Growing points in ethology

Growing points in ethology

Based on a conference sponsored by
St John's College and King's College, Cambridge

EDITED BY

P. P. G. BATESON

Lecturer im Zoology, University of Cambridge
Director of the Sub-Department of Animal Behaviour

R. A. HINDE

Royal Society Research Professor in the University of Cambridge
Honorary Director of the MRC Unit on the Development and Integration of Behaviour

CAMBRIDGE UNIVERSITY PRESS

CAMBRIDGE

LONDON · NEW YORK · MELBOURNE

Published by the Syndics of the Cambridge University Press
The Pitt Building, Trumpington Street, Cambridge CB2 1RP
Bentley House, 200 Euston Road, London NW1 2DB
32 East 57th Street, New York, NY 10022, USA
296 Beaconsfield Parade, Middle Park, Melbourne 3206, Australia

First published 1976

Printed in Great Britain
at the
University Printing House, Cambridge
(Euan Phillips, University Printer)

Library of Congress cataloguing in publication data

Main entry under title:
Growing points in ethology.

Includes bibliographical references and index.
1. Animals, Habits and behavior of – Congresses.
2. Psychology, Comparative – Congresses. I. Bateson,
Paul Patrick Gordon, 1938– II. Hinde, Robert A.
III. Cambridge. University. St John's College.
IV. Cambridge. University. King's College.
[DNLM: 1. Ethology – Congresses. QL751 G884]
QL750.G76 599'.05 76-8291
ISBN 0 521 21287 1 hard covers
ISBN 0 521 29086 4 paperback

Contents

List of contributors

R. J. Andrew, School of Biological Sciences, University of Sussex, Falmer, Brighton BN1 9QG, Sussex

P. P. G. Bateson, Sub-Department of Animal Behaviour, University of Cambridge, Madingley, Cambridge CB3 8AA

B. C. R. Bertram, Sub-Department of Animal Behaviour, University of Cambridge, Madingley, Cambridge CB3 8AA

N. G. Blurton Jones, Institute of Child Health, University of London, 30 Guilford Street, London, WC1N 1EH

T. H. Clutton-Brock, School of Biological Sciences, University of Sussex, Falmer, Brighton BN1 9QG, Sussex

R. Dawkins, Department of Zoology, University of Oxford, South Parks Road, Oxford OX1 3PS

J. F. Dunn, MRC Unit on the Development and Integration of Behaviour, University of Cambridge, Madingley, Cambridge CB3 8AA

J. C. Fentress, Department of Psychology, Dalhousie University, Halifax, Nova Scotia B3H 4JL, Canada

J. M. Hall-Craggs, Sub-Department of Animal Behaviour, University of Cambridge, Madingley, Cambridge CB3 8AA

P. H. Harvey, School of Biological Sciences, University of Sussex, Falmer, Brighton BN1 9QG, Sussex

R. A. Hinde, MRC Unit on the Development and Integration of Behaviour, University of Cambridge, Madingley, Cambridge CB3 8AA

N. K. Humphrey, Sub-Department of Animal Behaviour, University of Cambridge, Madingley, Cambridge CB3 8AA

A. Manning, Department of Zoology, University of Edinburgh, West Mains Road, Edinburgh EH9 3JT

P. Marler, Field Research Centre for Ecology and Ethology, The Rockefeller University, Tyrrell Road, Millbrook, New York 12545, USA

D. J. McFarland, Department of Zoology, University of Oxford, South Parks Road, Oxford OX1 3PS

P. B. Medawar, Clinical Research Centre, Watford Road, Harrow HA1 3UJ

J. Rosenblatt, Institute of Animal Behavior, Rutgers University, 101 Warren Street, Newark, New Jersey 07102, USA

M. J. A. Simpson, MRC Unit on the Development and Integration of Behaviour, University of Cambridge, Madingley, Cambridge CB3 8AA

J. Stevenson-Hinde, MRC Unit on the Development and Integration of Behaviour, University of Cambridge, Madingley, Cambridge CB3 8AA

N. Tinbergen, Department of Zoology, University of Oxford, South Parks Road, Oxford OX1 3PS

W. H. Thorpe, Sub-Department of Animal Behaviour, University of Cambridge, Madingley, Cambridge CB3 8AA

Introduction

Ethology is now well established, with ethologists making contributions to many disciplines ranging from neurophysiology to psychiatry and from genetics to ecology. Recently the growth of ethology has been accompanied by self-criticism, methodological improvement and a striving for conceptual rigour. In principle, the sharpened tools could be brought to bear on virtually all the behaviour of every living species. However, there would be serious shortcomings in thoughtlessly or mechanically embarking on such an enormous enterprise. Vast mounds of impeccably collected data could easily overwhelm creative thinking, and a preoccupation with method could displace a concern for problems. Such difficulties could be minimised if clear questions were formulated in advance – but clarity of purpose would not be enough. The vitality of ethology in the future will depend greatly on whether the questions that are being asked continue to be exciting and important.

As things stand, the roots ethology has put down seem healthy. But are they growing towards soil that will continue to nourish them? Are the shoots heading towards the light? In order to answer these questions we thought it might be worthwhile focussing on those growing points which, if tended carefully, are especially likely to develop strongly in the next few years. An opportunity presented itself with the twenty-fifth anniversary of the Sub-Department of Animal Behaviour at Madingley, or rather of the Ornithological Field Station, founded by W. H. Thorpe in 1950, from which the Sub-Department developed. We decided to commemorate the occasion with a small conference – not a conference to review past achievements or re-enact old battles, but to discuss some of the issues that seem to be emerging in ethology today. Obviously we could not cover the whole of ethology, but we invited contributions from representatives of widely different fields within the subject. Because of the occasion we were commemorating, many of the participants were past or present members of the Sub-Department and many of the others had had close links with it. We did not attempt to cover every controversial area. Our intention was that the papers should be formative rather than definitive. Therefore we asked the participants to prepare for circulation, in advance of the conference, papers concerned with some part of the framework of ideas which they expected would guide research in the next few years. The discussions were recorded and the participants were asked to produce final papers that took account of the discussion. In as much as they did so, the final product is a collective enterprise.

In editing the contributions, we have divided them largely on the basis of the questions traditional to ethological research: 'What were the immediate causes of this behaviour?'; 'What is its function?' and 'How did it evolve?'; and 'How did it develop in the individual?'. In addition we have included a section on 'Human Social Relationships', a special interest of several of the contributors: the chapters in this final section raise questions of all four types.

P.P.P.G.
R.A.H.

Acknowledgments

This volume contains the proceedings of a conference held in 1975, the 25th Anniversary of the founding of the Sub-Department of Animal Behaviour (University of Cambridge) at Madingley. The work of the laboratory would not have been possible without the financial help it has received over the years from research councils and foundations, especially the Rockefeller Foundation, The Nuffield Foundation, The Royal Society, Science Research Council, Medical Research Council, Mental Health Research Fund, Grant Foundation, Foundations' Fund for Research in Psychiatry, and Josiah Macy Foundation.

As editors of this volume, we are indebted to a number of people in addition to the contributors themselves. The conference was made possible by generous grants from St John's College and the F. P. Bedford Fund of King's College, Cambridge. King's allowed us to hold the meeting in their rooms. The discussions owed much to a number of people who attended the conference but did not contribute papers to the volume – especially Dr John Cowley, who chaired the meeting, and Dr J. M. Cullen. Mrs Cynthia Stott and Mrs Cilla Fuller were indefatigable in making the arrangements and in preparing the final manuscripts. To all these we express our thanks.

P.P.G.B.
R.A.H.

PART A

Motivation and perception

EDITORIAL: 1

Two basic concepts of classical ethology were the 'fixed action pattern' and 'sign stimulus'. The former was fundamental to evolutionary studies, providing a unit for comparison between species. It was also basic to Lorenz's (e.g. 1950) theory of motivation, which associated with each fixed action pattern the building up of a specific tension which sought release (see discussion in Thorpe, 1963). This issue of specificity was softened in Tinbergen's (1950) hierarchical model. Nevertheless, studies of motivation have been primarily focussed on such problems as why the frequency of a *particular* pattern of behaviour varies with time, and why the conditions necessary for its occurrence change. The importance of the relations between activities was undoubtedly recognised – for instance groups of responses were held to share causal factors, and mutual inhibition was often discussed (e.g. van Iersel, 1953; Hinde, 1970). Even so, the basic importance of such relationships for studies of motivation was not fully appreciated until relatively recently. The change in perspective is implicit in the first four chapters in this section which all focus primarily on relations between different patterns of behaviour.

Dawkins takes the most global view, examining the nature of hierarchical models and the diverse ways in which they have been and could be used in the study of behaviour. McFarland and Andrew are both concerned with the determination of the priorities for the performance of different activities within an animal's repertoire, but their approaches are very different. McFarland uses the traditional ethological units such as feeding and drinking, and derives the rules governing their priorities from consideration of their consequences. Andrew, on the other hand, is concerned with second-to-second changes in the direction of attention, whether or not a change in activity is involved. Finally, Fentress sounds a note of caution in emphasising that even the identification of units in the systems of behaviour is not so simple as it looks.

The concept of sign stimulus was linked with that of fixed action pattern in comparative studies, and was undoubtedly fertile, leading to numerous studies in which the stimulus characters selectively responsible for particular responses were carefully analysed (e.g. Tinbergen, 1948; Tinbergen & Perdeck, 1950). Nevertheless, the lock and key analogy implicit in the associated concept of the innate releasing mechanism had serious shortcomings (Marler, 1961). It became clear that some of the effective stimulus characteristics were not specific to a particular response but were common to many. This provided a new link with the general problems of perception which had concerned Lorenz in the thirties but were subsequently somewhat neglected by ethologists. Further links with problems of perception subsequently came from, for instance, studies of imprinting (e.g. Bateson, 1966). This general area of perception is represented by the chapter from Thorpe & Hall-Craggs, which discusses the utility of Gestalt theory in understanding the form of birds' songs.

REFERENCES

Bateson, P. P. G. (1966). The characteristics and context of imprinting. *Biological Reviews*, **41**, 177–220.

Hinde, R. A. (1970). *Animal Behaviour. A Synthesis of Ethology and Comparative Psychology*, 2nd edn. McGraw-Hill: New York.

Iersel, J. J. A. van. (1953). An analysis of the parental behaviour of the male three-spined stickleback (*Gasterosteus aculeatus* L.). *Behaviour Supplement*, **3**, 1–159.

Lorenz, N. (1950). The comparative method in studying innate behaviour patterns. *Symposia of the Society for experimental Biology*, **4**, 221–268.

Marler, P. (1961). The filtering of external stimuli during instinctive behaviour. In *Current Problems in Animal Behaviour*, ed. W. H. Thorpe & O. L. Zangwill. Cambridge University Press, London.

Thorpe, W. H. (1963). *Learning and Instinct in Animals*, 2nd edn (1st edn, 1956), Methuen: London.

Tinbergen, N. (1948). Social releasers and the experimental method required for their study. *Wilson Bulletin*, **60**, 6–51.

Tinbergen, N. (1950). The hierarchical organization of nervous mechanisms underlying instinctive behaviour. *Symposia of the Society for experimental Biology*, **4**, 305–312.

Tinbergen, N. & Perdeck, A. C. (1950). On the stimulus situation releasing the begging response in the newly hatched herring gull chick (*Larus argentatus argentatus* Pont.). *Behaviour*, **3**, 197–236.

1

Hierarchical organisation: a candidate principle for ethology

RICHARD DAWKINS

THE NEED FOR GENERAL PRINCIPLES: SOFTWARE EXPLANATION

If we look far into the future of our science, what will it mean to say we 'understand' the mechanism of behaviour? The obvious answer is what may be called the neurophysiologist's nirvana: the complete wiring diagram of the nervous system of a species, every synapse labelled as excitatory or inhibitory; presumably also a graph, for each axon, of nerve impulses as a function of time during the course of each behaviour pattern. This ideal is the logical end point of much contemporary neuroanatomical and neurophysiological endeavour, and because we are still in the early stages, the ultimate conclusion does not worry us. But it would not constitute understanding of how behaviour works in any real sense at all. No man could hold such a mass of detail in his head. Real understanding will only come from distillation of general principles at a higher level, to parallel for example the great principles of genetics – particulate inheritance, continuity of germ-line and non-inheritance of acquired characteristics, dominance, linkage, mutation, and so on.

Of course neurophysiology has been discovering principles for a long time, the all-or-none nerve impulse, temporal and spatial summation and other synaptic properties, γ-efferent servo-control and so on. But it seems possible that at higher levels some important principles may be anticipated from behavioural evidence alone. The major principles of genetics were all inferred from external evidence long before the internal molecular structure of the gene was even seriously thought about. Three computers with the same programming instruction set are in an important sense isomorphic in principle, even though their wiring diagrams may be utterly different, one employing valves, another transistors and the third integrated circuits; how all three work is best explained without reference to particular hardware at all (Simon, 1973). If a computer is doing something clever and life-like, say playing chess, and we ask how it is doing it, we do not want to hear about transistors, we simply accept them. The useful answer to the question is purely in terms of software; indeed the programme is likely to be written in such a way that it could easily be run with completely different hardware.

We need *software explanations* of behaviour. I do not mean that animals necessarily work like computers. They may be very different. But just as the lowest level of explanation is not always the most appropriate for a computer, no more is it for an animal. Animals and computers are both so complex that something on the level of software explanation must be appropriate for both of them.

Ethology has not lately produced many general explanatory principles. Its energy models of motivation were bold and aesthetically satisfying, but the predictions they made were too simple, and they were vulnerable to the first attacks of powerful critical intellect (Hinde, 1956, 1960). In the aftermath of their destruction grand general principles understandably became unfashionable. Some ethologists switched to other problems, such as the ecological roles of behaviour, and I have even heard the word 'ethologist' used to label a man who was *not* interested in mechanisms of behaviour!

At Madingley the broad balance which is the best feature of ethology (Tinbergen, 1963) has never wavered. It is characteristic that the Sub-Department of Animal Behaviour should celebrate its quarter century with an imaginative look forward rather than by dwelling on the past. I have tried to show that if we look forward far enough into the future, we are driven to seek general principles rather than detailed minutiae. We may as well start thinking now about likely *candidate principles*, and before looking at wholly new ideas it is worth dusting off some of the old ones. Hierarchical organisation is considered central to the whole of biology by the founder of the Sub-Department (Thorpe, 1974). Its ethological manifestation (Baerends, 1941; Tinbergen, 1950, 1951; Hinde, 1953; Kortlandt, 1955) came to grief in the general, deserved destruction of simplistic energy models, but its guilt was by association only. It really has nothing to do with energy models, but is a much more powerful principle in its own right. It is a particular pleasure to a pupil of Niko Tinbergen to try to point this out.

This paper will not be a review of the literature, nor will it use hard evidence to convince anyone. It will be an attempt to arouse the imagination of those more accomplished in research than I am, to persuade them to look again at the idea of hierarchical organisation, and use it in the future. But hierarchy is only one example of a principle of software explanation, and if it is eventually found wanting, other possibilities may be explored in a similar way.

DEFINITIONS AND CLASSIFICATION

ἱεραρχέω means to be supreme in sacred things (Liddell & Scott, 1883). The idea of supremacy or superiority is the basis of the following definitions, which are indented to distinguish them from plain English. They will be best understood with reference to Fig. 1. First the axioms:

> There exist elements: A, B, C etc., and a relation:
> '*is boss of*' (inverse: '*is bossed by*').

The elements can be represented by blobs, and the relations between them by arrows (Fig. 1). If A is boss of B then an arrow is drawn from A to B. We now define a more general relation:

> A is *superior* to B when either
> A is boss of B or
> A is boss of an element which is superior to B.

Thus an element can be superior to another element without being boss of it. Colloquially boss might mean 'immediate superior'. Using this preliminary definition we can now define a hierarchy. (I prefer to define it as a set of

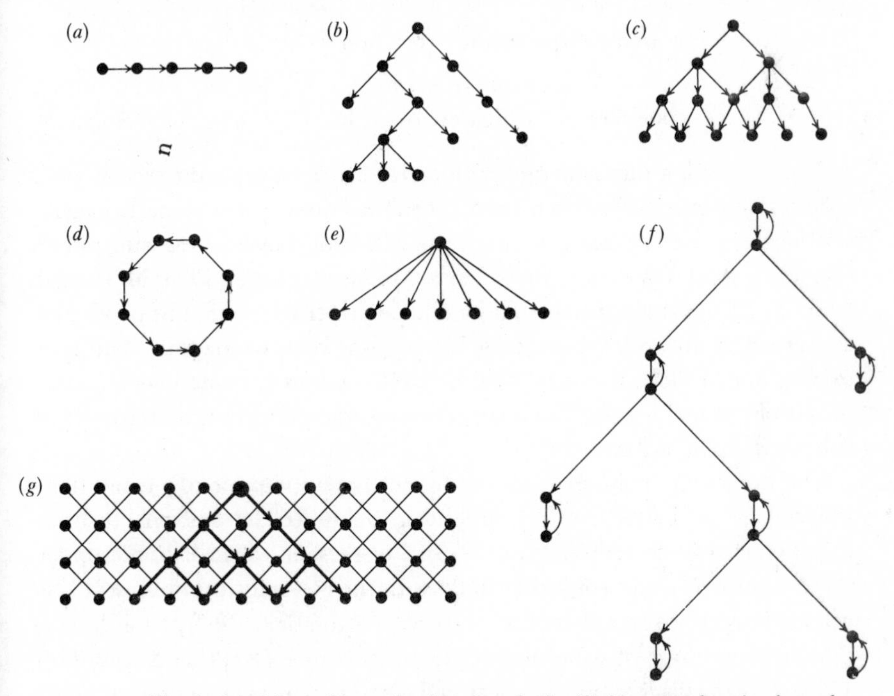

Fig. 1. (*a*) Linear hierarchy ('peck order'). (*b*) Non-overlapping branching hierarchy. (*c*) Overlapping hierarchy. (*d*) Not a hierarchy. (*e*) Shallow hierarchy. (*f*) Hierarchy of loops. (*g*) Network.

elements with a relation, rather than to follow Woodger (1937) in defining it as a class of relations. My usage is closer to that of the Church, in which the Hierarchy is the set of bishops.)

A *hierarchy* is a set which satisfies:

(i) There is no element in the set which is superior to itself, and

(ii) There is one element in the set, to be called the *hierarch*, which is superior to all the other elements in the set.

The first requirement (if strictly applied – see below) is equivalent to saying that the relation 'is superior to' must be transitive throughout a true hierarchy, i.e. there are no circular relationships as in Fig. 1*d*. The second requirement ensures that every hierarchy has a single overall superior or root. There is no reason why the hierarch should not be bossed by another element, but that element is by definition outside the hierarchy under discussion. Thus a hierarchy may be included in a larger hierarchy. Hierarchies may be classified into types as follows.

A hierarchy is *branching* if it includes at least one element which is boss of more than one element.

A hierarchy is *linear* if it is not branching.

A (necessarily branching) hierarchy is *overlapping* if it includes at least one element which is bossed by more than one element.

The model of a linear hierarchy (Fig. 1*a*) has been applied to social 'peck orders', and it has also been used, not always under the same name, for relationships within animals between different behavioural subsystems (Deutsch, 1960; Dawkins, 1969*b*; Davis, Mpitsos & Pinneo, 1974; McFarland & Sibly, 1975). It is a worthwhile candidate principle, but in this paper I am concerned mainly with branching hierarchies, both overlapping and non-overlapping. Indeed the very word hierarchy carries connotations for many people of a branching tree. Equivalent representations include nested brackets and overlapping areas.

The branching hierarchy idea has found its way into biology many times (Weiss, 1971; Pattee, 1973), and it is discussed by systems theorists (Mesarovic, Macko & Takahara, 1970). For reasons which I do not understand – perhaps it has somehing to do with the ecclesiastical meaning of the original Greek – it is a favourite of mystics (Koestler, 1967), or at least of those who are fond of using the word 'reductionism' (Koestler & Smythies, 1969). It is the basis of both Linnean and post-Darwinian taxonomy, although many people persist in seeing the animal kingdom as a linear hierarchy (Hodos & Campbell, 1969). It may have important applications in ontogeny at various levels (Manning, this volume; Simpson, this volume). As Kortlandt (1955) has shown in a brilliantly erudite, if somewhat outspoken review, branching hierarchy models have a long history in human and animal psychology, pre-dating the more familiar ethological ideas.

The nature of the relation 'is boss of' has beeen left unspecified so far. Interpret it as 'feeds motivational impulses to' and you have Tinbergen's (1950) hierarchical model. Change it to 'has a causal influence on', and you arrive at Hinde's (1953) less contentious formulation:

> ...of the causal factors controlling each activity, some are specific to it and others are also capable of influencing other activities. The causal factors are thus arranged in a hierarchical manner, those at the bottom influencing only one or a few activities whereas those at the top influence many. Ultimately, of course, there are factors which influence the occurrence of all possible behaviours – these are the conditions under which existence is possible...This hierarchy of causal factors gives us one way of classifying the various activities possessed by an animal into instincts and sub-instincts. All the activities belonging to a major instinct have certain causal factors which they share with other sub-instincts and some which are specific to them.

To list some other examples of hierarchies, 'is boss of' might mean 'controls' (as in a power station), 'is an ancestor of' (as in a family tree), 'is attached to' (as in an oak tree) or 'gives orders to' (as in the army). As we shall see, several interesting behavioural models are based on its meaning 'sets up a target or goal for'. In a computer programme it might mean 'calls up as a subroutine (procedure)'. A related meaning will be shown to give rise to another interesting class of models, those comparing behaviour organisation to human grammar.

What would *not* constitute a hierarchy? Within the context of sets of elements with a single relation 'is boss of', it follows from our definition that there are only two classes of non-hierarchical system, those which violate conditions (i) and (ii) on page 10. These are non-transitive systems and multi-hierarch systems respectively. To these might be added systems like Fig. 1*e*, in which there is a hierarch ruling over many subordinates, but all the subordinates are of equal status to each other. These 'shallow hierarchies' do not strictly violate the definition, but they are not hierarchical in an interesting sense, because their 'height' (Harding, 1971) is small compared with their 'span' (Simon, 1962).

Fig. 1*g* taken as a whole violates the single hierarch clause. Some people are reluctant to give the name hierarchy to overlapping hierarchies in general, because they feel that a logical extension of an overlapping hierarchy is a network like Fig. 1*g*, with no evident tree-like properties. However, although it is true that the set of all elements in that figure is not a hierarchy, it is possible to regard it as a set of hierarchies. Thus if all arrows are regarded as pointing downwards, each element in the top row becomes a hierarch superior to nine lower ones. One tree is emphasised in the figure for illustration. It is in this sense that a *Wirkungsgefüge* (von Holst & von Saint Paul, 1963) is hierarchical. Hinde (1952) pointed out long ago that even Tinbergen's model must really be overlapping, at least at the lower levels. The same is clearly implied in Weiss (1950) and in Lorenz's (1937) concept of 'tool-instinct'.

A network such as Fig. 1*g* functions as a collection of true hierarchies in the following sense. Although two hierarchs share many subordinates, each one has a unique set of them. Thus two cells in the visual cortex of a cat may have overlapping fields, both retinal and supra-retinal, but they will not overlap totally. Hence each one will be maximally stimulated by a unique pattern, though they share subpatterns, features, or at least spots of light. A similar point has been made for the motor side, for example with reference to invertebrate 'command interneurones' (Elsner, 1973; Kennedy, 1974).

Should systems with feedback be regarded as non-hierarchical because they violate the transitivity rule? Not necessarily. It depends how we interpret the word 'element' in the definition. In Fig. 1*f*, if the black blobs are the elements, the set is clearly not a strict hierarchy. But if we put each feedback loop into a box, and call each box an element, we then have a transitive hierarchy of feedback loops. Thus the TOTEs of Miller, Galanter & Pribram (1960) form true hierarchies. Ito (1974) postulates 'pyramid-like' hierarchies of control in the nervous system, with feedback units at lower levels, and feedforward and other kinds of control at higher levels, with an overall controller at the peak. Szentágothai & Arbib (1974) provide a stimulating discussion of feedback and feedforward in hierarchical systems.

On the other hand if feedbacks are not locally confined, but extend to more and more distant superiors it becomes less and less easy to regard the system as a hierarchy. There is no need to specify a definite cut-off point for the definition. This would be no more sensible than worrying about whether two animals belong in different genera or merely in different species.

However, simple rings such as Fig. 1*d* should definitely be ruled out. An example is a 'round-robin' computer time-sharing system. The different tasks are allocated a fixed share of the computer's time in strict rotation, like a chess master playing ten games 'simultaneously'. In less efficient round-robins, the timing is inflexible, even if particular tasks cannot make use of their time allocation for want of a busy peripheral device. R. H. McCleery (personal communication) suggests that *Arenicola* (Wells, 1966) may organise its behaviour like this. A more efficient method of time-sharing, a linear hierarchy by our definition, is 'priority-interrupt'. The top priority task retains control until it is held up, usually waiting for a slow peripheral device or human, whereupon the second-ranking task is disinhibited. Number two retains control until either it is held up and disinhibits number three, or it is interrupted by number one. McFarland (1974) has developed the analogy with animal behaviour.

Pure priority-interrupt systems have the potential disadvantage that very low ranking programmes may never be run at all. This problem can be overcome by a compromise with a round-robin, or by shifting priorities as a function of time, much as an animal gives higher priority to eating as the interval since its last meal increases.

Shifting of boss-ships among a community of potential equals is the basis of common ethological motivational ideas (e.g. van Iersel & Bol, 1957), in which everything inhibits everything else; the 'centre' which is for the moment most potent in inhibiting rivals, takes control of the final common path. By analogy with the 'pandemonium' system of sensory pattern recognition (Selfridge & Neisser, 1963), this may be called the 'motor pandemonium' model: whoever is shouting loudest at the moment is boss. Physiologically speaking, fluctuating factors such as hormonal and nutritional state, levels of skin irritation and other sensory factors, change the priorities among the competitors. Changes in observed behaviour are all directly caused by these changes in boss-ships. This is not true of interestingly hierarchical models such as McFarland's time-sharing model (see above), in which 'is boss of' means 'is capable of interrupting', and my 'attention threshold model' (Dawkins, 1969*b*), in which 'is boss of' means 'is attended to before'. In both these models, boss-ships remain stable while behaviour changes. The motor pandemonium model is not hierarchical except in a trivial sense. It is shallow (see above), and its slight hierarchical properties are not used to explain anything.

Nelson (1973) espouses 'distributed control', at times apparently setting it up as anti-hierarchical. Unfortunately, in a paper so rich in creative ideas, he does not find the space to define it. He is impressed with the power of simple networks of very few neurone-like elements to generate complex output patterns, and to switch abruptly from one stable pattern to another in response to simple quantitative changes in input. He goes on to visualise more complex networks in which 'the organization of connection would still be in a way hierarchical but lacking a hierarch. Control or dominance would shift from one to another part of the distributed perceptual-motor interface according to internal or external necessity....'

McCulloch (1945), also in the context of distributed control in networks, used the term 'heterarchy', and the word has recently been adopted by workers in artificial intelligence (Winston, 1972). Unfortunately it too has not been defined, as Winston admits. He goes on to provide a complicated list of five unconnected attributes of heterarchy, of which only the first two are easy to generalise from computer programming to biology. First, heterarchical programmes are 'goal-oriented'. We postpone discussion of goals until later. Secondly, 'executive control should be distributed throughout the system. In a heterarchical system, the modules interact not like master and slave, but more like a community of experts'.

Nelson's 'distributed control' and the 'heterarchy' of the artificial intelligence workers appear to have something in common. In both, as in the motor pandemonium model, boss-ships are capable of rapid reversal. Does this make them fundamentally anti-hierarchical? I prefer to think of changing boss-ships as an interesting complicating factor in systems which may or may not be hierarchical according to other criteria.

For example junior officers take orders from their superiors, but they also send information back in the reverse direction. There is flow of orders 'down' the pyramid, and flow of information 'up'. We could think of shifting control, depending on whether information or orders are being considered. But there is a more interesting sense in which the system remains a hierarchy, because information tends to converge towards an apex, and orders to diverge from the same apex. A captain can be identified as superior to a lieutenant whichever the direction of flow, because a captain interacts with several lieutenants, but a lieutenant only interacts with one captain.

Even a motor pandemonium model can be extended, so that each participant can be the head of a stable hierarchy. In a network such as Fig. 1*g*, control might shift from point to point, so that new hierarchs emerge. The hierarchies so brought into play may still have the attributes of a branching tree, with all that follows functionally (see below), even though they are temporary. Fig. 1*d* and 1*e*, however, are not interestingly hierarchical, whether they are regarded as temporary or permanent.

Hierarchical classification

Nelson (1973) makes an important distinction between hierarchies of embedment and hierarchies of connection. I think embedment is largely equivalent to classification, in which case the same distinction has been emphasised by others in the ethology literature (e.g. Kortlandt, 1955). The examples given above were hierarchies of connection. A hierarchy of classification is one in which 'is boss of' has some such meaning as 'contains' or 'includes'. Thus the general feature of a hierarchy of classification is that inferior elements actually make up the parts of higher elements. This is not so of hierarchies of connection. To paraphrase Nelson, the Curia is not a partitioning of the substance of the Pope. A soldier is a part of his platoon, but not of his platoon commander.

HIERARCHICAL ORGANISATION

Hierarchical classification is a vital convenience of everyday life, a universal means of organising information for easy access. A large postal system would be unworkable without hierarchical addressing. Even numerical ('zip code') addressing is really hierarchical, and in the same sense as the alphabetical arrangement of a dictionary (Longuet-Higgins & Ortony, 1968). Non-hierarchical methods of information retrieval such as random search, and sequential systematic search, are much too slow.

It is of course important not to muddle up hierarchies of classification and of connection. However, the muddling that has undoubtedly occurred in the ethological literaure is perhaps pardonable, because the two are frequently closely associated. Systems such as armies are hierarchically classifiable into functioning units, but these units are controlled by a hierarchical tree of command or connection. Similarly, biological taxonomy is hierarchical not only because an embedded structure is convenient, but also because of evolutionary connections. Each node in a taxonomic tree-diagram can be regarded as a taxon such as Mammalia which can be decomposed into subgroups, Carnivora, Rodentia, etc. It can also be regarded as a common ancestor.

In ethology it is clearly convenient to *classify* behaviour patterns hierarchically for the same reason as it is convenient to classify anything else in this way. But, at least until the concluding discussion, we shall be more concerned with the other question of whether there are hierarchies of connection between the subunits. This could be discussed from a neuroanatomical and neurophysiological point of view (Paillard, 1960: Szentágothai and Arbib, 1974; Eccles, 1975). In this paper I shall take two more indirect approaches. The first is a functional one – it might even be called a 'neuroeconomic' one – an attempt to show some of the economic pressures which might lead to the evolution of hierarchical systems. The arguments here are sufficiently general to apply to hardware or software. The second is a behavioural one, an examination of the temporal patterning of behaviour to see whether it appears to be governed by hierarchical pattern-generating (software) rules.

THE FUNCTIONAL SIGNIFICANCE OF HIERARCHICAL ORGANISATION

An interest in the functional significance or survival value of a biological feature is often regarded as confined to field-workers who see whole animals in their natural surroundings. This is obviously silly. Natural selection entitles us to expect with confidence that good, though not necessarily optimal, design principles will pervade the internal organisation of animals down to the smallest levels. What is less obvious is that it is good research strategy to think about design principles before, rather than after, attempting to find out what is actually the case. Good physiologists know this, and are guided by it straight to the most fruitful hypotheses for physiological testing.

Simon (1962), in a stimulating short paper suggesting that hierarchy is the basis of the 'Architecture of complexity', illustrates one reason why hierarchical design is good design by a parable about two watchmakers called Tempus and Hora. (Koestler in his re-telling (1967) changed the names to Mekhos and Bios – why?) Tempus's watches were as good as Hora's, but he took about 4000 times as long to make each one. The reason lay in a fundamental difference of design. Both types of watch had about 1000 components. Hora put these together first into 100 sub-assemblies of 10 components each. These in their turn were assembled into 10 larger units. Finally, the 10 larger units were combined to complete the watch. Thus if anything went wrong during assembly, Hora had only to go back and re-assemble the current subunit, and he did not lose much time. Tempus on the other hand tried to put together all 1000 components in a single large assembly operation. If there was a mistake or interruption the whole thing fell to bits, and he had to go right back to the start. He therefore very rarely completed a watch.

Parables should not be flogged to death, so I will refrain from relating stories about two computer programmers, or two sticklebacks, called Tempus and Hora, and turn to a listing of what seem to me the three main advantages of hierarchical organisation. When I wrote this I was unaware of the parallel but different discussion of Szentágothai & Arbib (1974).

The evolutionary rate advantage

This is the one illustrated by the watchmakers. In more abstract terms, the evolution of thermodynamically improbable assemblies proceeds more rapidly if there is a succession of intermediate stable sub-assemblies. Since the argument can be applied to the manufacture of each sub-assembly, it follows that highly complex systems which exist in the world are likely to have a hierarchical architecture (Simon, 1962), and nervous systems are presumably not exceptional. The word 'advantage' is perhaps misleading here, as indeed it often is in evolutionary discussion. 'Evolutionary stability' (Maynard Smith & Price, 1973) is nearly always better.

The local administration advantage

Consider the problem of controlling an automatic vehicle surveying Mars. The question is how much of the total electronic and computing power to put on Mars and how much to leave on Earth. One extreme, which has economic appeal because of the low rocket payload, is to leave almost all the decision-making circuitry on Earth – a general purpose computer could do the job – and equip the vehicle itself with little more than a two way radio set. That the economy was false would soon be apparent to anyone listening in to the radio messages. Every time the vehicle met a small local difficulty, a boulder say, it would relay the details to Earth, each bit of information taking four minutes to arrive. In a flash the giant computer would calculate the optimum tactic, but each bit of the returning instructions would take another four minutes to reach Mars, and the wretched robot would long since have ploughed into the boulder. Obviously detailed moment to moment radio-control from Earth is prohibited by the delays. Moreover, since much of the information necessary for control is all on Mars in the first place, it is a waste of the communication channel to refer it back to Earth where it is not used.

Clearly many detailed decisions based on local information are best taken locally, and this is a fundamental principle of far-flung organisations such as the late Roman and British Empires. On the other hand the main disadvantage of too much local responsibility is lack of coordination of different units towards a common purpose. The optimum balance between local responsibility and referring back to headquarters depends in a complex way on a number of factors, including the distance involved, measured in rate and cost of information transmission. Call this the 'information distance'.

Now suppose we need to control several vehicles on Mars. The information distance between each vehicle and Earth is the same. But the vehicles need to be coordinated together in a common plan, and the information distance between any two points on Mars is relatively small. Therefore the optimum balance will tend to shift towards setting up a local master computer on Mars which will take some decisions for all vehicles, and which will handle all communications with Earth. We have the beginnings of a branching tree, and the argument can be applied recursively to justify further, many-levelled branching.

The redundancy reduction advantage

A classic of the discipline which I am naming 'neuroeconomics', is Attneave's (1954) and (independently) Barlow's (1961*a*, *b*) analysis of principles underlying sensory systems. Most messages contain redundancy, that is they could be recoded more economically without loss of information. For example in most visual scenes there is a high correlation between the luminosity of neighbouring points. This means that if the intensity of the light falling on each retinal cell were simply mapped onto the visual cortex, the firing rate of any one central projection cell would be highly predictable from its immediate neighbours. This is not only wasteful of channel capacity; it is also unhelpful to the animal, which has to make practical decisions, to have information simply *reproduced*, however accurately, on yet another projection screen in its nervous system. As Barlow points out, nervous systems in fact recode so as to remove redundancy, lateral inhibition in the retina being the mechanism in this case. This ensures that maximal firing rates occur in cells bearing much information, which here are cells whose fields lie along *edges* in the scene. Similar arguments can be made for other aspects of sensory systems, and I will apply them also to the motor system. First we must make the link with the idea of hierarchy.

Many of the sensory patterns which an animal has to recognise – food, mates, obstacles, etc. – have features in common. This is another form of redundancy. Thus a straight line is redundant not only in Barlow's sense that all points along it are predictable from the two ends. It is also redundant in that it is a feature common to many ordinary objects. Rather than have entirely independent circuitry to recognise each important object it is therefore economical for pattern-recognising units to share subcomponents which recognise subfeatures which their key stimuli have in common. Recursive application of this argument leads to a system of overlapping hierarchical pyramids. This functionally sensible design is familiar from the work of neurophysiologists on visual mechanisms themselves, and of computer programmers facing the analogous problem of machine recognition of visual patterns (Sutherland, 1969; Barlow, Narasimhan & Rosenfeld, 1972).

Attneave and Barlow began their arguments independently by the same numerical thought-experiment. The human retina contains about four million light-sensitive cells. If we make the simplifying assumption that at any instant each cell is either signalling presence of light or not, the number of possible states of the system is $2^{4\,000\,000}$ which is not a small number. If there were one central cell tuned to each possible state of the retina, the volume of the brain would have to be measured in cubic light years. It was in this context that Barlow and Attneave postulated the pressing need for redundancy reducing mechanisms in the visual system.

Can we make a similar argument for the motor system? There are fewer muscle fibres in the human body than retinal cells, but there are enough to make difficulties. If we assume that at any instant each one is in a state of either contraction or relaxation we can arrive at a similar combinatorial explosion if we try to calculate the total number of possible states of the muscular system. In fact the potentially enormous number of states is greatly reduced by redundancy in the final motor output – large populations of muscle fibres contract and relax in a highly correlated way, and it is obviously sensible that they should do so.

The most perfect correlation is between members of the same 'motor unit', fibres which are all controlled by the same motoneurone. The correlation between motor units within one muscle is not perfect – if it were, graded contraction would be impossible. Nevertheless there is great redundancy in the behaviour of motor units within a single muscle, which is thus to some extent a unit of action. Then there are correlations between different muscles, both positive and negative, simultaneous and time-lagged. We are of course mounting the ladder of Weiss's (1941) well-known six levels of nervous organisation, which came to form the lower rungs of Tinbergen's hierarchical model. This arrangement confers the same economic advantage on the animal as subroutines give the computer programmer. Whole patterns of low-level coordination, programmed only once, may be called into the service of different high-level tasks.

Here then we have another economic argument in favour of hierarchical organisation from the animal's point of view. A similar, logical rather than functional argument can be advanced for the a priori plausibility of the nervous system's being hierarchically organised. This was briefly mentioned by Craik (1943) but it is fully set out by Bullock (Bullock, 1961; Bullock & Horridge, 1965). He considers 'the problem of recognition in an analyzer made of neurons'. He is concerned only with those cases where a definite all-or-none behavioural act emerges from the animal. How large this category is is open to empirical test (Dawkins & Dawkins, 1973), but many ethologists seem to assume in practice that it includes the behaviour they are studying. Bullock's point is that in such cases there must be, somewhere in the nervous system, a single unit which makes the decision. This unit may be regarded as the point of convergence of information from many sources, including sense organs, or it can be regarded as the starting point of outwardly radiating efferent information. In other words it is the confluence of two hierarchies. Bullock is much too cautious to call his units single cells, though this is one possibility he considers, and Barlow (1972) has recently argued provocatively for a 'neuron doctrine for perceptual psychology'. Certain large invertebrate neurones (Dorsett, Willows & Hoyle, 1973; Kennedy, 1974) would fit the bill. Bullock also considers multi-neurone decision-making units, which may go some way towards allaying the scepticism of Nottebohm (1970).

Hierarchy then seems to fulfil a major requirement of a good general candidate principle for the organisation of behaviour; it makes functional and logical sense. At present the evidence that it actually is an important principle is not convincing: not enough work has been done on it, which is one reason for writing this paper. I shall now turn to behavioural models.

SIMPLE CLUSTERING IN TIME

Eibl-Eibesfeldt (1975), who has given the most sympathetic recent treatment of Tinbergen's classic model, cites as evidence for it the fact that behaviour patterns tend to occur clustered in time, the clusters constituting functionally related groups. Tinbergen's (1950, 1951) own words can be interpreted to mean the same thing, although they have also been interpreted (and criticised – see above) as purely taxonomic in intent. Tinbergen of course also made use of other evidence such as that from electrical stimulation of the brain (see Vowles, 1970, for a more recent usage of the same kind of evidence in the service of another hierarchical model). In this section we are concerned with the relevance of the grouping of behaviour patterns in temporal clusters.

Simon (1962) considers a similar point more generally, under his heading of 'near-decomposability', for him a property of hierarchical systems generally, although he does not deal with animal behaviour. A small digression is needed to explain this.

Let the elements in a system be listed as the column and row headings of a matrix and let the body of the matrix contain numbers representing strengths of interaction between them. The order in which the elements are arranged as row and column headings is at first arbitrary. We now rearrange them so as to maximise the tendency for high interactions to be grouped in square submatrices around the major diagonal. The matrix is said to be decomposable if it is possible to arrange it so that all interaction scores lie in these square submatrices.

Fig. 2*a*, taken from Simon (1962), is an example of a *nearly* decomposable matrix. The elements around it refer to cubicles in a house, and the numbers in the table represent rate of heat flow between cubicles. Here it is possible to arrange the table so that all high numbers lie in three square submatrices along the diagonal, and all low numbers lie outside these submatrices. A rationale for the tendency to decompose into three submatrices is the following hierarchical scheme (Fig. 2*b*). The cubicles A1, A2 and A3 are in one room, B1 and B2 in another, and C1, C2 and C3 in a third. Insulation between rooms is good, while insulation between cubicles is poor.

More generally, low-level elements in a hierarchy are bound to each other by strong bonds, and their dynamic interactions are of high frequency (Bastin, 1969; Simon, 1973). Weaker bonds and slower dynamics characterise interactions between higher-level elements, which are clusters of low-level elements. If the differences between interaction strengths at different levels of organisation are large, the near-decomposability will show up in a matrix such as Fig. 2*a*, and the system can be treated as hierarchical.

Ethologists also represent strengths of interactions in square tables of this

(*a*)

	A1	A2	A3	B1	B2	C1	C2	C3
A1	—	100	—	2	—	—	—	—
A2	100	—	100	1	1	—	—	—
A3	—	100	—	—	2	—	—	—
B1	2	1	—	—	100	2	1	—
B2	—	1	2	100	—	—	1	2
C1	—	—	—	2	—	—	100	—
C2	—	—	—	1	—	100	—	100
C3	—	—	—	—	2	—	100	—

(*b*)

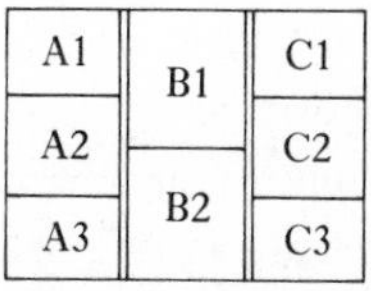

A1	B1	C1
A2		C2
A3	B2	C3

Fig. 2. Hypothetical nearly decomposable system. (*a*) Table of heat diffusion coefficients between cubicles, arranged so as to concentrate high entries in square submatrices along major diagonal. (*b*) Plan view of cubicles, to show their clustering in three rooms, A, B and C (from Simon, 1962).

type. Sometimes the figures in the table are temporal correlation coefficients between behaviour patterns, and sometimes they are probabilities or frequencies of transition. The ordering of rows and columns is usually arbitrary. However, Myrberg (1972) published transition matrices for behaviour patterns of the fish *Eupomacentrus partitus*, in which he rearranged the rows and columns so that the entries in the body of the table came to be clustered in square sub-tables, although he did not provide an explicitly hierarchical rationale. It is not obvious whether he achieved the optimal rearrangement, and he does not give in detail the algorithm he used. However, Fig. 32 of Myrberg's monograph is reminiscent of Simon's figure reproduced here, and there is a suggestion of the near-decomposability which Simon regards as a property of hierarchical systems. A more exacting test of the near-decomposability of Myrberg's table is given on p. 31.

This method of rearranging rows and columns of a behaviour transition-matrix is a form of cluster analysis, but rather a crude form. It treats behaviour patterns as arranged in a two-levelled hierarchy, with strong interactions between elements within a low-level cluster, and weak interactions

(*a*)

Preceding \ Following	FR	TG	HD	FM	NO	MV	BM	AB	WG	BF
FR		77	709	129	496	5				
TG	82				2					
HD	730	4			18					
FM	151		1		13					
NO	445	3	42	36		418	26	133	495	414
MV	7				412					1
BM					13					76
AB					93				3	195
WG					223	1	1	46		384
BF					741		62	92	156	

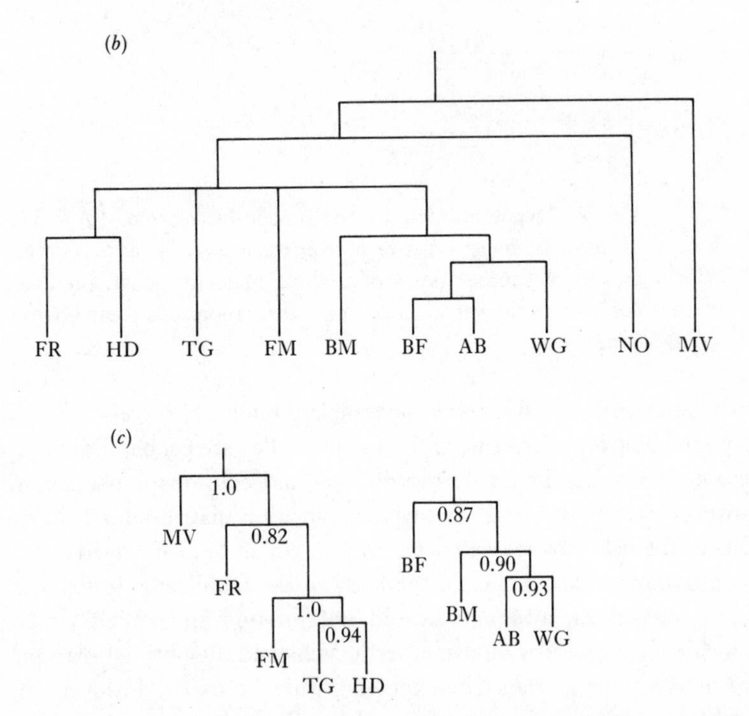

Fig. 3. Data on grooming behaviour in blowflies, *Calliphora erythrocephala* (from Dawkins & Dawkins, 1976). (*a*) Table of frequencies of transition between acts, arranged so as to concentrate high entries in two submatrices. (*b*) Single Linkage Cluster Analysis (Ross, 1969). Index of 'distance' between pairs of acts was $(1-r)$, where r is product moment correlation between time spent doing the two activities in successive five second periods. Two acts are clustered close together if they tend to occur in the same 5 second period. Two existing clusters are united, if any one

between elements of different clusters. Other kinds of cluster analysis are designed to group elements in multi-levelled hierarchical classifications (Everitt, 1974). Fig. 3*b* shows the application of one such, the Single Linkage method (Gower & Ross, 1969; Ross, 1969) to unpublished data on grooming in blowflies *Calliphora erythrocephala* (Dawkins & Dawkins, 1976). For comparison a table of transition frequencies is given with rows and columns rearranged in the same way as Myrberg's tables (Fig. 3*a*). Ignore Fig. 3*c* until later.

However, cluster analysis is only a technique of classification. As explained above, we want to know whether the underlying control is hierarchically organised. General cluster analysis techniques will tell us which of many alternative hierarchical clusterings best fit our data; they will not tell us whether our data really 'want' to be classified hierarchically at all, in the sense that the underlying mechanisms are hierarchically organised.

Similarly, we cannot necessarily regard the fact that behaviour patterns tend to be clustered in time, as evidence in itself for underlying hierarchical organisation. Behaviour patterns doubtless do form clusters in that each act is likely to be followed by another member of the same cluster, i.e. between-cluster transitions are rare. But, unless further conditions are met (see below), there is nothing here that could not be predicted by an ordinary one-levelled state-transition Markov model (references in Slater, 1973). Nothing is gained by speaking of clusterings of behaviour patterns. 'Rare transitions' and 'common transitions' are more parsimonious ways of expressing 'between-cluster' and 'within-cluster' transitions. We can indeed use such a Markov process as a kind of null-model, an example of a non-hierarchical model.

A behavioural model which allows us a truer test of near-decomposability is best introduced in terms of the concept of 'decision'. This must now be explained, since the word is used in a slightly unusual way.

---

of one cluster is close to any one of the other. (*c*) Mutual Replaceability Cluster Analysis (method explained later in text). Two acts are clustered together if they are mutually replaceable in behaviour sequences. Entries from existing clusters are added together for consideration for future clustering. Figs. 3(*b*) and 3(*c*) use the data in different ways, and are not simply mutually translatable. Key to behaviour patterns: 1. FR Rubbing front legs together; 2. TG Grooming proboscis with front legs; 3. HD Grooming head with front legs; 4. FM Grooming one middle leg with front legs; 5. BM Grooming one middle leg with rear legs; 6. BF Rubbing rear legs together; 7. AB Grooming abdomen with rear legs; 8. WG Grooming wing with rear legs; 9. MV Moving around, not grooming; 0. NO No grooming. Motionless.

Hierarchy of decisions

Dawkins & Dawkins (1973, 1974) analysed the temporal stream of events which is behaviour into a sequence of 'decisions'. A decision is defined as an event which itself could not easily be predicted, but from which future events can be predicted. Let letters of the alphabet represent behavioural events, and let the following be part of an observed sequence (ignore the underlinings for the moment):

VBQACVBQMFWACMFWMFWACVBQACVBQMFWACA-
CACVBQACMFWVBQVBQMFWACVBQACMFWMFWAC

Analysis of transition frequencies shows the following: V is always followed by B which is always followed by Q. However, Q may be followed by a variety of events. Therefore, the triplet VBQ may be regarded as a unit. The 'decision' to do BQ is said to be taken at the same moment as the decision to do V. V is called a 'decision-point'. All decision-points in the above sequence are underlined. Since BQ is redundant following V, C is redundant following A, and FW is redundant following M, the above sequence can be more economically represented without loss of information, as a sequence of decisions: VAVMAMMAVAVMAAAVAMVVMAVAMMA. This advantage of parsimony is enjoyed not only by the ethologist seeking to reduce the volume of his data to manageable proportions, but also by the animal itself struggling to control its many muscles in efficient temporal patterns.

This is a hypothetical extreme example. Real-life decisions are not absolute but relative. We tried to measure, in bits of information, the relative 'decisioniness' of successive frames of film of chicks drinking. We found not just two sorts of event, unpredictable decisions and predictable follow-ups of decisions, but rather a smear of intermediates with perhaps local modes. More interestingly from the present point of view, we speculated that there might be a hierarchy of decisions in the following sense. There might be some predictability between successive *decisions*, for instance VBQ and AC in the above example might be more likely than chance to alternate with each other. Then in the sequence MFWVBQACVBQACVBQACMFW, the first V would constitute a bigger decision than subsequent Vs or As, because it signals the onset of a new VBQAC cluster. Subsequent Vs and As are more predictable, hence smaller decisions; they are within-cluster decisions. They are still called decisions because they are less predictable than the Bs, Qs and Cs. Similarly there might be an even bigger, more global decision to enter the whole MFWVBQAC major cluster, as opposed to some other major cluster involving acts not yet mentioned, say XYZ.

A model along these lines is specified in the following assumptions. Like any non-trivial model this one is a piece of at least partially free invention which does not follow logically from known facts.

Assumptions of the model

The organisation of an animal's decisions is hierarchical if it is possible to group its behaviour patterns into clusters such that

(i) For each cluster there exists a state of the animal of being certain to do one element of the cluster, but still uncertain which. Thus A, B and C form a cluster if the animal is capable of entering a state in which it is definitely about to do (A or B or C) and nothing else, but this state still leaves open *which* of the three will be done (they are not necessarily equi-probable).

(ii) The elements within a cluster between which a choice is made may be single acts, or they may be subclusters defined in the same kind of way as the cluster under discussion. For example the animal might be capable of entering a state of being about to do (A or B or C or D or E) but nothing else. At other times it enters states of being certain to do (D or E) or of being certain to do (A or B or C), or of being certain to do D. However, there is no state of being certain to do (A or D) but not anything else, since A and D belong in different clusters at the same level; a state of being about to do A or D implies the possibility of doing B or C or E, the other members of the smallest cluster to which A and D both belong. The cluster of five acts can thus be represented as two subclusters:

((A or B or C) or (D or E)).

(iii) Choices may be influenced by previous choices only within clusters not between clusters, and only by previous choices during the current entry of the current cluster. Thus if a transition is observed between A and D this implies that the animal must have left Cluster 1 (A or B or C) and entered Cluster 2 (D or E). By assumption (iii), the choice of which member of the new cluster is performed is uninfluenced by which members of the old cluster had been chosen. Thus the transition A → D has a probability which is equal to the probability of the transition Cluster 1 → Cluster 2, with appropriate weighting for the overall rarity or commonness of A and D within their respective clusters. The same applies if A and D stand for subclusters rather than observed behaviour patterns.

A further assumption which is not essential but which it might be interesting to follow up is

(iv) Every decision the animal makes is a binary decision, a choice between two possibilities. This means that (A or B or C) must be dissectable into, for example ((A or B) or C). Any observed behaviour is then the consequence

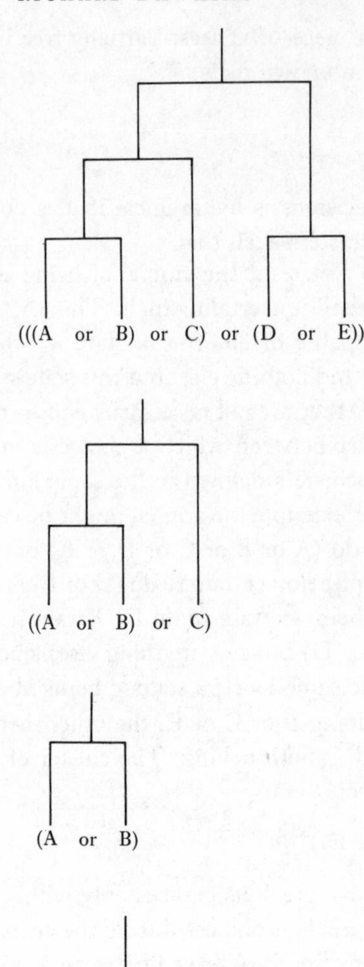

Fig. 4. From top to bottom, sequence of decisions leading to final choice of behaviour B, as explained in text. Illustrates equivalence between tree diagrams and bracket notation. Each inverted T-junction is a choice point.

of a particular series of binary decisions. In the case of B the decisions are

(((A or B) or C) or (D or E)) rather than anything else;
((A or B) or C) rather than (D or E);
(A or B) rather than C;
B rather than A.

Whether we adopt assumption (iv) or not, each successive decision is represented in this notation as entering a more deeply nested pair of brackets. It could equally well be thought of as dropping one level in a tree-diagram until a terminal branch (actual behaviour) is reached (Fig. 4). The equivalence of these two notations is familiar from 'list processing' computer techniques (Foster, 1967), techniques which form the basis of most of the programmes to be described below.

Evidence bearing on the assumptions

Assumptions (i) and (ii) were in a sense part of an earlier model which was rigorously tested through its quantitative predictions (Dawkins, 1969*a*; Dawkins & Dawkins, 1974 and references cited therein). This was a model of choice between external stimuli, for example coloured spots between which a chick might choose to peck. If a chick prefers red to blue, and blue to green, the following 'states' were admitted by the model: choice of red alone; choice of red or blue; choice of red or blue or green. The model was originally expressed in terms of 'thresholds', but in the bracket notation it can be represented as (((red) or blue) or green). The extra brackets round red indicate that there is a state in which only red can be chosen. An additional strong assumption was made about choices being exactly equally distributed among all colours eligible according to the hierarchical decision rule at any given time. This assumption flowed naturally from the 'threshold' way of picturing the model, but it will not be discussed here since it is not relevant to the idea of hierarchy itself. It led to the possibility of strong quantitative predictions being precisely deduced and tested. The success of these predictions gives some confidence in assumptions (i) and (ii) of the present model.

Assumption (iii) implies that there is not just one global set of transition rules governing all behaviour patterns of an animal, which can most economically be expressed in a single transition matrix. Rather it postulates nested sets of transition rules, each set of rules holding sway within a circumscribed cluster of elements. Transitions between elements which belong in different clusters at any particular level are predictable from the more global rules of transition between their respective clusters.

Fentress and Stilwell (Fentress, 1972; Fentress & Stilwell, 1973) counted frequencies of transition between different components of face-grooming in mice, and used an information measure of predictability in transition matrices of various order. The sequential structure revealed by these analyses fell far short of what the unaided human eye seemed to see. They therefore defined five higher-order units, recognised by eye, and given semi-precise definitions such as: 'Unit 4, repeated overhands with rare short licking interjected'. When frequencies of transition between these higher-order units were examined, a

auf \ folgt		Si	Fr	Ny	La	Fl	11Bo	11Lu	11	112	1f1	11R	11T	11K	1A1A	11Ar	11A1	33	332	33H	33F	3H3F	33T	3T3H	3T3F
		I							II									III							
I	Si		1	16	7	1	6	60	6		1					1		7		2	5				
	Fr	1		18	16	2	18	31	25			2			1	1		4	4						
	Ny	10	13		311	1	19	194	24	4	4	5			1	1	1	12	8	1	3		2	1	
	La	5	32	95		4	25	352	36	1	6	2			8	5	1	22	4	6	2				3
	Fl				10			13																	
	11Bo	8	19	50	14	2		110	41	3	2	3		1		5	3	11	1					1	
	11Lu	67	27	289	179	6	148		329	7	19	55	8	2	14	5	11	81	30	29	18		1	1	12
II	11	4	10	74	17		37	295		84	335	123	29	16	202	68	60	15	8	6	4				
	112		2	10	1			7	77		7		3		8			1	5						
	1f1		1	8	5		2	25	304	13		8	13	5	19	1	1		1	2					
	11R		2	4	1		4	11	156		3			1	7	5	1								
	11T							4	41	3	5				1										
	11K			1				2	14		2				1	6	2								
	1A1A						2		101	1		5	1	1		113	57								
	11Ar			4	3		1		187	1	1	1		1	17		3								
	11A1			1			2	4	124	1	1	1			1	3									
III	33	14	4	36	10		3	147	13	3	1				1	2	3		80	111	162	4	18	13	138
	332	1	1	8	2			21			1							109			3				1
	33H	2		5	1			11									1	168	2		53	13	1		5
	33F			2	3			13	2								1	268	2	68			5		6
	3H3F																	1		2	1		1		
	33T		1				1											12		1	9	1		1	4
	3T3H																	7		1	7				5
	3T3F			1	2			2										44		29	83	2	2	4	

auf \ folgt	I (Si, Fr, Ny, La, Fl, 11Bo, 11Lu)	II (11, 112, 1f1, 11R, 11T, 11K, 1A1A, 11Ar, 11A1)	III (33, 332, 33H, 33F, 3H3F, 33T, 3T3H, 3T3F)
I (Si, Fr, Ny, La, Fl, 11Bo, 11Lu)	$\chi^2 = 844.6$, d.f. = 25, ***, $C = 0.528$	$\chi^2 = 59.1$, d.f. = 48, N.S., $C = 0.290$	$\chi^2 = 34.4$, d.f. = 42, N.S., $C = 0.334$
II (11, 112, 1f1, 11R, 11T, 11K, 1A1A, 11Ar, 11A1)	$\chi^2 = 37.5$, d.f. = 48, N.S., $C = 0.254$	$\chi^2 = 1693.7$, d.f. = 49, ***, $C = 0.656$	$\chi^2 = 9.0$, d.f. = 56, N.S., $C = 0.421$
III (33, 332, 33H, 33F, 3H3F, 33T, 3T3H, 3T3F)	$\chi^2 = 10.9$, d.f. = 42, N.S., $C = 0.190$	$\chi^2 = 0.9$, d.f. = 56, N.S., $C = 0.174$	$\chi^2 = 730.5$, d.f. = 36, ***, $C = 0.579$

Fig. 5. Contingency tests on portions of Seibt's (1972) table of transition frequencies, as explained in text. Upper table gives frequencies with which acts listed as row headings were followed by acts listed as column

high degree of predictability was found. They conclude, 'These data provide a direct demonstration of the hierarchical structure in ongoing grooming behaviour which has often been postulated by workers in both the behavioural and neurological sciences.'

In our fly grooming study already cited, we made a half-successful attempt to test assumption (iii) directly (Dawkins & Dawkins, 1976). Here I shall illustrate the same method using data from other published literature.

Myrberg's method of arranging his tables of transition frequencies into sub-tables has already been mentioned. Seibt (1972) gives a similar table of frequencies of transition between grooming and other movements in diopsid flies, which she divides into three groups. Group 1 are non-grooming movements, group 2 grooming movements involving the front legs, and group 3 grooming movements involving the rear legs. As in our study of blowflies she found that group 2 and group 3 grooming movements tended to be separated from each other, and she concluded that they formed two self-contained complexes, each with its own 'grooming-programme'. She separated group 1 movements on the same kind of basis.

As explained above, mere temporal clustering is not in itself evidence for hierarchical organisation in Simon's (1962) sense or in the sense of the present model. A direct test of assumption (iii) would consist of a demonstration that, in the case of between-cluster transitions, which member of the second cluster is chosen is not influenced by which member of the first cluster had previously been performed. In the case of within-cluster transitions, however, considerable sequential influence may occur. For Seibt's data for instance, in those cases where a transition is observed from a member of group 2 to a member of group 3, the probability that the latter will be any one of the eight members of group 3 should be uninfluenced by which of the nine members of group 2 the former was. This may be tested by subjecting the sub-table involving transitions from any member of group 2 to any member of group 3 to a 9×8 χ^2 contingency test. χ^2 for 56 d.f. is only 9.043 which

headings. Lower table shows results of contingency test on each of the nine main sections. Stars indicate statistical significance. In general prediction is fulfilled that significant interactions occur in submatrices around main diagonal, but not in other portions. To correct for effects of large numbers, Pearson's coefficient of contingency $C = \sqrt{\chi^2/(N+\chi^2)}$ is calculated for all nine cells, where N = total number of contingencies. Again prediction supported. Original table has no entries along main diagonal – behaviour patterns by definition cannot follow themselves. Strictly speaking this necessitates special treatment in calculating χ^2 (Slater, 1973). However, the effect is so large that the precaution can be omitted. N.S. = not significant; d.f. = degrees of freedom.

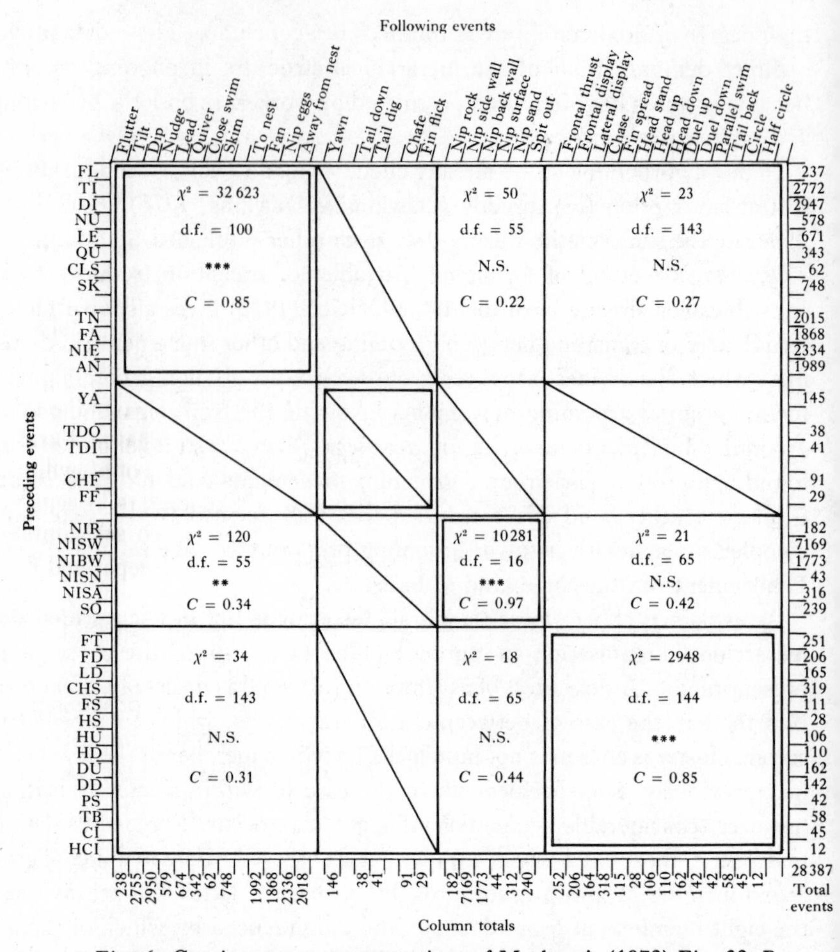

Fig. 6. Contingency tests on portions of Myrberg's (1972) Fig. 32. Row and column headings copied exactly from Myrberg. Three main submatrices around main diagonal are emphasised by a double line. All portions of the table not involving an interaction among these three ruled out (diagonal line). In remaining nine portions contingency tests were done as explained in text. Stars indicate statistical significance. In general prediction is fulfilled that significant interactions occur in the three square submatrices, but not in other portions (except one). To correct for effects of large numbers in square submatrices, Pearson's coefficient of contingency $C=\sqrt{\chi^2/(N+\chi^2)}$ is calculated for all cells. Again prediction supported. To eliminate effects due to behaviour patterns leading to themselves, all entries along main diagonal of original table have been eliminated.

accords with the prediction. On the other hand if the members of group 2 are subjected to a within-cluster 9×9 contingency test, χ^2 is 1693.690, showing highly significant within-cluster interaction again as predicted. Fig. 5 gives the equivalent χ^2 for all nine major subdivisions of Seibt's table. The prediction that the three entries along the major diagonal may be highly significant but all the rest will not be, is fulfilled.

The row and column headings of Fig. 6 are reproduced directly from Myrberg's (1972) Fig. 32. The body of the table is divided into the main regions discernible in Myrberg's arrangement ('partial decomposability analysis' described above). The detailed figures are replaced by a χ^2 value and contingency coefficient for each region of the table. The prediction of the model is fulfilled, with the exception of one cell.

*A cluster analysis based on Mutual Replaceability**

Normally a cluster analysis begins with a matrix of similarities or distances between elements (Everitt, 1974), and proceeds to group elements together which have high similarity. The single-linkage cluster analysis illustrated above (Fig. 3*b*) used temporal proximity as the index of similarity. Grooming acts were likely to be clustered together if they tended to occur in the same 5 second period. The method of rearranging transition matrices so as to maximise the concentration of entries in square submatrices is, as we saw, a form of two-levelled cluster analysis; in this case the index of similarity was probably of sequential contiguity.

But for the present model the appropriate index of 'clusteredness' of two behaviour patterns is not any form of temporal proximity, but rather *mutual replaceability* (cf. Kalmus, 1969) as far as between-cluster transitions are concerned. For example the blowfly head-grooming and proboscis-grooming movements belong in the same cluster, not because they tend to occur close to each other in time, but because their transition-relationships with members of other clusters are nearly the same: they are mutually substitutable in those parts of a transition matrix which do not involve their relationships with members of their own cluster. The χ^2 analysis above tested this kind of prediction using preconceptions about which behaviour patterns ought to be clustered together. The following method of cluster analysis was developed to discover which behaviour patterns are clustered together given no specific preconceptions other than the assumptions of the model itself. Assumption (iv), the one about all decisions being binary, was included for the mundane reason that it made computation easier.

* When I wrote this I was unaware that similar methods had been used before (Maurus & Pruscha, 1973).

The programme is provided with an ordinary first-order transition matrix as data. The arrangement of the rows and columns is irrelevant. It examines all possible pairs of behavioural acts in turn, calculating for each pair an 'index of mutual replaceability', which is the mean of two correlation (Spearman rank unless otherwise stated) coefficients r_r and r_c. r_r is the correlation between two rows, excluding the entries involving mutual interaction within the pair under investigation. r_c is the corresponding figure for the two columns. Having found that pair with the highest index of mutual replaceability, it designates them as members of the same cluster and prints out their names bound together in brackets. Following assumption (iii) of the model it then collapses the table so that no further distinction is made between these two behaviour patterns; their entries are lumped by simple addition. The whole operation is then repeated on the condensed table, and this continues until only two entries are left, or until no good correlation can be found, as defined by an arbitrary criterion. At each stage the pair of elements with the highest index of mutual replaceability is printed out; sometimes these two elements are single acts; sometimes they are already identified and lumped clusters, in which case a nested bracketting notation is used.

The method can be used on any transition-frequency data. Examples using published data are given in Fig. 7*a* and *b*. The clusterings of Myrberg's behaviour patterns can be compared with those which he arrived at (Fig. 6). Fig. 3*c* shows its application to blowfly grooming where it can be compared with a Single Linkage method already discussed. Do not expect that conventional methods of analysis based on temporal or sequential proximity will necessarily give similar clusterings to the Mutual Replaceability method. Two acts which are mutually replaceable would very probably *not* be sequentially close. To use Kalmus's (1969) analogy of a menu, two different fish dishes are mutually substitutable in the second slot of a four-course dinner; they are therefore unlikely to be served up in succession.

Fentress & Stilwell (1973) conclude by suggesting an analogy between mouse grooming and 'human grammar in which individual letters form different combinations in different words which in turn are sequentially arranged into phrases...'. As they point out similar suggestions have been made before. These will be discussed later. To anticipate, the main reason why I have preferred not to call the model discussed in this section grammatical is that a grammar is more than just a hierarchical system in which the higher units as well as the lower units have their own laws of transition. In addition, a sentence has a definite structure; at its crudest, it has a beginning, a middle and an end. A better way of putting this is in terms of 'correct nesting of brackets', and we will return to this later.

The behavioural analogy is with syntax not semantics, and it has nothing necessarily to do with the communicatory role of language. In some ways it

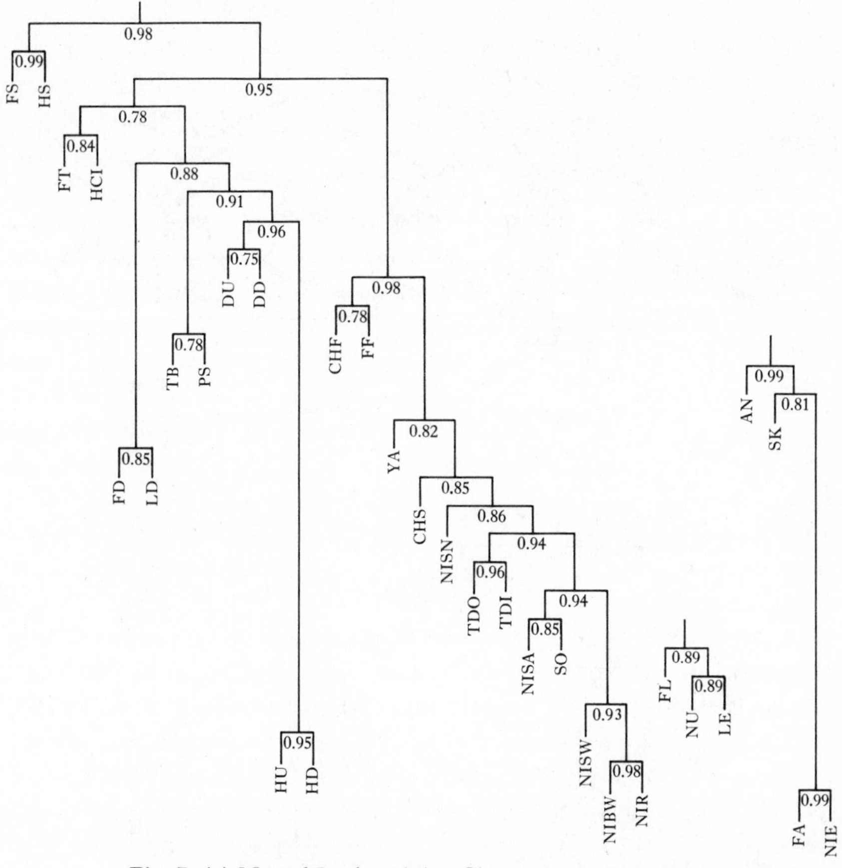

Fig. 7. (*a*) Mutual Replaceability Cluster Analysis on data from Myrberg (1972) on bicolour damsel-fish. Behaviour names as in Fig. 6. Each clustering of two subunits represented by one inverted T-junction. Members of earlier-formed clusters (*lower*), are lumped for later clustering (*higher*). Under each T-junction is Index of Replaceability for the two units joined. This is the mean of two correlation coefficients as explained in text. In this case product moment correlations used, as Spearman rank too costly of computer time.

is better to begin by comparing behaviour with simpler artificial 'languages' which were never designed for communication. These are the subject of the next section.

PATTERNS OF PATTERN

Several authors have developed what may be called pattern languages in the course of studies of serial pattern learning, the learning by human subjects of long sequences of symbols, digits, letters or responses. Simon (1972) has

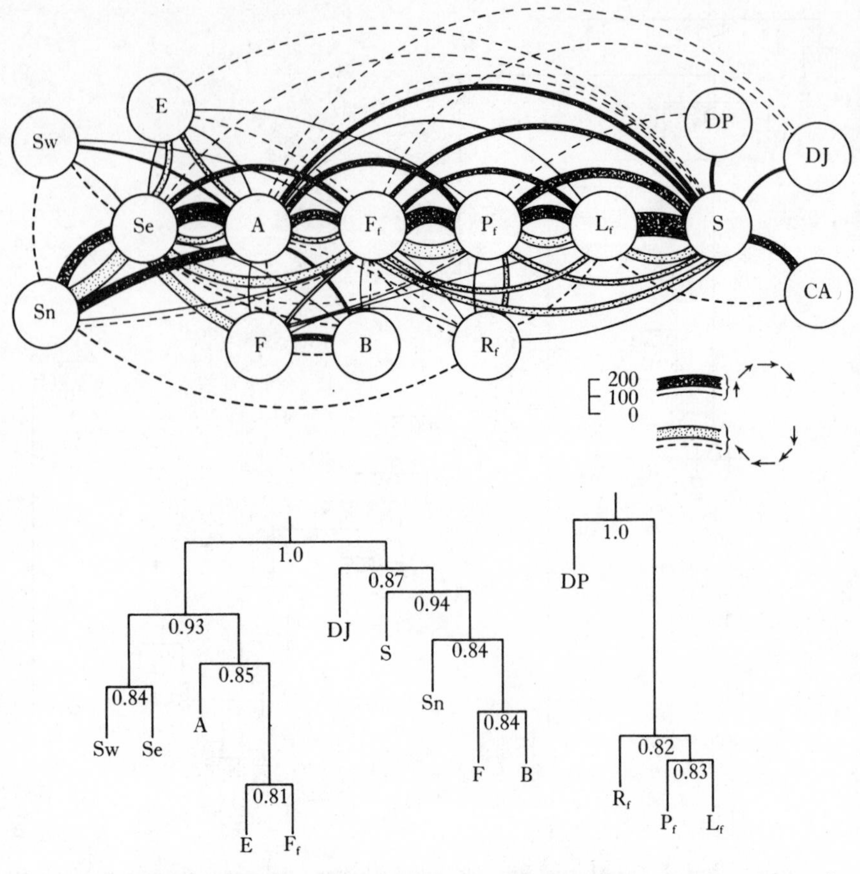

Fig. 7. (*b*) Mutual Replaceability Cluster Analysis on data from Baerends, Brouwer & Waterbolk (1955), on courtship behaviour of male guppies, *Lebistes reticulatus*. As Fig. 7*a*, except Spearman rank correlations used. Upper part of figure: representation of transition frequencies by original authors. Behaviour symbols are those of original authors.

shown that, although the formulations of the various authors seem different, they are mostly equivalent. I take the work of Restle (1970) as representative. He used a box with six numbered buttons and six corresponding lights. The lights flashed on in a non-random repeated pattern. The subjects' task was to anticipate each light by pressing its button first. Some serial patterns proved harder to learn than others, and mistakes occurred at certain points in the sequence more than others. The aim was to explain these findings in terms of theories about how the subjects encoded the sequences in their memories.

Naive 'stimulus–response chain' theories could quickly be dismissed, including more sophisticated versions in which each response was associated not only with its immediate predecessor, but with several predecessors. The subjects then were not behaving like a Markov process of any order. Nor were they simply storing the information in an ordered set of 'locations' as a computer might; if they were they would have learned a random sequence as readily as a patterned one. Restle used simple pattern languages to develop his theories of how they were doing it. These languages were ways of expressing a sequence using fewer symbols than the sequence itself. In each case the hypothesis was that the subjects were remembering, not the full sequence, but a set of *rules* for generating it, rules expressed in some formal equivalent of the 'language' under investigation.

A set of rules – I shall call it a programme – for generating a sequence would take the general form

(i) start with button x;

(ii) perform some transform on x to select the next button, for instance move to the next one on the right;

(iii) and subsequent steps. Perform some transform on the previous thing you did.

The important point is that in stage (iii), 'the previous thing' does not necessarily mean 'the last button'. The successful languages were all recursive, which means that 'previous thing' might be the last actual response, but it might be instead the last transform executed, either on a single response, or on an inner nested transform.

Particular transforms suggested by Restle were:

$T(x)$ move one to the right of x
$R(x)$ repeat x
$M(x)$ do the mirror of image of x

Thus $T(1) = 2$, $R(1) = 11$, $M(12) = 65$ since 6 is the mirror image of 1 on the button box, and 5 the mirror image of 2. $T(12) = 23$, $R(12) = 1212$, $R(R(12)) = 12121212$, $T(R(1)) = 1122$, $M(T(R(1))) = 11226655$. In all cases an equivalent tree-diagram can be drawn. For example the last 'programme' corresponds to Fig. 8*a*.

The sequence 1166225511662255225533442255334 4 is an example of one which proved easier to learn than one would naively suppose from its length. Restle suggested that this was because such sequences have a hierarchical structure, which in this case can be represented by $T(R(T(M(R(1)))))$. On the assumption that subjects represented the sequence internally by a set of rules equivalent to this formula, predictions were made about where in the sequence errors should be most likely to occur. They, and similar predictions for other sequences, were fulfilled. That particular 'programme' may be decoded into behaviour as follows: do behaviour 1; repeat it; do the mirror

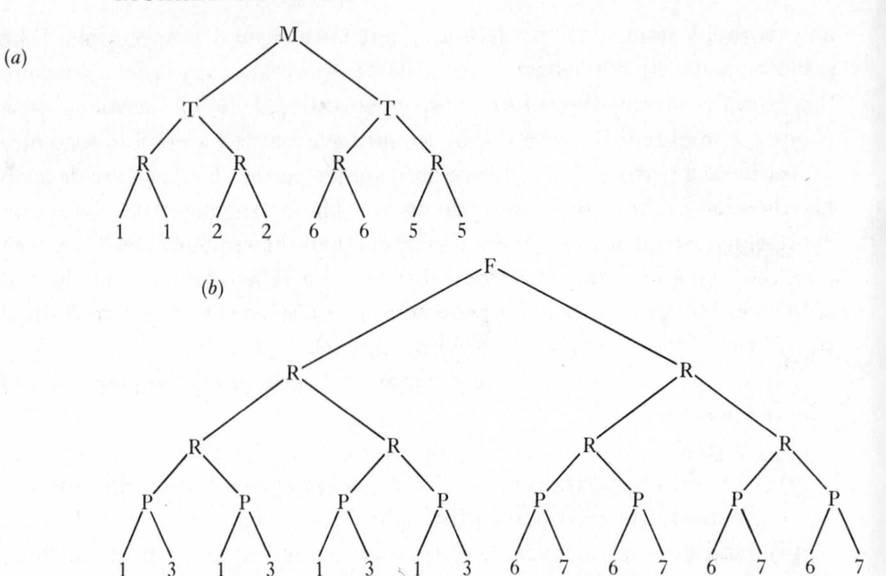

Fig. 8. (*a*) Tree-diagram illustrating 'programme' M(T(R(1))) for Restle's button box. (*b*) Equivalent diagram for hypothetical fly grooming programme F(R(R(P(1)))). Explanation in text.

image of the whole repeated unit; transpose both mirror halves; repeat everything done so far; transpose the whole repeated, transposed, mirrored, repeated unit.

Symmetrical binary trees have obvious limitations, for example they can only generate sequences whose number of elements is a power of two. Restle therefore also considers 'right-branching' asymmetrical trees. Using them he makes a spirited attempt to analyse J. S. Bach's Two-part Invention No. 1, and he suggests that the method is applicable to many forms of human behaviour, including such complex skills as playing the piano. I now leave human psychology, and return to animal behaviour.

The nature of the proposed analogy should be clear. The animal is supposed to generate sequences of behaviour by following economically stored sets of rules of the same type as those used by the subjects in the learning experiments. The following is a sequence of behaviour recorded from a grooming blowfly: 131313131313141414167676 (Dawkins & Dawkins, 1976). Each digit stands for one act, for example 1 is rubbing the front legs together, 3 is grooming the head with the front legs, 6 is rubbing the back legs together and 7 is grooming the abdomen. The most obvious patterns that can be seen are what Restle would call 'trills', long periods of alternation between say 1 and 3, and 1 and 4, However, before we rush into a direct search for hierarchical rules we must be careful. The numbers on Restle's box had some ordinal significance; thus a 121212 trill represented repeated pressing of a button and its neighbour. But the assignment of numbers to grooming movements is purely arbitrary. The operations 'transpose' and 'mirror' have no obvious meaning.

This is not to say that some sort of meaning for them could not be found. For example we might hypothesise the following transformations, some of them making use of the concept of 'postural facilitation' (Dawkins & Dawkins, 1976):

R(x) As in the case of Restle, repeat x

A(x) Groom that part of the body immediately anterior to x

P(x) Groom that part of the body immediately posterior to x

M(x) Groom that part of the body which is the left/right mirror image of x

F(x) Groom that part of the body which is the fore and aft mirror image of x, for instance 'rub rear legs together' might be the mirror image of 'rub front legs together'

As before, x might refer to a single grooming act, or more interestingly it might refer to a higher order unit as in F(R(R(P(front leg rubbing)))). This could be decoded as: rub the front legs together; move one stage posteriorly and groom the head; repeat the whole thing twice; then do everything again but fore/aft reversed, substituting back legs for front legs and abdomen for head (Fig. 8*b*). The whole sequence would be 131313136767676 7. Predictions could be made and tested about which sorts of sequences should and should not occur commonly if fly behaviour is generated by various programmes of calls of these procedures. We have not tried this yet, but it would be interesting to do so, and there may be other kinds of animal behaviour, bird song perhaps (cf. Nelson, 1973), for which a similar approach could be worthwhile. This is a suggestion for the future.

An algorithm for detecting patterns in behaviour – 'melodies'

The suggested transpositions given above are based on human preconceptions of what might be reasonable. Another approach is to scan the data for patterns which actually do occur, and for patterns of pattern. The following algorithm attempts to do this.

The programme is provided with raw data consisting of digits signifying acts, in the order in which they occurred. It scans through the data counting frequencies of doublet transitions. When it has found the commonest doublet it prints out the two behaviour names in order, bound together in brackets.

Frequency	Doublet
+83	(6 0)
+79	(1 3)
+65	((1 3)(1 3))
+51	(7 (6 0))
+45	(1 0)
+25	(1 2)
+23	(((1 3)(1 3))((1 3)(1 3)))
+21	(7 6)
+19	(1 4)
+16	((7 (6 0))8)
+15	((6 0)8)
+14	((1 2)(1 2))
+13	((1 0)(1 0))
+12	((1 4)(1 4))
+11	((7 (6 8))6)
+9	((7 (6 0))(7 (6 0)))
+7	(6 8)
+6	(5 (6 0))
+6	((1 3)(1 0))
+5	(0 (6 0))
+5	(6 (7 6))
+5	((1 0)(1 2))
+5	((((1 3)(1 3))((1 3)(1 3)))((1 3)(1 3)))
+5	((7 6)(7 6))
+5	(((1 2)(1 2))((1 2)(1 2)))
+4	(6 ((7 (6 0))6))
+4	(7 0)
+4	((1 0)4)
+4	(((6 0)8)((6 0)8))
+4	(((1 0)(1 0))(1 0))
+3	((7 (6 0))(7 6))
+3	((((1 3)(1 3))((1 3)(1 3)))((1 3)(1 0)))
+3	(((7 (6 0))8)(7 (6 0)))
+3	(((7 (6 0))8)((6 0)8))
+3	(((1 4)(1 4))(1 4)).

Fig. 9. Computer recognition of patterns in sequences of grooming movements by one blowfly 'April' (from Dawkins & Dawkins, 1976). Common doublet sequences of acts, and doublet sequences of already detected doublets (etc.), printed out in order of commonness. e.g. doublet $6 \rightarrow 0$ occurred 83 times and was commonest. When this was replaced by a single symbol (6 0), commonest doublet in altered record was $1 \rightarrow 3$. When this was replaced by (1 3), commonest remaining doublet was (1 3) $\rightarrow$ (1 3), which occurred 65 times, and so on until no doublets which occurred more than twice could be found. The two halves of each doublet may be identified from nesting of brackets.

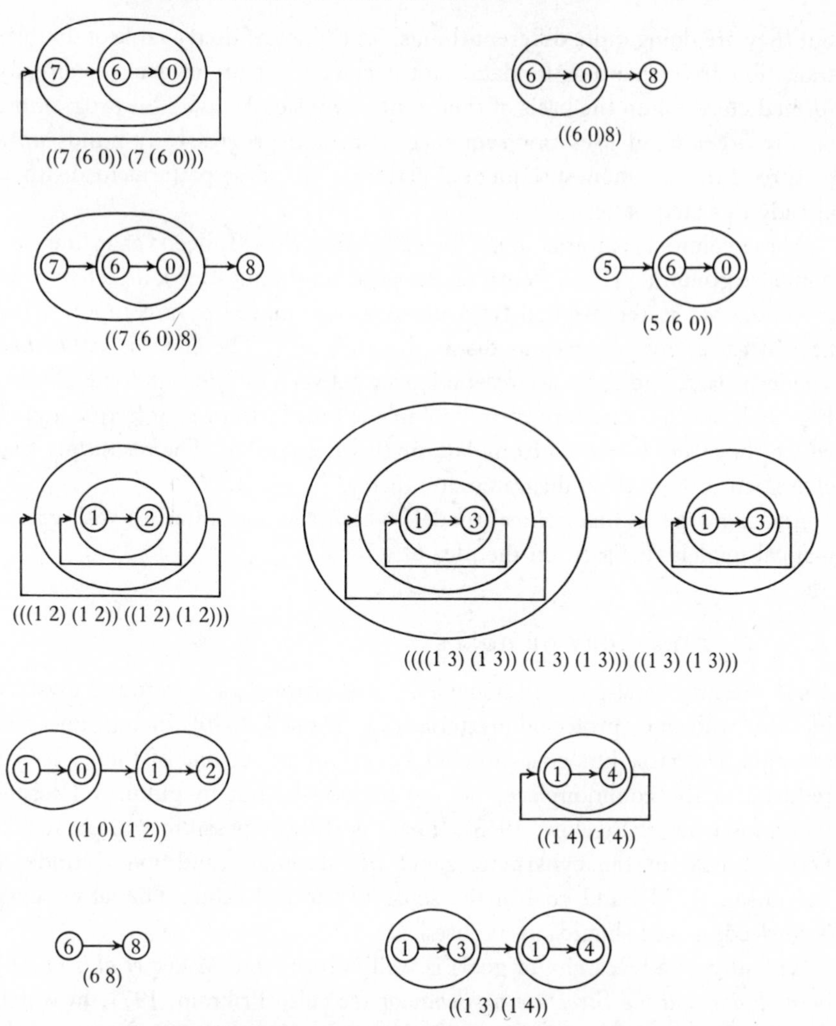

Fig. 10. Illustrations of some of the commoner 'doublets' from Fig. 9. Data from grooming sequences of a blowfly. Numbers refer to grooming acts as listed in caption to Fig. 3.

It then goes back over the data, replacing all occurrences of that doublet by a single symbol which stands for the doublet. It then repeats the process on the altered record. At each stage, one doublet is printed out, the elements of the doublet being either single behaviour names, or already identified patterns, represented in a nested bracket notation. Similar algorithms have been developed for detecting the natural segmentation of language (Wolff, 1975).

Superficially this programme may seem like the other one described above, but they are doing quite different things. The first one used a table of doublet transition frequencies as its data, and it clustered pairs of acts (or already formed clusters) on the basis of their mutual replaceability. This programme on the other hand uses raw sequence data, and progressively builds up a picture of the commonest sequential patterns, including patterns made up of already detected patterns.

These common patterns might be called melodies. Indeed, as in the case of mouse grooming (J. C. Fentress, personal communication), if each blowfly grooming act is represented by a musical note played by a computer, the human ear seems to pick out distinctive melodies. The effect is rather like 'modern' jazz, and there are resemblances between different individual flies. Fig. 9 shows the common 'melodies' in the order (decreasing frequency) in which they were extracted from data by the programme. The results are also expressed in Fig. 10 in diagrammatic form.

Another type of hierarchical model, which also has affinities with grammatical models, is based on the idea of a 'goal'.

HIERARCHY OF GOALS

Until recently 'goal-directed behaviour' was claimed as a profound mystery by those who enjoy profound mysteries (e.g. Russell, 1946). Biology does still present mysteries, but goal-directed behaviour is not one of them. It was reduced to the commonplace, not by philosophy but by gunnery (Rosenblueth, Wiener & Bigelow, 1943). There is still the possibility of confusion between goal in the cybernetic sense of 'stopping condition' (Hinde & Stevenson, 1970), and goal in the sense of survival value. The latter sense is misleading and should not be used.

The idea of a hierarchy of goals is well known from Miller *et al.*'s (1960) book *Plans and the Structure of Behavior* (see also Pribram, 1971, in which the idea is extended by the incorporation of feedforward). They are mainly concerned with human psychology, but their book has a chapter which calls attention to the similarity between their own hierarchical 'Plans', and the hierarchical 'Instincts' of Tinbergen, which they discuss with approval. They might have been even more pleased by Kortlandt's work (1955) in which, although it can be criticised (Hinde, 1957), the link with their own ideas was more direct.

Kortlandt's is a hierarchy of 'appetites', where an appetite is defined by the condition which brings it to an end. He saw the nest-building behaviour of his cormorants as mediated through the arousal of hierarchically subordinate appetites. Fig. 11 illustrates his term 'concentric purposiveness' in two equivalent ways. It could also be easily drawn by Miller *et al.* as a 'TOTE'

(*a*)

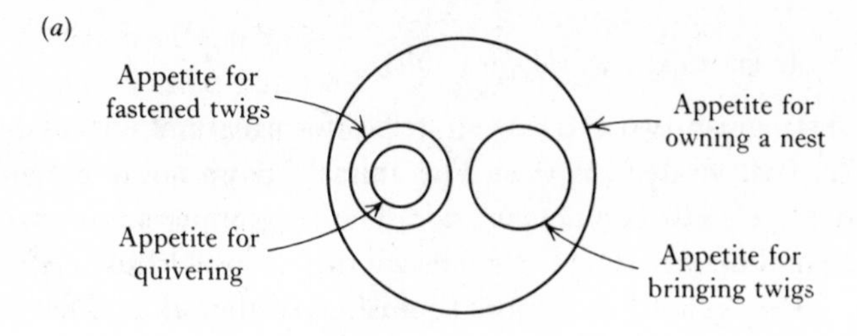

(*b*)

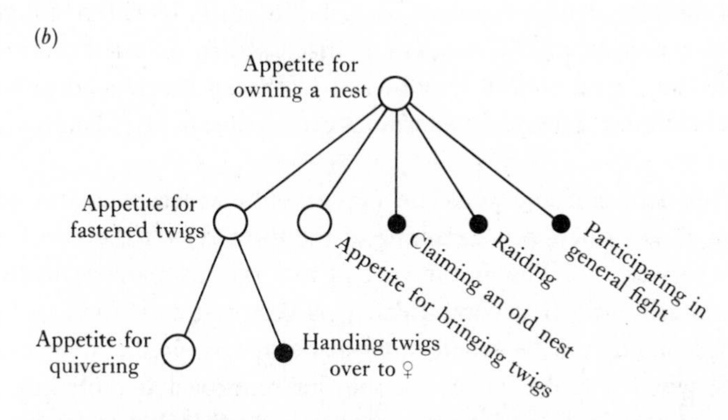

Fig. 11. 'Hierarchy of appetites' in the male cormorant *Phalacrocorax carbo* (from Kortlandt, 1955). Each appetite is defined by the state which brings it to an end. It may achieve its 'goal' by calling up subsidiary appetites. Thus the appetite for owning a nest is superior to the appetite for fastened twigs which in turn is superior to the appetite for quivering. (*a*) 'Concentric purposiveness'. (*b*) Equivalent, but extended, tree-diagram.

hierarchy. Thorpe (1963) gives an illuminating discussion of how nest-building and artifact construction in general can be understood in similar terms.

Hinde and Stevenson (1970), also in the context of nest-building, warn against over-enthusiastic interpretation of behaviour in terms of goals. They go on to provide an admirable classification of the rather diverse kinds of phenomena which might be called goal-directed, and suggest that for classification purposes the category may be too heterogeneous to be useful. Nevertheless I believe the following general functional or economic argument may be worth putting, since it lays particular stress on the hierarchical organisation of goals. It has affinities with Simon's watchmakers.

Action rules and stopping rules

We may distinguish two extreme strategies for programming adaptive behaviour. The goal strategy is this. The animal knows nothing about how to behave adaptively except a *stopping rule*. The programme simply says 'Thrash around at random until state G is achieved, then stop'. Orthokinesis (Fraenkel & Gunn, 1940) is not much more sophisticated than this. That part of the programme which may be called the *action rule* is as simple as it could be, 'thrash around'. The goal state G may be simple as in orthokinesis, or it may be very complex. Under the name 'British Museum Algorithm' a slightly more systematic version of this strategy is discussed in the artificial intelligence literature as a kind of null hypothesis. In theory it can solve any problem, however complex, but only at enormous cost in time (Feigenbaum & Feldman, 1963).

The opposite extreme would be a programme which put all the complexity into the action rules, a programme of the form 'Do A then do B then if X do C otherwise...' This is the type of programme conventionally (though not necessarily) given to computers. It is very fast and efficient, but only if the environment can be relied upon to be utterly predictable; otherwise it will fail when it stubs its toe on the first unexpected mole-hill. It is not possible to plan in detail for every contingency; there are too many of them.

Is there any way of combining the virtue of stopping rule programmes, imperturbability in the face of unpredictable conditions, with the virtue of action rule programmes, speed and efficiency? Yes. The solution lies in hierarchically nested stopping-rule programmes. This is the basis of the models of Kortlandt, and of Miller *et al.*

Stopping-rule programmes can be fast, provided the goal state is simple. This may be because it is sometimes possible to measure quantitatively the discrepancy between the goal and the present state, in which case the full power of negative feedback can be brought to bear (McFarland, 1971). But even if this is not so, simple stopping conditions may be achieved rapidly, because simple means not improbable. A complex goal state, like the particular permutation of letters which is any book in the British Museum, is inherently improbable. A goal state like 'stomach full of zebra meat' is too complex to achieve through random movements. A predator programmed with only such a stopping rule might take millions of years to achieve a square meal. It is obviously better to break down complex and improbable goals, into a series of simple goals which can be more rapidly achieved. For instance immediately subordinate subprogrammes might be: stop searching when zebra seen; stop pursuit when zebra very close; stop killing when zebra motionless; stop eating when stomach full. 'Searching', 'pursuit', 'killing' and 'eating' are deliberately left vague. Each of them would have its own programme, which might consist purely of action rules, but more probably, since they are all still quite complex, each one would call up its own subordinate stopping-rule programmes. Even at the very lowest level, there seem to be stopping-rule programmes, in the form of the γ-efferent servo loops. Action rules are perhaps mainly confined to determining the order in which stopping rules or targets are set up.

Another aspect of hierarchies of goals may be expressed as 'perfection of nesting', using the word in the sense of nesting of brackets! This is best discussed after considering grammatical models of behaviour.

GRAMMATICAL MODELS

The idea of some similarity between the principles underlying language and those underlying the serial organisation of behaviour in general is obviously of great interest from many points of view, including that of the evolution of language. It seems to have originated from Lashley (1951), but to have been first turned into an explicit model by Marshall in an unpublished paper in 1965. His 'phrase structure grammar' model did not become widely known until it was discussed in print by Hutt & Hutt (1970), and by Vowles (1970) who added some speculations about neurophysiological implications and about 'transformational grammar' (Chomsky, 1957). Meanwhile Kalmus (1969) had independently developed the analogy.

A major aim of grammarians at the time of Marshall's paper was to write down rules of the syntax of a particular language, in such a way that in theory a machine embodying these rules could generate all grammatical sentences recognised as correct by native speakers, and no ungrammatical sentences (Chomsky, 1957). Such a machine would only be concerned with syntactics, with deciding when to emit a noun, an adjective, a relative pronoun etc. The selection of *which* noun and which adjective is a semantic matter. It is not obvious what an analogy of semantics with animal behaviour would mean; perhaps something to do with the functional achievements of behaviour.

A computer programmed to follow some typical hierarchical and recursive rules of syntax, produced the following sentence (ignore the underlinings for the moment): 'The adjective noun of the adjective noun which adverbly adverbly verbed in noun of the noun which verbed adverbly verbed.' Dissect it carefully, and you will find that, although meaningless, it is syntactically correct English. The fundamental sentence is underlined: 'The adjective noun adverbly verbed'. All the rest consists of qualifications of the subject *noun*, in the form of hierarchically nested relative, prepositional and possessive clauses. The programme employed random numbers to make its successive choices, and it continued to generate prolific grammatical gibberish. However, the interest of it is not in its randomness, but in the nature of the *units* between which each random choice was made.

If the units had been *words*, with choices biased to follow transition probabilities actually found in English, the sentences generated would at first sight have looked acceptable. Adjectives would precede nouns and adverbs would precede verbs. If the Markov process simulated was one of high order, whole phrases would come out looking plausible. But there would be one thing fundamentally wrong: clauses which had been begun would not be properly finished. If we represent the opening of a new relative or possessive clause as equivalent to the opening of a bracket, the Markovian programme would not correctly close all brackets opened. In terms of the above example, the Markovian model would drift off into relative and possessive clauses, and would not 'remember' that there was an initial subject still waiting for its main verb.* Even the 'decision cluster' model discussed above would not perform any better.

Our phrase structure grammar programme on the other hand achieved this correct rounding off of 'brackets' effortlessly, and it would have done so no matter how many subsidiary clauses had been opened. This is because each random number determined the choice not of a word, nor of a fixed number of words, nor even of a cluster from which words might be chosen, but of a *procedure*. These procedures, with names like 'noun-phrase' and 'verb-phrase' themselves used random numbers to choose either words or other procedures (including themselves) and so on. For example the procedure noun-phrase might generate any of the following: 'noun', 'adjective noun', 'noun of' noun-phrase, 'noun which' verb-phrase, and many others. Since the procedures noun-phrase and verb-phrase are recursive (i.e. they call themselves), there is no limit to the depth of nesting of 'brackets' which will be correctly rounded off.

Marshall (I am referring to secondary sources only) proposed a grammar of this type to generate the sequences of pigeon courtship described by Fabricius & Jansson (1963). The 'words' are seven behaviour patterns with names like Bow and Copulate. The higher units (equivalent to noun-phrase, etc.) have names like Preparatory and Consummatory. The complete grammar is given in Fig. 12, both in tree-diagram form, and in Algol-60. The latter is preferred to Marshall's own metalinguistic symbolism, which it closely resembles, as it may be directly run on a computer as a simulation of a pigeon, and the results of one such run are given in the Figure. It should

* Compare the case of a woman with a severe lesion of the left frontal lobe, who wrote: 'Dear Professor, I want to tell you that I want to tell you that I want to tell you that I want to tell you...' and so on for several pages (Luria, 1970).

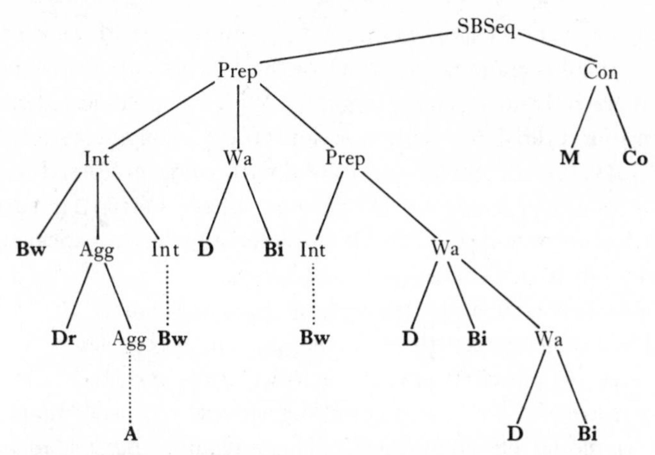


```
begin comment Marshall's pigeon grammar;
procedure SBSeq; begin Prep; Con end;
procedure Prep; begin Int; Wa; if p then Prep end;
procedure Int; begin "BW"; if p then Agg; if p then Int end;
procedure Agg; begin if p then "DR"; if p then "A"; if p then Agg
        end;
procedure Wa; begin "D"; "BI"; if p then Wa end;
procedure Con; begin "M"; "CO" end;
Boolean procedure p;
  begin comment true or false at random. Probability manipulated.
  end;
start: SBSeq; goto start
end of pigeon grammar;
```

Sample results of running the programme:
BW DR D M CO
BW A D BI BW DR D BW A D BW A D BI M CO
BW A D BI M CO
BW DR D BW DR D BI BW DR D BI BW A D BW A D M CO

Fig. 12. Marshall's pigeon grammar (recursive version) (from Hutt & Hutt, 1970, courtesy of C. Thomas Publ.) after Marshall (unpublished). Underneath is the same grammar written out in Algol-60, followed by some results of running the programme as a simulation of a pigeon.

be self-explanatory even to those who do not know the language (**begin** and **end** are brackets: the programme consists mostly of definitions of **procedures** in terms of other procedures, followed by repeated final 'calls' of the most global procedure SBSeq).

Marshall shows that his grammar accounts for many of the findings of Fabricius & Jansson. However, we have to remember that he chose one grammar out of a very large number of possible ones.* Moreover there is no indication that his grammar accounts for the data better than a Markov model, though it certainly does so more elegantly. With data predigested into the form of a transition probability table it is doubtful if a convincing test of the two models could be accomplished. Raw data or data processed as in Fig. 9 ('tunes') would probably be better, but I have not so far succeeded in framing definite testable predictions characteristic of grammatical models, and I can only put it as a challenge for the future.

The problem is this. In the case of human language the criterion for grammatical correctness is the judgement of a native speaker of the language. In the case of animal behaviour 'correct' and 'incorrect' have no such meaning (Altmann, 1965). It is not obvious for example what might constitute correct 'rounding off of brackets'. There seem to be two main ways of starting to look at it and it is best not to muddle them together. The first is in terms of pattern as discussed in the section on pattern languages. The second is in terms of goals as discussed in the previous section.

A 'correctly rounded off' serial pattern might show itself in observed behaviour in the following general way. Let A and B be two particular acts. Then we might observe that the sequence AXB is very common, where X is not one particular act, but any one of a set. In a sense then the performance of A opens a sequence which demands to be 'closed' by the occurrence of B, regardless of what X happens to be. Now let X stand for, not one single act but a series of unfixed length – a subsidiary pattern in fact. Thus we might see A12B, A12121212B and so on. In all of these the sequence AXB is still there, but no Markov chain analysis would ever detect it. It is a task for the future to find convincing examples of this, but a start has been made by Dawkins & Dawkins (1976).

Correct 'rounding-off' also has meaning with respect to goals. When a subsidiary goal is set up in the service of a more global one, the global one presumably can be still 'there' in whatever sense a goal is ever 'there'. Thus if a hyena sets up a subsidiary goal of 'the other side of the hill' in the service of the more global goal of 'catch zebra', it may be that in some sense the goal 'zebra' is still 'set up'. In the case of spotted hyenas (*Crocuta crocuta*) this is quite plausible, since Kruuk (1972, and personal communication) could tell in advance what prey hyenas were setting out to hunt, and they would not be distracted by the 'wrong' prey, even of a species which on another occasion they might set out to hunt. Correct 'rounding off' of goals then might

* Harding (1971) shows that the number of labelled binary tree shapes of degree n is $\frac{(2n-1)!}{2^{n-1}(n-1)!}$.
For seven types of pigeon behaviour, $n = 7$ and the number of possible grammars in 10395 discounting variants due to optional choices and recursion. The real number is very much larger, but many can be immediately rejected.

show itself as a tendency to return to a global appetitive pattern after subsidiary patterns have been initiated and satisfied, and also after distraction or interruption. Searching behaviour and the notion of a 'searching image' is of great interest to ethologists and ecologists (Dawkins, 1971; Krebs, 1973). The possibility that whatever is searched for may be hierarchically nested should be borne in mind in future.

If it is ever possible to attach meaning to 'correct' and 'incorrect' in animal behaviour, either in terms of statistical rarity, or in terms of functional appropriateness ('displacement activities'?), it might then be possible to analyse such 'mistakes' as do occur in the same kind of way as linguists have analysed speech errors to dissect underlying organisation (Fromkin, 1973). Spoonerism in song-birds would make a good subject for this centre of bird-song research (see Thorpe & Hall-Craggs, this volume).

SUMMARY COMPARISON OF BEHAVIOURAL MODELS DISCUSSED

Markov models of any order (our non-hierarchical 'null-models') have the following property. Events influence future events, and the degree of influence is less for the distant future than for the near future, a decreasing monotonic function of time. If behaviour sequences are regarded as the consequence of a corresponding sequence of decisions (as defined above), in the case of simple Markov models there is a one to one (or at least one to some fixed number) relationship between decision and observed act. In the case of all the hierarchical models this is not so.

In the 'decision cluster' model, the animal has to decide which cluster to enter, i.e. from which group of acts the final selection will be made. Then, after a series of subdecisions, it arrives at something the observer can actually see. As so far expressed, this is not different from a simple Markov model; it only sounds different. The crucial difference results from assumption (iii), that decisions in any cluster are not influenced by previous decisions within different clusters. This means that the decay of influence of present events on future events is not a constant function of time. It depends on whether a new cluster is chosen, in which case the decay is abrupt. Influence is still always less on the distant future than on the nearer future.

In the grammatical and 'pattern language' and goal models, not only is the number of acts per decision variable, but decisions about the distant future may be taken before decisions about the nearer future. This was expressed by the metaphor about brackets once opened having to be closed.

IS IT BIG ENOUGH FOR THE JOB?

At school we learnt the evidence for the theory of evolution, fossils, geographical distribution, hierarchical taxonomy and so on. Of course the evidence is very important – why otherwise should we remember Darwin more than Wallace? – but I confess it was not evidence that convinced me! The compelling thing about the theory of evolution is that it is big enough to do the job of explaining the otherwise inexplicable fact of our existence.

The nervous system will provide the last of the deep problems of biology. Anybody who has thought about it must be awed by what his own brain can do. Even its lesser accomplishments, the control of complex behaviour, the analysis of complex sensory data, the storage and rapid retrieval of voluminous memories, raise difficult enough problems. Perhaps their solution will never be anything but a mess of detail, but big problems invite big solutions. The question of how complexity in the world could come out of simplicity was answered in two words: natural selection. If we were forced to look into the future and guess which two words of our present meagre vocabulary might come closest to playing the same role for the understanding of complex behaviour, what would they be? Negative feedback? Powerful, but only in explaining simple behaviour: not big enough for the job. Hierarchical organisation?

I have emphasised the distinction between hierarchies of classification and hierarchies of connection. However, we have seen that some muddle between them is pardonable, and we can now see that the two may eventually come together. I think something like the principle we know as hierarchical organisation may turn out to be 'big enough for the job'. If so it will be for the same reason that it is an indispensable classificatory device. Whether it is complexity of stored information, complexity of pattern in incoming data, or complexity of controlled output, hierarchical organisation provides a way of making complexity manageable.

SUMMARY

(1) In the long term, general principles of Software Explanation of behaviour will be required. Hierarchical Organisation became unfashionable in ethology for the wrong reason.

(2) A hierarchy is defined semi-rigorously as a set of elements together with a relation called 'is boss of'. Linear, branching and overlapping hierarchies are also defined in the same way, and a distinction made between hierarchies of classification and hierarchies of connection. Possible examples of non-hierarchies are discussed.

(3) Three functional or 'neuroeconomic' arguments for regarding hierarchical design as good design are put forward. These are the Evolutionary Stability, the Local Administration, and the Redundancy Reduction advantages.

(4) In spite of statements common in the ethology literature, clustering of behavioural acts in time is not evidence for underlying hierarchical organisation.

(5) A truly hierarchical model is developed, based on the idea of 'decision' taken from previous papers. Animals are supposed to take global decisions initially, and to take progressively narrower subdecisions, ending up with an observed act. Published data are re-analysed to provide evidence bearing on the model. A method of cluster analysis based on the model, and called Mutual Replaceability Cluster Analysis is described, and tested.

(6) Special purpose 'languages' for describing serial patterns economically have been developed by human psychologists. These may be applicable to animal behaviour if we suppose that animals generate patterns by executing economical stored programs using hierarchical and recursive procedure calls. A hypothetical example is given, involving postural facilitation in grooming. An algorithm for extracting 'melodies', simple patterns, and patterns within patterns, is described and tested.

(7) Hierarchical models based on the idea of 'goal' are discussed. A distinction is made between two strategies of programming behaviour, based on 'action rules' and 'stopping rules'. Both have their disadvantages, and a good compromise is a set of hierarchically nested stopping rules.

(8) Models comparing animal behaviour with human syntax are discussed. An analogy with 'correct nesting of brackets' is developed. Recursive hierarchical rules of syntax have the interesting property that they effortlessly 'round off' main clauses even after subsidiary clauses, nested to an indefinite depth, have been initiated and concluded.

(9) The behavioural models used in the paper are compared according to two criteria: whether the number of acts per 'decision' is fixed or variable; and whether behaviour in the near future is necessarily easier to predict than behaviour in the more distant future.

(10) It is suggested that hierarchical organisation may be a generally powerful explanatory concept.

Many of the ideas in this paper were developed in discussions, over a long period, with Marian Dawkins. I am very grateful to her and also, for various reasons, to Bruce Anderson, Michael Arbib, Ted Burk, David McFarland, Sir Peter Medawar, Pat Searle, Richard Sibly, Wilson Sutherland, Niko Tinbergen, and the Editors of this book.

REFERENCES

Altmann, S. A. (1965). Sociobiology of rhesus monkeys. II. Stochastics of social communication. *Journal of theoretical Biology*, **8**, 490–622.

Attneave, F. (1954). Informational aspects of visual perception. *Psychological Review*, **61**, 183–193.

Baerends, G. P. (1941). Fortpflanzungsverhalten und Orientierung der *Ammophila campestris* Jur. *Tijdschrift voor entomologie* **84**, 68–275.

Baerends, G. P., Brouwer, R. & Waterbolk, H. T. (1955). Ethological studies on *Lebistes reticulatus* (Peters). I. An analysis of the male courtship pattern. *Behaviour*, **8**, 249–334.

Barlow, H. B. (1961*a*). The coding of sensory messages. In *Current Problems in Animal Behaviour*, ed. W. H. Thorpe & O. L. Zangwill. Cambridge University Press: London.

Barlow, H. B. (1961*b*). Possible principles underlying the transformations of sensory messages. In *Sensory Communication*, ed. W. A. Rosenblith. MIT Press: Cambridge, Mass.

Barlow, H. B. (1972). Single units and sensation: a neuron doctrine for perceptual psychology? *Perception*, **1**, 371–394.

Barlow, H. B., Narasimhan, R. & Rosenfeld, A. (1972). Visual pattern analysis in machines and animals. *Science, Washington*, **177**, 567–575.

Bastin, E. W. (1969). A general property of hierarchies. In *Towards a Theoretical Biology*, vol. **2**, *Sketches*, ed. C. H. Waddington. Edinburgh University Press: Edinburgh.

Bullock, T. H. (1961). The problem of recognition in an analyser made of neurons. In *Sensory Communication*, ed. W. A. Rosenblith. MIT Press: Cambridge, Mass.

Bullock, T. H. & Horridge, G. A. (1965). *Structure and function in the Nervous System of Invertebrates*. MIT Press: Cambridge, Mass.

Chomsky, N. (1957). *Syntactic Structures*. Mouton: The Hague.

Craik, K. J. W. (1943). *The Nature of Explanation*. Cambridge University Press: London.

Davis, W. J., Mpitsos, G. J. & Pinneo, J. M. (1974). The behavioral hierarchy of the mollusk *Pleurobranchaea*. I. The dominant position of the feeding behaviour. *Journal of Comparative Physiology*, **90**, 207–224. II. Hormonal suppression of feeding associated with egg-laying. *Journal of Comparative Physiology*, **90**, 225–243.

Dawkins, M. (1971). Perceptual changes in chicks: another look at the 'search image' concept. *Animal Behaviour*, **19**, 566–574.

Dawkins, R. (1969*a*). A threshold model of choice behaviour. *Animal Behaviour*, **17**, 120–133.

Dawkins, R. (1969*b*). The attention threshold model. *Animal Behaviour*, **17**, 134–141.

Dawkins, R. & Dawkins, M. (1973). Decisions and the uncertainty of behaviour. *Behaviour*, **45**, 83–103.

Dawkins, M. & Dawkins, R. (1974). Some descriptive and explanatory stochastic models of decision-making. In *Motivational Control Systems Analysis*, ed. D. J. McFarland. Academic Press: London.

Dawkins, R. & Dawkins, M. (1976). Hierarchical organisation and postural facilitation: rules for grooming in flies. *Animal Behaviour*, in press.

Deutsch, J. A. (1960). *The Structural Basis of Behavior*. Chicago University Press: Chicago.

Dorsett, D. A., Willows, A. O. D. & Hoyle, G. (1973). The neuronal basis of behaviour in *Tritonia*. IV. The central origin of a fixed action demonstrated in the isolated brain. *Journal of Neurobiology*, **4**, 287–300.

Eccles, J. C. (1975). In *The Creative Process in Science and Medicine*, ed. H. A. Krebs & J. H. Shelley. Excerpta Medica: Amsterdam.

Eibl-Eibesfeldt, I. (1975). *Ethology. The Biology of Behavior*, 2nd edn. Holt, Rhinehart & Winston: New York.

Elsner, N. (1973). The central nervous control of courtship behaviour in the grasshopper *Gomphocerippus rufus* L. (Orthoptera: Acrididae). In *Neurobiology of Invertebrates*, ed. J. Salanki. Tihany: Budapest.

Everitt, B. (1974). *Cluster Analysis*. Heinemann: London.

Fabricius, E. & Jansson, A.-M. (1963). Laboratory observation on the reproductive behaviour of the pigeon. *Animal Behaviour*, **11**, 534–547.

Feigenbaum, E. A. & Feldman, J. (1963). *Computers and Thought*. McGraw-Hill: New York.

Fentress, J. C. (1972). Development and patterning of movement sequences in inbred mice. In *The Biology of Behavior*, ed. J. A. Kiger. Oregon University Press: Eugene.

Fentress, J. C. & Stilwell, F. P. (1973). Grammar of a movement sequence in inbred mice. *Nature, London*, **244**, 52–53.

Foster, J. M. (1967). *List Processing*. Macdonald/Elsevier: London.

Fraenkel, G. S. & Gunn, D. L. (1940). *The Orientation of Animals*. Oxford University Press: Oxford.

Fromkin, V. A. (1973). Slips of the tongue. *Scientific American*, **229** (6), 110–117.

Gower, J. C. & Ross, G. J. S. (1969). Minimum spanning trees and single linkage cluster analysis. *Applied Statistics*, **18**, 54–64.

Harding, E. F. (1971). The probabilities of rooted tree-shapes generated by random bifurcation. *Advances in applied Probability*, **3**, 44–77.

Hinde, R. A. (1952). The behaviour of the great tit and some other related species. *Behaviour Supplement*, **2**, 1–201.

Hinde, R. A. (1953). Appetitive behaviour, consummatory act, and the hierarchical organisation of behaviour – with special reference to the great tit (*Parus major*). *Behaviour*, **5**, 189–224.

Hinde, R. A. (1956). Ethological models and the concept of drive. *British Journal for the Philosophy of Science*, **6**, 321–331.

Hinde, R. A. (1957). Consequences and goals. Some issues raised by Dr. A. Kortlandt's paper on 'Aspects and prospects of the concept of instinct'. *British Journal of Animal Behaviour*, **5**, 116–118.

Hinde, R. A. (1960). Energy models of motivation. *Symposia of the Society for experimental Biology*, **14**, 199–213.

Hinde, R. A. & Stevenson, J. G. (1970). Goals and response control. In *Development and Evolution of Behavior*, ed. L. R. Aronson, E. Tobach, D. S. Lehrman & J. S. Rosenblatt. Freeman: San Francisco.

Hodos, W. & Campbell, C. B. G. (1969). Scala naturae: why there is no theory in comparative psychology. *Psychological Review*, **76**, 337–350.

Holst, E. von & Saint Paul, U. von (1963). On the functional organisation of drives. *Animal Behaviour*, **11**, 1–20.

Hutt, S. J. & Hutt, C. (1970). *Direct Observation and Measurement of Behavior*. Thomas: Springfield, Ill.

Iersel, J. J. A. van & Bol, A. C. (1957). Preening of two tern species. A study on displacement activities. *Behaviour*, **13**, 1–87.

Ito, N. (1974). The control mechanism of cerebellar motor systems. In *The Neurosciences – Third Study Program*, ed. F. O. Schmitt & F. G. Worden. MIT: Cambridge, Mass.

Kalmus, H. (1969). Animal behaviour and theories of games and of language. *Animal Behaviour*, **17**, 607–617.

Kennedy, D. (1974). Connections among neurones of different types in crustacean nervous systems. In *The Neurosciences – Third Study Program*, ed. F. O. Schmitt & F. G. Worden. MIT: Cambridge, Mass.

Koestler, A. (1967). *The Ghost in the Machine*. Hutchinson: London.

Koestler, A. & Smythies, J. R. (1969). *Beyond Reductionism*. Hutchinson: London.

Kortlandt, A. (1955). Aspects and prospects of the concept of instinct (vicissitudes of the hierarchy theory). *Archives néerlandaises de zoologie*, **11**, 155–284.

Krebs, J. R. (1973). Behavioral aspects of predation. In *Perspectives in Ethology*, ed. P. P. G. Bateson & P. H. Klopfer. Plenum Press: New York & London.

Kruuk, H. (1972). *The Spotted Hyena*. Chicago University Press: Chicago.

Lashley, K. S. (1951). The problem of serial order in behavior. In *Cerebral Mechanisms in Behavior*, ed. L. A. Jeffres. Wiley: New York.

Liddell, H. G. & Scott, R. (1883). *A Greek–English Lexicon*. Oxford University Press: Oxford.

Longuet-Higgins, H. C. & Ortony, A. (1968). The adaptive memorization of sequences. In *Machine Intelligence*, vol. **3**, ed D. Michie. Edinburgh University Press: Edinburgh.

Lorenz, K. Z. (1937). Uber die Bildung des Instinktbegriffes. *Naturwissenschaften*, **25**, 289–331. Reprinted (1970) as 'The establishment of the instinct concept', in *Studies in Animal and Human Behaviour*, vol. **1**. Methuen: London.

Luria, A. R. (1970). The functional organisation of the brain. *Scientific American*, **222** (3), 66–79.

Maurus, M. & Pruscha, H. (1973). Classification of social signals in squirrel monkeys by means of cluster analysis. *Behaviour*, **47**, 106–128.

Maynard Smith, J. & Price, G. R. (1973). The logic of animal conflict. *Nature, London*, **246**, 15–18.

McCulloch, W. S. (1945). A heterarchy of values determined by the topology of nervous nets. *Bulletin of mathematical Biophysics*, **7**, 89–93.

McFarland, D. J. (1971). *Feedback Mechanisms in Animal Behaviour*. Academic Press: London.

McFarland, D. J. (1974). Time sharing as a behavioral phenomenon. In *Advances in the Study of Behavior*, **5**, 201–225.

McFarland, D. J. & Sibly, R. M. (1975). The behavioural final common path. *Philosophical Transactions of the Royal Society of London* B, **270**, 265–293.

Mesarovic, M. D., Macko, D. & Takahara, Y. (1970). *Theory of Hierarchical Multilevel Systems*. Academic Press: New York.

Miller, G. A., Galanter, E. & Pribram, K. H. (1960). *Plans and the Structure of Behavior*. Holt, Rhinehart & Winston: New York.

Myrberg, A. A. (1972). Ethology of the bicolor damselfish *Eupomacentrus partitus* (Pisces: Pomacentridae): a comparative analysis of laboratory and field behaviour. *Animal Behaviour Monographs*, **5**, 197–283.

Nelson, K. (1964). The temporal pattern of courtship behaviour in the glandulocaudine fishes (Ostariophysi, Characidae). *Behaviour*, **24**, 90–146.

Nelson, K. (1973). Does the holistic study of behavior have a future? In *Perspectives in Ethology*, ed. P. P. G. Bateson & P. H. Klopfer. Plenum Press: New York & London.

Nottebohm, F. (1970). In *Auditory Processing of Biologically Significant Sounds*, N. R. P. Bulletin, ed. F. G. Worden & R. Galambos, page 76.

Paillard, J. (1960). The patterning of skilled movements. In *Handbook of Physiology*, ed. J. Field, H. W. Magoun & V. E. Hall, Sect. I, vol. **3**. American Physiological Society: Washington.

Pattee, H. H. (1973). *Hierarchy Theory. The Challenge of Complex Systems*. Braziller: New York.

Pribram, K. H. (1971). *Languages of the Brain*. Prentice–Hall: Englewood Cliffs, N.J.

Restle, F. (1970). Theory of serial pattern learning: structural trees. *Psychological Review*, **77**, 471–495.

Rosenblueth, A., Wiener, N. & Bigelow, J. (1943). Behavior, purpose and teleology. *Philosophy of Science*, **10**, 18–24.

Ross, G. J. S. (1969). Algorithms AS 13, AS 14, AS 15. *Applied Statistics*, **18**, 103–106.

Russell, E. S. (1946). *The Directiveness of Organic Activities*. Cambridge University Press: London.

Seibt, U. (1972). Beschreibung und Zusammenspiel einselner Verhaltenswiesen von Stielaugenfliegen (Gattung *Diopsis*) unter besonderer Berücksichtigung des Putzverhaltens. *Zeitschrift für Tierpsychologie*, **31**, 225–239.

Selfridge, O. G. & Neisser, U. (1963). Pattern recognition by machine. In *Computers and Thought*, ed. E. A. Feigenbaum & J. Feldman. McGraw-Hill: New York.

Simon, H. A. (1962). The architecture of complexity. *Proceedings of the American Philosophical Society*, **106**, 467–482. Reprinted in Simon, H.A. (1970), *The Sciences of the Artificial*. MIT Press: Cambridge, Mass.

Simon, H. A. (1972). Complexity and the representation of patterned sequences of symbols. *Psychological Review*, **79**, 369–382.

Simon, H. A. (1973). The organization of complex systems. In *Hierarchy Theory*, ed. H. H. Pattee. Braziller: New York.

Slater, P. J. B. (1973). Describing sequences of behavior. In *Perspectives in Ethology*, ed. P. P. G. Bateson & P. H. Klopfer. Plenum Press: New York & London.

Sutherland, N. S. (1969). Outlines of a theory of visual pattern recognition in animals and man. In *Animal Discrimination Learning*, ed. R. M. Gilbert & N. S. Sutherland. Academic Press: London.

Szentágothai, J. & Arbib, M. A. (1974). Conceptual models of neural organization. *Neurosciences Research Program Bulletin*, **12**, 307–510.

Thorpe, W. H. (1963). *Learning and Instinct in Animals.* Methuen: London.

Thorpe, W. H. (1974). *Animal Nature and Human Nature.* Methuen: London.

Tinbergen, N. (1950). The hierarchical organisation of nervous mechanisms underlying instinctive behaviour. *Symposia of the Society for experimental Biology*, **4**, 305–312.

Tinbergen, N. (1951). *The Study of Instinct.* Oxford University Press: London.

Tinbergen, N. (1963). On aims and methods of ethology. *Zeitschrift für Tierpsychologie*, **20**, 410–433.

Vowles, D. M. (1970). Neuroethology, evolution and grammer. In *Development and Evolution of Behavior*, ed. L. R. Aronson, E. Tobach, D. S. Lehrman & J. S. Rosenblatt. Freeman: San Francisco.

Weiss, P. (1941). Self-differentiation of the basic patterns of coordination. *Comparative Psychology Monographs*, **17**, 1–96.

Weiss, P. (1950). Experimental analysis of coordination by the disarrangement of central-peripheral relations. *Symposia of the Society for experimental Biology*, **4**, 92–110.

Weiss, P. (1971). *Hierarchically Organised Systems in Theory and Practice.* Hafner: New York.

Wells, G. P. (1966). The lugworm (*Arenicola*) – a study in adaptation. *Netherlands Journal of Sea Research*, **3**, 294–313.

Winston, P. H. (1972). The M.I.T. Robot. In *Machine Intelligence*, **7**, ed. B. Meltzer & D. Michie. Edinburgh University Press: Edinburgh.

Wolff, J. G. (1975). An algorithm for the segmentation of an artificial language analogue. *British Journal of Psychology*, **66**, 79–90.

Woodger, J. H. (1937). *The Axiomatic Method of Biology.* Cambridge University Press: London.

2

Form and function in the temporal organisation of behaviour

D. J. McFARLAND

Ethologists are familiar with the idea that behaviour patterns are subject to natural selection in the same manner as the morphological characteristics of animals. In this paper, I wish to discuss the view that, not only are the species-typical characteristics of behaviour patterns subject to selection, but so also is the order in which they are performed. Moreover, a quantitative study of the survival value attached to the order in which behaviour patterns occur is essential for a proper study of the mechanisms responsible for the temporal organisation of behaviour.

This paper is not an attempt to argue the case in a rigorous manner, but rather an attempt to introduce various concepts for discussion, and to outline the way in which the problem of temporal organisation of behaviour might be tackled, both theoretically and empirically.

FORM AND FUNCTION

That form and function go hand-in-hand is the essence of Darwinism. However, to identify *a* function with any particular characteristic of an animal is an over-simplification, since any modern form must be the outcome of competition between various selective pressures. For example, the functions of vision in nocturnal animals differ from those in diurnal species. For nocturnal animals sensitivity is at a premium, and the eye must collect as much light as possible. The pupil should be large relative to the size of the eye as a whole, and this entails a big lens, to avoid spherical aberration of the lens periphery. To maintain the image in focus on the retina, increased lens size must go hand-in-hand with increased lens curvature, producing a smaller image on the retina. Thus many nocturnal animals have a large lens, producing a small bright image, whereas diurnal animals have a smaller lens, producing a larger image with consequent improved acuity. Typical nocturnal eyes are found in the opossum, house mouse and lynx (Fig. 1) and also in deep-sea

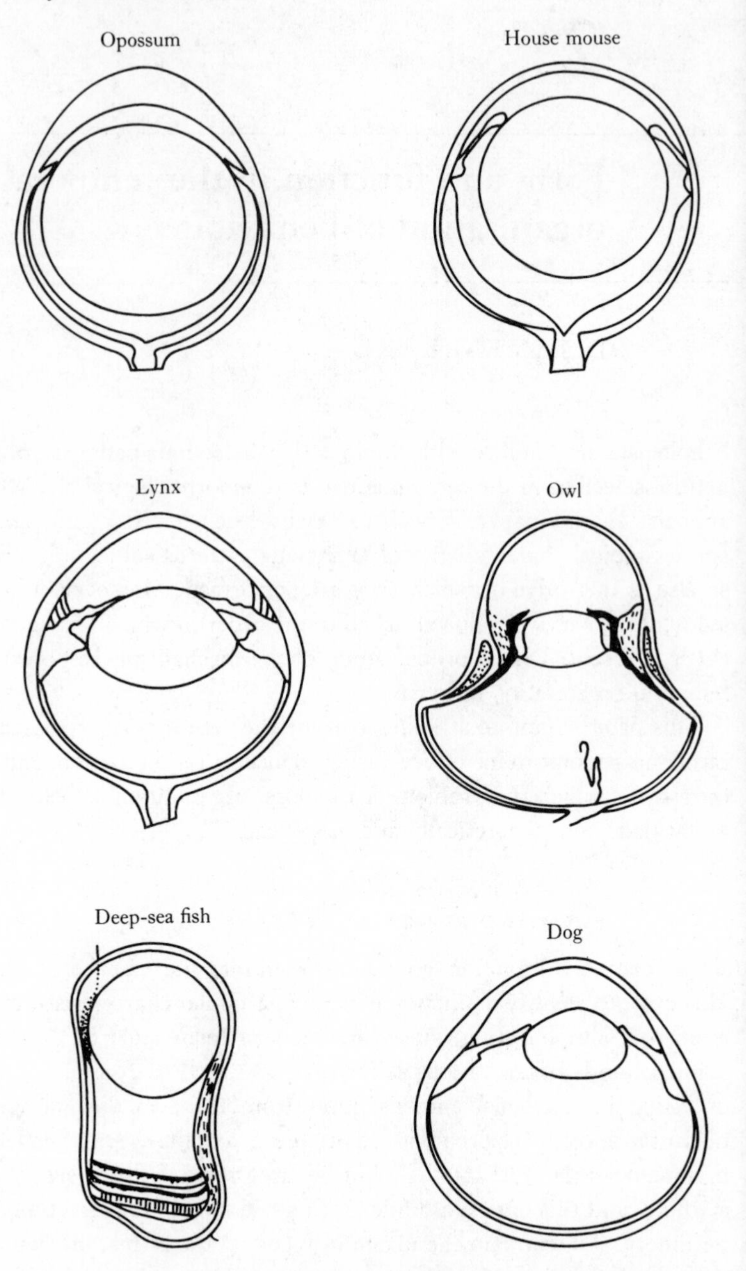

Fig. 1. Form and function in the vertebrate eye. The eye structure of various nocturnal animals, and of a deep-sea fish, are compared with that of the diurnal dog (after Tansley, 1965).

fish. In other nocturnal animals, such as owls and bush babies, the limitation on lateral expansion by the skull has led to tubular extension towards the back of the head. These animals have very restricted eye movement and have to turn the head instead (Tansley, 1965).

By comparing the eye structure of species from differing habitats, we can see that during the course of evolution there has been an interplay of competing selective pressures, resulting in compromise between the various alternative advantages. Thus acuity generally has been sacrificed to obtain increased sensitivity, and there may have been a trade-off between sensitivity and the ability to move the eyes. The 'best' form of eye for a particular species will be that for which the compromise with highest survival value has been achieved. This view implies a 'principle of optimal design', such that biological structures, subjected for a sufficiently long time to a specific set of selective pressures, will tend to assume characteristics which are optimal with respect to those circumstances. The notion of optimal design is an old one, and inherent in much of the older literature on the form of biological structures (e.g. D'Arcy Thompson, 1924; Huxley, 1932). More recently, a number of authors have been more explicit in their formulation of optimality principles as applied to biological structures (Cohn, 1954, 1955; Rashevsky, 1960; Rosen, 1967). In particular, it has been suggested that biological structures, which are optimal in the context of natural selection, are also optimal in the sense that they minimise some cost function derived from the engineering characteristics of the situation (Rosen, 1967, 1970). Thus the form of a biological structure is seen as the outcome of a design operation, which aims at the optimal form on the basis of certain criteria. This idea is well illustrated by a study of blood-vessel design, derived from the work of Murray (1926*a*, *b*), who was probably the first to present a theoretical–experimental analysis of optimal design in a physiological system.

In considering the design of a simple blood vessel, we might ask what radius the vessel should have. Since all fluid flow involves a frictional loss of energy, the pumping mechanism must maintain an appropriate power output if a steady flow is to be achieved. We might expect blood vessels to be designed to minimise power loss by presenting the least possible resistance to the flow of blood. Since resistance to flow rate falls as vessel radius increases, it would appear that the radius should be as large as possible.

A large vessel radius implies a large volume of blood to be maintained, and a large area of vessel wall to be maintained. We might, therefore, expect the benefit of large vessels to be offset by high maintenance costs. We now come up against a perennial problem in optimal design considerations: how should the benefits of low resistance-to-flow be quantitatively evaluated and offset against the maintenance costs of the blood vessels?

A general feature of optimisation is that it results from a 'trade-off' process, in which the costs and benefits of different aspects of the process are counterbalanced. In analytical optimality theory the costs and benefits have to be quantitatively assessed, and then combined into a single *cost function* which represents the precise manner in which the various costs and benefits are counterbalanced. We may call this the 'trade-off problem'.

In the present example, the quantitative relationship between the radius of the vessel and the frictional resistive power can be calculated from simple physical considerations, as can that of the maintenance power (Milsum & Roberge, 1973). The conceptual difficulty involved in referring to these cost components in terms of a common metric is ameliorated by the observation that the metabolic energy cost involved in maintaining the blood vessels must be obtained from the same source as that which supplies the heart's pumping energy. Thus Milsum & Roberge (1973) postulate a cost function P for a circular vessel of radius r, given length L and flow rate Q, as the sum of the frictional resistive power P_1 and the maintenance power P_2.

$$P = P_1(R(r)) + P_2(V(r)) = \beta L Q^2 R(r) + \alpha V(r), \tag{1}$$

where R is resistance, pressure drop per unit length per unit flow rate; P_1 is hydraulic power, pressure drop$\times$flow rate; Q is the flow rate; V is the volume of the vessel; β and α are appropriate conversion factors, for example, in milligrams of glucose per second, per hydraulic power and per vessel volume respectively. This cost function is illustrated in Fig. 2. From this function it is possible to calculate the optimal radius r^* for the minimal cost function P^*.

On the basis of this type of formulation, a number of authors have calculated and verified empirically the optimal angles and radii involved in vascular branching and bifurcation (Murray, 1926*a,b*; Cohn, 1954, 1955; Milsum & Roberge, 1973) thus providing a truly quantitative consideration of form and function which has considerable everyday importance.

In general, we can expect from the Theory of Natural Selection that the form of a biological structure will tend towards the optimal form in relation to its functions. Thus the form of a blood vessel, in terms of its radius, branching pattern etc., will tend to be that which best serves the function of transporting blood with minimal energetic cost. Whenever we try to specify an optimal condition we have to make clear the criteria by which the degree of optimality is to be judged. This is the purpose of introducing the concept of the cost function into considerations of optimal design. In the case of blood vessels, the cost function represents the combination of factors, such as resistance-to-flow, which we suppose to be most subject to selective pressure. Thus to state that the function of a blood vessel is to transport blood

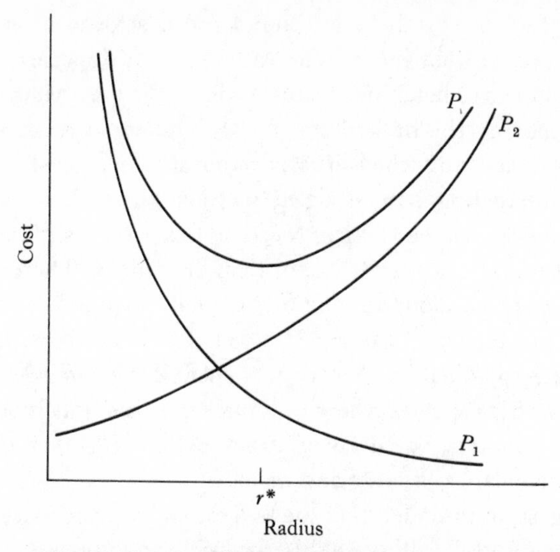

Fig. 2. Cost function for single blood vessel of given length, as a function of radius *r*. *r** is the optimal radius for the minimal cost function. P_1 = frictional resistive power, P_2 = maintenance power, P = total power (i.e. total cost in terms of energy) (from Milsum & Roberge, 1973).

with minimal energetic cost, is indirectly to specify the selective pressures which influenced the evolution of the vessel into its present form. In considering form and function in animal behaviour, the same general considerations should apply, but we must remember that optimisation of structures is a static problem, whereas optimisation of behaviour involves consideration of dynamics, because behaviour is continually changing in time.

COST BENEFIT ANALYSIS OF BEHAVIOUR

Cost–benefit analysis is a way of setting out the factors which need to be taken into account in making decisions. In economics, the decisions may concern investment projects, price schemes, etc. By using cost–benefit analysis, the economist hopes to achieve a formulation, in which the aim is to maximise the present value of all benefits less that of all costs (Prest & Turvey, 1965).

Animals take decisions almost every minute of their lives. The decisions concern what the animal ought to do in a given situation, where many courses of action are possible. The various possible activities differ in their consequences, and have different costs and benefits attached to them. A distinctive feature of decisions made by animals is that much of the cost–benefit analysis has been done by natural selection. For example, Tinbergen and co-workers (1962) demonstrated by field experiment that black-headed gull (*Larus ridi-*

bundus) nests with broken egg-shells are found and destroyed by predators more readily than nests without shells. Whereas in some bird species the shells are removed very soon after hatching, black-headed gulls may delay a matter of hours before removing the broken egg-shells. Tinbergen reasoned that, despite the obvious benefits attached to early removal of the egg-shells, there must be some counteracting cost attached to their removal. Observation revealed that wet newly hatched chicks were at risk from cannibalism by neighbouring adult gulls, to a greater extent than dry chicks. Therefore, the cost of leaving a wet chick alone for the few seconds required for removal of the egg-shell is greater when the chick is wet than when it is dry. The removal of the broken egg-shell has greater net benefit if it is carried out when the chicks are dry in those species where cannibalism exists, but is otherwise better carried out promptly. The decision to remove the egg-shell therefore rests upon a delicate balance of costs and benefits.

There is no suggestion intended that the individual animal carries out the cost–benefit analysis for itself, although this may occur to some extent in more 'intelligent' animals. The animal is designed, by natural selection, to behave in such a manner that the greatest net benefit is attained. In other words, the form of the behaviour is the outcome of a design operation which aims at the optimal compromise between the competing selective pressures characteristic of the environment. Just as the structure of the eye can be studied in terms of both form and function, so too can the mechanisms controlling behaviour. This is no more than a re-statement of Tinbergen's long-standing view 'that behaviour can, and should, be studied in the same way as any other group of life processes' (Tinbergen, 1951).

The work of Tinbergen and his co-workers, which has done much to establish an understanding of the relationship between form and function in behaviour, has been based upon three main methods of approach to the problem of establishing the selective pressures operating in a particular behavioural situation. The first, in terms of ethological chronology, compares the behaviour of genetically related populations, and involves the assumption that the observed behavioural differences between populations are the result of selective pressures due to observed environmental differences between the populations. This approach is not without its difficulties (Hailman, 1965) but it has, nevertheless, proved to be a cornerstone of the ethological approach (Lorenz, 1950; Cullen, 1957; Tinbergen, 1959).

The second method involves experimental demonstration that a particular feature of behaviour is likely to have consequences that affect the survival of the species. This approach is exemplified by the study of Tinbergen *et al.* (1962) on egg-shell removal in the black-headed gull. It was shown that the natural camouflage of the gulls' eggs was reduced in effectiveness by the proximity of a conspicuous broken egg-shell. By laying out eggs in the gull colony, half of them with a broken egg-shell nearby, and half without, it was found that the former were taken by predators much more readily than the latter.

Experiments on the survival value of behaviour, which are necessarily conducted in somewhat artificial conditions, can never reveal the exact way in which natural selection operates. This is because, in the complex natural situation, it is the overall effect of selection that is important. In nature there are many conflicting pressures, so that the design effected by natural selection is inevitably a compromise. For example, as mentioned above the black-headed gull does not remove the broken egg-shell soon after hatching, as do some other ground-nesting birds, but delays one or two hours. This is probably due to the fact that the gull colony is closely packed, and neighbouring black-headed gulls prey on wet chicks, but not on dry ones. There may, therefore, be survival value in delaying egg-shell removal until the chicks have dried, and the parents' absence does not expose them to predation from neighbours.

The problem of precise estimation of selective pressures is largely a problem of evaluating the selective pressures characteristic of each of the many dimensions of the niche. For example, although it has been demonstrated that removal of egg-shells by black-headed gull parents has an anti-predator function, this does not mean that it is the only important function. Tinbergen and his co-workers (1962) realised that there were a number of other possibilities, viz. (1) the sharp edges of the shell might injure the chicks, and (2) the chicks might become trapped in the empty shell. (3) The shells might interfere with brooding, and (4) the broken shell might provide a breeding ground for bacteria, and thus increase the danger of infection amongst the chicks. Tinbergen argues that because the kittiwake (*Rissa tridactyla*), a highly nidicolous species, lacks the egg-shell removal response (Cullen, 1957), neither avoidance of injury, nor of infection, nor interference with brooding, are the main functions of egg-shell removal. However, it remains possible that while the prime function is maintenance of the camouflage of the brood, other factors may also exert a small selective pressure.

The third method involves direct comparison of mortality, or breeding success, for animals differing in some behavioural characteristic. For example, Kruuk (1964) showed that black-headed gulls, nesting near the periphery of a colony, were more heavily predated by foxes than gulls nesting in the centre of the colony. Patterson (1965) showed that the breeding success of black-headed gulls was reduced, the further from the centre of the colony they nested. It was also reduced if the time of nesting was advanced or retarded, compared with the average time for the colony as a whole. While this method is satisfactory for the purpose of demonstrating that a behavioural character is under the influence of a specific selective pressure, it becomes a gigantic open-ended commitment, when the aim is to account for the behaviour in terms of the totality of selective pressures responsible.

The problem of quantitatively evaluating the numerous, and often conflicting, selective pressures that shape the evolution of behaviour is a formidable one. To some extent it can be tackled by direct measurement of mortality and reproductive success. For example, the great tit (*Parus major*) has been studied in Marley Wood on the Wytham estate, near Oxford, since 1947. Basic census data, consisting of the number of breeding birds, the number of eggs laid, the number of eggs hatching, and the number of young fledging, have been collected annually (Lack, 1954, 1966; Perrins, 1965). Analysis of these data (Krebs, 1970) suggests that clutch size and hatching success are density dependent, as found previously in Holland (Kluijver, 1951). The main factors responsible for population regulation appear to be availability of food at the time of egg-laying, predation of eggs, and spring territorial behaviour. For the student of behaviour the relevance of such studies is not always obvious, particularly if he is primarily interested in the causal mechanisms underlying behaviour.

DECISION-MAKING CRITERIA

I wish to argue that, not only is the functional aspect of behaviour important and interesting in its own right, but it is essential for a proper study of the causal mechanisms responsible for the temporal organisation of behaviour. However much we know about the processes which determine the motivational state of an animal, such as body temperature, hormonal balance, degree of fear, or readiness to attack, we shall never be able to achieve precise predictions about its behaviour, unless we also understand the organisation of the decision-making mechanisms of the animal. Moreover, we cannot understand the decision-making processes without also appreciating the selective pressures that shaped them. I propose to illustrate this point by means of a simple everyday example.

Suppose we consider a university committee, set up to review applications for a lecturership. Suppose the committee arrives at an agreed short list of six applicants, *a*, *b*, *c*, *d*, *e*, *f*. The task is to decide which is the best applicant, taking both teaching suitability and research promise into account. We assume the abilities to teach and to do research are independent, and have to be assessed separately. A possible approach would be for the committee to arrive at an agreed rating of teaching ability for each applicant, scored on a 10-point scale. A separate rating, on a similar basis, could be agreed for research ability. Possible results of this dual exercise are illustrated in Fig. 3. The question is, which is the best applicant?

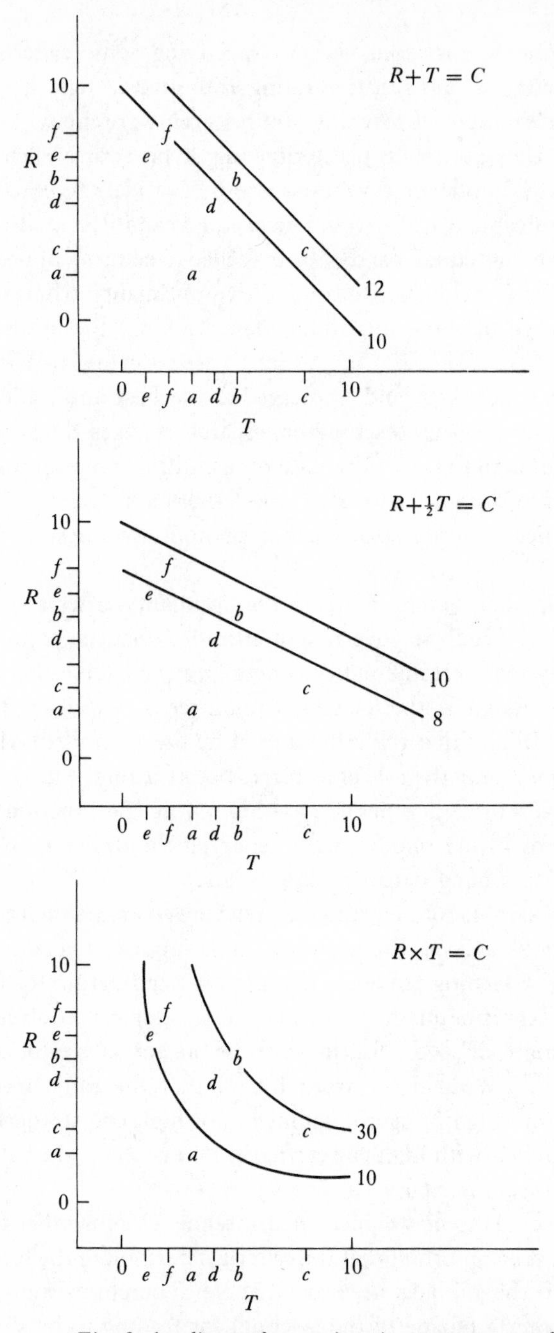

Fig. 3. Applicants for a university post are scored separately for estimated teaching (T) and research (R) ability. The scores obtained by applicants *a*, *b*, *c*, *d*, *e*, and *f*, are shown as labelled points on the graph. Lines joining points of equal candidature (isoclines) are labelled according to the candidature value calculated on the basis of optimality criteria indicated in the inset formulae.

This question cannot be answered without specifying some criteria for combining teaching rating T and research rating R to give a single strength of candidature C. For example, if teaching and research were thought to be of equal importance, the (optimality) criterion might be additive, so that $C = R + T$. In this case, applicant b would score $6+5 = 11$, $c = 3+8 = 11$, $f = 8+2 = 10$, and applicants b and c would have equal candidature. In Fig. 3*a* lines joining points of equal candidature (called isoclines) appear as straight lines, which is characteristic of an additive optimality criterion. If research were thought to be more important than teaching, the optimality criterion might be $C = R + \frac{1}{2}T$, in which case $b = 6 + \frac{1}{2}(5) = 8\frac{1}{2}$, $c = 3 + \frac{1}{2}(8) = 7$, $f = 8 + \frac{1}{2}(2) = 9$, and f emerges as the best applicant. Alteration of the weighting given to teaching or research changes the slope of the isoclines, as illustrated in Fig. 3*b*. Alternatively a multiplicative optimality criterion might be more suitable, so that $b = 6 \times 5 = 30$, $c = 3 \times 8 = 24$, and $f = 8 \times 2 = 16$. Multiplicative criteria produce hyperbolic isoclines, as illustrated in Fig. 3*c*.

We now have to ask what factors influence the optimality criteria. In the university context, factors such as government attitude, financial pressures, and so forth, are likely to determine policy concerning the balance between teaching and research. As far as the selection committee is concerned, these are ecological factors. In addition the criteria used by the committee will be influenced by ideas concerning the role of lecturers both within the university and in the educational world in general. In other words the function of a university lecturer is of prime importance in shaping the decision criteria involved in the selection among particular applicants.

Translation of this example into terminology relevant to animal behaviour is not difficult at a superficial level. For example, let us suppose that two main factors contribute to the feeding tendency (strength of candidature for feeding) of an animal: (1) degree of hunger, in terms of physiological requirement for food; and (2) strength of food cues, in terms of the animal's estimate of the availability of food. We can see from Fig. 4 that the same feeding tendency can result from high hunger combined with high cue strength, as from low hunger combined with high cue strength. Points of equal tendency are joined by a *motivational isocline*.

Any tendency to feed has to be weighed against tendencies for other types of behaviour, because it is important that the criteria determining the feeding tendency be related to the animal's needs as a whole. Therefore, we would expect the decision criteria (shape of the isocline) for feeding to be shaped by natural selection in accordance with the animal's ecological circumstances. For instance, where food availability is erratic, more weight should be given to cue strength, while the weight attached to hunger should be related to the animal's physiological tolerance.

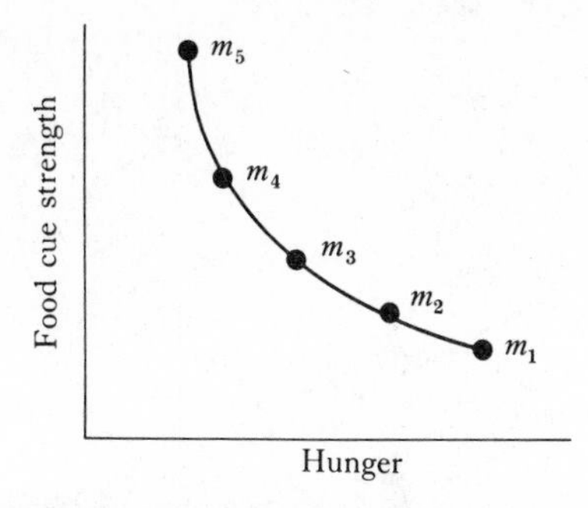

Fig. 4. Feeding tendency as a function of cue strength and hunger. The feeding tendency is the same for points m_1, ..., m_5, and the line joining these points is a motivational isocline (from McFarland & Sibly, 1975).

In real life, the animal has to decide whether to sleep, feed, groom etc. at any particular time. The present argument is that, in addition to the strength of motivational factors, the decisions reached will be heavily influenced by the decision criteria. If I believe that animal behaviour is subject to natural selection, I must conclude that decision criteria are shaped by natural selection, because the decision criteria strongly influence the order in which animals go about their daily tasks.

It has been argued by McFarland & Sibly (1975) that *any* model of the motivational (i.e. reversible) processes governing the behaviour of an animal can be represented by means of isoclines in a suitable multi-dimensional space. This argument is axiomatic, based upon the two prime assumptions: that (1) it is always possible to classify the behavioural repertoire of a species in such a way that the classes are mutually exclusive in the sense that the members of different classes cannot occur simultaneously, and (2) these incompatible actions are uniquely determined by a particular set of causal factors. The isoclines join all points in the space which represent a given 'degree of competitiveness' of a particular 'candidate' for overt behavioural expression. The competition between candidates is an inevitable consequence of the fact that animals cannot 'do more than one thing at a time'.

When we come to the problem of how the various incompatible activities are related to the motivational state of the animal, we can see that each is controlled by a set of causal factors. The causal factors will include variables describing the animal's 'estimate' of the stimuli present in the external environment (e.g. cues to the availability of food), and variables relevant to the animal's internal state. McFarland & Sibly (1975) represent the total motivational state in a *causal factor space*, in which there is an axis corresponding to each class of causal factor, with the classes defined in terms of some suitable arbitrary criterion.

A simple example is provided by the work of Baerends, Brouwer &

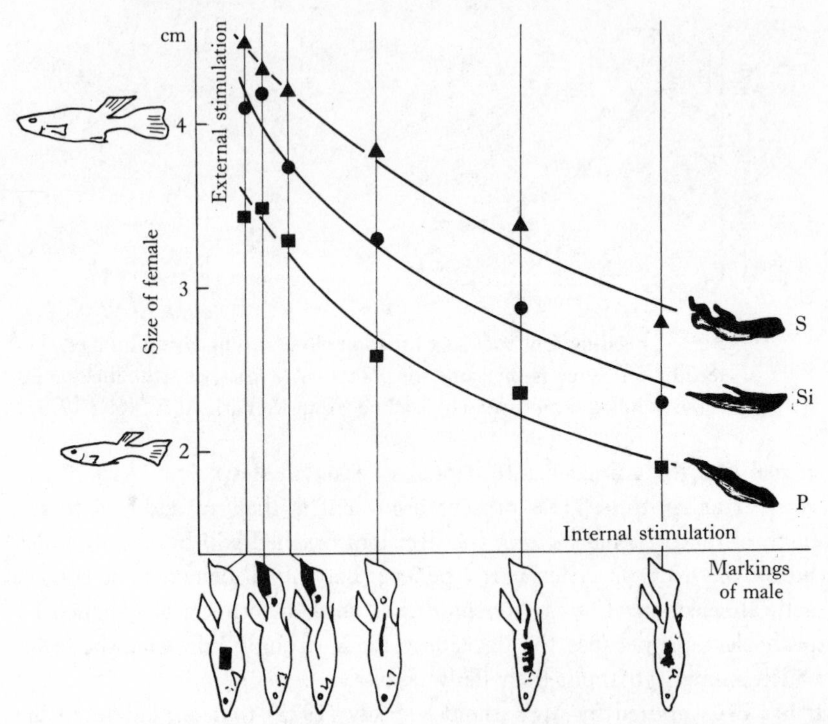

Fig. 5. The influence of the strength of external stimulation (measured by the size of the female) and the internal state (measured by the colour pattern of the male) in determining the courtship behaviour of male guppies. Each curve represents the combination of external stimulus and internal state producing posturing (P), sigmoid intention movements (SI) and fully developed sigmoid (S), respectively (from Baerends *et al.*, 1955).

Waterbolk (1955) on the courtship of the male guppy *Lebistes reticulatus*. The tendency of the male to attack, flee from, and behave sexually towards the female can be gauged from the colour patterns characteristic of each motivational state. In Fig. 5, increasing sexual motivation is plotted as an index of colour change along the abscissa. The effectiveness of the female in eliciting courtship increases with her size and is plotted on the ordinate. The points plotted on the graph represent the relationship between the measures of internal state and external stimulation at which particular patterns of behaviour are observed. If the patterns P, Si, and S are taken to represent increasing values of response strength, and the scaling of the ordinate is taken at face value, then the isoclines obtained represent closely those that would result from multiplication of internal and external factors. In this case the method of quantification is somewhat arbitrary, the scaling on the abscissa depending on the association of the different colour patterns with the relative frequency of activities characteristic of sexual tendency and that on the ordinate being arbitrarily linear. Nevertheless, Fig. 5 is a good example of the type of representation that we have in mind. McFarland & Sibly (1975) discuss various ways in which arbitrary scaling can be overcome, but for our present purpose it is sufficient to note only the principle involved.

The importance of the motivational isoclines is that their shape is a design feature, which is intimately related to the animal's ecological circumstances. Any decision-making process, in which there is 'totally decidable logical preference', can be expressed as a set of isoclines, or 'indifference curves' (Kaufman, 1968). The essence of the argument is that, because animal behaviour can always be classified into mutually exclusive and exhaustive categories (i.e. the animal is always doing something), a totally decidable logical preference always exists, albeit inherent in the design of the decision-making machinery (McFarland & Sibly, 1975). Knowledge of the shape of motivational isoclines, or equivalent knowledge of the decision-making machinery, expressed in terms of decision rules, neural coding, or any suitable and equivalent alternative, is a necessary condition for obtaining a full understanding of the relationship between form and function in behaviour.

It should not be thought, however, that knowledge of motivational isoclines, or the decision-rules which underlie them, is a sufficient condition for precise quantitative formulation of a behavioural cost function, equivalent to that obtained for the design of blood vessels (Fig. 2), described above. To achieve this degree of precision, we should have to know how the animal 'calibrates' the various internal and external stimuli that determine its motivational state.

At this stage, in order to formulate the concepts involved in more concrete terms, it may be useful to consider a particular example in some detail. This example has been chosen because of its relative simplicity and amenability to reasonable conjecture. It should be understood at the outset that the evidence available is quite insufficient to justify the speculations involved, and that these are introduced purely for the sake of illustration.

Ecology and behaviour of the Namib desert lizard Aporosaura anchietae

Of the many desert lizards which have been studied few have been found to have specific physiological adaptations to the desert environment. The majority escape unfavourable conditions by well-defined adaptive behaviour (Mayhew, 1968). Escape from thermal stress and desiccation by behavioural means usually involves hiding beneath sand or rocks. This behaviour has the disadvantage that it severely limits opportunities for feeding. This is particu-

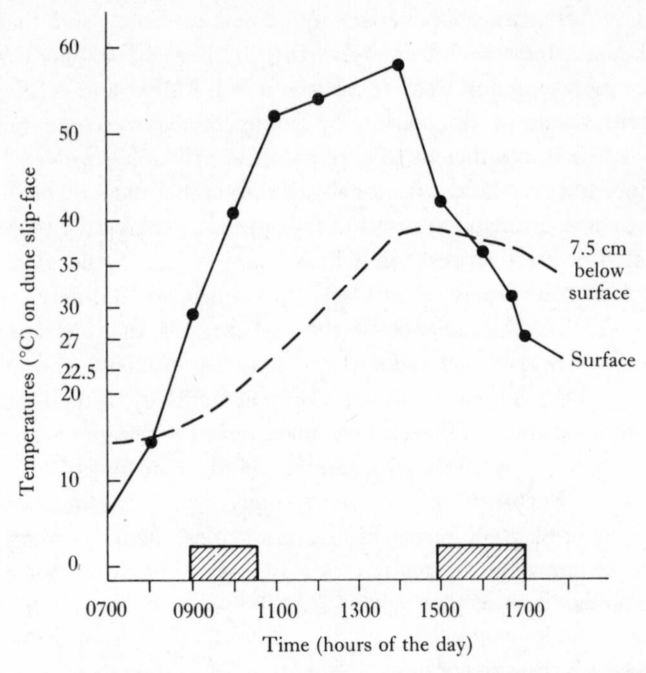

Fig. 6. Typical temperature conditions on the dune slip-face, which constitutes the micro-habitat of *Aporosaura anchietae*. Hatched blocks indicate the periods for which the lizard is above the surface of the sand (after Louw & Holm, 1972).

larly true of the lizard *Aporosaura anchietae* found in the Namib desert in South-west Africa. It typically inhabits the slip face on the leeward side of sand dunes. The leeward side of the dune acts as a trap for wind-blown organic matter which congregates at the foot of the slip face, where *Aporosaura* feeds upon grass seeds and insects. The physiological and behavioural adaptations of this lizard have been described in some detail by Louw & Holm (1972).

The leeside of the dunes is virtually wind free and frequently faces the rising sun. Consequently the surface temperatures rise very rapidly, as illustrated in Fig. 6. When the surface temperature exceeds 30 °C the lizard emerges from the sand and presses its ventral surface against the substrate. It takes a special 'dished' posture which achieves the maximum contact between its body and the substrate. By this behaviour the animal achieves a rapid rise in body temperature and is soon able to move about and forage on the dune face. As substrate temperature approaches 40 °C the animal straightens its limbs in a characteristic 'stilt-walk', which raises the body as far as possible above the substrate. It periodically raises the diagonally opposite limbs while the base of the tail is used for support. At about 40 °C the lizards dive beneath the soft sand to reach the cooler depths. The behaviour of *Aporosaura* thus appears to be very much dictated by its thermo-regulatory requirements. These generally result in a biphasic period of activity during which the animal is able to forage on the slip face, as illustrated in Fig. 6. The work of Louw & Holm (1972) shows that these lizards are able to consume large amounts of food in a short period of time. After feeding in this manner, captive animals exhibit little or no interest in feeding and remain submerged beneath the sand for longer periods than usual. These workers also found that *Aporosaura* is able to store water in the digestive tract. It is capable of ingesting water amounting to 11% of its body weight within three minutes and can subsequently survive for more than eight weeks without further opportunity to drink. These lizards obtain water primarily through eating insects; though they also have occasional opportunity to drink condensed water from periodic advective sea fog, which is typical of the area.

When the lizards are under the surface of the sand, they are protected from climatic changes and from predation. However, they have no opportunity to obtain food or water. When weather conditions are favourable during the day, the animals emerge upon the slip face and feed upon detritus and arthropods. Here they suffer from a much increased rate of dehydration and are vulnerable to predation by the sidewinding adder (*Bitis peringueyi*), the black-backed jackal (*Canis mesomelas*), the chanting goshawk (*Melierax musica*) and the rock kestrel (*Falco tinnunculus*). The lizard's normal reaction to danger is a fast sprint followed by emergency diving beneath the loose sand. It is capable of this rapid anti-predator response only when its body temperature is in the region of 30–40 °C. Below these temperatures the lizards are less mobile, and on cold days they generally remain below the surface of the sand.

On the basis of these observed behaviour patterns of *Aporosaura*, and of their probable survival value, we can make a guess about the optimum survival strategy for this species. When the animal is replete with food and water we would expect it to spend less time on the surface of the sand and thus avoid predation and dehydration. When it has a negative energy or water balance, it can avoid eventual death only by venturing on to the surface of the sand to search for food and water. In cases of extreme hunger or thirst, we would expect the animal to run a greater risk of predation by emerging from the sand at a lower temperature than normal. In such circumstances a period of specialised warming behaviour would be necessary. Similarly, we would expect the animal to run greater risks of dehydration by remaining on the surface at ever-increasing temperatures and indulging in bouts of specialised cooling behaviour. These hypotheses are summarised in Fig. 7, which can be seen as a map of the behaviour upon a basis of physiological coordinates.

We now have the problem of accounting for the mechanisms by which

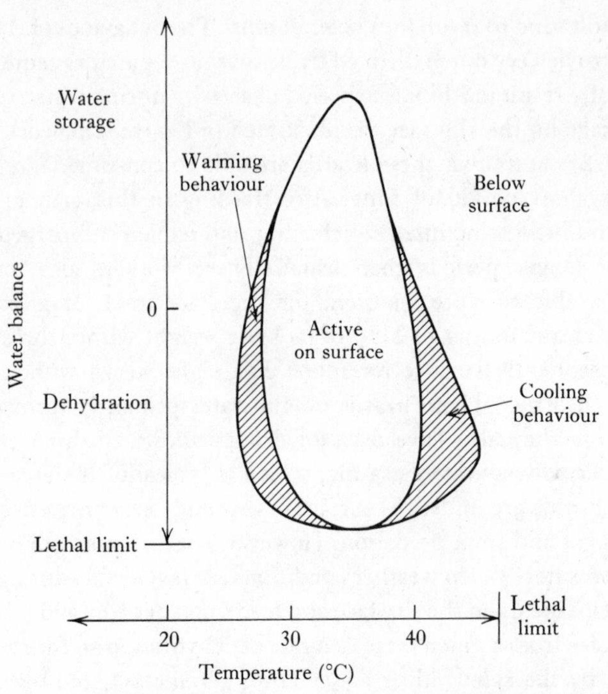

Fig. 7. Two-dimensional representation of the physiological states at which *Aporosaura* might be expected to be below the surface of the sand, or above the surface and involved in warming or cooling behaviour, or actively foraging. Boundary lines indicate transitions from one type of behaviour to the other. Water balance in arbitrary units.

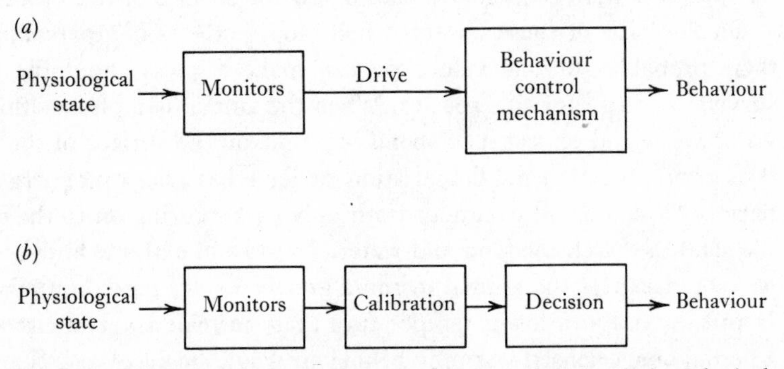

Fig. 8. (*a*) Conventional representation of the influence of physiological state upon behaviour. (*b*) Modification of (*a*) to allow for division of the behaviour control mechanism into calibrations and decision subsystems.

the animal achieves such appropriate behaviour in changing environmental circumstances. The traditional approach is to look for the sense organs through which information about the external environment, and about internal physiological state, reaches the animal's brain. On the basis of work on other lizards, we have good reason to suppose that central thermoreceptors (Whitfield & Livezey, 1973) and dehydration monitors (Fitzsimons & Kaufman, 1974) might exist in *Aporosaura*. The representation of such information in the brain is commonly thought of as some kind of drive, or potentiality, to perform appropriate behaviour. The drives compete, or interact, in such a way that the observed behaviour is controlled by the strongest drive. A summary of this situation is illustrated in Fig. 8*a*, in which the range of behaviour is restricted for the sake of simplicity.

Let us pretend that the animal merely has to make a decision to remain above, or below the surface. In accounting for the adaptiveness of the lizard's behaviour, knowledge of the nature and strength of the relevant drives would not be sufficient. It is obvious that there must be some calibration of the drives, in terms of their relative importance, as indicated in Fig. 8*b*. We would then have to specify how calibrated drives combine and interact. The necessity for both calibration and combination can be illustrated by reference to the stylised *Aporosaura* example, illustrated in Fig. 7.

In investigating calibration of the monitored internal states relevant to temperature and dehydration, we would expect to find that the calibrated drives were related to the survival value of the physiological states. For example, the probability of death resulting from thermal stress might well be a function of temperature such as that illustrated in Fig. 9*a*, and the probability of death by predation might also be a function of temperature, for an animal on the surface, because the lizards are less mobile at lower temperatures (Fig. 9*b*). On this basis we might expect the calibration of the thermal drive to be somewhat like that illustrated in Fig. 9*c*. Similarly, we might expect the calibration of dehydration to be something like the function illustrated in Fig. 10: the tendency to surface increases with dehydration, thus increasing the animal's chances of finding water; but at extreme dehydration this advantage is offset by the increased loss of water that results from being on the surface and exposed to the sun and wind. We now have a calibrated tendency to surface, based on temperature and a similar tendency based on dehydration. The question is – how do these combine to produce a single tendency to surface?

In order to illustrate the difficulty of this aspect of the problem, I have calculated the consequences of assuming that the combination law is additive, and have separately calculated the consequences of assuming that the law is multiplicative. The additive rule implies that the isocline representing a

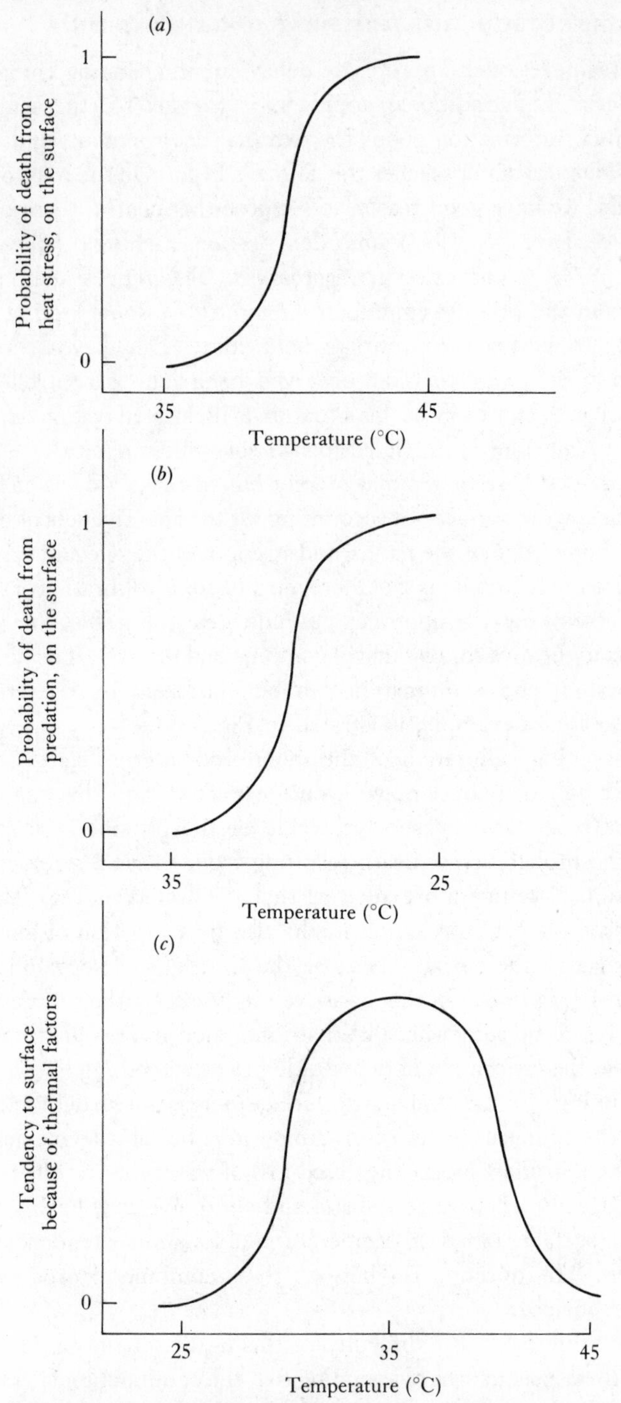

Fig. 9. Hypothetical functions relating temperature and survival for *Aporosaura*. (*a*) and (*b*) are cumulative probability functions. (*c*) A calibration function derived from (*a*) and (*b*).

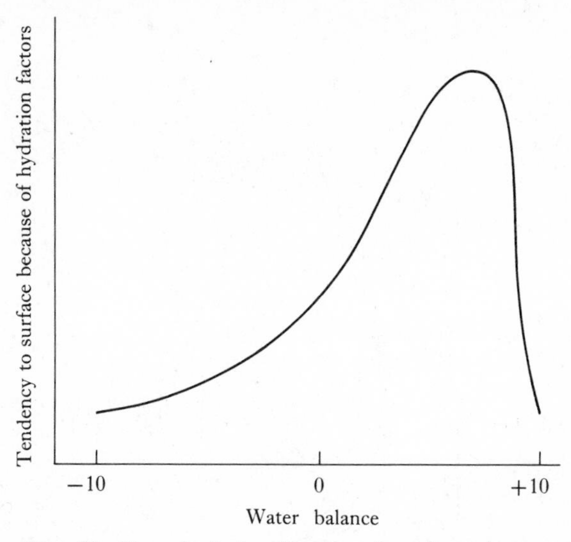

Fig. 10. Hypothetical calibration function relating water balance and tendency to surface. Water balance in arbitrary units.

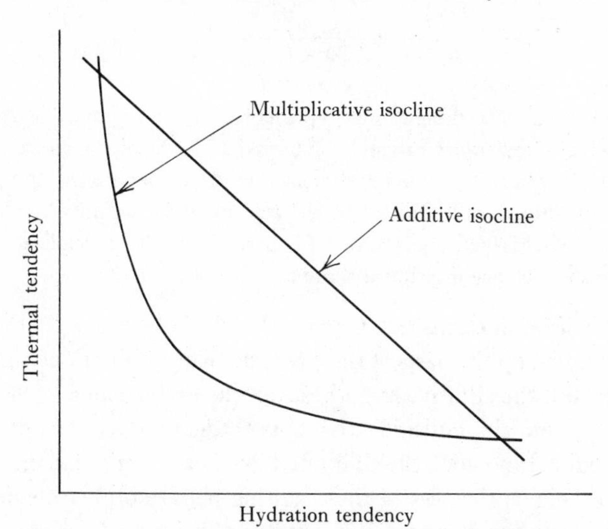

Fig. 11. Hypothetical motivational isoclines for *Aporosaura*, joining points of equal tendency to surface as a result of the combined thermal and hydration tendencies.

particular tendency to surface is linear, while the multiplication rule implies that the isocline is a hyperbola, as illustrated in Fig. 11. If we take these isoclines to represent the threshold tendency to surface (i.e. the tendency which, on average, will be large enough to cause the animal to be above the surface), then we can represent the consequences of the hypotheses embodied

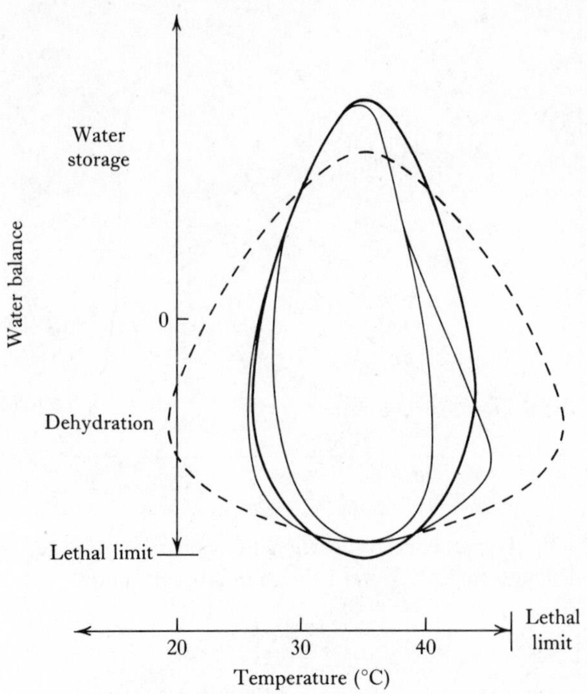

Fig. 12. Two dimensional representation of the physiological states at which *Aporosaura* might be expected to be below or above the surface of the sand. The thin continuous boundary lines are transposed directly from Fig. 7. The thick line is based upon the additive isocline of Fig. 11. The dashed boundary line is based upon the multiplicative isocline. Water balance in arbitrary units.

in the two isoclines in terms of a map on the physiological coordinates (Fig. 7). In other words, on the basis of the (hypothetical) calibrations summarised in Figs. 9 and 10, the additive hypothesis produces the map shown by a thick line in Fig. 12, and the multiplicative hypothesis produces the dashed one. Clearly the manner in which the calibrated drives are combined makes a large difference to the adaptiveness of the resulting behavioural strategies.

The problem with this approach is that it is difficult to foresee the day when the physiological analysis of behavioural control mechanisms will have advanced to such an extent that it will be possible to measure the way in which drives are calibrated, and combine to produce decision rules appropriate to the animal's way of life. It is equally difficult to see how traditional motivational analysis, with its view of behaviour as driven from within, can ever succeed in this task. The traditional approach to motivation is to attempt to specify the links in the causal chain between the animal's physiological state, its perception of the environment, its calibration of these factors in terms of their importance for survival, and its formulation of appropriate decisions, which are translated into behaviour. As I have argued above, a knowledge of the decision criteria is an essential ingredient of this recipe, and it is possible to gain this only through consideration of the function of the mechanisms involved. An alternative approach to the problem is to admit that we will never be able to establish all the details of the causal network controlling behaviour, and to emphasise the consequences of behaviour, rather than its causes (see Fig. 13*a*). The relevant consequences of behaviour can be divided into those which are perceived by the animal, and alter its motivational state; and those which are not relevant to the behaviour of the animal in the short-term, but nevertheless are subject to natural selection. For example, an animal may eat a poisonous plant without perceiving the consequences as being any different from those of eating a non-poisonous plant. Even if the poison does not not kill the animal, but merely makes it sick, ingestion of the plant may affect the animal's chances of survival by temporarily lowering its general efficiency, ability to avoid predators, etc. Naturally, we would expect that in the long term the animal would develop a suitable avoidance mechanism, such as that responsible for bait shyness in rats (Rozin & Kalat, 1970; McFarland, 1973). However, we can never assume that there is always, or even mostly, a match between the consequences of behaviour as perceived by the animal, and those acted on by natural selection. Nevertheless, every behaviour sequence will cause changes in the internal state of the animal, that can be related to the survival value of that same behaviour sequence, because every behaviour sequence has both of the types of consequence discussed above, and these are highly çorrelated. The correspondence between changes in the internal state of the animal, and the survival value of the behaviour responsible for some of those changes, forms the basis of a new approach to the analysis of motivational mechanisms in terms of both form and function.

THE ANALYSIS OF BEHAVIOUR BY MEANS OF OPTIMALITY THEORY

This approach has been formally described elsewhere (Sibly & McFarland, 1976), and all I wish to do here is to outline the approach and indicate how it might be applied in practice. Whenever we try to specify an optimal condition, we have to make clear the criteria by which the degree of optimality is to be judged. In applying optimality theory to behaviour, the costs and benefits of behaviour have to be quantitatively assessed, and then combined into a single cost function, which represents the precise manner in which the various costs and benefits are counterbalanced. A specified cost function can be used to calculate an optimal control law, which will generate a (supposedly optimal) behaviour sequence in any given set of circumstances.

If all the causal factors of behaviour are known and represented in an n-dimensional vector space (called a causal factor space by McFarland & Sibly, 1976), then the internal state of the animal, and the state of the environment as it affects the animal, can be represented at any moment in time as a point in this causal factor space. Changes in the environment as perceived by the animal, or changes in the internal state of the animal owing to hormonal or other changes and consequences of behaviour (McFarland & Sibly, 1972; Sibly & McFarland, 1976), combine to move the point in causal factor space along a trajectory. We shall designate the state in causal factor space as the vector $\mathbf{x}(t)$, being a function of time. We shall suppose that the behavioural repertoire of the animal is also described in vector space form so that the action chosen by the animal may be represented by a vector $\mathbf{u}(t)$. We are now in a position to state the problem rigorously. A behavioural sequence (described by $\mathbf{u}(t)$ in a period T) may be said to be *optimal* if it maximises fitness. Sibly & McFarland have shown that this is equivalent to saying that a behaviour sequence is optimal if it minimises the cost

$$\int_0^T C(\mathbf{x}, \mathbf{u})\mathrm{d}t,$$

where C is a cost function, similar to that described above in relation to the design of blood vessels. The problem for the experimenter is to know which $\mathbf{u}(t)$ the animal should choose at each time in order to maximise fitness. In other words, given the information that the animal has at its disposal, and the repertoire of actions which it is capable of performing, what is the control law, or set of decision rules, which will produce the optimal behaviour?

The situation is summarised in Fig. 13*b*. The state of the system $\mathbf{x}$, which can be represented as a point in a causal factor space, is determined by the system 'plant',* which includes the interaction of all the factors influencing $\mathbf{x}$. Knowledge of the system plant and the control law is sufficient to characterise the system completely, and to permit calculation of the cost function that the system is designed to fulfil. Alternatively, knowledge of the system plant, together with an explicit hypothesis about the cost function, is sufficient to permit calculation of a control law, which will produce a supposedly optimal behaviour sequence that can be compared with observed behaviour sequences.

* This term is borrowed from the field of process control. It has become thoroughly incorporated into the vocabulary of control theory, and is particularly appropriate in behaviour studies, because it makes intuitive sense to distinguish between the bodily processes (plant) and their control by the brain.

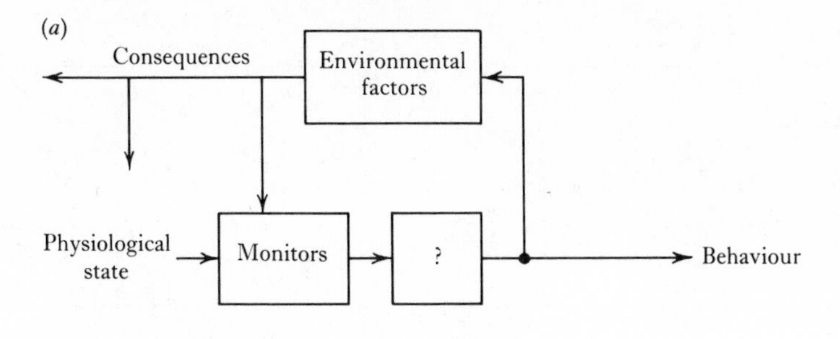

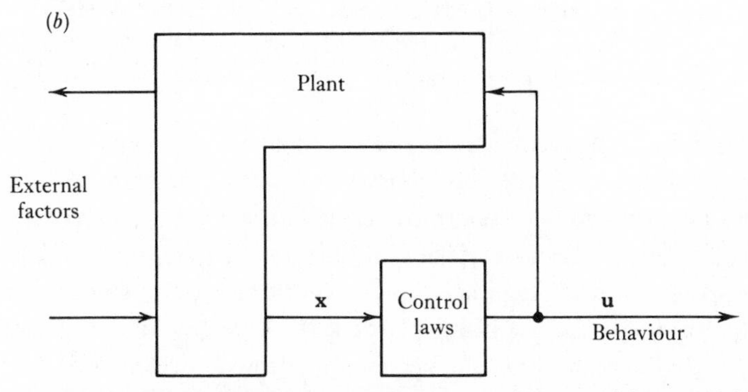

Fig. 13. Relationship between behaviour and its causal factors. (*a*) View of the problem, in which the consequences rather than the causes of behaviour are emphasised. In (*b*) the environmental factors, the consequences of behaviour, the monitoring processes, and the systems controlling the physiological state of the animal, are all incorporated into the system 'plant'. The changes in state of the system, represented by the vector **x**, are determined by the plant equations. The behaviour, represented by the vector **u**, is determined by the system state **x**, in accordance with the 'control laws' or 'decision-rules'.

The system plant

The first problem is that we are not in a position to know all the causal factors of the behaviour of any individual animal. However, we can often narrow down the problem to manageable proportions by considering the constraints acting upon the system. Constraints may stem from the nature of the environment, and its roles in determining the 'climate' in which the animal finds itself (Sibly & McFarland, 1974) and in determining the consequences of the animal's behaviour (McFarland & Sibly, 1972). Another form of constraint is provided by the fact that the animal must use a limited set of muscles to carry out a wide range of activities (McFarland & Sibly, 1976). A second line of attack can be provided by computer simulation, designed to take account

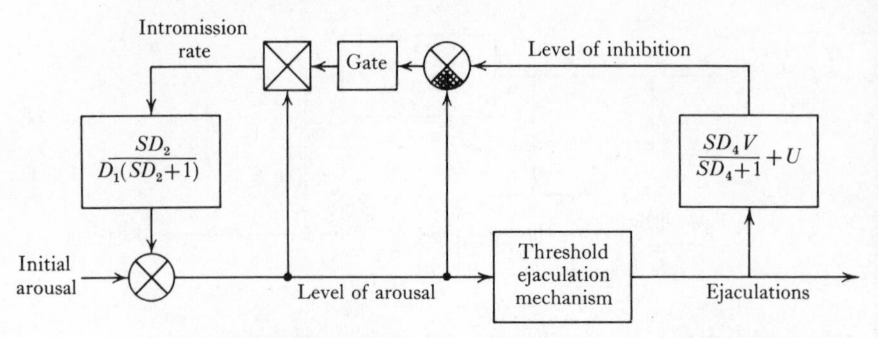

Fig. 14. Simplified version of the model of Freeman & McFarland (1974) for the copulatory behaviour of the male rat. S is the Laplace operator; D, V and U are constants.

of the results of experiments carried out by many workers over long periods of time. For example, the copulatory behaviour of the male rat has been the subject of intensive behavioural investigation since the pioneering work of Beach (1956) and Larsson (1956). By means of computer simulation it has been possible to draw together the diverse threads of much of this research, and to formulate in quantitative terms the changes in causal factors that underly this aspect of sexual behaviour (Brown, Freeman & McFarland, 1974). The dynamics of the copulatory behaviour of the male rat seem to depend on two consequences of the behaviour (Fig. 14). The consequence of intromissions is a positive feedback, with a built-in decay, which tends to increase level of arousal. As a consequence of ejaculation there is a rapid build-up of inhibition, which has both a rapid and a slow decay term, and which interacts with level of arousal in determining the duration of the refractory period. Results obtained from the simulation are compared with observed data in Fig. 15.

This rather gross approach can form a basis for more sophisticated methods of data analysis, designed to elucidate the causation of behaviour sequences in individual animal subjects (McFarland & Nunez, 1976). A further promising development is the elegant use of the techniques of stochastic analysis, which have been successfully deployed by Heiligenberg (1973, 1974) to determine fluctuations in the behavioural states of readiness to attack in cichlid fish.

In view of the rapid progress made in recent years, there are grounds for optimism, and it seems likely that a good understanding of the causal factors relevant in certain situations will be forthcoming in the foreseeable future. From the viewpoint of optimality theory, it is important that these investigations be conducted by behavioural means, rather than by physiological means. As can be seen from Fig. 13*b*, the system plant, which determines all changes

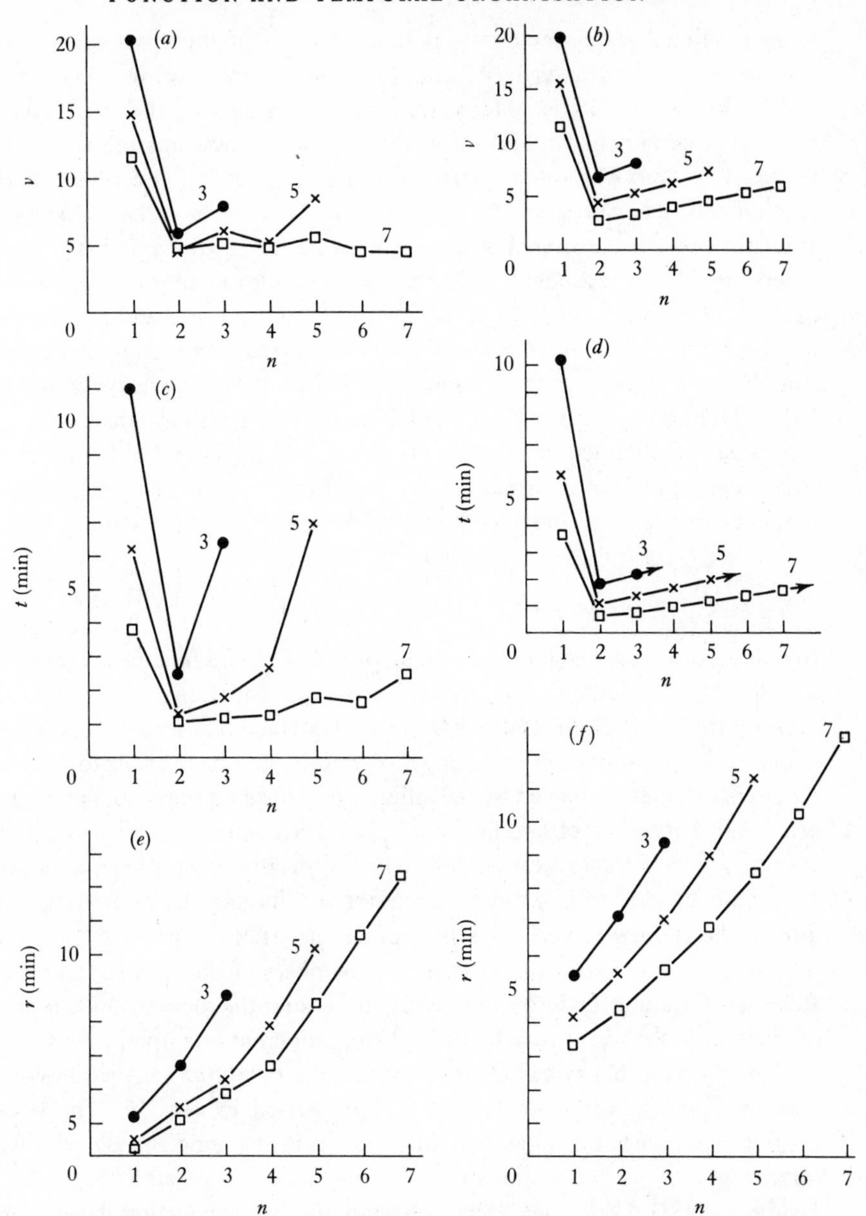

Fig. 15. Aspects of the copulatory behaviour of the male rat: observed results. In all cases, numbers on curves indicate number of ejaculations occurring in one hour of observation. (*a*) Intromission frequency (*v*) versus ejaculation number (*n*). Data from Larsson (1956). (*b*) Simulated version of (*a*) (from Freeman & McFarland, 1974). (*c*) Ejaculatory latency (*t*) versus ejaculation number. Data from Larsson (1956). (*d*) Simulated version of (*c*), from Freeman & McFarland (1974). (*e*) Durations of successive refractory periods (*r*). Data from Larsson (1956). (*f*) Simulated version of (*e*), from Freeman & McFarland (1974).

in motivational state, includes (*a*) those aspects of the environment that determine the consequence of behaviour, (*b*) the monitoring processes by which the brain obtains information about changes in the internal and external environment, and (*c*) the calibration of the incoming information in terms of its survival value. What is required is an understanding of the relationship between **u** and **x**, and this can only be done by behavioural studies, in which the animal is made to indicate the relationship between the independent and dependent (motivational) variables in an unambiguous yet quantifiable manner. Success in this respect has been achieved in a few cases in which the animal has been made to demonstrate its scaling of external stimuli (e.g. Heiligenberg, Kramer & Schultz, 1972; Baerends & Kruijt, 1973). McFarland & Sibly (1976) and Sibly (1975) have shown that analogous results can be obtained in experiments in which animals are made to demonstrate their scaling of internal factors, quantitative results being achieved by reference to a 'motivational landmark' (McFarland, 1974).

The control law

In many cases, knowledge of the 'system plant' will be fragmentary. However, it will often be sufficient to permit determination of the control law, or decision rules (Fig. 13*b*). If the changes in causal factors relevant to one aspect of behaviour are relatively well known, then it should be possible to discover the decision rules employed by the animal, by opposing the tendency for the behaviour with that of another less well-known aspect of behaviour. For example, if the sexual readiness of a rat can be measured with some accuracy, we should be able to find out about its relative hunger, by presenting food during the refractory period. This type of approach is currently being employed in a wide variety of situations by members of the Animal Behaviour Research Group at Oxford, and before discussing the more detailed aspects of optimality theory, it may be helpful to outline one example.

The courtship behaviour of the smooth newt (*Triturus vulgaris*) has been described by Halliday (1974), and is summarised in Fig. 16. The whole courtship sequence takes place on the substratum of a pond and is fairly often interrupted by ascent to the surface of the water to breathe. Spurway & Haldane (1953) report that 'ascent certainly interrupts courting, but it is not clear that courting inhibits ascent'. The latter question is being studied in some detail by Halliday & Sweatman (1976). By restraining the female in a strait-jacket (Fig. 17), Halliday (1976) is able to investigate the courtship tendency (libido) of the male. Preliminary results from Halliday's work (Fig. 18) show clearly that breathing is influenced by courtship. It appears that the male is more likely to ascend and breathe if the courtship is 'going badly'.

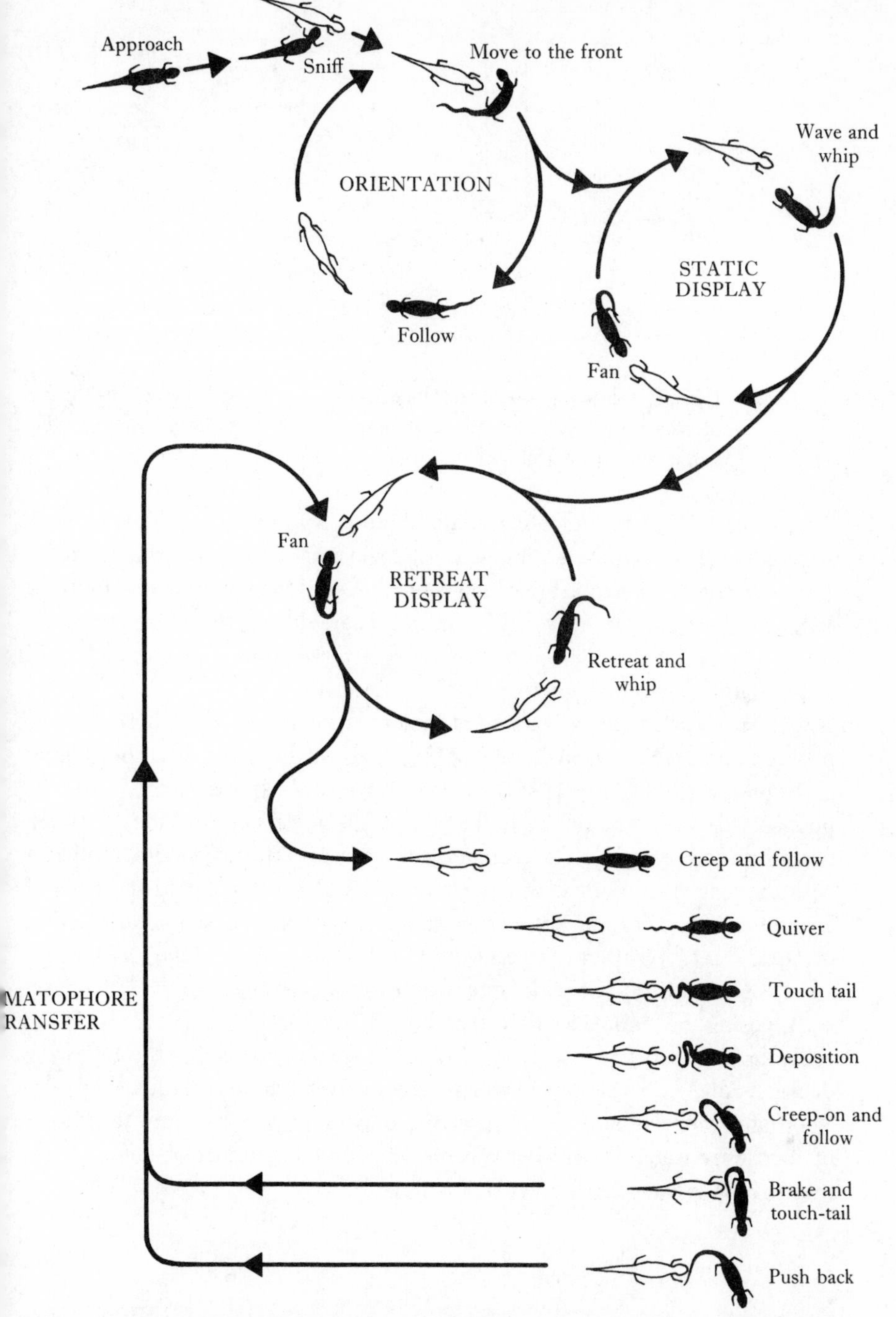

Fig. 16. Schematic representation of the courtship sequence of the smooth newt *Triturus vulgaris* (from Halliday, 1974). The male is shown black. The orientation, static display and retreat display phases may involve repetition, but the spermatophore transfer phase is considerably more stereotyped.

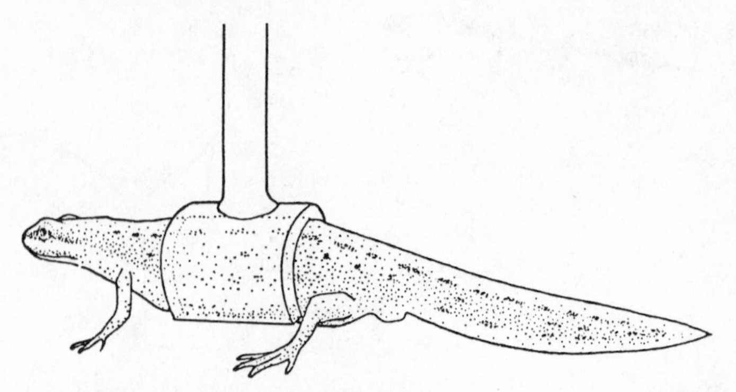

Fig. 17. A drawing of a female smooth newt in a 'straight-jacket', by means of which the experimenter can manipulate her stationary and approach behaviour (from Halliday, 1976).

Thus breathing occurs sooner if male libido is low, or if the female remains stationary. If the female is removed, the male ascends and breathes immediately. On the other hand, breathing is postponed if the female is manipulated so as to approach the male, and it never occurs during the spermatophore-transfer phase of the courtship. Breathing is very common on completion of a spermatophore-transfer sequence.

Current experiments are aimed at manipulation of the breathing tendency by alteration of the oxygen content of the gas above the water, after the manner of Spurway & Haldane (1953). In this way it is expected that it will be possible to work out the rules by which the male newt decides between courting and breathing. The technique of opposing one behavioural tendency with another, as a method of determining decision-rules, forms the basis of a number of studies currently in progress at Oxford. These include studies of sticklebacks (*Gasterosteus aculeatus*), great tits (*Parus major*) and doves (*Streptopelia risoria*). Considerable progress has been made in the latter case, by opposing the tendencies to feed and drink (McFarland, 1974; Sibly, 1975; Sibly & McCleery, 1976). Sibly (1975) obtained evidence for the following decision-rule, for situations in which food and water are not readily available: 'food and water should be taken at the maximum possible rate, the choice of food over water should depend on whether the product of deficit times incentive is greater for hunger or for thirst'.

The cost function

Ideally, the cost function represents the precise manner in which the costs and benefits of the structure of a behaviour sequence, or of an organ, are counterbalanced. Whereas the 'plant' represents the actual controlled system,

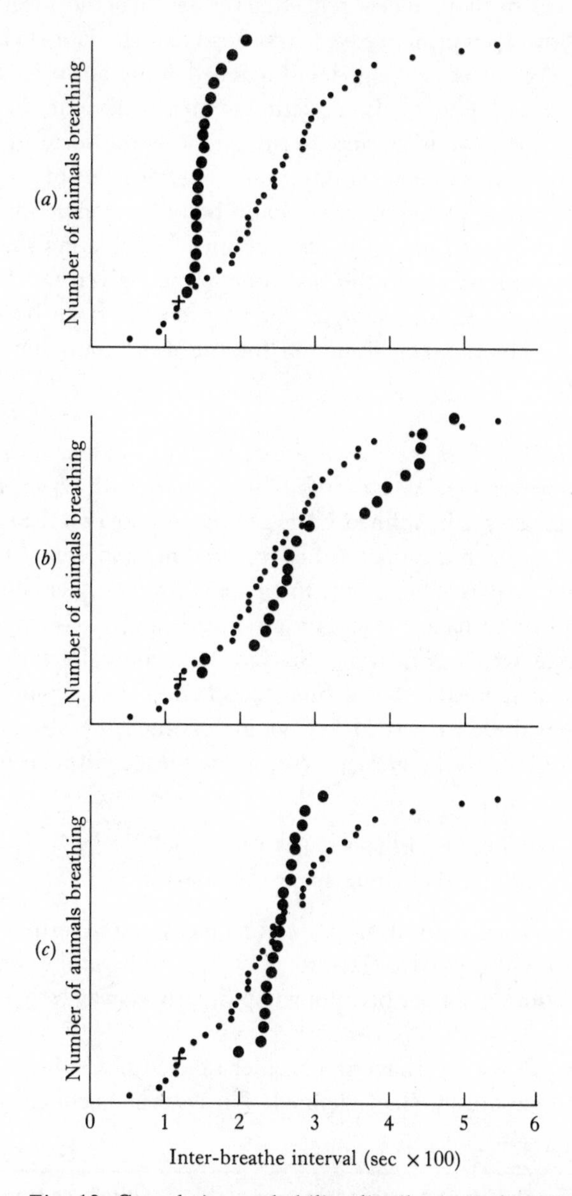

Fig. 18. Cumulative probability distributions of intervals since previous breathing episode (time zero) of male newts during courtship. The control situation is the distribution obtained in the presence of a static female. (*a*) The female is removed (cross), and the male goes quickly to breathe. (*b*) The female is made to approach the male slowly, thus becoming slightly more stimulating than when static. (*c*) The female is made to approach the male quickly (very stimulating), so that the male

the cost function can be thought of as reflecting the design of the 'ideal' system, with maximum fitness. In principle we might hope to formulate a behavioural cost function as a result of prolonged and detailed study of the survival and reproductive success of individuals in relation to their behaviour. In practice, the best we can hope for is an intelligent guess at the form of the cost function, based upon our biological intuition and knowledge of the animal's way of life. This guess, or hypothesis, would be open to experimental test in accordance with predictions made as a result of calculating the optimal strategies that would result from the cost function in a given set of environmental circumstances. As an example, let us proceed, stage by stage, to formulate a guess about a cost function for the newt courtship situation described above.

Stage I. The first task is to prescribe the problem in a suitably precise manner, without being too ambitious to start with. A crude initial approximation can always be refined later. Let us suppose that the male newt is in a situation in which a female is always present, and one of two types of male behaviour is always observed: the male is always either courting the female (courting) or surfacing to obtain air (breathing). It is important to define the possible activities in terms that are both mutually exclusive and exhaustive, even if it means that a category of 'null' behaviour has to be devised to contain those aspects of behaviour in which the observer is not particularly interested. In the present case, we define a control vector $\mathbf{u}$ with two elements:

u_1 = courting rate (displays per unit time),
u_2 = breathing rate (time spent not courting).

Here we are assuming that courting rate can be measured in terms of display elements observed per unit time (Halliday, 1975), and breathing rate can be approximated by time spent per breathing trip (i.e. time away from female).

Stage II. We now have to consider how costs and benefits might be ascribed to the activities we have defined. The costs and benefits that result

---

can complete the courtship sequence within a short space of time. These results show that breathing is disinhibited when the courtship is terminated by the female (*a*), or by the male (*c*). During courtship, breathing is suppressed (*b*). Essentially similar results are obtained when the male is tested at a greater oxygen debt, induced by making the male court a stationary female for a longer period before the female is moved. (Data from Halliday (1976).)

from behaviour depend largely upon the consequences of the behaviour; i.e. upon the way in which the state of the system is affected by the behaviour. In the present case, two states are of importance: (*a*) the state of oxygen 'debt' of the male will obviously affect his survival probability and reproductive success; (*b*) the state of sexual readiness of the female, in terms of picking up spermatophores, will also affect the outcome of the courtship. The sexual state of the male will also have an influence on the success of the courtship, because his courting is much more vigorous when his libido is high, and he deposits more spermatophores per courtship encounter (Halliday, 1976). However, it can be argued that the state of the female is so largely dependent upon that of the male, that the two variables are not independent and only one is necessary in a preliminary formulation, such as this. On this basis we define the state vector $\mathbf{x}$ in terms of two state variables:

x_1 = the state of sexual readiness of the female,
x_2 = the state of oxygen 'debt' of the male.

Stage III. We now need to have some idea of how the state of the system is altered as a consequence of behaviour. In general, we can say that $\dot{\mathbf{x}} = \mathbf{K}\mathbf{u}$, where $\dot{\mathbf{x}} = \mathrm{d}\mathbf{x}/\mathrm{d}t$, and $\mathbf{K}$ is a matrix representing the dynamic relationships between the behaviour $\mathbf{u}$ and its consequences, expressed in terms of changing the state of the system. Our knowledge of the system (plant) controlling the courtship behaviour of the male newt is very limited (Houston, Halliday & McFarland, 1976), so we cannot specify $\mathbf{K}$ in detail. We can however, make reasonable guesses based upon common-sense considerations.

The probable consequences of male behaviour (u_1, u_2) are four-fold: The state of sexual readiness of the female (x_1) is increased by male courtship, and there is also likely to be an increase in oxygen debt (x_2), as a consequence of the increased activity. We can expect breathing (u_2) to decrease female readiness (x_1), because the male has to abandon her in order to visit the surface of the water, and female responsiveness probably declines in the absence of a male (Halliday, 1975). We can also reasonably expect that breathing (u_2) will decrease the oxygen debt (u_1). We can now summarise the effects of the male newt's behaviour in terms of changes in the state of the system, viz.

$$\dot{\mathbf{x}} = \mathbf{K}\mathbf{u}, \quad \begin{bmatrix} \dot{x}_1 \\ \dot{x}_2 \end{bmatrix} = \begin{bmatrix} k_1 & -k_2 \\ k_3 & -k_4 \end{bmatrix} \begin{bmatrix} u_1 \\ u_2 \end{bmatrix}, \tag{2}$$

$$\begin{aligned} \dot{x}_1 &= k_1 u_1 - k_2 u_2, \\ \dot{x}_2 &= k_3 u_1 - k_4 u_2. \end{aligned} \tag{3}$$

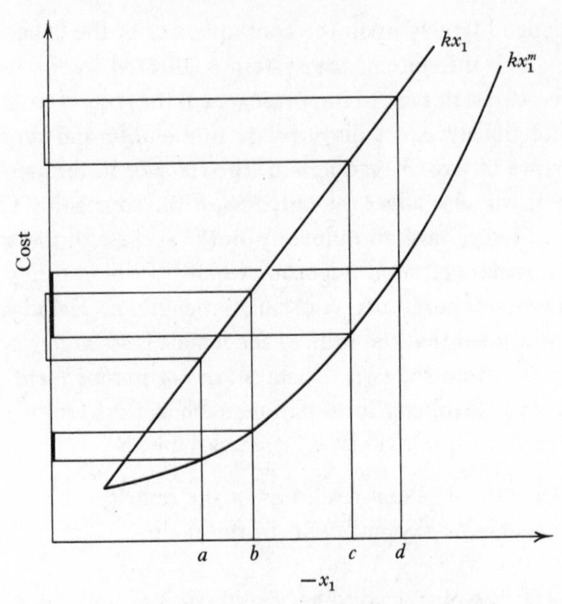

Fig. 19. Possible relations between cost and sexual state. When the relationship is linear, a given change in sexual state $(a-b)$ has the same effect on cost as a similar sized change $(c-d)$ at a higher level of x_1. When the relationship is convex, the former change in x_1 has a smaller effect on cost than the latter.

Stage IV. We have now specified six variables upon which the costs and benefits of newt courtship are likely to depend. These have to be combined into a single cost function C, so that $C = C(\mathbf{x}, \mathbf{u}, \dot{\mathbf{x}})$. However, we know that $\mathbf{u}$ and $\dot{\mathbf{x}}$ are not independent, since $\dot{\mathbf{x}} = \mathbf{Ku}$ (above). Therefore, only $\mathbf{u}$ or $\dot{\mathbf{x}}$ need appear in the cost function, and we have a choice as to which this should be. It is more convenient to choose $\mathbf{u}$, so that $C = C(\mathbf{x}, \mathbf{u})$. The next step is to consider each of the possible components of the cost function in turn, and assess its probable functional importance.

Sexual state (x_1). Successful fertilisation is more likely when the female's state of sexual readiness is high, as she is more likely to follow the male and pick up the spermatophore successfully (Fig. 16). Thus greater fitness is likely to be associated with high values of x_1. The relationship between fitness and sexual state might be a linear one, in which case a given increment in x_1 will have the same effect on fitness, at a high value of x_1 as at a low value (see Fig. 19). On the other hand the relationship might be convex, such that a given change in x_1 makes a smaller contribution to fitness at a low value of x_1 than at a high value (Fig. 19). If we do not know how to assess this particular aspect of the problem, we can leave the question open for the time being, by expressing this component of the cost function as $-x_1^m$. This means that cost is reduced (i.e. fitness is increased) as x_1 is increased, but that the exponent m remains unknown. Values of m substituted into the final cost function will make a profound difference to the outcome, so that the most probable values for m will soon become apparent, once optimality calculations are undertaken (see below).

Oxygen debt (x_2). The survival of being in a low state of oxygen debt will obviously be higher than that of having a high debt, but here, as above, it is difficult to know whether the relationship should be linear or convex, and we provisionally write the contribution to the cost function as x_2^n.

Courting (u_1). Here we have to remember that both courtship *per se*, and the consequences of courtship, may affect fitness. Courting may increase the oxygen debt, the probability of predation, and use up time that could usefully be spent feeding, etc. On the other hand, the consequences of courting, expressed in terms of alteration of the state of sexual readiness of the female, may greatly increase fitness. On the whole, we must assume that there are costs associated with courtship, which the system as a whole is designed to minimise. We can provisionally write the contribution to the cost function as u_1^p.

Breathing (u_2). Although breathing has obvious beneficial effects, these are not gained without considerable cost, especially when breathing occurs during courtship. The quicker the animal can make a breathing trip, the less cost is incurred without there being much change in the benefit gained. On this basis it is reasonable to express the contribution of breathing to be the inverse of the rate, viz $1/u_2^q$.

Separability. In the foregoing discussion, there is an implicit assumption that the cost function is separable into the component costs associated with performing particular activities (**u**), and component costs resulting from particular values of state variables (**x**). Such separability means that the cost function $C(\mathbf{x}, \mathbf{u})$ is the sum of a term in u_1, a term in u_2, a term in x_1 and a term in x_2. Sibly & McFarland (1976) show that this separability is equivalent to probabilistic independence of the various factors. Independence in this context means that the contribution to fitness owing to the value of one variable (e.g. courting rate u_1) is unaffected by the values of the other variables. This would be the case, for example, if the chance of successful fertilisation resulting from a particular level of female readiness (x_1) were the same, however great the male's oxygen debt. This is not to say that the overall chance of success is not affected by the oxygen debt, but that the chance *owing to* the female readiness is not affected. In the case of newt courtship where male courtship vigour changes little with oxygen debt, separability seems a reasonable assumption to make, and on the basis of the foregoing discussion, the cost function could be formulated as follows:

$$C = x_2^{\,n} - x_1^{\,m} + u_1^{\,p} + 1/u_2^{\,q} \tag{4}$$

Another example of a cost function, for which there is some evidence (Sibly & McFarland, 1976), is the following:

$$C = x_1^2 + x_2^2 + u_1^2 + u_2^2, \tag{5}$$

where x_1 is hunger state, x_2 is thirst state, u_1 is feeding rate and u_2 is drinking rate in the Barbary dove (*Streptopelia risoria*).

The optimal behaviour

An hypothesis about the cost function, formulated in suitably rigorous terms, can be used to calculate an optimal behaviour sequence, for a given set of conditions. In very general terms the procedure is to find the Hamiltonian, a function that should always be maximised by the animal to obtain the optimal behaviour sequence. Whereas the cost function summarises the cost and benefits that should be minimised in general terms, the Hamiltonian takes account of the specific situation in which the animal finds itself at any particular time. In general

$$\text{Hamiltonian} = \text{plant equations} - \text{cost function}$$
$$H = \mathbf{p}\dot{\mathbf{x}} - C$$

where $\mathbf{p}$ is a 'co-state vector', $\mathbf{p} = \partial H / \partial \mathbf{x}$. In terms of the simple situation illustrated in Fig. 13*b*, the Hamiltonian can be used to find the decision rules, or control law. For example, in the case of the cost function given in equation (5), the corresponding control law (under certain constraints, Sibly & McFarland, 1976) was found to be

eat at rate k_1 if $x_1 r_1 k_1 > x_2 r_2 k_2$
drink at rate k_2 if $x_1 r_1 k_1 < x_2 r_2 k_2$

where x_1 is hunger state, x_2 is thirst state, r_1 is food availability, r_2 is water availability, k_1 is the limiting rate of eating, k_2 is the limiting rate of drinking. This control law is equivalent to the decision-rules obtained experimentally by Sibly (1975), for the case of feeding and drinking in doves discussed above. In this case the dynamics of the interaction of hunger and thirst was fairly well known from previous work (McFarland, 1970, 1971, 1974) so that computation of the Hamiltonian was not too difficult. In cases where the plant equations are less well known, it may be possible to make intelligent guesses, and this has been done for the thermoregulatory behaviour of *Aporosaura*.

In other cases, such as the courtship behaviour of newts, there may be little hope of obtaining the plant equations (Houston *et al.*, 1976). However, as indicated above, it may be possible to determine the decision-rules directly by experiment. Knowledge of the decision-rules might then be combined with a hypothesis about the cost function to calculate the optimal behaviour.

Calculation of the optimal behaviour is the main objective, because it provides a test of any hypothesis purporting to relate form and function in the organisation of behaviour sequences. The testing procedure may be summarised as follows:

(1) Observation from ecological studies of the animal's environment, and estimates of the consequences of its behaviour are used to produce an adequate state-variable description of the behaviour in particular situations, and to discover roughly the dependence of cost upon these variables.

(2) An hypothesis about the cost function is formulated on the assumption that behaviour is optimal. In the example outlined by Sibly & McFarland (1976), a sufficient number of observations were taken to indicate the plausibility of a quadratic cost function.

(3) The hypothesised cost function is used in optimality computations to predict sequences of behaviour that should occur under the observed circumstances and conditions.

(4) The predicted behaviour sequences are compared with those in the natural, or laboratory situation.

(5) The discrepancy between predicted and observed behaviour sequences is used to reformulate the cost function, redefine the constraints, etc. so that a new model emerges.

This approach embodies the normal scientific procedure of hypothesis, prediction and test. It is to be hoped that this approach will be useful in analysing both the form and the function of the behavioural strategies of individual animals. The form of the strategy, expressed in terms of the state variables x_i and u_i, is embodied in decision-rules. The function, or survival value, of the strategy can be expressed in terms of a cost function or in terms of its contribution to fitness.

SUMMARY

This paper discusses the view that not only are species-typical characteristics of behaviour subject to natural selection, but so also is the order and timing of their performance. Observational and ecological studies of an animal can be used to formulate a cost function, representing the balance of costs and benefits of the various aspects of the behaviour of the animal.

The hypothesised cost function can be used in optimality computations to predict sequences of behaviour that should occur under observed circumstances and conditions. These predictions can be compared with observed behaviour sequences, thus completing the normal scientific procedure of hypothesis, prediction and test.

The author wishes to express his appreciation to John Krebs and Alasdair Houston for their helpful comments on the manuscript, and to the Science Research Council for their support of this research.

REFERENCES

Baerends, G. P., Brouwer, R. & Waterbolk, H. Tj. (1955). Ethological studies of *Lebistes reticulatus* (Peters): I. An analysis of the male courtship pattern. *Behaviour*, **8**, 249–334.

Baerends, G. P. & Kruijt, J. P. (1973). Stimulus selection. In *Constraints on learning: Limitations and Predispositions*, ed. R. A. Hinde & J. Stevenson-Hinde, pp. 23–50. Academic Press: London & New York.

Beach, F. A. (1956). Characteristics of masculine 'sex drive'. In *Nebraska Symposium on Motivation*, ed. M. R. Jones, pp. 1–38. University of Nebraska Press: Lincoln.

Brown, R., Freeman, S. & McFarland, D. J. (1974). Towards a model for the copulatory behaviour of the male rat. In *Motivational Control Systems Analysis*, ed. D. J. McFarland, pp. 461–510. Academic Press: London.

Cohn, D. L. (1954). Optimal systems: I. The vascular system. *Bulletin of mathematical Biophysics*, **16**, 59–74.

Cohn, D. L. (1955). Optimal systems: II. The vascular system. *Bulletin of mathematical Biophysics*, **17**, 219–227.

Cullen, E. (1957). Adaptations in the kittiwake to cliff-nesting. *Ibis*, **99**, 275–302.

Fitzsimons, J. T. & Kaufman, S. E. (1974). Osmometric thirst in the iguana. *Journal of Physiology*, **242**, 112p–114p.

Freeman, S. & McFarland, D. J. (1974). RATSEX – an exercise in simulation. In *Motivational Control Systems Analysis*, ed. D. J. McFarland, pp. 479–510. Academic Press: London.

Hailman, J. P. (1965). Cliff-nesting adaptations of the Galapagos swallow-tailed gull. *Wilson Bulletin*, **77**, 346–362.

Halliday, T. R. (1974). Sexual behaviour of the smooth newt, *Triturus vulgaris* (Urodela, Salamandridae). *Journal of Herpetology*, **8**, 277–292.

Halliday, T. R. (1975). An observational and experimental study of sexual behaviour in the smooth newt, *Triturus vulgaris* (Amphibia, Urodela). *Animal Behaviour*, **23**, 291–322.

Halliday, T. R. (1976). The libidinous newt. An analysis of variations in the sexual behaviour of the male smooth newt, *Triturus vulgaris*. *Animal Behaviour*, **24**, 398–414.

Halliday, T. R. & Sweatman, H. P. A. (1976). To breathe or not to breathe; the newt's problem. *Animal Behaviour*, **24**, in press.

Heiligenberg, W. (1973). Random processes describing the occurrence of behavioural patterns in cichlid fish. *Animal Behaviour*, **21**, 169–182.

Heiligenberg, W. (1974). A stochastic analysis of fish behaviour. In *Motivational Control Systems Analysis*, ed. D. J. McFarland, pp. 87–118. Academic Press: London.

Heiligenberg, W., Kramer, U. & Schultz, V. (1972). The angular orientation of the black eye-bar in *Haplochromis burtoni* (Cichlidae, Pisces) and its relevance to aggressivity. *Zeitschrift vergleichende Physiologie*, **76**, 168–176.

Houston, A., Halliday, T. & McFarland, D. J. (1976). Towards a model of the courtship of the smooth newt *Triturus vulgaris*, with special reference to problems of observability in the stimulation of behaviour. *Medical and biological Engineering*, in press.

Huxley, J. (1932). *Problems of Relative Growth*. Methuen: London.

Kaufman, A. (1968). *The Science of Decision-making*. Weidenfeld & Nicolson: London.

Kluijver, H. N. (1951). The population ecology of the great tit, *Parus m. major* L. *Ardea*, **39**, 1–135.

Krebs, J. (1970). Regulation of numbers in the great tit (Aves, Passeriformes). *Journal of the zoological Society of London*, **162**, 317–333.

Kruuk, H. (1964). Predators and anti-predator behaviour in the black-headed gull (*Larus ridibundus* L.). *Behaviour Supplement*, **11**, 1–30.

Lack, D. (1954). *The Natural Regulation of Animal Numbers*. Oxford University Press: London.

Lack, D. (1966). *Population Studies of Birds*. Oxford University Press: London.

Larsson, K. (1956). *Conditioning and Sexual Behaviour in the Male Albino Rat*. Almkvist & Wiksell: Stockholm.

Lorenz, K. Z. (1950). The comparative method of studying innate behaviour patterns. *Symposia of the Society for experimental Biology*, **4**, 221–268.

Louw, G. N. & Holm, E. (1972). Physiological, morphological and behavioural adaptations of the ultrapsammophilous Namib Desert lizard *Aporosaura anchietae* (Bocage). *Madoqua*, **1**, 67–85.

Mayhew, W. W. (1968). Biology of desert amphibians and reptiles. In *Desert Biology*, ed. G. W. Brown, pp. 196–356. Academic Press: New York.

McFarland, D. J. (1970). Recent developments in the study of feeding and drinking in animals. *Journals of psychosomatic Research*, **14**, 229–237.

McFarland, D. J. (1971). *Feedback Mechanisms in Animal Behaviour*. Academic Press: London.

McFarland, D. J. (1973). Stimulus relevance and homeostasis. In *Constraints on Learning: Limitations and Predispositions*, ed. R. A. Hinde & J. Stevenson-Hinde, pp. 141–155. Academic Press: London & New York.

McFarland, D. J. (1974). Experimental investigation of motivational state. In *Motivational Control Systems Analysis*, ed. D. J. McFarland, pp. 251–282. Academic Press: London.

McFarland, D. J. & Nunez, A. (1976). Systems analysis and sexual behaviour. In *Biological Determinants of Sexual Behaviour*, ed. J. Hutchinson. Wiley: London, in press.

McFarland, D. J. & Sibly, R. (1972). 'Unitary drives' revisited. *Animal Behaviour*, **20**, 548–563.

McFarland, D. J. & Sibly, R. (1975). The behavioural final common path. *Philosophical Transactions of the Royal Society*, B, **270**, 265–293.

Milsum, J. H. & Roberge, F. A. (1973). Physiological regulation and control. In *Foundations of Mathematical Biology*, vol. **3**, ed. R. Rosen, pp. 1–95. Academic Press: London & New York.

Murray, C. D. (1926*a*). The physiological principle of minimal work. I. *Proceedings of the National Academy of Sciences USA*, **12**, 207–214; and II, **12**, 299–304.

Murray, C. D. (1926*b*). The physiological principle of minimum work applied to the angle of branching of arteries. *Journal of general Physiology*, **9**, 835–844.

Patterson, I. J. (1965). Timing and spacing of broods in the black-headed gull *Larus ridibundus* L. *Ibis*, **107**, 433–459.

Perrins, C. M. (1965). Populations and clutch size in the great tit *Parus major* L. *Journal of Animal Ecology*, **34**, 601–647.

Prest, A. R. & Turvey, R. (1965). Cost–benefit analysis: a survey. *Economic Journal*, **75**, 685–705.

Rashevsky, N. (1960). *Mathematical Biophysics*. University of Toronto Press: Toronto.

Rosen, R. (1967). *Optimality Principles in Biology*. Butterworth: London & New York.

Rosen, R. (1970). *Dynamical System Theory in Biology*. Wiley: New York.

Rozin, P. & Kalat, J. W. (1970). Specific hungers and poison avoidance as adaptive specialisations in learning. *Psychological Review*, **78**, 459–486.

Schultz, D. G. & Melsa, J. L. (1967). *State Functions and Linear Control Systems*. McGraw-Hill: New York & London.

Sibly, R. M. (1975). How incentive and deficit determine feeding tendency. *Animal Behaviour*, **23**, 437–446.

Sibley, R. M. & McCleery, R. (1976). The dominance boundary method of determining motivational state. *Animal Behaviour*, **24**, 108–124.

Sibly, R. M. & McFarland, D. J. (1974). A state–space approach to motivation. In *Motivational Control Systems Analysis*, ed. D. J. McFarland, pp. 213–250. Academic Press: London.

Sibly, R. M. & McFarland, D. J. (1976). On the fitness of behaviour sequences. *American Naturalist*, in press.

Spurway, H. & Haldane, J. B. S. (1953). The comparative ethology of vertebrate breathing. I. Breathing in newts, with a general survey. *Behaviour*, **4**, 8–34.

Tansley, K. (1965). *Vision in Vertebrates*. Science Paperbacks no. **11**. Associated Book Publishers: London.

Thompson, D'Arcy W. (1924). *On Growth and Form*. Cambridge University Press: London.

Tinbergen, N. (1951). *The Study of Instinct*. Clarendon Press: Oxford.

Tinbergen, N. (1959). Comparative studies of the behaviour of gulls (Laridae): a progress report. *Behaviour*, **15**, 1–70.

Tinbergen, N., Broekhuysen, G. J., Feekes, F., Houghton, J. C. W., Kruuk, H. & Szulc, E. (1962). Egg-shell removal by the black-headed gull *Larus ridibundus* L.: a behaviour component of camouflage. *Behaviour*, **19**, 74–188.

Whitfield, C. L. & Livezey, R. L. (1973). Thermoregulatory patterns in lizards. *Physiological Zoology*, **40**, 285–296.

3

Attentional processes and animal behaviour

R. J. ANDREW

This paper represents a first attempt to explore the ways in which attentional processes may affect and even control the behaviour of animals.

Since human studies of attention are far more advanced than animal studies, I begin by summarising the theoretical principles which have been developed as a result of such studies. I then attempt to extend these principles to explain a wider range of situations, including ones in which there is mismatch between the stimulus to which the animal is attending and the centrally held specifications which have been activated during recognition. A number of specific assumptions are required if this is to be done, and these are listed and described, so that a relatively precise model of attentional processes can be employed in the remainder of the paper.

The third section presents evidence from studies of chicks and mice which show that (i) the model may be usefully applied to data from animals, (ii) a particular parameter of attention (the persistence in use of the rules by which stimuli are selected for examination) is directly affected by testosterone, suggesting that a single central mechanism is involved in the application of such rules in a variety of situations. The situations considered are tests involving search, distractibility, presentation of a novel stimulus and extinction. Other possible explanations of the testosterone effect are discussed.

The last section discusses some general applications of attention theory to animal studies. It begins with a consideration of possible functions of changes in the persistence with which rules of selection are applied. The application of attention theory to search by animals and to processes which occur at low and high arousal, is discussed next. Finally, the factors controlling the occurrence of mismatch are taken as a special example of the processes underlying the classification and recognition of stimuli by animals.

THEORETICAL BACKGROUND

Selection and recognition of stimuli

Many of the assumptions made in the model of attention processes used in this article are taken from human attention theory. Others derive from different bodies of theory, such as that concerned with response to novelty. A brief summary of such theories will serve both to acknowledge some of the more important of my sources, and to introduce the main concepts which will be of importance here.

Broadbent (1971) has provided an authoritative and comprehensive account of studies of human attention. The central problem of such studies is the way in which certain stimuli are selected for examination and response whilst others are excluded. Tasks which are used to study such processes aim typically to present many stimuli rapidly or simultaneously so as to occupy attention as fully as possible. The classical task of this type is perhaps that of dichotic acoustic shadowing. Here the subject has to repeat a message which he hears in one ear whilst an irrelevant message (i.e. one which he has been instructed to ignore) is presented to the other. A crucial observation was that although most practised subjects can carry out such a task with few errors, and with no recollection of most of the content of the irrelevant message, certain words of great significance such as the subject's own name are likely to be noticed even if they are embedded in the irrelevant message. In order to cope with such findings, Treisman (1960) and Broadbent (1971) proposed that irrelevant signals are not fully excluded from analysis, but are attenuated so that they are less likely to set in motion the full processes leading to recognition and response. Attenuation involves a variety of mechanisms, ranging from receptor reflexes, such as direction of gaze, to central processes which distinguish between complex stimuli on elaborate criteria.

Broadbent (1970) distinguishes two types of selection: 'filtering' (or 'stimulus set') and 'response set'. In filtering a single easily perceived characteristic is used such as which ear receives the sound, or, in vision, the colour of the stimulus. In response set, a number of characteristics have to be used, no one of which is sufficient to allow selection by itself. Clearly these two strategies are extremes, and intermediate strategies can be imagined. However, the distinction is a real and useful one in animal studies as well as humans.

Rules of selection are often set up in human subjects by direct instruction, but it is equally possible to impose them by the conditions of the test: thus the use of response set can be forced on a subject by demanding search for a stimulus against a background of stimuli which can only be discriminated by a careful use of a number of parameters. I will try to show that similar processes are possible in animals. Finally, rules of selection do not work perfectly, which is another way of saying attenuation of other stimuli is not complete.

Selection of stimuli affects recognition processes in that it determines whether an input will occur from a stimulus or not, and for how long it continues. When an input does occur, the basic event in recognition is assumed to be the activation of one of a number of central mechanisms which each correspond to a particular type of previously encountered stimulus. These mechanisms have been variously termed 'dictionary units' (Treisman, 1960), 'category states' (Broadbent, 1971) and 'recognition units' (Kahneman, 1973); the last term will be used here as having most general application to animal as well as human studies. In the situation usually considered by attention theory, activation is assumed to result in, and be followed by, a response (e.g. verbal identification of a recognised word) which is used as a reliable index that recognition has occurred.

The simplest such case is that of tests (e.g. vigilance tests), in which the subject must detect the moment at which one particular (expected) stimulus occurs. In this instance, signal detection theory allows quantitative treatment (Green & Swets, 1966). The input to the recognition unit corresponding to the stimulus, which is the only unit whose activation need be considered, is supposed to derive from two sources: random inputs provided by the environment (or the experimenter), which can be treated as noise, and inputs resulting from the stimuli themselves. Both will vary, and so it is assumed that the subject sets a criterion value (β), inputs above which are treated as representing the onset of a stimulus; put in other words, inputs greater than β activate the recognition unit.

Changes in β would clearly allow the subject to adjust the proportions of the four possible outcomes of a decision: hits, misses, false alarms and correct rejections. In experiments with human subjects, the ability to change β can be tested by instructing the subject to use several different levels of confidence when identifying target stimuli. Thus in one run he might be told to respond only when absolutely certain, and in another whenever there is any hint of the stimulus. Results from such experiments show that β does indeed vary appropriately.

The second important variable in tests of this sort is signal detectibility. If it is assumed that inputs to the central nervous system from the target stimuli, and from the background stimuli from which the first are to be distinguished, both vary along a Gaussian distribution, then it is possible to estimate detectability (d^1) as the distance in standard deviations between the peaks of the two distributions. In most psychophysical studies d^1 is the measure of chief interest. It also is of considerable use in studies of attention, since stimuli which are not selected for examination will have low or zero values of d^1.

Methods for the estimation of β and d^1, and the assumptions upon which these methods are based, are presented with great clarity by Green & Swets (1966). The important point for the purposes of this paper is that in order to do this it is necessary to be able to measure not only hits and misses but also false alarms and correct rejections.

So far, we have only considered the case of inputs which vary along a single parameter. Broadbent (1971, p. 251 *et seq.*) has shown that the same methods of analysis can be applied when stimuli involve several parameters, at least when the parameters each contribute independently to recognition. The final expressions for the multiplicative interaction of each such contribution then reduce to the same form as those for a stimulus involving a single parameter. Strategies of recognition may well exist for which such a simplifying assumption is not possible. Thus if a hierarchically organised key were in use, in which the answer to each question determined the next question, some relevant parameters might not contribute to recognition, if the stimulus were rejected on evidence from another parameter examined early in recognition. As a result changes in β would tend always to be linked with changes in d^1. However, in practice the application of signal detection theory to the recognition of complex stimuli has been justified in that it has allowed problems to be attacked which are not otherwise amenable to quantitative examination.

An important finding as a result of such an approach (Broadbent, 1971, p. 251 *et seq.*) has been that different recognition units differ in their criterion of activation even in the same individual or during the same period of test. Such differences are both long term and short term. Thus to take examples from verbal material, recognition units for common words usually have low criteria, whilst those for rarer words may have low criteria at moments when their occurrence is made likely by the preceding portion of a sentence.

Now that the properties of both rules of selection and recognition units have been described, it is necessary to consider briefly how attention theory treats periods of sustained attention and shifts of attention. During the first, a particular set of rules of attention are in operation. Inputs from stimuli which are not selected under these rules may nevertheless activate recognition units. This can occur for two different kinds of reason. First, input from non-selected stimuli may be adequate for activation when a recognition unit has a low criterion of activation. Secondly, the rules of selection may effectively be briefly changed. Conspicuous stimuli (e.g. with sudden onset, or ones which contrast markedly with their background) will tend to be targetted and at least briefly fixated under most circumstances; this could be regarded as a basic and over-riding rule of selection. Moray (1969) postulated, in addition, that the absence of any stimulus acceptable to the current rules of selection allows other stimuli to be examined. Such examination is assumed to be brief, since it is terminated as soon as any acceptable stimulus is perceived.

Shifts of attention are equivalent to a change in the rules of selection. Such a change is particularly likely at a point at which a recognition unit has been activated by a stimulus which was not selected by the current rules. In order to explain this, Broadbent (1971, p. 164 *et seq.*) postulates that stimulus properties specified by an activated recognition unit may become incorporated under some circumstances into subsequent rules of selection.

An example may help to clarify the sort of sequence of events which have been observed. Shifts of attention from one ear to the other may occur in the dichotic shadowing task, when the sense of the message on the shadowed ear is continued on the other ear. Thus if 'England expects' is heard on the first ear, and 'that every man' follows on the second, the subject may shift to the second ear and continue for a little to tend to repeat the subsequent message presented to that ear.

Circumstances under which more permanent shifts of attention occur have been little studied in work on human attention. However, it seems likely that shifts are short and infrequent in tests such as the dichotic shadowing task precisely because of initial instructions to concentrate on one 'input channel' alone. When attention is not so controlled, shifts of the sort which have just been described are likely to occur much more freely.

It will be seen that we have now reached a point considerably removed from the signal detection theory which provides the basic theoretical underpinnings of attention theory. The reader is referred to Kahneman (1973) for discussion of more complex aspects of human attention, such as the control of the processes which have so far been considered by higher-level mechanisms (e.g. 'momentary intentions').

It is important to note that ideas derived from signal detection have been used in contexts where quantitative techniques of analysis such as the estimation of β may be impossible. It may be useful, for example, to postulate differences in ease of activation of recognition units, when under some circumstances no response (or no response that is easily measured) follows activation. When this is the case, I shall follow Treisman (1960) in speaking of 'threshold' of activation.

Mismatch

A number of influential theories have postulated a process of comparison of input due to a stimulus with central specifications which describe a former state of that stimulus (or of a stimulus which might have been expected to occur at that point in time). The degree of mismatch between stimulus and central specifications then determines the type of behaviour which appears. Hebb (1946) and later Salzen (1962) postulated that fear responses are evoked by large mismatch. McClelland, Atkinson, Clark & Lowell (1953) supposed that smaller mismatch evokes examination, and may be sought after.

The most concretely formulated of such hypotheses is that of Sokolov (1963) who postulated that repeated exposure to a stimulus, which is initially capable of entire fixation, allows elaboration of a 'neuronal model'. Once this has occurred, the stimulus no longer evokes fixation and other components of the orientation reflex because inputs from the stimulus are blocked or attenuated centrally. If, however, the stimulus is changed in some way so that it no longer corresponds to its neuronal model, then the orientation reflex once more appears.

The resemblances between the concepts of 'neuronal model' and 'recognition unit' are obvious. Indeed, Kahneman (1973) accepts their equivalence. He interprets the receptor reflexes of the orientation reaction which follow 'violation of expectation' (i.e. a mismatch between model and input) as an allocation of further attentional capacity for information processing (Kahneman, 1973, p. 44 *et seq.*).

It is important to be clear that such an assumption represents a considerable extension of attention theory as so far summarised. It amounts to the assumption that activation of a recognition unit is followed by a further phase of comparison of stimulus input and information.

It will be seen that, as the concepts of attention theory are applied to a wider range of situations, it is necessary to make a number of specific and interrelated assumptions. In the next section, the assumptions which have been made in the model of attentional processes used in this article, are listed with some comments.

ATTENTIONAL PROCESSES: ASSUMPTIONS OF MODEL

(1) A 'recognition unit' (Treisman, 1960) is one of a population of central mechanisms, each of which corresponds to a particular category of stimulus, in the sense that the activation of the unit must occur before responses normally evoked by that category of stimulus are evoked by its presentation. A recognition unit specifies the properties of the category of stimulus in two senses. First, activation of the unit is possible only by a stimulus which, in some parameters at least, resembles the properties of the category (see assumption 2). Secondly, the information which is used in a decision that a stimulus resembles an established category but differs from it in such a way as to represent a significant transformation of it, is associated with the recognition unit and available after its activation.

(2) Recognition units differ in their threshold of activation, and the threshold of a particular unit may change. When it can be assumed that activation is always followed by an identifiable response specific to the unit and when test conditions allow the measurement of hits, misses, false alarms and correct rejections, changes in threshold can be estimated as changes in a criterion value β (see previous section). It is assumed that changes also occur in situations where this is not possible. The factors which affect threshold are as follows:

(i) The likelihood of occurrence of such a category of stimulus. This may be a short term effect associated with a particular contextual cue which indicates that the category has become likely, or more long term, in that a frequently encountered category will have a recognition unit of lowered threshold.

(ii) The relative weightings attached to hits, misses, false alarms and correct rejection. These might be derived from the positive and negative reinforcements derived from each type of outcome.

Clearly if threshold is low, the initial decision that a particular category of stimulus is present (i.e. activation of recognition unit) does not involve all information potentially available concerning the category.

(3) A number of different events may follow activation:

(i) Response specific to the category of stimulus. This may be delayed despite activation for reasons such as adjustment of position relative to the stimulus, or because the nature of the response calls for further comparison.

(ii) No response, amounting to the decision that the stimulus is familiar and does not call for a response of higher priority than that in hand.

(iii) A phase of comparison during which the animal is affected by mismatch between the stimulus and the category of stimulus, the recognition unit for which has been activated.

(4) Mismatch specifically evokes a range of behaviour. The duration and degree of mismatch determines which type of behaviour is evoked (e.g. extended intense mismatch may evoke fleeing). The degree of mismatch depends on the number of dimensions in which stimulus and recognition unit differ, and on the magnitude of the difference along each dimension. It also depends on the accuracy of specification of each dimension (thus a wide or narrow range of variation may be permissible), and the stability of the specifications associated with the unit (see (ii) p. 127).

(5) If a novel stimulus fails to activate a recognition unit, then no mismatch is possible. This assumption is particularly difficult to test, since it is impossible to be sure that even a very novel stimulus may not activate a recognition unit with generalised properties (e.g. specifying absence of any object at a particular point in the environment). The assumption has one particular

extension of importance: it is assumed that it is possible for elaboration of a new recognition unit to occur during examination of a novel stimulus, without responses associated with mismatch being evoked.

(6) Stimuli are selected for examination using particular cues. Stimuli which are not selected provide inputs which are reduced or abolished, i.e. their detectability is reduced.

(7) The rules of selection may involve one or few simple parameters ('filtering'), or at the other extreme ('response set'), most or all of the parameters of a stimulus category with complex properties. When filtering is in use, some stimulus parameters will be irrelevant to selection, even though they are important in activation of the appropriate recognition unit. Human data (see Search situations, p. 120) indicate that certain cues permit selection by filtering to be as quick and efficient with many irrelevant alternative targets as with few.

(8) Rules of selection commonly differ at different times in the same individual, although certain rules probably always hold. The latter include those concerned with (i) the manner in which a conspicuous stimulus (i.e. one with a sudden onset, or which contrasts markedly with background stimulation) evokes targetting; and (ii) the manner in which, if there is mismatch between an activated recognition unit and a stimulus, examination of the stimulus is enhanced (e.g. fixation may occur) and prolonged.

(9) Rules of selection take their character from several different effects:

(i) The parameters which are used in selection are likely to be amongst those specified by a recognition unit which has recently been activated, or possessed by the stimulus responsible for that activation.

(ii) The parameters used will be those which allow efficient selection of appropriate stimuli under the conditions in operation. Thus when only one stimulus, which has been detected, is of importance, specification of position may suffice; whereas, in search over a complex background, response set may be necessary.

(iii) Other effects are likely. Thus a rule of selection might perhaps come into operation as a learned strategy without detecting any stimulus which might activate an appropriate recognition unit.

(10) Rules of selection will be maintained when the stimuli selected evoke responses which have the highest current priority. This condition will tend to be stable when there are no stimuli present which, if recognised despite exclusion, would change the priority of response (e.g. by evoking escape in place of feeding). Attentional processes can be (and are) neglected in motivational analysis when this is true.

(11) Rules of selection tend to continue in use for some time and/or over a number of examinations, independently of the effect postulated in (10)

above. This will be clearest when changes in motivation can be largely neglected (e.g. in choice amongst different types of food subject which are nutritionally equivalent).

(12) Stimuli which are excluded by the current rules of selection may nevertheless provide enough input to activate recognition units with low thresholds. Examination (or even brief response appropriate to the stimulus) may follow, and must be regarded as involving a change in rules of selection. This is often temporary, being followed by the immediate reinstatement of the original rules, but such a change may also be more permanent (assumption 14).

(13) Stimuli which do not fully meet the current rules of selection become more likely to be selected the longer a period lasts, in which appropriate stimuli do not occur. The simplest example is momentary shifts of gaze away from the area of fixation. If a stimulus which does meet the rules occurs at such a time it is still likely to be preferentially selected.

(14) Changes in rules of selection tend to occur when a stimulus which does not meet the current rules, succeeds in activating a recognition unit. As postulated in assumption 9(i) parameters belonging to the stimulus or specified by the recognition unit may then be incorporated into the rules. This will be more likely if a high priority response is called for following activation; a change in response is not necessarily involved (e.g. feeding might now be evoked by a second type of food object). The simplest example is a change of specified position following the location of a target in a new search area. Another important case is the activation of a recognition unit which results in mismatch, evoking a range of high priority behaviour.

(15) Persistence in use of rules of selection is assumed to be capable of adjustment independently of any factors already discussed. Such increased persistence is assumed to affect the reinstatement of current rules after brief interruption of the character described in assumption 12. The example with which I shall chiefly be concerned here is increase in such persistence in certain male birds and mammals (domesic chicks, mice and perhaps man) when circulating levels of testosterone are high.

In a previous discussion of increased persistence caused by testosterone (Andrew, 1972*a*), it was postulated that this is caused by a more persistent 'activation' once in use of 'central specifications' of stimuli. Such 'activation' was used to explain the selection of stimuli. Such a formulation has considerable disadvantages: it leaves vague the question of which stimulus parameters included in the 'specifications' are used in selection. More important, it differs in a confusing way from the use of the term 'activation' in human attention theory. The earlier terminology has therefore been abandoned.

R. J. ANDREW

APPLICATION OF ATTENTION THEORY TO DATA FROM CHICK AND MOUSE

Introduction

The assumptions which were listed in the previous section were intended to include, as far as is possible, all of the attentional processes which have to be postulated when developing a model which can be applied to a wide range of situations. Theories which are intended to deal with a narrower range of situations need (and rightly make) fewer assumptions. One example which has been dealt with elsewhere (Andrew, 1972*a*) is the question of the circumstances under which it may be useful to postulate mechanisms with properties such as those of recognition units. Clearly, other things being equal, the simpler the model which can adequately describe the phenomena under study the better. On the other hand, processes which are ignored as a useful initial approximation may never be studied.

I shall attempt two approaches here as a first step towards the justification of a model with general applicability, and an investigation of some of its predictions. First, I shall try to show for particular situations that the application to animal studies of a model of attentional processes which is derived from human attention theory is useful; in the case of search tests, comparison of empirical data from human and animal studies is possible. Secondly, I shall use the changes in persistence produced by testosterone to show that an effect, which is best explained as an increased persistence in use of rules of selection, occurs in all the situations so far examined. I shall argue that this suggests that at least one of the processes postulated by the model is mediated by a particular central mechanism, which is real rather than a convenient feature of a theoretical model, and which is indeed important in a wide range of situations. In the remaining parts of this section both of the above approaches will be applied to data taken in turn from tests involving search, distractibility, novelty and extinction.

It is clearly important to consider alternative explanations of these effects of testosterone. This can be done most conveniently after each situation has been described.

Finally, for the sake of brevity, the following abbreviations will be used throughout this section: C = intact control male chick in the first two weeks of life, in which endogenous levels of testosterone are very low, T = intact male chick receiving exogenous testosterone (details of dosage, Andrew, 1972*b*, 1975*a*, *b*).

Search tests

Many of the features of tests used in studies of human selective attention can be provided by search tests designed for administration to animals. If the animal attempts to identify and respond to targets at high speed (which is certainly true of chicks after even brief deprivation of food), then it is likely that more potential targets (suitable and unsuitable) are usually available than can be examined before the next response is initiated. As a result, just as in auditory shadowing with two or more simultaneous messages or in tachistoscopic presentation of a complex visual display, the attentional processes which control the selection of stimuli are likely to dominate the pattern of responses which is shown.

It is possible in animal studies to control which out of two or more alternative types of targets are sought. Thus brief exposure to one out of two or more acceptable food objects ('priming') may cause search for the object given at priming to continue over an area on which other types of food objects are also present (e.g. domestic chick, Dawkins, 1971; Andrew & Rogers, 1972). At its most complex, such search may commence with matching to sample (Croze, 1970), thus providing a parallel to the role of instructions from the experimenter in human tests. In the chick experiments, data from which are about to be discussed, rather longer-term preferences were also used: thus a chick tends to choose the type of food grain on which it has been feeding for the last few days when two types of food grain are presented simultaneously.

Two test floors were used in the chick studies (Rogers, 1971, 1974; Andrew & Rogers, 1972, Andrew, 1972*a*).* Two types of food grain, differing only in colour (i.e. values of hue, saturation and contrast) were scattered in equal numbers over the floors. The chicks had had extensive experience of both, but it was arranged that one (P food) should be preferred to the other (NP food), either because of longer exposure in the home cage or because of a priming experience immediately before test. In the first test (plain floor) no other grains were present, whilst in the second (pebble floor) pebbles of a size and general colour similar to P food were present in numbers approximately equal to P food. Chicks rapidly learnt to avoid pecking pebbles, in part because they were glued down and could not be swallowed.

The data which have been obtained from such tests support two of the most basic features of the present model: the distinction between filtering and response set in selection of stimuli, and the assumption that selection and threshold changes in recognition units are independent processes.

* Unless otherwise indicated, findings mentioned in this section are described in these references.

Broadbent (1970) notes the empirical finding from human studies that intrusions of single incorrect choices occur when filtering is in use, whereas they are absent when response set is in use. These intrusive errors involve recognition units of low threshold. The most obvious explanation is that for such recognition units selection plays the predominant role in determining whether activation will occur, and failures of selection are more likely when only one or two simple parameters are in use than when a variety of requirements must all be met.

The conditions of search on the plain floor make filtering using colour a possible strategy, whilst response set is necessary on the pebble floor if P food is to be taken without pecking pebbles in error. Intrusive pecks on NP food indeed occur on the plain floor and not on the pebble floor. They may be defined as follows: a single (rarely double) peck at NP food which occurs as the last peck before shift to a new search area, where P food is at once again chosen. (Search areas were studied by providing food grains in clumps separated by bare floor.) Such an intrusive peck thus reflects a partial failure of filtering, since a stimulus of correct position but incorrect colour is taken.

This interpretation is strengthened by the fact that failures of filtering, as shown by intrusive pecks, are made more likely if past experience has been with grains which are somewhat variable in colour. This is usual if dye is applied as a solution to the grains, since small pieces may later flake from a dyed grain, leaving patches of the original colour showing. Grains of an entirely uniform colour can be produced by dying the food as powder and forming it into grains after dying. If now two groups of chicks are compared in a test in which both P and NP grains are of uniform appearance, intrusive pecks occur only in the birds with past experience of grains of slightly variable colour (L. J. Rogers and R. J. Andrew, unpublished).

Evidence for changes in the threshold of a recognition unit, independent of what stimuli are being selected, comes from tests in which birds with a clear preference for one type of food (i.e. which will consistently select P food when P and NP food are both present) are allowed to feed briefly on NP food alone before test on a plain floor. Such preliminary priming greatly increases the number of intrusive NP pecks even in birds which otherwise consistently select P food. It has been assumed here that immediately prior experience with a stimulus is likely to lower the threshold of the corresponding recognition unit; the effect of priming which has been just described thus agrees with prediction.

It remains to be established for search situation that testosterone increases the persistence in use of rules of selection. In pebble floor tests the sign of a change of rule is simple, namely the end of a run of choice of one type of food and the beginning of choice of the other. In plain floor tests the sign is a change from a run in which one colour is chosen on entry into a new search area, and the second is taken, if at all, as an intrusive choice just before shifting to a new area, to a run in which the second colour is now chosen as the first was before. It is in fact the case that changes are markedly and significantly less frequent in Ts than in Cs in both types of test. This is true whether the colour treated as P food at the beginning of the test is determined by feeding experience immediately before or considerably earlier (e.g. up to two days earlier, the birds being force-fed in the meantime).

The use of position as a rule of selection requires special comment. In general, position too is more persistently specified by Ts. This is true of plain floor tests and of feeding on a floor with only one type of food present (Rogers, 1971): in both cases, Ts emit more pecks within a clump before moving. In tests in which groups of identical light caps, some of which cover a food grain and some do not, form the areas of search, Ts turn more caps before moving than do Cs (Messent, 1973).

The interpretation of the pebble floor test is more complicated. Here Cs emit more pecks in a particular area before moving. The explanation is probably that, once a shift to choose NP food has occurred (which occurs much more frequently in Cs), location of targets becomes very much easier, since NP food differs in colour from both P food and pebbles. It is then more likely that all of the targets of the type sought will be readily found in any particular area of search.

Distractibility tests

The basic design of such tests is as follows: the animal's attention is concentrated, as a result of prior training, on stimuli associated with an intermediate goal (e.g. a food dish or a key whose operation delivers a food reward). It may be assumed that, at such a time, rules of selection are in use which specify properties of these stimuli, and tend to exclude all others. This assumption has been tested by making available a food dish in an unaccustomed location and showing that the animal at once responds to it. At test a change is induced either in a stimulus which would be expected to be excluded or in one which is associated with the secondary goal, and so should be selected. If input from the changed stimulus succeeds in activating the recognition unit corresponding to its original state, the resulting mismatch should cause sustained attention. This is measurable as an increased latency of performance of the final response.

The present hypothesis would predict that Ts would be less likely than Cs to be affected by changes in excluded stimuli, but that this should not be the case following changes in a selected stimulus. The exact prediction in this second instance is slightly different in the two different versions of the test which have been used (below).

In the first (Archer, 1974), chicks were trained to run down a runway to a food dish. When this task was well learned, the appearance of the runway wall was changed. Cs were markedly and significantly more delayed by such a change. They were more likely to stop and examine the change for some time; often they showed no signs of continuing on to the food dish. Ts, on the other hand, were often delayed only briefly, if at all. A change in the food dish would be predicted to produce the reverse difference, on the following argument: in both Ts and Cs, the changed stimulus is not excluded but selected. The resulting mismatch will delay feeding, until rules of selection are changed to select stimuli presented by food grains and exclude those associated with the food dish. Such a change would be expected to take longer in Ts. Prediction is in fact once more confirmed since Ts take longer to feed from a food dish of changed appearance. They are also disturbed for longer by other similar changes, such as a shift in position of the food dish.

A very similar result has been obtained with this test in male mice (J. Archer, unpublished.) The effect of testosterone was to decrease the delay produced by change in the runway wall and to increase that produced by change in the food dish. Both castrates which received exogenous testosterone and normal intact males differed in this way from castrates, the difference being significant in both cases, but larger in the first group, perhaps because endogenous levels were not maximal in all the intact males.

The second type of test (Messent, 1973) consisted of a visual discrimination task, in which the positive and negative stimuli were presented simultaneously on the two sides of a split key, between which they alternated regularly. Pecks to the positive stimulus brought food reward, whilst pecks to the negative stimulus produced a brief time-out. The effect measured was the length of time for which pecks ceased following a change in the test chamber. Once again, Ts were less affected than Cs by changes in stimuli which were likely to be excluded by rules of selection. It is worth noting that this was true of changes in the negative stimulus, which was certainly always in full view, as well as of more remote changes (e.g. in the colour of the house lights).

Ts and Cs, on the other hand, were equally affected by change in the positive stimulus. This is consistent with the present model, although no strong prediction is possible in this instance. The positive stimulus would be expected to be selected in both Ts and Cs. However, in this case, unlike that of the first test, a change in the rules of selection is likely to have no very clear effect on the time taken to resume pecking at the positive stimulus. This can only occur when the bird is attending to the positive stimulus, and prepared to peck despite mismatch: the present hypothesis gives no reason to suppose that the processes leading to such a condition are directly affected by testosterone.

Presentation of novel stimulus

This test differed in a number of ways from the distractibility tests which have just been discussed. The novel stimulus consisted of either the introduction or removal of a small object within the isolation chamber in which the chick had been living for some days; as a result the change was treated by the chick as a marked one. Further, the change began with a movement which almost invariably attracted the attention of the chick and, finally, it occurred at a time when the chick (which had food and water continuously available) was not strongly motivated to perform any other response. As a result of all these factors the behaviour of all chicks was affected markedly and for some time: the resulting changes in behaviour were recorded in as much detail as possible (D. Clayton and R. J. Andrew, unpublished).

All chicks should thus begin the period after the presentation of the novel stimulus with rules of selection which specify the novel stimulus. Once again, the hypothesis that such rules are more persistent in Ts, once they have come into use, is confirmed. Ts take considerably and significantly longer to start once more to show the usual behaviour of an undisturbed chick. Further when they do begin feeding (for example), they commonly interrupt this almost at once and resume locomotion and examination of the object. Cs, on the other hand, usually complete a normal feeding bout, once they begin. The resemblance to normal feeding extends to such details as stretching at the end of the bout, as both Ts and Cs do when undisturbed, but only Cs do when feeding in the presence of a strange stimulus. The overall impression gained from observation, just as from these quantitative measures, is that Ts find it difficult to shift attention away from the strange stimulus, and do so only partially and briefly even when they begin to respond to other stimuli again.

Certain responses (e.g. head-shaking) which occur also in extinction situations are commoner in Ts than Cs. It is possible that this difference is associated with a more continuous and protracted exposure to mismatch, much as is argued in the next section; this is not, however, a firm prediction from the present model.

Finally, there is one further difference between Ts and Cs which is interesting enough to deserve mention, even though its interpretation is still speculative. During the period of fairly continuous observation of the strange stimulus, two phases of behaviour can be distinguished. These are as follows: (*a*) A phase ('calling and response') in which calling accompanies frequent locomotion: observation of the novel object appears to continue throughout. (*b*) A phase ('silent investigation') in which the chick remains still, but not immobile since investigatory head movements continue; such periods are often terminated by stretching, perhaps because the chick remains still for sufficiently long as to be slightly cramped. Silent investigation takes up significantly more time in Cs than in Ts; the associated stretching is commoner in Cs.

Evidence has been presented elsewhere (Andrew, 1974*b*) which shows that chick calls characteristically occur as one of the consequences of activation of a midbrain mechanism, which is responsible for periods of enhanced readiness to respond: a typical example is the initial long bout of excited feeding, accompanied by calling which occurs on first catching sight of food. It is probable that the phase of calling and response in the present test corresponds to periods in which mismatch is directly affecting behaviour by evoking responses. An interesting hypothesis for further study is, therefore, that silent investigation corresponds to periods in which the recognition unit for the previous state of the stimulus is no longer activated, and a new unit is being elaborated in the absence of mismatch.

Whatever its applicability to the present test, the distinction between two types of examination of, and response to, a changed stimulus is potentially useful, and is examined further in later sections.

Extinction tests

Two types of extinction tests have been used. In the first, the chick was trained to approach and feed from a dish, which was empty during extinction: the responses whose extinctions were measured were pecking and looking in the dish. The second involved the extinction of a key pecking response, previously rewarded by food on a continuous reinforcement schedule.

It will be useful to distinguish five possible consequences which may follow the failure of a response to produce an expected reinforcement. These are (i) the repetition (often more vigorously) of the same response to the same stimulus; (ii) inhibition of response, without shift of attention to another stimulus; (iii) sustained shift of attention to an alternative target of the same general type (e.g. another potential food source, if the animal was attempting to obtain food); (iv) shift of attention and response to a quite different type of stimulus, with return to the first type immediately afterwards; (v) as (iv), but involving a sustained shift.

The last category is likely to be rare or absent, when there is a single dominant system of responses (e.g. feeding following food deprivation), and the present model would predict that this should be true of both Ts and Cs. It can be neglected, therefore, in the present section.

Before predictions can be made about the remaining categories, it is necessary to note that there is no obvious effect of testosterone on the ability to withold a response to a particular stimulus, when that response is ineffective in obtaining reinforcement: this is true both of key operation (Messent, 1973) and time spent pecking and looking into an empty food dish (Andrew, 1972*a*). This finding should be carefully distinguished from another finding, namely that Ts will respond to more stimuli in a particular search area before moving to another area, even if the response to each stimulus is usually unsuccessful (p. 107). In the latter case, Ts and Cs differ in how likely rules of selection are to be changed in choosing a new stimulus.

Given that inhibition of response to the stimulus associated with reinforcement develops more or less as quickly in Ts and Cs, two predictions can be made. First, inhibition of response without shift of attention (ii) is more likely to develop in Ts than in Cs. Features of the behaviour of Ts which support this prediction include staying near the stimulus for longer without responding to it (Andrew, 1972*a*) and the appearance of periods of brief immobility (Andrew, 1972*a*; Archer, 1974). Secondly, shift of attention to potential alternative targets (iii), involving some change in rules of selection, is more likely in Cs than in Ts. Under the conditions of the test, with only one food source available, the only identifiable response which might indicate such shifts is looking and pecking at points on the floor of the test chamber away from the food source. Cs do show more of such behaviour than Ts (Andrew, 1972*a*).

If attention is sustained for longer and more continuously on the food source in Ts, it would be expected that any responses caused by non-reinforcement and frustration would be commoner in Ts. Falk (1972) reviews evidence from mammals which shows that drinking, which is unrelated to (and often in considerable excess of) any physiological deficit, is typically performed when food reinforcement is withheld. Such 'adjunctive' drinking is commoner in Ts under such circumstances (R. J. Andrew, unpublished), as are certain responses such as head- and body-shaking which might be regarded as displacement grooming. The occurrence of responses like drinking requires a shift of attention to a quite different type of stimulus, but there are no data from the tests so far performed which allow a proper test of the prediction that in Ts such shifts should commonly be followed by a return to the original stimulus (iv). Drinking was in fact usually very brief but this might be expected on a variety of theories.

Other explanations

Explanations other than the hypothesis that testosterone increases the persistence in use of rules of selection have been discussed elsewhere (Andrew, 1972*a*, 1976). However, it is only fair to the reader to summarise here the evidence indicating that these other explanations are inadequate, particularly as two explanations which I had not previously discussed (see (6) below) were suggested during discussion at the conference out of which this book developed.

(1) *Arousal level.* The proposition that testosterone changes the distribution or likelihood of periods of high or low arousal must be framed more specifically before it is testable, since 'arousal' has been given a variety of incompatible meanings (Andrew, 1974). Three of these are effectively considered under other headings: different responses occur at different levels of arousal, so that fear responses, for example, become possible only at high arousal; all responses become likelier at high arousal, so that inhibition of responding is less likely; attention changes with arousal. One possibility which is not otherwise considered is a change in the likelihood, duration or completeness of periods of drowsiness and inattention. As a result, the overall detectibility of stimuli should change.

There is no evidence from direct observation for such a change in Ts. Ts and Cs do not differ in the distribution or duration of components of free-running behaviour including sleep over long periods of time in isolation cages (P. G. Clifton, personal communication). In the tests which have been discussed above, Ts are less likely to be affected by changes which are not associated with the goal. On the other hand, they are more affected by changes in the goal, and by a markedly novel stimulus. In the pebble test they are better able to locate scattered stimuli in a series of difficult discriminations.

(2) *Effectiveness of food deprivation and of reinforcement associated with feeding.* The changes in the food search tests which are produced by testosterone are present (Andrew & Rogers, 1972) over, and unaffected by, a range of deprivation levels (0–5 hours), which is wide for the young chicks, and which includes the levels used in tests. Increased effectiveness of food deprivation might make Ts more likely to ignore a change in the runway wall, but it should make them quicker rather than slower to return to feeding after the introduction of a novel object.

(3) *Ease of evocation of fear responses.* No consistent (and certainly no significant) differences between Ts and Cs in the frequency of escape responses were found in experiments using three very different novel environments and open fields (Archer, 1973*a*). There were differences in the time to onset of locomotion, but these were in opposite direction under different conditions (Archer, 1973*b*, and unpublished data) and are best explained as indirect consequences of attentional changes (Andrew, 1976). In the novel object test (p. 109), there were no differences in the frequency of overt escape

or freezing. The differences between Ts and Cs in the distractibility tests cannot be explained by a change in the threshold of low intensity escape or freezing responses since that should affect the effectiveness of change wherever it is situated. This is confirmed in the case of mice where two strains were examined (J. Archer, unpublished). Both showed the testosterone effect as already described, but in addition one (BALB/C) was more affected than the other (Porton) by any change, whether in runway or food dish.

(4) *Changes in rates of learning.* An effect of testosterone on the rapidity into which stimulus–response connections are formed would presumably mean greater habit-strengths after the same number of training trials in Ts, when the trials follow administration of the hormone. Higher habit-strength in Ts could explain reduced distractibility to changes in the runway wall. Indeed Archer (unpublished) has found in the runway tests with mice (p. 108) that distractibility is lower after a larger number of training trials, whether testosterone is present or not. However, in the runway tests discussed, training preceded hormone administration, so that it is difficult to see how an increased rate of learning in Ts is a possible explanation for the results which were obtained.

The same is true in the case of search tests, where the differences between Ts and Cs hold whether injection precedes or follows training (Rogers, 1971; Andrew & Rogers, 1972). It is in any case difficult to see how increased habit strength could explain either intrusive NP pecks in Ts on the plain floor or the fact that the differences in pattern of search between Ts and Cs still hold when it is possible to change which food is preferred by a brief priming experience immediately before testing. Finally, Rogers (1971) showed that inexperienced Ts and Cs do not differ in the rate at which they learn to discriminate pebbles and food grains. Messent (1973) presents similar evidence for a pattern discrimination.

The explanation is difficult to apply to the other tests. Thus, one might perhaps argue in the case of the novel object test that the stronger the stimulus–response connection the more likely it would be that the bird would continue to respond in the usual way to familiar cues (e.g. feed to the sight of the food hopper: that is, the reverse of what happens in Ts). However, no very firm prediction seems possible.

(5) *Inability to withhold response.* It has already been noted that this does not differ between Ts and Cs (see also Andrew, 1972*a*).

(6) *Other attentional processes specified by present model.* Increases in the readiness with which conspicuous or novel stimuli evoke targetting and

fixation would not explain the results of the distractibility tests nor of the search tests, although they might explain those of the novel stimulus test.

A more persistent lowering in Ts of threshold of recognition units following activation might explain the results of the novel stimulus tests. However, if persistence in search for P food on the pebble floor were explained by lowering of the threshold of the recognition unit for P food, it would be expected that pebbles (which closely resemble P food in this test) should activate the same recognition unit more readily in Ts than in Cs. In fact, pecks at pebbles are not more frequent in Ts.

SOME GENERAL APPLICATIONS OF ATTENTION THEORY TO ANIMAL STUDIES

Persistence of attention

So far the increased persistence of attention which results from testosterone has been discussed as a convenient means of studying attentional mechanisms. There is, however, evidence that the effect is a physiological one in domestic fowl, some strains of laboratory mice and perhaps man.

Relatively high dosages of testosterone are required to obtain the effect in domestic chick. However, this is also true of other effects of the hormone, which are known to be androgen dependent in the adult (Andrew 1975*a*, *b*): thus the dosages required for increased persistence and for the facilitation of aggression are very close, and both effects occur only in males (Andrew, 1972*b*). Rogers (1974) found in adult fowl that castrates or cocks receiving androgens behaved in search tests like Cs, whereas intact cocks or castrates receiving exogenous testosterone behaved like Ts. J. Archer (unpublished) found in two strains of laboratory mice that castrates receiving testosterone and intact males showed the typical effects of testosterone in distractibility tests, whereas castrates behaved like Cs.

Relevant human studies have employed very different tests. However, results have been obtained suggesting increased ability to sustain attention on a particular stimulus or type of stimulus and, conversely, decreased likelihood of shifting attention (review, Andrew, 1972*a*), in men with chronically high levels of circulating androgens (but within normal physiological range) or following the administration of additional testosterone (Klaiber *et al.*, 1971).

It thus seems reasonable to enquire for what reasons general changes in persistence of attention, particularly in association with the onset of breeding in males, might be adaptive.

The main potential function for increased persistence of attention (i.e. a decreased likelihood of changing rules of selection) which will be considered here is to increase the tendency to sustain response to a particular stimulus, despite difficulties and distractions, when alternative equivalent stimuli are available.

It is important to distinguish such a change from short- or long-term changes in the priority of a particular system of responses. When feeding is facilitated after deprivation of food, or copulation is facilitated during the breeding season, changes occur such as an increased ease of evocation by appropriate stimuli. However, such changes do not directly or very effectively alter the balance between, on the one hand, a strategy of persisting in response to one type of stimulus and on the other shifting readily between other types of stimuli which all evoke the same response.

It is also worth noting that a decreased likelihood of changing rules of selection is not the same as a decreased ability to inhibit responses, when they fail to evoke reward or result in punishment (see pp. 110, 111). The first type of change leads to adaptive strategies of behaviour. Thus it will tend to sustain repeated response to a stimulus when this does not result in negative reinforcement. Further, when response is inhibited because of punishment or lack of success, sustained attention to the stimulus will allow immediate resumption of response, if the obstacle or opponent is removed. If we make the further assumption that when changes in rule of selection occur in persistent animals they tend to be small and partial (and there is evidence suggesting this is the case of search tests, p. 105), then persistence will result in a number of slight variations of approach being tried to a problem such as blocked access, as each in turn proves unsuccessful.

The second type of change is likely to be maladaptive in that repetition of an unsuccessful response at best wastes metabolic energy, and at worst may provoke an opponent. Further, and more importantly, unless changes in persistence of attention are also assumed, decreased ability to inhibit response will not be associated with sustained attention, once inhibition finally does occur. Under some circumstances a specific change in the effectiveness of punishment might be adaptive, such as insensitivity to pain during intense defensive fighting; but I am not concerned with such circumstances here.

Increased persistence of attention is likely to be adaptive when there is a need (1) to exclude distracting stimuli, (2) to secure a particular resource, (3) to establish an attachment to a particular place or conspecific as quickly as possible, or (4) to adhere to established strategies rather than try new ones. These categories overlap a good deal but they provide a useful working classification:

(1) *Distracting stimuli.* Examples have already been discussed above of the way in which rules of selection which specify properties of stimuli serve to maintain approach and response to the goal, despite the presence of novel and distracting stimuli. It will be remembered that it was

necessary to assume (see assumptions 12 and 15, p. 103) that these rules of selection were not necessarily lost during inspection of a novel stimulus but could re-assert control, and that this was particularly true of persistent animals. With these points in mind, it can be seen that the extent to which irrelevant and distracting stimuli, response to which should be avoided, are excluded depends on the character of the rules (which is not directly affected by changes in persistence). On the other hand, the likelihood that rules will change, at any particular frequency of examination and recognition of distracting stimuli (which itself will depend on the character of the rules), is decreased in persistent animals.

This distinction is important when considering the two main types of situation in which distracting stimuli are very frequent and likely to affect behaviour. In the first, near total exclusion of distracting stimuli involves negligible disadvantage. This is likely to be the case only if it is very unlikely that stimuli will be excluded which, if they were to activate a recognition unit, would evoke a response of very high priority. Thus if predators are likely to be encountered, marked exclusion of sudden sounds or movements would be dangerous. If marked exclusion is a sensible strategy, then human studies of the effects of such background stimulation as loud noise (e.g. evidence reviewed in Broadbent, 1971, p. 428 *et seq.*) suggest that exposure to many distracting stimuli (e.g. during work in an environment full of unpredictable loud noises or movement) leads to the adoption of rules of selection which are extremely effective in exclusion. It is difficult to predict whether increased persistence is likely to be useful during exposure to such conditions. On the one hand, it might be that no degree of exclusion by the nature of the rules of selection can be sufficient to prevent some possibility of distraction; impairment of selection by filtering is certainly caused by noise in human studies (e.g. Broadbent, 1971, p. 430). A predator, that is chasing one out of a confusing crowd of prey that are dashing in all directions, may well be faced with this problem and need to be as persistent as possible in the application of its current rule of selection. On the other hand, if rules yielding adequate exclusion can be applied, changes in persistence may be irrelevant; indeed, in a noisy or visually distracting environment, it might be better at intervals to change rules of selection in order to scan the environment.

The second situation is more straightforward and may be exemplified (hypothetically) from the problems which are likely to face a young male at puberty in a social group containing older males. An increase in the likelihood that he will be attacked means that each change in behaviour of an older male requires evaluation in a way not necessary when enjoying the privileges of infancy.

A general increase in persistence of attention would make it easier for subadult males to sustain behaviour directed towards an important goal, despite repeated potentially distracting examination of older males. The increased ability of male chicks receiving testosterone to pass, on the way to a goal, a distracting stimulus which they nevertheless examine carefully, provides an exact parallel. It should be noted that in this instance persistence of attention shows itself as the reinstatement of the original rules of selection after an intervening brief examination (see assumption 15, p. 103). At the same time and most importantly, increased persistence of attention acting in this way would not reduce the ability of the young male to avoid or flee without delay when necessary. A general increase in persistence of attention around puberty would be an appropriate way of meeting this problem.

(2) *Secure particular resource*. Under certain circumstances, it is important to pick one out of a number of equivalent targets, and persistently attend to it, despite difficulties and obstacles. This will be true when a certain minimum amount of time and effort is required to obtain any target, so that if the animal shifts readily between targets, it is unlikely to obtain access to any one. This is probably true of predators like wolves and hunting dogs which commonly run down a particular prey.

An important and general case, which deserves consideration at greater length, is that of disputes between conspecifics for resources which are rare and must be acquired quickly. Such resources are likely to be disputed repeatedly.

Two central changes which might lead to an increased ability to win disputes are possible. One is a facilitation of attack, the other an increased readiness to persist in a dispute. Facilitation of attack (i.e. lowering the threshold for the evocation of attack) might sometimes lead to victory by causing an animal to attack unexpectedly and earlier in a potential dispute than usual. Maynard Smith & Price (1973) have argued, both from direct evidence and from computer simulation, that the strategy of pre-emptive strikes and rapid escalation of disputes is likely to be evolutionarily unstable. In part this is because individuals employing such a strategy will suffer greatly from injury once they become common.

This argument has been disputed (using similar evidence from computer simulation of a quantitative model) by Gale & Evans (1975). However, they too agree that direct evidence strongly suggests that, in many vertebrates, disputes between males are stabilised at a level at which physical injury is unlikely. In such disputes, victory is likely to go to the animal who continues for longest.

It is likely but not certain that increased persistence of attention during a dispute would favour such a victory. The argument in favour of such a proposition is briefly as follows: a prolonged dispute may be treated as a special case of the situaton in which an animal is confining its attention to a particular stimulus (here the opponent). A special difficulty is introduced by the fact that it is difficult to decide whether the steps in a balanced dispute should be regarded as cumulatively punishing or not. However, the simplifying assumption will be made that cessation of threat or attack for this reason will be as likely in a persistent as a non-persistent animal (since increased persistence caused by testosterone does not affect the inhibition of an unsuccessful response, pp. 110–111).

There is evidence for attentional changes which would tend to lead to the termination of even a continuously successful series of responses if these involved continuous attention to the same stimulus (Broadbent, 1971, p. 82 *et seq.*). Unless 'effort' is used to sustain attention (Kahneman, 1973), shift to new stimuli becomes more likely under such circumstances. Such effort (which is measurable by increased muscular tension and even overt movements, Broadbent, 1971, p. 48 *et seq.*) is necessary not only in prolonged observation of a single stimulus but also over relatively short periods in such tests as the Stroop Test (Jerison & Rohwer, 1966) in which there is continuous difficulty in sustaining rules of selection (the names of colours are printed in different colours, and it appears to be difficult to select names whilst ignoring colours).

If it is assumed that a progressive increase in the likelihood of a change in the rules of selection occurs during continuous attention, which can be opposed only partially and with difficulty, then an opposite effect (e.g. caused by testosterone) should postpone the point at which attention is terminated. It is encouraging that Broverman *et al.* (1964) have shown that, in man at least, males who are persistent because of high levels of androgen, are indeed better able to sustain attention over long periods to the same small number of stimuli presented repeatedly at high speed.

Finally, it is important to remember that victory is only a means to an end. Persistence of attention should also be effective in gaining access to the disputed resource between disputes. Thus if it is assumed that both disputants break-off dispute without a clear decision, then if one of them is able to return at once to response to the disputed target, it will have an advantage over the other.

(3) *Establishment of attachment.* An animal often has an attachment to one particular member of a class of important resource. The bond to a mate, and the attachment to resources within a territory are obvious examples. Commonly such an attachment gives an advantage in disputes with conspecifics for the resource: when this is so, it is useful to speak of an 'owner' of the resource. A male passerine bird such as a robin (*Erithacus rubecula*; Lack, 1939) in this sense 'owns' song posts and potential nest sites within his territory.

An owner might be expected to have an advantage over an otherwise equally matched opponent, on the following general argument. Other things being equal, it is an advantage to both disputants to curtail a dispute in order to avoid waste of time, metabolic energy and possible injury (Parker, 1974). The possession of a particular resource may be of more importance to the owner since he is likely to own other resources nearby which he can best use if he retains the disputed resource, whereas his opponent would be as well off with any equivalent alternative. In other cases, retention by the owner may be more important to him because he can continue to profit by the resource until the dispute is resolved (e.g. he may be able to snatch bites at food). In either case, it would be sensible for the non-owner to be the one to break off a protracted dispute.

It is thus likely often to be advantageous to establish as rapidly as possible attachments to particular resources which are capable of being owned (which will include the proviso that disputes must not be potentially so frequent as to make defence unprofitable). If it is assumed that the longer such a resource is attended and responded to at first encounter, the quicker a strong attachment is established, then increased persistence of attention will produce this desirable effect.

(4) *Adherence to established strategies.* In any situation in which the current rules of selection are successful in obtaining sought-after targets, there will be alternative advantage to be obtained in changing the rules from time to time so as to sample other targets, whose exploitation may be even more profitable. The balance of advantage will shift with increase of experience, since the more that is known of the likely range of targets (e.g. the likely food sources), the greater the probability that the animal has already chosen to search for the most profitable type of target. The increase in efficiency obtained by persistently applying rules of selection which are successful is likely to become progressively of greater relative importance with increasing experience.

In some of the instances which are discussed above, it might be enough to have an increase in persistence of attention which occurs only during the performance of a particular system of responses. Thus, for example, persistence of attention might be specifically increased during threat and attack directed at a conspecific. A general change in persistence might be expected, on the other hand, when persistence in response to a wide variety of stimuli or a general resistance to distraction is needed. The factors which may have put an increase in persistence of attention under control of testosterone in males of some species may be taken as a convenient example.

In species in which, as is usually the case, it is males who are chiefly involved in competition to own resources and to win disputes and fights, all of the first three categories of factors considered above are likely at the onset of breeding to favour increased persistence: thus males with elevated levels of androgens are likely to have to continue initiated behaviour, despite distraction caused by proximity of potentially hostile males, to win disputes and to attach rapidly to particular resources. The final category of factors (increases in persistence with increased experience) will tend to act in the same direction in that breeding animals are older than juveniles. In one particular instance this factor too may make increased persistence appropriate at puberty: Lack (1966, pp. 270 and 310) has suggested that long-lived species may postpone breeding until they are fully experienced in the exploitation of the environment.

Search situations

In general, when a great many potential targets have to be scanned, a predator should aim at the use of filtering, since this has the great advantage (Broadbent, 1971, p. 181 *et seq.*) that the speed of selection is little affected, and in some cases not at all, by increases in the total number of irrelevant targets. In man, cues of different types differ in effectiveness for this purpose. Thus a colour cue allows filtering which appears to be unaffected by the number of irrelevant cues (see e.g. Green & Anderson, 1956); the pitch of a human voice appears to be a cue which is very effective in picking out one voice from another in a medley of speech (see e.g. Broadbent & Ladefoged, 1957).

The evolution of perceptual mechanisms allowing very effective filtering of this sort is a largely unexplored field. A first step would be an investigation of the effect of varying the number of irrelevant stimuli when various different cues were available for use in selection, so as to establish which could be used in filtering. One cue which is almost certainly available in many tetrapods for use in filtering is movement. In Anura, retinal and tectal mechanisms lead to targetting to a moving stimulus presented in any part of the visual field (see e.g. Ewert, 1970; Ewert & von Wietersheim, 1974). Effectively, this is equivalent to filtering using movement: indeed, really efficient filtering might be expected often to involve relatively peripheral perceptual mechanisms. It should be noted that even in Anura more complex processes are also involved in response to prey: thus naive toads will target on to a wasp and seize it, but after one sting they are able to avoid catching wasps (Brower, van Zandt Brower & Westcott, 1960).

The selection of potential prey on the cue of movement must be highly effective, since it apparently persists in some mammalian predators. Thus

Leyhausen (1973) notes that some viverrids will only attack moving prey. The advantages conferred by filtering may be responsible for this surprising inflexibility.

In general, selection on complex characteristics (response set) would be expected to be slow in the presence of many other targets. However, human studies show that extensive experience of a particular complex stimulus may allow the development of the ability to select it at high speed, even if it can only be distinguished from other targets by the use of a complex of characteristics. It is not certain what is the exact mechanism of selection. Kahneman (1973, p. 81) argues that it may involve activation of the appropriate recognition unit, followed by a recursive effect which then locks attention on to the target involved; one piece of evidence suggesting this, which is quoted by Kahneman, is that, when subjects are looking for several targets at once, they may be aware that they have located one before they can tell which it is (Neisser, 1967). The suggested strategy would thus seem to be to scan rapidly with some very simple rule and with one recognition unit at low threshold, and to follow its activation by further investigation, so as to reduce or prevent mistaken responses. The development of such an ability takes considerable time (Rabbitt, 1964, 1967).

Tinbergen (1960) postulated a rather similar process, when he introduced the concept of search image into discussion of search for prey. Royama (1970) has shown that the specific use made by Tinbergen of the concept was probably an inappropriate one. The initially slow development of exploitation by titmice of a new food species as it first appears during the summer, probably usually depends on changes in the length of time which individual titmice spend hunting in different ecological niches, rather than on the time taken to elaborate a search image for the new prey. However, it seems possible, in view of the human data, that the ability to select a cryptic prey efficiently and quickly (which is certainly shown by titmice, Royama, 1970) may be acquired only slowly during first encounter with the prey. It would be interesting to compare the foraging patterns of naive (first season) and experienced birds during the time of appearance of a particular cryptic prey. The prediction would be that crypticity would delay the appearance of full exploitation in first season birds, but that once exploitation did develop it might be as marked and efficient as for less well-concealed species; experienced birds might show no greater delay for a highly cryptic prey than for any other.

The most important reason for general changes in persistence in food search has already been briefly discussed in general terms (p. 119): the more experienced the animal the more persistent it should be in the use of rules of selection which are proving successful.

It has been assumed here that position of search, if specified, is a rule of selection, and so affected by changes in persistence. As a result it would be expected that, other things being equal, a persistent animal would spend longer in a particular search area, and move less far between points of search in a uniform environment. Interestingly, a result suggesting an effect of the latter type has been described recently in the field by Smith (1974), who found that male blackbirds (*Turdus merula*) move less far in each move than do females, when both are foraging for the same (artificial) prey over a uniform area of grassland. It will be noted that, perhaps by coincidence, the sex difference is in the direction which would be expected if androgens were to affect persistence in the same way in this species as in the domestic fowl.

It is more difficult to predict what might happen when it is possible for an animal to compute and use an optimum value for such variables as give-up time following last successful choice (Krebs, Ryan & Charnov, 1974), or distance to be moved between points of search. The most likely stable outcome is that such values should come to be used whatever the level of persistence of the animal. The question could readily be investigated in the chick and deserves study.

Finally, the application of signal detection techniques to search situations is likely to be valuable in a variety of ways. A major obstacle to such application is the need to measure (see pp. 97–98) the stimuli which are effectively presented to the animal but do not evoke response (misses, and correct rejections). One possible approach would be to present all types of target on a uniform background in discrete and limited areas, and to assume that the background itself did not present any stimuli, and that all of the stimuli in each area examined were presented for detection. If such simplifying assumptions could be empirically justified, measurements of β and d^1 would be possible. Changes in β under different conditions would provide evidence for changes in the weightings assigned by the animal to the different possible outcomes of stimulus presentation (hits, misses, false alarms and correct rejections). Examples of the sort of condition which might usefully be varied are (*a*) nutritional value of each (food) target, energy expended in actions of search, level of deprivation; (*b*) time since last success, ratio of targets of the type sought to other targets; (*c*) degree of difference between the different targets.

Measurements of detectability (d^1) have one potentially important application: they would allow comparative estimates of the ease with which an animal can distinguish a cryptic target from various objects into which it might be confused, which would be separate from possible accompanying changes of criterion of match (β).

Arousal and attention

General changes in attentiveness (to use a deliberately vague word) are very important in behaviour. The most usual way of discussing such changes has been to assume that attentional processes are in some way related to level of arousal. Thus it is often assumed that rate of entry of sensory information is (or should be!) higher at higher levels of arousal (e.g. Hinde, 1970, p. 216). It has proved to be very difficult to frame such a proposition in terms which allow it to be tested: Delius (1970) suggested that arousal level could be equated with the rate at which decisions were being taken by the central nervous system.

The problem is made worse by the wide variety of other meanings which have been attached to the term 'arousal' (review, Andrew, 1974*a*). Here, it will simply be assumed that general changes occur in behaviour over sleep–activity cycles, and that periods before (and to a lesser extent immediately after) sleep or drowsiness will have special properties, which can for brevity be termed those of low arousal. The first problem is thus to define the properties of such periods; it can then be asked whether such properties also occur at times when the temporal proximity of overt sleep or drowsiness is not available as an index.

Human attention studies have tested and confirmed the commonsense assumption that periods of inattention (which may be brief and unaccompanied by external signs) will be more frequent when arousal levels are low. Since all signals tend to be missed during such periods, their presence may be inferred from an overall decrease in detectability of signals.

In human studies (Broadbent, 1971, pp. 81 *et seq.* and 418 *et seq.*), the incentives, overt and covert, to follow the instruction of the experimenter are apparently usually sufficient to allow the subject to oppose such effects on detectability. Only after prolonged deprivation of sleep, for example, do clear decreases in d^1 occur. Situations in which low arousal states are not opposed by the efforts of the subject have been little studied. However, it is just such situations that require study when investigating the nature of normal low arousal states that occur as part of natural cycles of behaviour.

The first step is clearly to establish whether brief periods of marked inattention really occur, as opposed to an overall and sustained change. Direct observation of the distribution and duration of overt signs of slight drowsiness may be useful. However, measurements of d^1 are essential, despite the difficulties of sustaining response under such conditions. This is unlikely always to be an insuperable problem: chicks, for example, will often continue to peck slowly at targets, even when breaks for brief periods of drowsiness repeatedly intervene (Andrew & Rogers, 1972). When response does continue, it should be possible to apply variants of standard human tasks (e.g. one in which the subject has to judge whether a target is present or not in each presentation interval) and measure d^1.

An approach of this sort should make it possible to decide whether periods of inattention occur sufficiently often to affect behaviour over substantial parts of cycles of waking behaviour. No doubt such inattention would affect behaviour in a variety of ways, but relatively concrete predictions could be made and tested in a way which would be impossible using a vague intervening variable such as arousal.

Human attention studies also suggest new ways of looking at high arousal states. Broadbent (1971, p. 440 *et seq.*) has argued that the similar changes produced by increased incentives or loud background noise in tests involving stimulus selection and detection reflect heightened arousal. They can be explained as a lowering of criteria so that response becomes more likely; as a result, to take one example, intrusive errors become more likely during filtering. Such test situations, in which attention is restricted to a particular small population of stimuli, and the vast majority of motivational states are excluded, cannot be assumed to be representative of all behavioural states in which it would commonly be supposed that arousal is high. Indeed there is no reason to suppose that such states can be ordered on any single continuum, attentional or otherwise (Andrew, 1974*a*). However, changes in β are likely to suggest processes which may underlie general changes in behaviour in the fully awake animal.

Mismatch

The discussion has so far been largely concerned with the application of assumptions concerning rules of selection. I shall take the effects on behaviour of mismatch between recognition unit and input as one example of the way in which a theoretical framework can be developed based on the postulated population of recognition units. It should perhaps be noted for the sake of clarity that mismatch caused by non-reinforcement will not be discussed here: some of the principles which will be developed for perceptual mismatch may well apply to non-reinforcement, but the importance of other factors make the discussion of this too complex to be broached here.

A variety of effects of mismatch have been postulated by different theorists. It has been assumed to be an important cause of fear responses (Hebb, 1946; Salzen, 1962) of a variety of components of the orientation reflex (Sokolov, 1963) of examination (Berlyne, 1967) and of reward or punishment (McClelland *et al.* 1953). A more extensive review of the range of effects which may be caused by mismatch is given in Andrew (1972*a*). Evidence is also considered there, which tends to further widen the applicability of theories concerned with mismatch, in that it suggests that a brief period of comparison, with some consequences equivalent to those of mismatch, may sometimes follow the activation of a recognition unit which proves eventually to match input. This may be the case when activation of the unit tends to initiate a response, whose nature calls for further examination of the stimulus, and effectively sets further requirements of match before performance can occur. This special case will not be pursued further here (but see p. 101).

Any precise formulation of the factors which might affect the magnitude and duration of mismatch would thus find application in a variety of theories. Such factors will here be considered under three heads: (i) those that determine whether a recognition unit, comparison with which will yield mismatch, will be activated; (ii) those that determine the degree and duration of mismatch; (iii) the special role of mismatch during the elaboration of a new recognition unit.

(i) Kagan (1970) formulated an important principle governing the activation of recognition units, in a discussion of the inverted-U relationship between the intensity of response shown by a baby, and the discrepancy between a presented stimulus and stimuli previously encountered in the same situation. He suggested that very large discrepancies meant that stimuli were no longer treated as transformations of the previously presented stimulus. In terms of the present model, such a stimulus would activate either a quite different recognition unit, which matches it relatively closely, or no recognition unit at all. The achievement of substantial mismatch thus cannot be guaranteed by the presentation of a marked transformation of a previously encountered stimulus. It is likely to result only if there is a recognition unit with a sufficiently low threshold to be activated by input from the stimulus, despite the fact that input departs markedly from the properties specified by the recognition unit (assumption 2, p. 101). One obvious but important example is provided by the fact that in the case of a novel object the cues which define its position in space also serve to specify what should or used to be in its place, and thereby simultaneously lower the threshold of a recognition unit and provide an input with at least one parameter (position) appropriate to activate it.

The principles involved may be illustrated by the following hypothetical explanation of how jokes work. Kagan (1970) has shown that smiles and laughter in the human infant can be evoked by mismatch. The evocation of laughter by jokes can be explained if adult laughter can be similarly caused. A good joke would then be assumed to depend on the establishment by a series of powerful cues of an expectancy (i.e. the lowering of the threshold of a particular recognition unit) which is able to cause the activation of what proves to be an inappropriate recognition unit. As a result substantial mismatch results.

(ii) The effects of mismatch will depend upon its degree and its duration. These no doubt interact in ways yet to be defined. They will here be considered separately.

(ii*a*) It seems likely that the degree of mismatch will be greater when many rather than few parameters are changed: the question is open to direct test so long as other factors such as likelihood of activation of the appropriate recognition unit can be held approximately constant.

The accuracy with which parameters are specified will also be important and should be largely controlled by past experience. It seems likely that recognition units are available under some circumstances which specify stimulus parameters only within broad limits and so can accept a broad range of objects as adequate matches. Thus variable past experience of objects such as stones or pebbles on a beach might well result in the elaboration of a generalised recognition unit, whereas experience of a series of billiard balls would be expected to produce a recognition unit with precise specification of parameters such as size and shape. Thus a rounded white object may either activate a 'pebble' recognition unit, with likelihood of no mismatch over a wide range of characteristics, or a 'billiard ball' recognition unit, with near certainty of mismatch (unless the object *is* a billiard ball)! Which happens may again depend on contextual cues (e.g. whether the object is lying on the ground or on a billiard table).

An important recent study which has directly investigated problems of this sort is that of Bateson & Chantrey (1972). They found that when chicks were exposed in close proximity in time to two imprinting objects which differed in colour the chicks had difficulty in learning to respond separately to one of the two objects as a positive discriminating stimulus. No such difficulty resulted when the objects were presented for exposures of the same duration but more widely separated in time. Bateson & Chantry argued that under the first conditions the objects had come to be classified together. This is clearly close to the idea that a single recognition unit is elaborated which is activated by both.

Psychologists have taken a different approach to the problems raised by inter-individual differences in the ease with which different parameters of discriminating stimuli are used during learning. Mackintosh (1965) and Sutherland (1966) have argued that learning proceeds in two stages. The first consists of learning which parameters to use, whilst, in the second, values of these parameters appropriate to the current discrimination are conditioned to positive and negative responses. The first phase has much in common with the elaboration of rules of selection, with the interesting extension that these are applied as a strategy in appropriate situations prior to the elaboration of what might be regarded as recognition units. No assumption has been made here as to the degree to which the use of particular rules of selection may be learned independently of the elaboration of recognition units.

There are thus two possible approaches to data such as those provided by Bateson & Chantry. One is to assume the elaboration of a single recognition unit for both stimuli. This would result in these two types of stimuli being classified together in the sense that a response conditioned to one would be evoked by the other. It would be interesting to know whether objects of the same shape but of any colour would be treated as equivalent or whether only the two original colours would be acceptable. The use of stimuli of other shapes and dimensions, which differ in colour, should not be affected on this first hypothesis.

The second approach is to assume that rules of selection excluding colour have been learned. This hypothesis would suggest that the relative ineffectiveness of colour as a discriminating cue would extend to stimuli differing in a variety of ways from the original pair of stimuli used in training.

(ii*b*) The duration of mismatch is likely to depend on at least two factors. The first is the availability of other recognition units which may be activated by the input. One instance in which mismatch is likely to be brief is that holding when a mismatching recognition unit has been activated as a result of its low threshold, but another and matching unit of higher threshold exists.

A second factor, which is likely to be both important and difficult to study, is that of the stability of the parameter specifications associated with the recognition unit. It is well established from psychological studies that repeated exposure to stimuli which are grouped towards one end of the scale of values of a particular parameter (e.g. heavy as opposed to light weights) tends to shift subsequent judgements of that parameter (e.g. review, Helson, 1964).

McClelland *et al.* (1953) discussed the problems raised by such possible shifts in central representations of stimuli for the application of a theory which related degree of mismatch to judgements of pleasantness. This proposed that small deviations of input from central representations of the current state of affairs (which included simple stimuli such as temperature or sweetness, as well as more complicated representations of stimuli) were pleasant, whilst large deviations were unpleasant. Haber (1958) provided data showing this to be true for judgements based on changes in skin temperature. McClelland *et al.* (1953) noted that stability of the central representation is essential if deviations are to have affective consequences, and assume that the longer the prior exposure to an unchanging stimulus the greater stability is likely to be.

An experimental study of the effects of duration, spacing (e.g. massed or distributed) and recency of past exposure on stability is badly needed. It seems likely that the conditions under which the changed stimulus is presented will also be important. Thus in situations comparable to human psychophysical tests, in which a particular value of a stimulus parameter has to be chosen, a shift in the absolute value chosen, caused by exposure to stimuli with a marked bias towards one end of the scale of parameter values, is likely to leave the parameter value changed but still precise. On the other hand, in a situation in which alarmed investigation is being evoked by novelty, the final result, if the recognition unit concerned is unstable, might be a considerable widening of the range over which parameter variation is specified to occur.

One immediately useful outcome of such a study would be the ability to cause the elaboration of recognition units of high stability. One of the major obstacles to testing theories which postulate differing effects of mismatch of different degrees, is the likelihood that each exposure during tests tends to change the central specifications from which discrepancy must be calculated. Ideally, one would hope for a manipulation which could give very high stability independently of training conditions. Effectively, this means blocking at least one type of perceptual learning. The discovery by Rogers, Drennen & Mark (1974) that the administration of cycloheximide to young chicks during an early sensitive period (in which drug administration must coincide with exposure to visual stimuli) disturbs visual learning abilities, and prevents habituation to repeated visual changes, offers some hope of providing an appropriate technique. The crucial question is whether visual learning occurring prior to drug administration survives or not: H. D. Drennan and I have preliminary indications that this is so.

If such a technique were to be discovered, one could hope for stable and persistent rates of self-presentation of stimulus change (to take only one example) which would allow proper investigation of this phenomenon and its implications for behaviour such as exploration and play.

(iii) Particularly complex problems are posed by the possibility that mismatch plays a part in the elaboration of new recognition units. A specific example may be helpful. Bateson (1973) has argued that there is a phase during imprinting when the chick prefers to examine slight variants on the original imprinting object. Initially, when it has learned very little about the object, this effect cannot occur; later, as the central specification of the object become more accurate (or perhaps more stable), the absolute magnitudes of parameter change which will produce the effect will become less and less, because small changes will now be treated as a marked discrepancy. Bateson makes the very interesting suggestion that only in this way can it be made certain that the chick will learn about all aspects (i.e. front, back and sides) of its mother.

This hypothesis can be generalised to cover exploration and perceptual learning in general: it amounts to the assumption that mismatch may guide the elaboration of a recognition unit. On such a hypothesis, perceptual learning might be divided into three overlapping phases. In the first, with no recognition unit or an extremely labile one, mismatch would be absent or so brief as to be unimportant. In the second, with a partially elaborated recognition unit, mismatch as a result of a particular parameter would serve to sustain attention (as postulated in the case of a fully elaborated and relatively stable recognition unit: assumption 8, p. 102); as a result of such examination the representation of the parameter would change towards a matching value. In the final phase, the recognition unit would be relatively stable, and mismatch would begin to have the sort of consequences discussed earlier in this section.

Such a model, although complex, can yield precise predictions, as Bateson has shown. It has the additional merit that the development of the final phase should be independently measurable by the appearance of responses associated with mismatch.

CONCLUSION

In the preceding pages, I have attempted to develop a theory which accounts for results from experiments of a sort which are not often done, and which asks questions about processes which are often neglected as a reasonable and necessary simplification of an otherwise too complex model. Some justification for embarking on such an enterprise and, worse still, expecting readers to accompany me, seems only fair.

Studies of animal behaviour have had great success in a number of areas. Any list of such areas would certainly include the causation of behaviour such as hunger and thirst, mechanisms of perception and conditioning processes. Very much less attention has been given to other processes, the understanding of which is at least as important and interesting to human beings. These include on the one hand, the moment-by-moment changes in the type of stimulus which an animal notices, and in its intentions and strategies in relation to such stimuli; on the other, the way in which it categorises and classifies the stimuli which make up its world and the sort of information which it possesses about each.

I would argue that the most challenging goal in brain research is the identification and study of the structures which directly mediate conscious thought: that is, exactly such moment-by-moment changes in interest and attention as are modelled in a very simple way by rules of selection and activation of recognition units. Studies of brain function relevant to such processes can only proceed in animals if a theoretical framework and a body of experimental results relating to them can be developed from animal studies.

SUMMARY

Concepts from human attention theory are applied to a number of problems in animal behaviour. First, a distinction is drawn between rules of selection of stimuli for examination and the readiness with which input from stimuli will be treated as acceptable evidence of the presence of a particular stimulus or category of stimuli.

Secondly, rules of selection differ greatly in their character. At one extreme, 'filtering' allows the rapid selection of stimuli on some simple characteristic despite the presence of a wide variety and number of targets; errors of selection are likely to be common. At the other, 'response set' will usually not allow errors but is slow in the presence of alternative targets.

This model of attention is applied to the increased persistence of certain aspects of behaviour produced by testosterone in male chicks, mice and possibly man. It is argued that such persistence is best described as representing a greater persistence in use of particular rules of selection. Evidence is taken from

(i) search tests involving situations where the test conditions constrain the animal to use either filtering or response set. Testosterone sustains search for a particular type of stimulus in the presence of alternative targets.

(ii) Runway tests in which a distracting change is introduced either in the runway wall or in the goal dish. Here, testosterone reduces the effect of change in stimuli which are not selected and prolongs it for change in ones that are.

(iii) Presentation of a very novel object in the home cage, followed by observation of return to maintenance behaviour. Testosterone prolongs the period of attention to the object.

These effects of testosterone are shown not to depend on changes in the ability to withhold responses, in the distribution and duration of drowsy periods, in the ease of evocation of specific systems of response such as feeding or fear, in the rate of learning, or in the readiness with which novel or conspicuous stimuli evoke examination.

Since increased persistence of attention is produced by the normal secretion of testicular androgens in cocks and male mice, it is justifiable to consider the biological functions of such an effect. Some possibilities are discussed.

Finally, a number of other possible applications of attention theory are discussed. These include search for food items, changes in attention in high and low arousal states, and the way in which novelty affects behaviour. In this last instance, the duration and intensity of the effects of a novel stimulus are likely to depend on the stability and detail of the recognition unit which is activated, as well as on the factors leading to its activation.

REFERENCES

Andrew, R. J. (1972*a*). Recognition processes and behaviour, with special reference to effects of testosterone on persistence. *Advances in the Study of Behavior*, **44**, 175–208.

Andrew, R. J. (1972*b*). Changes in search behaviour in male and female chicks, following different doses of testosterone. *Animal Behaviour*, **20**, 741–750.

Andrew, R. J. (1974*a*). Arousal and the causation of behaviour. *Behaviour*, **51**, 135–165.

Andrew, R. J. (1974*b*). Changes in visual responsiveness following intercollicular lesions and their effect on avoidance and attack. *Brain Behaviour and Evolution*, **10**, 400–424.

Andrew, R. J. (1975*a*). Effects of testosterone on the behaviour of the domestic chick. I. Effects present in males but not in females. *Animal Behaviour*, **23**, 139–155.

Andrew, R. J. (1975*b*). Effects of testosterone on the behaviour of the domestic chick. II. Effects present in both sexes. *Animal Behaviour*, **23**, 156–168.

Andrew, R. J. (1976). Increased persistence produced by testosterone, and its implications for the study of sexual behaviour. In *Biological Determinants of Sexual Behaviour*, ed. J. Hutchison. Academic Press: London, in press.

Andrew, R. J. & Rogers, L. (1972). Testosterone, search behaviour and persistence. *Nature, London*, **237**, 343–346.

Archer, J. (1973*a*). The influence of testosterone on chick behaviour in novel environments. *Behavioral Biology*, **8**, 93–108.

Archer, J. (1973*b*). Effects of testosterone on immobility responses in the young male chick. *Behavioral Biology*, **8**, 551–556.

Archer, J. (1974). The effects of testosterone on the distractibility of chicks by irrelevant and relevant stimuli. *Animal Behaviour*, **22**, 397–404.

Bateson, P. P. G. (1973). Preferences for familiarity and novelty: a model for the simultaneous development of both. *Journal of theoretical Biology*, **41**, 249–259.

Bateson, P. P. G. & Chantrey, D. F. (1971). Retardation of discrimination learning in monkeys and chicks previously exposed to both stimuli. *Nature, London*, **237**, 173–174.

Berlyne, D. E. (1967). Arousal and reinforcement. *Nebraska Symposium on Motivation*, **15**, 1–110.

Broadbent, D. E. (1970). Stimulus set and response set: two kinds of selective attention. In *Attention: Contemporary Theories and Analyses*, ed. D. Mostofsky. Appleton-Century-Crofts: New York.

Broadbent, D. E. (1971). *Decision and Stress*. Academic Press: London.

Broadbent, D. E. & Ladefoged, P. (1957). Vowel judgements and adaptation level. *Proceedings of the Royal Society of London*, B, **151**, 384–399.

Broverman, D. M., Broverman, I. K., Vogel, W., Palmer, R. D. & Klaiber, E. L. (1964). The automatization cognitive style and physical development. *Child Development*, **35**, 1343–1359.

Brower, L. P., van Zandt Brower, J. & Westcott, P. W. (1960). Experimental studies of mimicry. 5. The reaction of toads (*Bufo terrestris*) to bumblebees (*Bombus americanorum*) and their robberfly mimics (*Mallophora bomboides*), with a discussion of aggressive mimicry. *American Naturalist*, **94**, 343–355.

Croze, H. (1970). Searching image in carrion crows. Hunting strategy in a predator and some anti-predator devices in camouflaged prey. *Zietschrift für Tierpsychologie*, **5**, 1–86.

Dawkins, M. (1971). Perceptual changes in chicks: another look at the 'searching image' concept. *Animal Behaviour*, **19**, 556–574.

Delius, J. D. (1970). Irrelevant behaviour, information processing and arousal homeostasis. *Psycholgische forschung*, **33**, 165–188.

Ewert, J. P. (1970). Neural mechanisms of prey catching and avoidance behavior in the toad (*Bufo bufo* L.). *Brain Behaviour and Evolution*, **3**, 36–56.

Ewert, J. P. & von Wietersheim, A. (1974). Musterauswertung durch Tectum und Thalamus/Praetectum-Neuronen in visuellen System der Kröte *Bufo bufo* (L.). *Journal of comparative Physiology*, **92**, 131–148.

Falk, J. L. (1972). The nature and determinants of adjunctive behaviour. In *Schedule Effects: Drugs, Drinking, Aggression*, ed. R. M. Gilbert & J. D. Keehn. University Toronto Press: Toronto.

Gale, J. S. & Evans, L. J. (1975). Logic of animal conflict. *Nature, London*, **254**, 403–464.

Green, B. F. & Anderson, L. K. (1956). Colour coding in a visual search task. *Journal of experimental Psychology*, **51**, 19–24.

Green, D. M. & Swets, J. A. (1966). *Signal Detection Theory and Psychophysics*. Wiley: New York.

Haber, R. N. (1958). Discrepancy from adaptation level as a source of affect. *Journal of experimental Psychology*, **56**, 370–373.

Hebb, D. O. (1946). On the nature of fear. *Psychological Review*, **53**, 259–276.

Helson, H. (1964). *Adaptation-level Theory*. Harper & Row: New York.

Hinde, R. A. (1970). *Animal Behaviour*, 2nd edn. McGraw-Hill: New York.

Jerison, A. R. & Rohwer, W. D. (1966). The Stroop color-word test: a review. *Acta psychologica*, **25**, 36–93.

Kagan, J. (1970). Attention and psychological change in the young child. *Science, Washington*, **170**, 826–832.

Kahneman, D. (1973). *Attention and Effort*. Prentice-Hall: Englewood Cliffs, N.J.

Klaiber, E. L., Broverman, D. M., Vogel, W., Abraham, G. E. & Cone, F. L. (1971). Effects of infused testosterone on mental performances and ICSH. *Journal of clinical Endocrinology and Metabolism*, **32**, 341–349.

Krebs, J. R., Ryan, J. C. & Charnov, E. L. (1974). Hunting by expectation or optimal foraging? A study of patch use by chickadees. *Animal Behaviour*, **22**, 953–964.

Lack, D. (1939). The behaviour of the Robin: I and II. *Proceedings of the zoological Society of London*, A, **109**, 169–178.

Lack, D. (1966). *Population Studies of Birds*. Oxford University Press: London.

Leyhausen, P. (1973). In *Motivation of Human and Animal Behaviour: an Ethological View*, ed. K. Lorenz & P. Leyhausen. Holt, Reinhart & Winston: New York.

Mackintosh, N. J. (1965). Selective attention in animal discrimination learning. *Psychological Bulletin*, **64**, 124–150.

Maynard-Smith, J. & Price, G. R. (1973). The logic of animal conflict. *Nature, London*, **246**, 15–18.

McClelland, D. C., Atkinson, J. W., Clark, R. A. & Lowell, E. C. (1953). *The Achievement Motive*. Appleton-Century-Crofts: New York.

Messent, P. R. (1973). Distractibility and persistence of chicks. D.Phil. thesis, University of Sussex.

Moray, N. (1969). *Attention: Selective Processes in Vision and Hearing*. Hutchinson: London.

Neisser, U. (1967). *Cognitive Psychology*. Appleton-Century-Crofts: New York.

Parker, G. A. (1974). Courtship persistence and female-guarding as male time investment strategies. *Behaviour*, **48**, 157–184.

Rabbitt, P. M. A. (1964). Ignoring irrelevant information. *British Journal of Psychology*, **55**, 403–414.

Rabbitt, P. M. A. (1967). Learning to ignore irrelevant information. *American Journal of Psychology*, **80**, 1–13.

Rogers, L. J. (1971). Testosterone, isthmo-optic lesions and visual search in chickens. D.Phil. thesis, University of Sussex.

Rogers, L. J. (1974). Persistence and search influenced by natural levels of androgens in young and adult chickens. *Physiology and Behavior*, **12**, 197–204.

Rogers, L. J., Drennen, H. D. & Mark, R. F. (1974). Inhibition of memory formation in the imprinting period: irreversible action of cycloheximide in young chickens. *Brain Research*, **79**, 213–233.

Royama, T. (1970). Factors governing the hunting behaviour and selection of food by the great tit (*Parus major* L.). *Journal of Animal Ecology*, **39**, 619–668.

Salzen, E. A. (1962). Imprinting and fear. *Symposium of the zoological Society of London*, **8**, 199–217.

Smith, J. M. N. (1974). The food searching behaviour of two European thrushes. I. Description and analysis of search paths. *Behaviour*, **59**, 276–302.

Sokolov, E. N. (1963). *Perception and the Conditioned Reflex*. Pergamon Press: Oxford.

Sutherland, N. S. (1966). Successive reversals involving two cues. *Quarterly Journal of experimental Psychology*, **18**, 97–102.

Tinbergen, L. (1960). The natural control of insects in pine woods. *Archives néerlandaises de zoologie*, **13**, 265–379.

Treisman, A. M. (1960). Contextual cues in selective listening. *Quarterly Journal of experimental Psychology*, **12**, 242–248.

4

Dynamic boundaries of patterned behaviour: interaction and self-organization

JOHN C. FENTRESS

STATEMENT OF FRAMEWORK

Overview

The basic idea explored in this paper is that integrative systems in behaviour commonly display two fundamental principles of operation: interaction and self-organization. The balance between these two principles may depend upon the degree to which the systems are activated. The particular hypothesis is that as activity within an integrative system increases, its intrinsic organization tightens, with the result that extrinsic factors become less important to the control of expressed behaviour. This basic postulate helps account for dynamic shifts in the balance between central and peripheral mechanisms plus changing rules of interaction between different behavioural systems.

The focus of the argument is to provide a means for thinking about the problem of dynamic pattern in behaviour. Our models are frequently couched in what might be viewed as unrealistically static terms. Further, these models do not deal in an adequate way with the possibility of shifting overlap between separately defined systems of behavioural control. To help illustrate these issues I shall draw from relevant literature on motivational and motor systems, and shall subsequently attempt to link these arguments with data on phenotypically derived potentialities and constraints that underlie observed patterns of behavioural expression. First, some more general issues will be examined.

Boundaries

The concept of boundary is fundamental to all natural science. It refers to the way we segregate the universe under discussion into component dimensions. The subsequent relations we seek depend upon these initial divisions (Brown, 1969).

In the study of patterned behaviour one approach to the question of control boundaries is seen in the literature on motivation. The concept of motivation refers to the relations between factors intrinsic and extrinsic to the organism, as well as to the relations between defined classes of behaviour. Where does one draw the boundary between intrinsic and extrinsic contributions to behaviour, or between one dimension of behaviour and another?

One reason that these questions become difficult is that behaviour systems, like organisms, appear both to interact with extrinsic factors and display patterns of self-organization. It is this dual theme that I wish to pursue in the present paper, for it permits the formulation of dynamic principles that may have considerable generality.

Systems

The term 'system' refers to a definable set of elements, or dimensions, which are associated with one another by specified, or at least specifiable, rules. Thus the system's operation is presumed to depend both on the properties of component parts and on the rules that link these parts. This basic tenet can be found, for example, in the analysis of neural networks of varying degrees of complexity (Fentress, 1976). The distinction between systems is made primarily upon the dual criteria of relatively tight linkage between subcomponents within the system and relatively marked discontinuities between subcomponents in different systems. More formal definitions of systems can and have been proposed (e.g. McFarland, this volume), but this general framework is sufficiently precise to pursue the major arguments of the present paper. The emphasis is comparable to that proposed by Reiner for adaptive control systems in biology: '. . . it is of the utmost importance to keep in mind that analysis analyzes something which is a *system* of parts, and that one must, after isolating and studying the parts, also study the *relations* that hold the system together' (Reiner, 1968).

The precise definition of a *behaviour system* is made complex for several reasons. First, multiple criteria can be employed at the descriptive level. For example, classification of behaviour by movement, consequence, or antecedent conditions can give different results (Hinde, 1970; Fentress, 1973). Secondly, it is often difficult to determine precisely where to separate one dimension of behaviour from another. This appears particularly true for mammals where different acts often appear to blend together without any obvious single line of demarcation (Fentress, 1967). Thirdly, underlying control processes often cannot be viewed directly, at least in satisfactory detail at the same time that integrated behaviour patterns at the organism level are examined. They must therefore be inferred on the basis of specified manipulations and observations of input and output. In this way, intervening 'transfer functions' are derived. Measurements of the divergent effects of a given input and the convergence of antecedent factors on a given output can give somewhat different pictures (Fentress, 1973). Fourthly, initial classifications of behavioural control systems typically can be fractionated upon finer analysis (Hinde, 1959). Once this happens, other classifications may be possible. Fifthly, control analyses frequently indicate varying degrees of operational overlap between systems classified separately at the level of behavioural output, with the result that these systems must be viewed as interdigitated rather than independent. Questions of independence versus suppression versus facilitation must be carefully separated (Fentress, 1973). Finally, it appears possible that the boundaries of a given system, defined through specified input and output relations, may vary in time. One consequence of this is that there can be a shift in the operational overlap or interconnection between systems as a function of the internal dynamics of the organism. This last point is emphasized in the present paper within the general framework of interaction and self-organization. (See also Bateson, this volume, in which related issues from a developmental perspective are discussed.)

Dynamics

The argument stressed here is that systems are both interactive and self-organizing, but that the balance between these two modes of organization can shift as a function of dynamic principles which have not been sufficiently explored. The basic idea of an interactive/self-organizing system is that the system can be activated by a variety of factors normally defined as extrinsic to the system, but once activated the system generates patterns of activity that are to a large extent independent of these extrinsic factors. In one sense, therefore, we can say that the system redefines its control boundaries as a function of its state of activation.

One of the interesting corollaries of this approach is that systems of behavioural integration may be definable partially in terms of their interactions with other systems. This is because the interactions with other systems change the boundaries of a given system. Such a suggestion is clearly the reverse of more traditional procedures which first define several motivational systems independently, and then seek rules of interaction between these systems. However, if the systems overlap with one another then they are not mutually independent, and if they shift their individual boundaries as a function of both interaction and self-organization then each must be defined in reference to the current state of others. This point should become clear as the discussion proceeds.

To develop the idea of interactive/self-organizing systems I shall divide the subsequent material into three sections. The first looks at the interplay between factors intrinsic and extrinsic to the organism as a whole. The second carries the same basic arguments to the level of factors intrinsic and extrinsic to a given control system. The third section explores the role of phenotypic substrates and capacities in the production of dynamic patterns of behaviour.

Implications

It should be clear that the present approach offers an alternative to the traditional dichotomy and debate between specific and non-specific factors in behaviour. As I have pointed out previously the concept of specific factors in its strict form implies complete separation between control systems in the sense that they do not interact at any level. Conversely, the strict definition of a non-specific factor is that it mediates all possible input–output combinations equally (Fentress, 1973). It seems reasonable, perhaps particularly in the study of behaviour in higher vertebrates, to view control pathways that mediate different classes of behaviour as being partially shared with each other under certain circumstances. It is this that I mean by the term overlap. Possible asymmetries and specificities in overlap between differently classified dimensions of behaviour can subsequently be evaluated in this way. Further, since the models we make must adequately account for dynamic fluctuations in behaviour, it seems reasonable that we also consider the underlying control systems as having dynamic properties in the sense that their boundaries of operation may shift as a function of the internal dynamics of the organism.

Such a view is not complicated, but it can be difficult to think about. The reasons for this, I think, are that we tend to view categories we construct in unitary, mutually exclusive, and static terms. It is apparent that in other branches of natural science, particularly in the physical sciences, simple models of unitary, mutually exclusive, and static mechanisms have had to be replaced with a more dynamic and even relativistic framework (Bohm, 1969). I think that it is possible that the same will hold true for integrated patterns of behaviour. It is the exploration of this possibility that has led to the tentative particular formulations in the present paper. I say this so that even if the particular concept outlined here proves on further analysis to need modification, the general strategy of looking for more dynamic principles of interacting subsystems will not be lost sight of (cf. Goodwin, 1970; Katchalsky, Rowland & Blumenthal, 1974, for similar discussion of biological systems). Since definitions of behavioural systems in ethology involve abstractions that are typically based upon inferred as well as directly observed relations between events, we must adopt a sufficiently flexible approach to integration to allow exploration and test of various possible modes of organization.

Obviously in some sense our models of integrated behaviour must be formulated within a framework that is not only empirically testable but static (which it must be to be empirically testable). This can be done if we look for consistencies in the rules of interaction between variables which together produce the picture of dynamic flux. In biological systems we can be aided in this by attempting to superimpose principles of dynamic organization upon the more stable phenotypic substrates of the organism which set the potentialities and constraints of performance. These potentialities and constraints on the one hand produce the diversity in detailed patterns of organization which place important limitations on our generalizations (Hinde, 1972). At the same time there may be deeper commonalities of organization which will be missed if we concentrate solely on individual and particular details of behaviour or on individual and particular mechanisms. An approach that will be explored here is that each organism has phenotypically determined capacity limitations, and that these limitations transcend the execution of particular forms of behaviour in the sense that qualitatively different classes of behaviour can in part be arranged in terms of their overall demands on the processing abilities of the organism as an intact entity. This would suggest that models of motivational fluctuations in behaviour cannot be entirely separated from considerations of the more permanent and deeper phenotypic structures of the organism.

THE ORGANISM AND ITS ENVIRONMENT: THE QUESTION OF INTRINSIC AND EXTRINSIC FACTORS IN BEHAVIOUR

The first lesson of the literature on motivation is that factors intrinsic and extrinsic to the organism act in unison to produce the integrated patterns of behaviour that we observe. Thus, for example, animals may display quite different degrees and forms of response to the same external situation from one occasion to the next. When these fluctuations in response are reversible and cannot be attributed to direct changes at the sensory or motor level, we infer higher order changes of internal state.

The phenomenological situation can be illustrated by examining changes in an animal's response to a given stimulus as a function of the animal's ongoing behaviour. For example, it has been found that voles in an experimental alley were much more likely to flee rather than freeze in response to an overhead moving object if they were moving at the time that the stimulus was presented. Conversely, they were more likely to freeze than flee if they were engaged in non-locomotor activities.

Even here, however, one must go further than to isolate a momentary expression of ongoing behaviour from its broader temporal context. For example, it was found that animals which were sitting still at the instant of the presentation of the overhead stimulus, but which had been locomoting within the previous 10 seconds, were just as likely to flee as they were during actual locomotion (Fentress, 1968*a*, *b*). This not only illustrates a certain inertia in motivational systems which can outlast overt behavioural expression, but at the same time indicates that processes which underlie one dimension of behaviour may be active during the expression of another.

The basic idea is that we cannot account for fluctuations and constancies in behaviour solely through reference to events external to the organism. This is an important principle, but on finer analysis it refers to the basic properties of *all* causal statements within a systems context: i.e. every 'cause' must 'affect' something to produce an 'effect', and the state of the thing (or process) affected is just as important to the effect as is the cause. *Every* causal statement involves a relation between two or more states. Thus the organization of spinal reflexes depends as much upon the state of spinal interneurones as it does upon the activation of sensory afferents (Thompson, 1967), and the contractions of a muscle depend as much upon the muscle's initial state as upon neuronal inputs to the muscle (Kennedy, 1976). Simple stimulus–response (S–R) models of behaviour are blatently illogical, for they omit reference to the state variables that lie between S and R. This criticism would hold for any S–R models, whether in the physical or biological realm. What motivational models stress is that state variables imposed between S and R can change, and this in turn alters the relation between S and R. Looked at in this way, motivational constructs lose their mystique but also emphasize a point of general significance with reference to the problem of control.

This general point is the importance of examining *relations* between separately defined dimensions that underlie behaviour. It is the basic idea of relations within a dynamic framework that I wish to emphasize here. Other chapters in the present volume examine similar issues from slightly different perspectives (e.g. Bateson, Hinde & Stevenson-Hinde, and Simpson). In this paper the question of relations is viewed primarily within the context of systems which appear to be both interactive and self-organizing.

Species behaviour

The fact that fluctuations and constancies intrinsic to an animal interplay with environmental factors to produce the patterns of species behaviour that we observe does not mean that the relative contribution of these intrinsic and extrinsic factors cannot be determined for a given situation, nor that the respective contributions must remain constant. When one examines face grooming sequences in rodents, for example, an obvious explanation for both the occurrence and patterning of grooming is that the animal is responding to external irritation. The importance of peripheral irritants in grooming is easy to verify experimentally, but this certainly is not the whole story.

For example, when sensory branches of the trigeminal nerve were sectioned in mice to denervate the face, characteristic patterns of face-grooming could still be elicited. More importantly, the degree to which the trigeminal nerve lesions modified the expression of grooming was a function of the circumstances under which the animals were tested as well as the animals' behavioural state. When the animals were tested in their home cages considerable distortions of normal grooming patterns were observed, whereas when the animals were tested in small novel environments it was not possible to distinguish between operated animals and control animals (Fentress, 1972). Under the latter circumstances mice frequently display particularly prolonged, high frequency, and stereotyped movement sequences in grooming. This correlates with a shift toward central as opposed to peripheral control as measured by reduced responsiveness to changes in facial input. In this sense one can say that the control boundaries for a given class of behaviour have shifted from one behavioural situation to another. That is, the peripheral inputs that contribute importantly under one set of circumstances are effectively excluded in another, and central mechanisms begin to predominate. The output mechanisms that produce the observed sequences of grooming become self-organized.*

One can also legitimately argue that when mice are in a novel environment as compared to their home cages they are subjected to increased attentional demands which in turn reduce the effectiveness with which they can process specific peripheral information from the face. This argument cannot be discounted at the present time, but one must recognize that a necessary corollary is that the animals can only process a limited amount of information from the environment which is the definition of limited capacity. I shall return to this consideration in a later section.

* Two important and related points were raised by conference participants in preliminary discussion of these ideas. The first is that the relative importance of different elements might shift without necessitating the postulate of shifting boundaris *per se*. The second is that one must be careful to indicate that a given element may have an altered threshold for expression rather than being simply operative or inoperative. The first question gets directly at the question of *where* we draw boundaries while the second gets at the question of whether boundaries are thought of on a simple inclusion versus exclusion basis. As long as we are careful to recognize that the idea of boundaries is used here as an abstract concept to help us *think about* dimensions of behavioural control these questions can be regarded as clues to possible alternate and refined formulations. Any abstract concept must be viewed in this light. The latter point of shifting relative thresholds should certainly be born in mind throughout the subsequent discussion which has been simplified for the sake of clarity in exposition. A third point which one might raise is that the organism may become particularly sensitive to some classes of input at the same time it becomes relatively insensitive to others. This point has been examined previously (Fentress, 1973), but is not directly relevant to the more broadly defined possibilities I wish to focus upon here.

A related impression from preliminary data, however, is that actual execution of rapid movements during grooming can serve to isolate the 'grooming system' from both dependence upon and sensitivity to interruption by sensory factors. First, rapid and stereotyped phases of a grooming sequence are less influenced by disruptions of both peripheral and proprioceptive inputs than are slower and more complex movements that occur in the same context (Fentress, 1972). Secondly, when one attempts to disrupt ongoing grooming with peripheral stimulation, such as a click or mild electric shock, the rapid and stereotyped movement sequences appear more autonomous (difficult to interrupt) than do slower and more complex movement sequences. Analogous results have been obtained independently by Dawkins & Dawkins (1974) on the phases of pecking and drinking sequences in chicks.

Movement stereotypes

Perseverant motor stereotypes provide another example at the behavioural level: during rapid execution of a movement sequence animals are relatively uninfluenced by sensory cues that play a more important role during less vigorous performance of the same basic motor pattern. I report here two previously unpublished sets of observations.

The first example involves a captive Cape hunting dog (*Lycaon pictus*) which had developed a figure eight locomotor pattern in its cage. Once the pattern was well established a chain, approximately 0.8 metres high at the centre, was placed at the crossing point of the figure eight. The animal initially changed its path to avoid the chain. However, when external disturbances generated a higher speed of pacing the animal reverted back to its previous path. The animal then stumbled several times over the chain. Soon, however, it began to clear the chain in an apparently mechanical manner. The chain was left in place for a period of three months. It was then lowered to the ground. During subsequent pacing at an undisturbed rate the animal jumped several times as if the chain were still raised then continued the basic figure eight movements without jumping. When various external disturbances such as noise from the zoo keepers occurred, however, the animal both increased its speed of locomotion and reverted back to its jumping pattern as if the chain were still raised. In these latter circumstances one obtained the impression that the animal was no longer processing detailed sensory information about the chain's location, but had reverted to a previously established central motor programme. There are limitations to this statement, however, which indicate that it is the *relative* balance between central and peripheral mechanisms that has changed. For example, when the chain was removed completely from the pen, the animal was observed to jump on only two occasions (cf. footnote, p. 141).

The second illustration involves weaving and jumping movements that developed around a water spout in voles isolated for several months in enclosed biscuit tins. As a parenthetical statement it is interesting that *Microtus agrestis* developed a species characteristic weaving movement and *Clethrionomys britannicus* developed a jumping pattern. The basic point is that when the water spouts were subsequently removed each species continued its previously established motor profile during conditions that led to rapid execution of the stereotyped behaviour pattern, whereas during phases of less rapid execution the animals compensated for the absent water spout.

In each of these cases, as with the control of more normal species characteristic movement patterns such as self grooming, one gains the impression that during rapid execution of a motor sequence the animals rely primarily upon previously established central programmes. The movement sequence becomes self-organizing in the sense that the animals no longer appear sensitive to sensory cues that under other circumstances modulate both the probability and form of behaviour. A related point is that during these phases of rapid execution various external disturbances that would normally block the performance of the stereotyped motor sequences can be presented without obvious effect.

Other movement sequences

There is also evidence that well established and rapidly performed human movement sequences are relatively independent of sensory feedback that is necessary for slower, more complex, and less well-established movements (e.g. Keele, 1968; Konorski, 1967). Further, when the same basic movement sequence is performed at different speeds, both the dependence upon sensory feedback and the ease of disruption by extraneous stimuli are reduced during rapid execution. Thus it is reasonable to postulate that under these conditions of rapid execution of a well-established movement the control boundaries become centrally constricted.

Classic criteria for 'intensity' of motivated behaviour are the ease with which the behaviour is interrupted by extraneous factors and the dependence of the behaviour upon external cues (Fentress, 1973). The control systems appear to sculpture themselves away from extraneous influences. A single and relatively extreme example illustrates the main point I wish to make.

When I was a research student at Cambridge I kept groups of voles in large alleyways. In several of these alleys I had placed a variety of bricks, sticks, and small stones to give the animals a relatively diverse and normal environment. The animals soon established preferred paths which entailed both going over and around certain of the objects in their alleyways. At the conclusion of my experiments (Fentress, 1968*a*, *b*) the animals were caught and released. For two of the alleys I moved several of the objects and observed the animals prior to catching and releasing them. At the points where the objects had been removed and added the animals spent considerable time sniffing, digging, chewing, and grooming. The animals also spent a large percentage of their time off their pathways as if exploring the changes in their environment. When I subsequently proceeded to remove the animals, however, individuals fled along previously established pathways. Several jumped at places where they had previously crawled over objects (now removed) and repeatedly ran into objects that were placed in previously unobstructed paths. Again it was apparent that the animals had become unresponsive to sensory factors that normally guided their behaviour. Once the escape behaviour was triggered by my movements it appeared to organize itself relatively independently of further environmental guidance.

Such observations are striking at a phenomenological level, but are obviously far removed from precise consideration of mechanism. There are similarly a variety of observations in human behaviour that indicate 'attentional demands' in the rapid execution of performance skills can reduce sensitivity to sensory information that might otherwise be effective in the guidance of these skills (Broadbent, 1971; Kornhuber, 1974). Although the detailed mechanisms obviously differ in these different cases the basic operational principle that systems become both self-organizing and immune from extrinsic inputs during rapid execution appears widespread in a variety of species and for a variety of particular activities. Are there any data on how such operational principles could be mediated at the physiological level?

Systems and mechanisms

One approach to this question is to examine simpler integrative networks for possible similarities in operation. In this manner the individual elements of operation can often be studied directly. With the diversity of biological material one would obviously expect parallel diversity in specific mechanisms even when the overall principles of operation are analogous. However, it is the basic principles of operation that are selected for, and here we might expect nature to place similar demands on performance. For example, both vertebrate and invertebrate motor systems utilize similar principles of load compensation although the particular mechanisms have evolved independently (Kennedy, 1976). Further, when operational principles prove effective at one level of organization it is reasonable to postulate that nature will conserve similar operational principles at other levels of organization. We might therefore expect to gain some insights into principles of organization in motivational systems through an examination of integration at the sensory and motor levels of organization. Although the importance of different levels of organization cannot be denied, the examination of simpler networks may provide insights into principles of control in patterned behaviour that might be both relevant to, and obscured in, more complex systems (Fentress, 1976).

It has been recognized for many years that activation of motor pathways in mammals can block the processing of specific sources of sensory information (Jasper, 1963). One striking illustration that has received considerable attention in recent years is the phenomenon of saccadic suppression where visual information is momentarily blocked during rapid movement of the eyes (Wurtz, 1969). This is obviously a specialized case which prevents both blurring the visual image plus sensation of movement of the external environment when the point of fixation jumps from one place to another. However, other cases which have more obvious relevance to complex patterns of behaviour can also be cited. For example, high levels of pyramidal tract activity can block the transmission of sensory information, such as that from trigeminal afferents (e.g. Wall, 1968; Somjen, 1972). It is possible that similar mechanisms participate in the reduced sensitivity of mice to alterations in sensory information from the face during rapid phases of self-grooming (above). Through such a mechanism one could think of the 'grooming system' being triggered by sensory information from the face, and then isolating itself from this same sensory information.

Analogous principles have now been reported for a variety of species. Russell (1971) has shown that rapid, but not slow, body movements in the amphibian *Xenopus* are closely preceded by inhibition of lateral-line mechanoreceptor afferents. One consequence is that reflexes which might interfere with the completed execution of these rapid movements are prevented. Similarly, Carpenter & Rudomin (1973) have demonstrated a powerful inhibition of skin afferents in the frog (*Rana* sp.) by antidromic stimulation of the extensor muscle nerves. It is reasonable to postulate that similar inhibition of sensory information occurs during the course of a normal jump. In this sense we could think of a 'jumping system' which interacts with, and indeed may be triggered by, skin afferents, but which subsequently becomes self-organizing in part through the blockage of information from these same afferents. Murphey & Palka (1973) have demonstrated central inhibition of synapses from caudal cerci in crickets during periods of locomotion. During periods of quiescence stimulation of these same receptors of the caudal cerci by a puff of air produces rapid running. Again, the output system in behaviour appears first to interact with sensory information and then to isolate itself from this same sensory information. In this sense one can think about the behavioural system as constricting its boundaries (becoming self-organized) during periods of strong activation.

A particularly elegant recent illustration of these basic operational principles has been provided by Krasne & Bryan (1973) for the lateral giant fibre system in crayfish. Stimulation of this giant fibre system produces a rapid sculling stroke that moves the animal upwards, often to the point of producing a complete somersault (Wine & Krasne, 1972). The vigour of this movement might be expected to elicit a variety of conflicting reflexes mediated by the same mechanoreceptor hair system that triggers activity in the lateral giant fibre. Another potential problem for the animal is that the synaptic sites between tactile afferents and their interneurones are very susceptible to habituation (Zucker, 1972). What Krasne & Bryan demonstrated is that activity in the lateral giant neurone produces powerful inhibition upon these synapses. This has subsequently been demonstrated to be due to presynaptic inhibition via depolarization of the afferent terminals (Kennedy, Calabrese & Wine, 1974). The important dimension for the present discussion is that, once activated, the lateral giant fibre escape system isolates itself from potentially disruptive sources of input. In this sense the behavioural output becomes self-organizing even though it is triggered by interactions with exogenous inputs. (I thank Donald Kennedy for bringing these last examples to my attention.)

There is evidence that behavioural output systems not only become more immune from extrinsic sensory input during periods of activation, but also less dependent upon sensory information during rapid performance. For example, removal of proprioceptive information by dorsal root lesions in mice primarily affects movements that are both complex and slow (Fentress, 1972). Similarly, Northup (1973) has shown that mice with cerebellar defects of genetic origin are most affected during the production of slow grooming movements in contrast to fast grooming movements. One could reasonably postulate that these slow movements involve cerebellar processing of proprioceptive information whereas faster movements are more centrally controlled (e.g. Eccles, Ito & Szentagothai, 1967). Northup also showed that the modification of grooming movements by weights added to the limbs was maximal during slow movements. Similarly, removal of proprioceptive information in insects can block the execution of slow locomotor patterns while more rapid locomotor patterns are relatively little affected (Pearson, 1972). The relative importance of sensory input via the trigeminal system in mice during face grooming at different speeds makes a similar point (above).

A variety of recent data from mammalian neurophysiology supports the basic idea of system isolation during activation. Doty (1976) reviews the concept of control centres (defined primarily by functional rather than localized anatomical criteria), and points out that these 'centres' may isolate themselves upon activation through processes of collateral inhibition. Similarly, Llinás & Walton (1976) provide evidence that the inferior olive system may act to isolate intrinsic pattern generating mechanisms of cerebellar origin that underlie integrated motor output. Obviously there are marked differences in detail between individual mechanisms within a species and for different species, but the basic theme of shifts toward intrinsic control upon system activation appears with sufficient frequency to deserve further consideration as an operational principle of some generality (cf. Dawkins, this volume, on search for general principles).

These ideas are also supported by much of the earlier behavioural literature. For example, isolation of a triggered movement from further external control is part of the classic definition of fixed action pattern in ethology (e.g. Tinbergen, 1951; Hinde, 1970; Schleidt, 1974). Indeed, this has led to a variety of neurobiological studies of intrinsic pattern generation (e.g. Hoyle, 1976; Ikeda, 1976). Here it is important to point out the behavioural evidence indicates that isolation may be highly selective; e.g. most orienting stimuli are not filtered and releasing stimuli are. Further, it is often useful to distinguish between more variable skilled movements which depend upon sensory modulation and less variable stereotyped components (e.g. Fentress, 1972). Finally, the idea that rapid production of movement can occur at rates that do not allow for feedback control was recognized many years ago by Lashley (e.g. 1951). (My thanks to Robert Hinde for suggesting that I bring these examples into the present context.)

In this discussion I have lumped together considerations of isolation of sensory information with extrinsic and proprioceptive origin. I do this because similar broad principles appear to apply in each case. This does not imply that distinctions cannot be made. For the most part, however, distinctions in detail must await further analysis, and their relevance to the more general level of discussion employed here is at present uncertain.

THE RELATIONSHIP BETWEEN DIFFERENT BEHAVIOURAL SYSTEMS

The next issue concerns the relations between behavioural systems classified as different. Unlike the situation with intrinsic and extrinsic factors defined for the organism *in toto*, there are no clear (e.g. morphological) landmarks that separate one behavioural system from another. It is useful to distinguish between the performance characteristics of the system once it is activated and the factors which activate and/or block the system (Fentress, 1973). A related question concerns the influence of an activated behaviour upon the performance of others. Each of these dimensions inevitably involves questions of the relations between one behavioural system and another as well as the boundary conditions that exist for a single behavioural system (above). As each organism lives in the context of its environment so too does each dimension of behaviour that we classify occur in the context of others (preceding, subsequent, simultaneous, and/or alternating). It makes no more sense to ignore the relationships between one dimension of behaviour and another than it does to ignore the relationships between individual notes and pauses in a musical composition. How do these dimensions of behaviour fit together, and what does this indicate about underlying mechanisms?

Performance characteristics and consequences

As the speed of a behavioural performance increases, the output dimensions frequently become more tightly coupled. This often occurs at the same time that the performance of the behaviour becomes (1) more independent from modulation (e.g. disruption) by extrinsic inputs, (2) less dependent upon specific sources of sensory information, and (3) more effective in blocking other dimensions of behaviour. These factors together provide one measure of the concept of behavioural intensity (Fentress, 1973).

Coupling of output components. Von Holst (e.g. 1948) demonstrated that individual fins in teleosts can beat at different frequencies, but as the vigour and/or frequency of movements increases the fins tend to become entrained to a common or dominant rhythm. This 'magnet effect' involves a simplification of the animal's motor output in the sense that a single set of rules can now be applied to all the fins. Mates (1973) has recently demonstrated an analogous phenomenon in his quantitative analysis of eye movements in chameleons. When the mean frequency of eye movements is low, the intermovement interval for a given eye shows considerable variability and the two eyes are independent in the sense that observation of one eye alone is a poor predictor of movements in the other eye. As movement frequency increases, however, each eye moves in a more stereotyped way with respect to time intervals, and the two eyes become phase-locked in their movements. It is as if a single set of rules for movement both becomes more rigid and applies to each eye. Similarly, I have found that face-grooming movements in mice become more rhythmical, follow a more stereotyped course, and involve a tighter coupling of the forelimbs as frequency and apparent vigour of the movements increases (Fentress, 1972 and unpublished data). Information measures of the tightness of sequential coupling between paired grooming movements, for example, dropped from an H_2 value of 1.71 (predictability approximately 1 in 4) during the performance of eight low frequency movements to 0.86 (predictability approximately 1 in 2) for the same eight movement components performed at a high rate.* It is at these latter times that the animals are less easily interrupted by extraneous disturbances.

A plausible model is that as movement frequency increases, the output demands upon the organism also increase, which in turn produces both a simplification of output rules plus a reduced sensitivity to extraneous influences. A similar relationship between frequency, coupling, and stereotyping of movements can be demonstrated for human performance. For example, if one moves the index fingers on each hand slowly and with a low mean frequency it is possible to generate complex patterns where each of the fingers follows somewhat independent rules. As the frequency and speed of movements increases, however, the intermovement intervals become more regular and the two fingers become phase-locked. The most frequent movements can be produced when the two fingers move identically (J. C. Fentress, unpublished). Miller & Frick (1949) demonstrated that a person's ability to press a series of buttons in random order decreases as a function of the mean frequency of presses required. In each of these cases increases in the 'gain' of the output system results in a simplification of output rules.

I grant that these observations fall outside the usual literature on motivation. Can similar phenomena be demonstrated in the sequential articulation of species-characteristic patterns of behaviour as investigated within the framework of motivational control? I mentioned previously that when voles were being caught in home pen alleys for subsequent release they reverted back to previously established pathways, often with the result that they would bump into obstacles repeatedly and in a stereotyped manner. One can find numerous examples where increase in inferred motivation tightens the temporal organization of functionally integrated sequences of behaviour. Two examples of the influence of food deprivation on the sequencing of attack behaviour will be given here. Zack (1975) demonstrated that moderate levels of food deprivation both increased the probability of intra-specific agonistic encounters in the marine mollusc *Hermissenda crassicornis* and simplified the temporal organization of component movements. Similarly, R. Adamec (unpublished)

* H_2 refers to the overall predictability of acts in a behavioural sequence given knowledge both of individual probabilities and of sequence pairs. For convenience the figures are calculated on a logarithmic scale with the base 2, where H is defined by the exponent. Thus eight equally probable and independent behavioural acts would generate an H value of 3 ($2^3 = 8$), and a sequence which was perfectly predictable would give an H value of 0 ($2^0 = 1$). For the present purposes it is sufficient to recognize that lower H values indicate less uncertainty in the sequence; cf. Fentress, 1972, and Fentress & Stilwell, 1973, for further discussion.

has found that food deprivation in rats both increases their killing of mice and reduces the complexity of the sequence of motor acts involved.

I suggest these and numerous related observations support the hypothesis that increased activation of a motivational or output system frequently increases the tightness of its internal organization, with the common result that movement components become more frequent, more stereotyped, more tightly coupled, and both less dependent upon and less influenced by extraneous factors. The central control system in a sense sculptures itself into a position of autonomous organization where the various subcomponents of control increasingly share common operational principles which are limited by the organism's overall processing capacity.

Responsiveness of system A to system B. Evidence that performance of a given class of behaviour (A) reduces responsiveness to factors that normally underlie another class of behaviour (B) is well known and has been reviewed in some detail previously (e.g. Hinde, 1970; Fentress, 1973). The point I wish to make explicit here is that this phenomenon is analogous to reduced responsiveness to extraneous factors defined in terms of the organism *in toto* (above). Thus animals which are strongly engaged in a behaviour such as feeding often become relatively insensitive to factors that might otherwise produce sexual behaviour.

Two further examples will be given for the point of illustration. The first concerns the reduced responsiveness of rodents to specific peripheral irritants that are relevant to face grooming when the animals are actively engaged in other forms of ongoing behaviour. We have recently conducted a series of experiments in which air puffs and water drops were applied to the head and body of mice during various phases of behaviour. To summarize these results briefly it was found that the mice were most likely to switch from ongoing behaviour to grooming during periods that might be described as alert quietness. The probability of switching to grooming within 60 seconds after application of a peripheral irritant during active feeding, for example, was less than 0.17 whereas during the quiet–alert stage the probability of switching to grooming was 0.81. When the same peripheral irritants were applied during agonistic or fleeing behaviour the probability of switching to grooming within 60 seconds averaged less than 0.06. Further, it was possible to demonstrate that the probability of switching from feeding to grooming was reduced by a factor of more than 3 after the animals had previously been food deprived or were presented with particularly palatable food. Similarly, during periods of particularly vigorous bouts of fighting and fleeing the probability of switching to grooming was less than ¼ that observed during encounters judged as less intense by independent criteria (J. C. Fentress, unpublished). A similar drop in the responsiveness to grooming stimuli was previously found during the rapid execution of perseverant movement stereotypes (Fentress, 1965).

The second example concerns the threshold for centrally elicited grooming movements as a function of ongoing behaviour. We have found that it is possible to elicit face-washing sequences in mice through stimulating electrodes implanted in various extrapyramidal loci; e.g., globus pallidus, putamen (Fentress, 1972 and unpublished data). Again the threshold for elicited grooming is lowest during periods of quiet–alert. During periods of specifically oriented behavioural sequences, for example, the mean threshold for eliciting grooming was raised from 0.48 mA during quiet alert to a mean of 1.79 mA (50 c.p.s., 0.5 msec duration). Further, these figures discount trials during active ongoing behaviour where grooming could not be elicited by central stimulation at all.

The basic suggestion of these and related observations is that when a given behavioural system becomes strongly activated it increases its immunity from influence by extraneous factors defined in terms of other behavioural systems. This is further evidence that activated behavioural systems generate properties of self-organization.

Blocking effects of system A on system B. A complementary way to view the previous material is to ask the extent to which activation of system A blocks the expression of behaviour patterns defined in terms of system B. The issues turn out to be relatively complex and dependent upon the degree of system A activation as will be indicated more fully below. Here we can confine ourselves to situations in which independent measures indicate a strong activation of system A.

It is obvious that animals cannot do everything at once, and that the performance of one class of behaviour often precludes the expression of another. One cannot, for example, stand up and sit down simultaneously. The more fundamental issue, however, is that when independent evidence is obtained for strong activation of factors relevant to behavioural class A the probability of producing behavioural class B is frequently reduced even in the presence of factors that might otherwise be sufficient for the production of B.

Because this basic relationship between the activation of a given class of behaviour and the reduced probability of expression of a different class of behaviour has been reviewed extensively (e.g. Hinde, 1970; Fentress, 1973) a single illustration is sufficient here. In our work with timber wolves (*Canis lupus*) we have had several occasions to observe the influence of active exploratory behaviour upon food consumption. When the animals are placed in a novel environment they go through prolonged periods of active investigation. If food is placed in this environment it will be ignored until the investigatory phase of behaviour has been completed, a process which often takes several hours. This is true even when the animals are previously food deprived and are offered highly preferred dietary items (Fentress, 1967, and unpublished observations). One can think of this in terms of broadly expressed behavioural inhibition of other systems when a given class of behaviour is strongly activated (cf. above and Fentress, 1973). In this sense the expressed behaviour not only becomes self-organized from the perspective that it isolates itself from extraneous factors, but it also actively blocks the performance of activities normally associated with these extrinsic factors. The next question is whether there can also be positive associations between different classes of behaviour.

Activating factors

I have previously argued (Fentress, 1973) that the question of whether behavioural system A is facilitated or blocked by factors primarily related to behavioural system B can only be determined precisely when qualitative variables are paired systematically with both temporal and quantitative variables. In particular the model which I have suggested argues that moderate levels of activity in system B may facilitate the expression of behavioural class A, whereas higher levels of activation in B block the performance of A. This could be understood if low levels of activation of a given behavioural system produce relatively diffuse excitatory effects while higher levels of activation produce a tighter focus of activity, with the result that other forms of behaviour are blocked. Temporal variables can to a large extent be placed in the same framework if one assumes that the level of activation of a given behavioural system takes some period of time to reach its maximum and some period of time after the overt expression of behaviour to become dissipated (Fentress, 1972, 1973).

The essence of this position is that factors that are defined in terms of one behavioural system can be partially shared with factors defined in terms of another behavioural system. Whether the operational overlap between these systems is positive or negative depends in part upon the inferred level of activation within a system. At relatively low levels of activation the relationship between systems is more likely to be positive than at high levels of activation. This dynamic relationship between systems could be accounted for if (*a*) the excitatory boundaries of a given system (as measured by the range of behavioural outputs that are facilitated) are relatively broad during low periods of activation and more tightly focussed during high periods of activation and (*b*) the inhibitory boundaries of a given system tend to expand with increasing levels of activation.*

The model is designed to provide a conceptual orientation to problems of dynamic relations between different classes of behaviour, and obviously should not be equated with patterns of excitation and inhibition at physiological level of analysis. At the same time direct physiological analogues of these basic operational properties at the behavioural level can be found, which indicates that the basic concepts are compatible with information available at the physiological level (Fentress, 1973).

The issue of activating one class of behaviour by factors defined as primarily intrinsic to a different class of behaviour can provide complex statistical considerations in the normal performance of higher intact organisms (e.g. Fentress, 1967, 1968*a*, *b*, 1972, 1973). Therefore, for the purpose of illustration it is convenient to concentrate our attention upon simplified levels of organization. I shall therefore emphasize illustrative data on the control of (*a*) perseverant movement stereotypes, (*b*) sequential recovery of movement patterns after anaesthesia, and (*c*) the control of behaviour patterns following damage to the central nervous system.

Perseverant movement stereotypes. When animals are confined for long periods of time under restricted environmental conditions they frequently develop very stereotyped and perseverant forms of motor behaviour. Three sets of observations on the activation of these stereotypes are reported here. The first involves perseverant pacing behaviour by a captive Cape hunting dog housed at the Zoological Society of London. Observations over a period of six months indicated that this perseverant pacing could be facilitated by a variety of inputs normally associated with other dimensions of behaviour. For example, when zoo keepers approached with food the pacing behaviour increased prior to the execution of specifically oriented approach behaviour. Similarly, when groups of school children approached the pen, pacing in-

* At the behavioural level (*a*), this model can be tested by first determining the range of factors that contribute to different classes of behaviour. Once this is done experiments are performed where the animal is placed in circumstances in which one of these classes of behaviour is most likely to occur, but is, for example, not yet expressed, e.g. placement of food in front of a moderately 'hungry' animal. Then factors not normally associated with feeding can be introduced in different amounts and at different times. For example, one could apply a light pinch of the tail. The basic prediction is that a light tail pinch would facilitate feeding whereas a strong tail pinch would block feeding. This could then be repeated for a variety of classes of behaviour (e.g. tail pinch and copulation or drinking). Part (*b*) of the model can be tested with the prediction that animals who are actively engaged in one class of activity, such as feeding, will become less responsive to factors which normally underlie a different class of behaviour (e.g. turning in response to light tail pinch). In that sense it is possible to determine the extent to which performance of one behaviour 'inhibits' mechanisms underlying the performance of another.

creased prior to specifically oriented avoidance reactions. Various noises and visual stimuli presented experimentally by the author had a similar effect. If one postulates that the proximity of an extrinsic disturbance correlates with its effective intensity and thus its effectiveness in producing specifically oriented responses, then the conclusion which follows is that intermediate levels of the various disturbances have relatively diffuse excitatory effects which can facilitate a predominant behaviour such as pacing, whereas higher intensity and/or more proximal stimuli block pacing. When pacing is blocked by more intense stimuli specifically oriented behaviour patterns are observed. One can thus think of the facilitatory foci of behavioural systems as becoming more restricted in their expression as the inferred activation of these systems increases. Under these circumstances behaviour patterns which are facilitated at low levels of activation are blocked and replaced by more specifically oriented responses.

Two additional observations are of related interest. The first is that 'high intensity' (e.g. close proximity, loud) inputs frequently produced a momentary acceleration in pacing prior to suppression. The second is that *after* 'high intensity' inputs (which blocked pacing during their presentation) the perseverant pacing movements were frequently enhanced. It is from such observations that one can argue that (*a*) during moderate levels of activation behavioural systems can have relatively diffuse excitatory effects which can 'feed into' predominant output systems as in perseverant pacing, and (*b*) the effective intensity of activation of a given behavioural system follows definable temporal rules which are analogous to the phenomena of 'warm up' and 'after discharge' in the neurophysiological literature (cf. Fentress, 1973).

The second set of observations was made on weaving and jumping movement stereotypes in captive voles (Fentress, 1965). Here again a variety of factors which by independent assay could be shown to be primarily associated with other classes of behaviour facilitated the movement stereotypes if they were presented within moderate limits of intensity. Similarly, it was possible to observe both momentary facilitation of the weaving and jumping movements if a sudden strong 'irrelevant' stimulus was presented, and a more protracted facilitation of these stereotypes following the cessation of intense stimuli that produced specific orientation during their presentation.

My third set of observations on the activation of perseverant movement stereotypes involves electrical stimulation of the mesencephalic reticular formation in squirrel monkeys. Monkeys which had been implanted for other purposes (e.g. Fentress & Doty, 1971) were confined individually for long periods of time in small primate cages. During the course of experimentation two of the monkeys developed cage stereotypes. Prior to the development of these stereotypes stimulation of the mesencephalic reticular formation produced a variety of locomotor patterns, visual inspection of the hands, and feeding as a function of stimulus parameters and ongoing behaviour. Once the perseverant locomotor movements had become well established, however, stimulation at moderate intensities through the same leads elicited the perseverant movement patterns. Stronger stimulation blocked the movement stereotypes with subsequent rebound. Once again the impression gained was that moderate levels of activation could 'feed into' and facilitate the movement stereotypes whereas higher levels of stimulation blocked these same behaviour patterns. This is compatible with the hypothesis that low levels of activation of a given behavioural system can have relatively diffuse facilitatory effects whereas higher levels of activation result in a more tightly focussed range of excitation and an increased inhibitory surround (cf. Fentress, 1973).

Recovery of behaviour following anaesthesia. I shall give a single example here for it illustrates with particular clarity the major issues of activation and suppression of one behavioural system by another. When C57 mice recover from anaesthetic doses of a barbiturate (Nembutal) they can be made to display a sequence of three motor patterns: (1) hind leg scratch reflex, (2) gnawing on a food pellet placed in front of the face, and (3) face-grooming. The hind leg scratch reflex occurs first during recovery and can be elicited by placing the animal on its back and gently twisting the spine. The probability of this behaviour is increased markedly by a light pinch on the tail. A similar pinch on the tail increases the probability of gnawing and face washing during appropriate stages of the recovery sequence, whereas stronger pinches block these same behaviour patterns. It is obvious that tail pinch has no specific association with any of these motor patterns. The suggestion, therefore, is that the activation which results from moderate, but not strong, tail pinches has relatively diffuse excitatory properties which can facilitate different behavioural outputs that happen to be predominant at a particular time. With stronger tail pinches specific orienting responses are observed which block the performance of these predominant behaviour patterns. After the cessation of stronger tail pinches the predominant behaviour patterns are again facilitated (cf. Fentress, 1972).

Recovery of behaviour following lesions in the central nervous system. Teitelbaum and his associates have performed an extensive series of observations on the sequence of recovery from lateral and posterior hypothalamic lesions. One major consequence of these lesions in both rats and cats is the production of aphagia. Wolgin, Cytawa & Teitelbaum (1975) found that feeding could be elicited in aphagic animals by either pinching the tail or injection of amphetamine. They attribute this to a relatively non-specific activation. The amphetamine effects are of particular interest since the same 2 mg/kg dose in cats without lesions causes a sharp decrease in feeding. They argue that the hyperphagic effect of amphetamine is normally blocked by more specific anorexigenic effects which in turn prevent feeding. In rats with lesions, feeding could also be facilitated by placing drops of water on the animals' nose. This manipulation at first elicits grooming which subsequently spreads to feeding. Similarly the authors report that activation by a variety of inputs can reinstate the attack of cats with hypothalamic lesions upon rats.

The lesson of these observations is that factors normally associated with the production of one class of behaviour can under specified circumstances facilitate a different class of behaviour. The authors also note that during recovery from hypothalamic lesions 'stereotyped automatic movements' appear prior to more complex forms of behaviour. For example, grooming behaviour can be observed during early stages of recovery whereas more complex appetitive behaviour patterns recover much later. One striking finding is that rats recovering from lesions of the posterior hypothalamus can be made to groom by sinking them into a tank of warm water. Immersion in cold water, on the other hand, elicits specific escape responses. The suggestion of this latter observation is again that moderate activation of a behavioural system can lead to relatively diffuse excitation (which here feeds into the predominant behaviour of grooming) while more intense activation focusses the animals' response more specifically (which here is represented by escape).

Neuropharmacological data. Antelman and his colleagues (e.g. Antelman & Szechtman, 1975) have confirmed that tail pinches of moderate pressure can facilitate relatively stereotyped movement patterns such as feeding, gnawing, and licking. When appropriate objects are removed a more generalized hyperactivity or escape behaviour is observed. These workers suggest that the nigrostriatal dopamine system may play an important role in mediating the relatively non-specific activation by tail pinch since dopamine antagonists either block or reduce the observed effects. Valenstein (1976) reviews related data that support this hypothesis. He also reports that a greater variety of activated behaviour patterns can be evoked with tail pinching by manipulating the animals' environmental context (e.g. by removing goal objects). It should also be pointed out that dopaminergic pathways have been implicated in modulating sensitivity to sensory input (e.g. Ungerstedt, 1974). Since the major effect of dopaminergic activation reported is an increase in sensitivity to sensory (e.g. tactile) input this suggests a *positive* correlation with facilitation of stereotyped motor output. At one level this appears to contradict the *negative* correlation between movement and sensory responsiveness suggested in this paper. Further research is needed, however, for the consequences of motor performance *per se* upon receptivity to extrinsic factors has not been adequately studied from this perspective. The limitations to the generalization that 'output interferes with input' might be usefully approached at the neurobiological level through further evaluation of the nigrostriatal system in species behaviour (cf. Fentress, 1972).*

At moderate doses, and against an appropriate background, amphetamines can also increase the probability as well as vigour of relatively stereotyped movement sequences, such as grooming in rodents (Fentress, 1965 and unpublished observations). At higher doses even more highly stereotyped behaviour patterns such as sniffing, gnawing, and head turning predominate. Behaviour patterns such as grooming then occur during the recovery phase. These results have been supported by the more recent studies of Randrup & Munkuad (1970). Amphetamine releases both dopamine and norepinephrine. The former catecholamine appears most important, however, since direct DOPA injections can produce the stereotypes while inhibition of norepinephrine synthesis does not block the effects of amphetamine injections.

Conclusion. The data summarized above support the suggestion that factors which normally might be viewed as specifically relevant to one class of behaviour can under certain circumstances 'activate' another class of behaviour. This does not force the conclusion that these factors are entirely 'non-specific' (cf. Fentress, 1973). A more cautious view is that certain facilitatory factors can be shared between more than one behavioural system under appropriate circumstances. In this sense the behavioural systems overlap. Several of the data also support the hypothesis that behavioural systems become more tightly organized as they are activated beyond a minimal level. This is measured by an increased tendency to block rather than to facilitate other classes of behaviour. One can postulate, therefore, that with increased 'activation' (e.g. measured by stimulus strength), the *relative* specificity of a given system, as indicated by its range of facilitatory action, increases; and thus the facilitatory overlap between behavioural systems decreases: i.e. a change in the relationship of behaviourally measured boundaries of facilitation and suppression. In summary, with increased 'activation' a behavioural system becomes progressively more constrained, self-organized, and effective in blocking the expression of other behavioural systems.

I have argued previously that it is often valuable to consider temporal factors in conjunction with models of 'activation levels' and the 'tightness'

* Complex relations between efferent excitation and inhibition of afferent informaton are reviewed in Somjen, 1972.

(specificity, overlap) of integrative processes. For example, both anticipation of and recovery from relatively intense forms of stimulation may result in the facilitation of an increasingly broad band of behaviour patterns as one moves away in time from the actual stimulus presentation (Fentress, 1972, 1973). To illustrate, it has recently been found that aggressive behaviour (Archer, 1976) and mounting behaviour (Hanby, 1974) can be facilitated some time *after* a variety of environmental disturbances. Similarly, Killeen (1975) reports that 'general activity' initially increases during inter-reinforcement intervals then drops with the advent of more specifically oriented competing behaviour.

It may prove useful to compare these data to the distinction made by Staddon & Simmelhag (1971) on the inter-reinforcement interval relationship between 'interim behaviours' and 'terminal behaviours'. These latter observations are compatible with the idea that temporal as well as intensity factors may contribute to the operational boundaries of integrative systems. The question of the generality of these properties as well as the extent to which temporal, quantitative, and qualitative variables can be interchanged in a formalized quantitative model must, however, await further experimentation on a variety of species, a variety of behaviour patterns, and a variety of situations. The question of *symmetry* in relations between two or more separately classified behaviour systems is not explored explicitly in the present paper. It is safest to assume that the effects of 'system A' upon 'system B' may be quite distinct from the effects of 'system B' upon 'system A'. Within these limits, however, similar organizational principles can be postulated. The basic arguments to this point are summarized in Table 1.

PHENOTYPIC SUBSTRATES AND CAPACITIES

One approach to the problem of generalization is to seek dimensions of behavioural organization that can be applied to different specific activities. With reference to the question of integrative boundaries that underlie patterned behaviour, two such dimensions are *stereotyping* and *processing capacity*. I think that these can be related to each other as well as to the problems of dynamic organization in behaviour outlined previously.

The basic arguments are these: (1) Frequently repeated and stereotyped behaviour patterns are to a large extent centrally programmed ('pre-organized') and require less processing capacity than do more complex and less often repeated patterns of behaviour. (2) These stereotyped behaviour patterns frequently occur during transitions between more complex goal directed activities, during and after relatively extreme levels of 'arousal', and following various types of neurological insult – in part because of their minimal process-

Table 1. *Summary perspective on dynamic boundaries*

1. General statement
   - A. Behavioural boundaries can appear blurred (indistinct) at descriptive level
   - B. Boundaries are abstracted by criteria and analysis levels employed
   - C. Control boundaries may overlap with one another to varying degrees
   - D. Boundaries may display dynamic shifts of interaction and self-organization

2. Interactive/self-organizing systems
   - A. During low level, initial, or terminal activation, systems tend toward diffuse organization
     1. A wide variety of behavioural outputs may be facilitated (neither total non-specificity nor symmetry among systems is implied)
     2. The system is responsive to a relatively broad range of influences, which in turn further define the system's boundaries
     3. Behavioural outputs are loosely organized and easily disrupted
   - B. As activation (organization) of system increases, its dynamic structure becomes more tightly constrained
     1. Afferent information less necessary to performance
     2. Performance less easily disrupted by extrinsic factors
     3. Sequential structure of system outputs increases
     4. Increased tendency to block rather than facilitate behaviour patterns associated with other systems

3. Some common (not invariable) correlations

   Speed→amplitude→sequential structure of system outputs→reduced dependence upon system inputs→increased immunity from disruption by extrinsic factors→increased blocking of extrinsic inputs and outputs

ing demands. (3) The tightness of sequential structuring of behaviour in a more general sense is also enhanced by overall processing demands and capacity limitations. (4) During these periods underlying integrative processes tighten their intrinsic organization in the sense that they become both more immune from influence by and less dependent upon extrinsic factors (which may be either occluded or actively inhibited). One might also tentatively postulate that in 'higher' vertebrates subcortical (including brain stem) mechanisms are largely responsible for the articulation of stereotyped output sequences (cf. Antelman & Szechtman, 1975; Vanderwolf, 1975; Wolgin *et al.*, 1975; Valenstein, 1976), and that integration via these structures involves a less complex form of processing (defined in terms of the behavioural capacities of the organism as a whole) than does the mediation of behaviour via cortical mechanisms. I recognize that statements of this general nature may have important exceptions. For example, certain species-characteristic movement sequences may be performed relatively infrequently and still be centrally programmed (cf. point 1). However, exceptions test the generality of the rule, and are often clarified if a broad framework is provided initially.

It should be clear that these formulations are both exploratory and interconnected. They are presented here as a 'growing point' (which I suppose becomes a blob!) in my own thinking as requested by the organizers of this conference. It should be made explicit that such general statements do not in any way negate the importance of particular mechanisms that differentiate the specific control of one dimension of behaviour from another. The suggestion, however, is that in addition to these specific mechanisms there may be common features at another, and admittedly more abstract, level of organization.

Elicitation and control of stereotyped behaviour patterns

It is well established in the ethological literature that specific high probability stereotyped behaviour patterns frequently occur during periods of 'conflict' or 'thwarting'. Because these stereotyped movements often bear no obvious functional relationship to other dimensions of the overall behavioural context they have been given a special descriptive label, i.e. 'displacement activity' (reviewed in Fentress, 1973). The same behaviour patterns can often be observed also during or after periods of inferred 'arousal' or drowsiness (e.g. Delius, 1967; Fentress, 1968*a*, *b*; Andrew, 1974). Similarly, stereotyped or 'automatic' behaviour patterns often appear during electrophysiologically inferred periods of cortical deactivation (e.g. Caspers, 1961; Vanderwolf, 1975), after direct blocking of cortical activity as by topical application of KCl (e.g. Huston & Bures, 1970; Andrew, 1974; J. C. Fentress, unpublished), and during early stages of recovery from surgical insult of the brain (e.g. lesions of the hypothalamus; Wolgin *et al.*, 1975). It is reasonable to postulate that behavioural situations that are variously defined in terms of 'conflict', 'thwarting', 'stress', 'overload', etc., reflect a processing capacity which is reduced with respect to any particular dimension of behaviour (cf. Broadbent, 1971) and therefore bears some analogy to direct physiological or surgical disruption of brain function. If this makes sense, then there is circumstantial evidence for the suggestion that stereotyped behaviour patterns appear when they do *partially because* these stereotyped behaviour patterns place fewer processing demands on the organism. In this sense we can think of stereotyped behaviour patterns as being 'pre-organized'.

There is indeed a large body of data on human performance that supports this basic contention (e.g. Broadbent, 1971; Kornhuber, 1974). When we try to perform two skills simultaneously that are either complex or newly acquired we find that they tax our capacity in the sense that the performance of one interferes with the performance of another. It is not wise to carry on an esoteric conversation with a novice driver! Further, complex and newly acquired skills are easily disrupted by conflict, stress, fatigue, neurological insult, etc. An interesting fact is that during these situations one frequently observes (*a*) an increased probability of stereotyped and well-established behavioural patterns, (*b*) an overall increase in the stereotypy of sequentially articulated activities, and (*c*) a reduced sensitivity to extraneous inputs (cf. Luria, 1966; Broadbent, 1971; Fentress, 1973; and previous sections).

A simple way to summarize much of the literature on the elicitation of 'functionally irrelevant' stereotyped behaviour patterns in animals is that they often occur when the animals act as if they 'feel' they should do something but do not know what to do. Granted the vagueness of such terminology this general picture might be expected if the processing demands for stereotyped movements are less than for more complex forms of behaviour. This could occur if the performance of stereotyped behaviour patterns in minimally dependent upon sensory regulation or modulation by 'higher' regions of the central nervous system (cf. Dawkins, this volume).

Obviously specific factors are relevant to the question of why certain stereotyped movements occur in one situation and others in another (cf. Tinbergen's (1952) discussion of these issues in the context of displacement behaviour). Still, as we have seen above, once stereotyped movements are elicited through interactions among different integrative systems they often display a remarkable degree of intrinsic organization. If tight intrinsic structuring of behaviour involves less complex processing, then situations which increase demands for processing – such as increased speed of sequentially articulated movements – should be expected to reduce the organism's sensitivity to extraneous variables as well as to produce simpler rules of simultaneous and sequential coupling of component acts. We have seen above that this commonly occurs. We would also expect to find that reduction in processing capacity as through neurological insult often has similar effects, which is indeed the case (e.g. Luria, 1966).

Conclusion

The simplest way to summarize the suggestions of this section is that when the processing demands of the organism begin to outstrip the organism's processing capacities there is an increased probability that (*a*) ongoing sequences of behaviour will become more stereotyped, tightly organized, less flexible, and more autonomous, and/or (*b*) phenotypically well established and 'pre-organized' stereotyped movement sequences will occur. The rules underlying (*a*) also apply to (*b*) as we have seen, for example, in the changes

Table 2. *Well-established and stereotyped behaviour patterns*

1. Demand less processing capacity for performance
2. Relatively independent from sensory guidance (i.e. centrally controlled)
3. Often appear when processing capacity reduced or approaching limit, and/or when sources of sensory modulation reduced
4. Provide convenient 'simplified' model system for study of behavioural integration
5. Emphasize importance of capacity limitations and phenotypic structure in studies of integrated behaviour

of sequential structuring and sensitivity to extrinsic factors (defined either in terms of exogenous stimuli or other ongoing behavioural processes) with changes in speed of performance, attentional demands, etc. Processing demands versus capacities can be related to such variables as number of responses required per unit time (e.g. performance speed), number of tasks present at any given time (e.g. attentional demands), levels of 'activation', 'drowsiness', etc., and neurological insult. This indicates further that quite diverse dimensions of behaviour may compete for the same (limited capacity) mechanism(s), and thus emphasizes the necessity for examining the dynamic interplay between behavioural processes at the organism level if we are to understand the operation of individual components.

I fully appreciate that the details of these arguments may have to undergo revisions as a consequence of future research, and I suspect that even then there will be limitations of their general applicability to problems of patterned behaviour. Yet I am confident on the fundamental point that we must combine considerations of phenotypic substrates and processing capacity in our models of dynamic patterns of behavioural integration. Indeed, this may be one important way to anchor our models so that we can do better justice to the dynamic properties of integrative systems, and the dynamic principles of relationship among these systems. The boundaries of behaviour may be more fluid than we suspect in terms of the underlying phenotypic capacities that constitute the diverse species we both investigate and appreciate. Further detailed studies on the context of occurrence and control of well-established species-typical and acquired stereotyped movement sequences can help us to clarify further both the integrative dynamics of organization in motivational systems and the relationships between transient expressions of behaviour and more permanent rules of phenotypic organization. Five basic points about stereotyped behaviour and phenotypic capacities are summarized in Table 2.

DISCUSSION FOOTNOTES

Function

An issue raised at the conference was 'why' (functionally) animals might be built this way, particularly with reference to stereotyped movements. Such questions are of obvious interest but are difficult to confront with precision. The first argument for shifting boundaries in a general sense is that this would provide an economy of underlying (e.g. neuronal) machinery. If the same element can participate with other elements in different configurations, and if the behavioural output depends upon configurational properties of the 'system as a whole', then considerable savings could be obtained if a given element participates in more than one class of behaviour. This basic principle of a given element participating in a variety of higher order systems is supported at the neurophysiological level in both vertebrate and invertebrate preparations (e.g. Doty, 1976; Hoyle, 1976; Kennedy, 1976). Secondly, if the behavioural demands of an organism typically fall within a restricted range, limitations in overall processing capacity would balance favourably with the economy of biological encoding that such processing limitations make possible. Laboratory experiments which 'overload' the animal are convenient for determination of organizational principles, but it is important to recognize that these limitations may be of little practical import in the animal's normal environment. Thirdly, if it is indeed true that animals revert to stereotyped behaviour patterns under conditions of 'processing overload', then these behaviour patterns themselves provide a means for coping with a potentially disadvantageous situation. From a functional perspective this viewpoint gains added respectability when one assumes that often repeated motor acts themselves are likely to have worked effectively in the past. Fourthly, if the performance of a behavioural act itself blocks effective processing of certain forms of input this can prevent disruption of the act before it reaches its functional end-point. Here the reader is expected to realize that the blocking of potentially irrelevant or disruptive inputs does not imply necessary blocking of relevant information. It should be clear that this distinction was not emphasized above in order to simplify the main line argument. Valenstein (1976) has independently argued that the performance of stereotyped behaviour patterns (which he relates to activity in the nigrostriatal dopamine system) may help to reduce the variety of stimuli that an organism would otherwise process, thus providing a potential mechanism for coping with stimulus events which might otherwise block effective action.

Epilogue on dynamic pattern

The basic thoughts I have expressed about interactive/self-organizing systems in integration appear similar to the discussions of dynamic pattern recently developed in biophysics (e.g. Katchalsky, in Katchalsky *et al.*, 1974). To be properly understood, the construction of dynamic pattern must be evaluated in an appropriately abstract form in which the main emphasis is upon 'information space' as opposed to geometric space (e.g. Eigen, 1971). In Katchalsky's terms we are dealing with an issue of 'symmetry breaking' in which initially 'homogeneous' fields are redefined by successive discontinuities (specificities). The model proposed in this paper stresses the properties of self-organization among a subset of elements within the field as a function of initial interactions between these elements in a manner strikingly similar to (since it was derived independently of) the models of self-organization at the molecular level discussed by Eigen. The initial 'homogeneous' phase becomes structured, but the structure itself is defined in terms of dynamic stabilities of the 'information space'. What the present model does is to refer the issue of symmetry breaking via self-organization to hypothesized involvement of the participating units, with concominant coupling among some elements and self-isolation among others. This is the essence of the terminology of activation levels. While it is obviously imperfect and incomplete in detail, the construct of activation as used here does permit several explicit experimental referents by which further refinements can be obtained. The basic idea is that the system properties ('boundaries') become constrained as a function of interactions among participating elements, which in turn depend upon actions within elements.

For future progress in this line of thinking it will be necessary not only to work on the question of dynamic patterning within the framework of integration, but to relate these features of behavioural expression to the more stable boundaries provided by the organism's phenotypic structure. This can obviously be done in a trivial way, such as by noting that birds of one species articulate a different song than birds of another, and that individual experience may play a greater or lesser role in different species in the formulation of these song 'templates'. The question is what generality such specific statements can be expected to have. Thus I have attempted a complementary and more abstract formulation in terms of phenotypic capacities which are viewed as setting the ultimate constraints upon performance regardless of the particulars of expression at any given instant of time. To make this more concrete, I have given several examples of stereotyped and well-established motor activities (i.e. highly structured dynamic patterns).

I was interested to learn that Katchalsky also emphasized the importance of 'context' in defining system or element properties in contrast to the strict 'reductionist' view in which subsets of systems or elements are considered to display their basic properties ('boundaries') in a void, i.e. independent of extrinsic factors with which they are actually or potentially interconnected (cf. p. 137).

An apparent difference between the concept offered here and the models by Katchalsky is the greater emphasis he placed upon sudden transitions or 'jumps' from one multistable state to another. I have placed my arguments within a more continuous spatio-temporal framework, and it is possible that this will demand correction as analysis proceeds. Certainly in terms of overt behavioural expression we perceive discontinuities in order. These are indeed the essence of our ability to distinguish one act or set of acts from another. First, however, apparent discontinuities in overt expression do not necessarily imply discontinuities of underlying cause; after continuous heating, water may 'break into' a boil. Secondly, multi 'discontinuous' events may act together at the level of mechanism, producing higher-level phenomena that appear more continuous; molecules of black paint mixed with molecules of white paint may appear homogeneously grey (cf. Fentress, 1973, on 'multispecificity' as an alternative to non-specific models of behaviour). Thirdly, even at the descriptive level in behaviour, such as in the notation of movement, individual dimensions often interleaf in such a way that it is difficult to separate one act clearly from another (cf. Golani, 1976, for an excellent discussion of this issue).

SUMMARY

(1) Integrative processes in patterned behaviour can be thought of as having both changing and overlapping boundaries that lead to systematic principles of interaction and self-organization.

(2) These dynamic processes of integration are superimposed upon and constrained by more stable phenotypic dimensions of response coding and processing capacity.

I acknowledge with pleasure that my early thinking about problems of dynamic pattern in species behaviour and its relation to the potentialities and constraints of phenotypic structure was to a great extent inspired and guided by my teachers and colleagues at Madingley. It was particularly nice to have the opportunity to 'try again' in the context of the present conference. In spite of all its imperfections I would like to dedicate this paper to Bill Thorpe, a man who as a scientist, teacher, and human being has been a lasting inspiration to me.

The ideas I attempt to express here were collated in the stimulating atmosphere of my colleagues at Dalhousie University, and I owe particular thanks to Bob Adamec, Graham Goddard, Ilan Golani and Heather Parr for their critical discussion. I also thank each of the participants of the present conference who joined together in the spirit of co-operative exploration and debate. The research reported here was supported by National Institutes of Health grant MH 16955 and National Research Council of Canada grant A9787.

REFERENCES

Andrew, R. J. (1974). Arousal and the causation of behaviour. *Behaviour*, **51**, 135–165.

Antelman, S. M. & Szechtman, H. (1975). Tail pinch induces eating in sated rats which appears to depend on nigrostriatal dopamine. *Science, Washington*, **189**, 731–733.

Archer, J. (1976). The organization of aggression and fear in vertebrates. In *Perspectives in Ethology*, vol. **2**, ed. P. P. G. Bateson & P. H. Klopfer. Plenum Press: New York, in press.

Bohm. D. (1969). Some remarks on the notion of order. Further remarks on order. In *Towards a Theoretical Biology*, vol. **2**, ed. C. H. Waddington, pp. 18–60. Aldine: Chicago.

Broadbent, D. E. (1971). *Decision and Stress*. Academic Press: New York.

Brown, G. Spencer (1969). *Laws of Form*. George Allen & Unwin: London.

Carpenter, D. O. & Rudomin, P. (1973). The organization of primary afferent depolarization in the isolated spinal cord of the frog. *Journal of Physiology*, **229**, 471–493.

Caspers, H. (1961). Changes of cortical D.C. potentials in the sleep–wakefulness cycle. In *The Nature of Sleep*, ed. G. E. W. Wolstenholme & M. O'Connor, pp. 237–253. Churchill: London.

Dawkins, M. & Dawkins, R. (1974). Some descriptive and explanatory stochastic models of decision-making. In *Motivational Control Systems Analysis*, ed. D. J. McFarland, pp. 119–168. Academic Press: New York.

Delius, J. D. (1967). Displacement activities and arousal. *Nature, London*, **214**, 1259–1260.

Doty, R. W. (1976). The concept of neural centers. In *Simpler Networks and Behavior*, ed. J. C. Fentress, pp. 251–265. Sinauer Associates: Sunderland, Mass.

Eccles, J. C., Ito, M. & Szentagothai, J. (1967). *The Cerebellum as a Neuronal Machine*. Springer-Verlag: New York.

Eigen, M. (1971). Molecular self-organization and the early stages of evolution. *Quarterly Review of Biophysics*, **4**, 149–212.

Fentress, J. C. (1965). Aspects of arousal and control in the behaviour of voles. Ph.D. thesis, Cambridge University.

Fentress, J. C. (1967). Observations on the behavioral development of a hand-reared male timber wolf. *American Zoologist*, **7**, 338–351.

Fentress, J. C. (1968*a*). Interrupted ongoing behaviour in two species of vole (*Microtus agrestis* and *Clethrionomys britannicus*). I. Response as a function of preceding activity and the context of an apparently 'irrelevant' motor pattern. *Animal Behaviour*, **16**, 135–153.

Fentress, J. C. (1968*b*). Interrupted ongoing behaviour in two species of vole (*Microtus agrestis* and *Clethrionomys britannicus*). II. Extended analysis of motivational variables underlying fleeing and grooming behaviour. *Animal Behaviour*, **16**, 154–167.

Fentress, J. C. (1972). Development and patterning of movement sequences in inbred mice. In *The Biology of Behavior*, ed. J. Kiger, pp. 83–132. Oregon State University Press: Corvallis.

Fentress, J. C. (1973). Specific and nonspecific factors in the causation of behavior. In *Perspectives in Ethology*, ed. P. P. G. Bateson & P. H. Klopfer, pp. 155–224. Plenum Press: New York & London.

Fentress, J. C. (ed.) (1976). *Simpler Networks and Behavior*. Sinauer Associates: Sunderland, Mass.

Fentress, J. C. & Doty, R. W. (1971). Effect of tetanization and enucleation upon excitability of visual pathways in squirrel monkeys and cats. Experimental *Neurology*, **30**, 535–554.

Fentress, J. C. & Stilwell, F. P. (1973). Grammar of a movement sequence in inbred mice. *Nature, London*, **24**, 52–53.

Golani, I. (1976). Homeostatic motor processes in mammalian interactions – a choreography of display. In *Perspectives in Ethology*, vol. **2**, ed. P. P. G. Bateson & P. H. Klopfer, pp. 69–134. Plenum Press: New York.

Goodwin, B. C. (1970). Biological stability. In *Towards a Theoretical Biology*, **3** *Drafts*, ed. C. H. Waddington, pp. 1–18. Aldine: Chicago.

Hanby, J. (1974). Male–male mounting in Japanese monkeys (*Macaca fuscata*). *Animal Behaviour*, **22**, 836–849.

Hinde, R. A. (1959). Unitary drives. *Animal Behaviour*, **7**, 130–141.

Hinde, R. A. (1970). *Animal Behaviour: A Synthesis of Ethology and Comparative Psychology*, 2nd edn. McGraw–Hill: New York.

Hinde, R. A. (1972). *Social Behavior and its Development in Subhuman Primates*. University of Oregon Press: Eugene.

Holst, E. von (1948). Von der Mathematik der nervösen Ordnungsleistung. *Experientia*, **4**, 374–381.

Hoyle, G. (1976). Approaches to understanding the neurophysiological bases of behavior. In *Simpler Networks and Behavior*, ed. J. C. Fentress, pp. 21–38. Sinauer Associates: Sunderland, Mass.

Huston, J. & Bures, J. (1970). Drinking and eating elicited by cortical spreading depression. *Science, Washington*, **169**, 702–704.

Ikeda, K. (1976). Genetically patterned neural activity. In *Simpler Networks and Behavior*, ed. J. C. Fentress, pp. 140–152. Sinauer Associates: Sunderland, Mass.

Jasper, H. H. (1963). Studies of non-specific effects upon electrical responses in sensory systems. *Progress in Brain Research*, **1**, 272–293.

Katchalsky, A. K., Rowland, V. & Blumenthal, R. (1974). *Dynamic Patterns of Brain Cell Assemblies. Neurosciences Research Program Bulletin*, **12**, no. 1.

Keele, S. W. (1968). Movement control in skilled motor performance. *Psychological Bulletin*, **70**, 387–403.

Kennedy, D. (1976). Properties of neural elements in relation to network function. In *Simpler Networks and Behavior*, ed. J. C. Fentress, pp. 65–81. Sinauer Associates: Sunderland, Mass.

Kennedy, D., Calabrese, R. L. & Wine, J. J. (1974). Presynaptic inhibition; primary afferent depolarization in crayfish neurons. *Science, Washington*, **186**, 451–454.

Killeen, P. (1975). On the temporal control of behavior. *Psychological Review*, **82**, 89–115.

Konorski, J. (1967). *Integrative Activity of the Brain*. University of Chicago Press: Chicago.

Kornhuber, H. H. (1974). Cerebral cortex, cerebellum, and basal ganglia: an introduction to their motor functions. In *The Neurosciences: Third Study Program*, ed. F. O. Schmitt & F. G. Worden, pp. 267–280. MIT Press: Cambridge, Mass.

Krasne, F. B. & Bryan, J. S. (1973). Habituation: regulation through presynaptic inhibition. *Science, Washington*, **182**, 590–592.

Lashley, K. S. (1951). The problem of serial order in behavior. In *Cerebral Mechanisms in Behavior*, ed. L. A. Jeffress, pp. 112–136. Wiley: New York.

Llinás, R. & Walton, K. (1976). A simple neuronal system in vertebrate brain. In *Simpler Networks and Behavior*, ed. J. C. Fentress, pp. 274–279. Sinauer Associates: Sunderland, Mass.

Luria, A. R. (1966). *Higher Cortical Functions in Man*. Basic Books: New York.

Mates, J. (1973). Patterning of eye movements in the chameleon. Ph.D. thesis, University of Oregon.

Miller, G. A. & Frick, F. C. (1949). Statistical behavioristics and sequences of responses. *Psychological Review*, **56**, 311–324.

Murphey, R. J. & Palka, J. (1973). Efferent control of cricket giant fibres. *Nature, London*, **248**, 249–251.

Northup, L. (1973). A study of temporal structure of behavior in inbred mice, with special reference to the organization of face grooming. Ph.D. thesis, University of Oregon.

Pearson, K. G. (1972). Central programming and reflex control of walking in the cockroach. *Journal of Experimental Biology*, **56**, 173–193.

Randrup, A. & Munkuad, I. (1970). Biochemical, anatomical and psychological investigations of stereotyped behavior induced by amphetamines. In *Amphetamines and Related Compounds*, ed. E. Costa & S. Garratini, pp. 695–713. Raven Press: New York.

Reiner, J. M. (1968). *The Organism as an Adaptive Control System*. Prentice-Hall: Englewood Cliffs, N.J.

Russell, I. J. (1971). The role of the lateral-line efferent system in *Xenopus laevis*. *Journal of Experimental Biology*, **54**, 621–641.

Schleidt, W. M. (1974). How 'fixed' is the fixed action pattern? *Zeitschrift für Tierpsychologie*, **36**, 184–211.

Somjen, G. (1972). *Sensory Coding in the Mammalian Nervous System*. Appleton-Century-Crofts: New York.

Staddon, J. E. R. & Simmelhag, V. L. (1971). The 'superstition' experiment: a re-examination of its implications for the principles of adaptive behavior. *Psychological Review*, **78**, 3–34.

Thompson, R. F. (1967). *Foundations of Physiological Psychology*. Harper & Row: New York, Evanston & London.

Tinbergen, N. (1951). *The Study of Instinct*. Clarendon Press: Oxford.

Tinbergen, N. (1952). 'Derived' activities; their causation, biological significance, origin, and emancipation during evolution. *Quarterly Review of Biology*, **27**, 1–32.

Ungerstedt, U. (1973). Brain dopamine neurons and behavior. In *The Neurosciences: Third Study Program*, ed. F. O. Schmitt & F. G. Worden, pp. 695–703. MIT Press: Cambridge, Mass.

Valenstein, E. S. (1976). Stereotyped behavior and stress. In *Psychopathology of Human Adaptation*, ed. G. Serban & J. W. Mason. Plenum Press: New York, in press.

Vanderwolf, C. H. (1975). Neocortical and hippocampal activation in relation to behavior: effects of atropine, eserine, phenothiazines, and amphetamine. *Journal of comparative and physiological Psychology*, **88**, 300–323.

Wall, P. D. (1968). Organization of cord cells which transmit sensory cutaneous information. In *The Skin Senses*, Proceedings of the 1st International Symposium, 1966, ed. D. R. Kenshalo, pp. 512–533. Thomas: Springfield, Ill.

Wine, J. J. & Krasne, F. B. (1972). The organization of escape behavior in the crayfish. *Journal of experimental Biology*, **56**, 1–18.

Wolgin, D. L., Cytawa, J. & Teitelbaum, P. (1975). The role of activation in the regulation of food intake. (Paper presented at a conference entitled: *Hunger: Basic Mechanisms and Clinical Implications*). Raven Press: New York, in press.

Wurtz, R. H. (1969). Response of striate cortex neurons to stimuli during rapid eye movements in the monkey. *Journal of Neurophysiology*, **32**, 975–986.

Zack, S. (1975). A description and analysis of agonistic behavior patterns in an opisthobranch mollusc, *Hermissenda crassicornis*. *Behaviour*, **53**, 238–267.

Zucker, R. S. (1972). Crayfish escape behavior and central synapses. II. Physiological mechanisms underlying behavioral habituation. *Journal of Neurophysiology*, **35**, 621–637.

5

Sound production and perception in birds as related to the general principles of pattern perception

W. H. THORPE AND J. HALL-CRAGGS

Recent studies on vocal communication in birds leads us to the conviction that many bird utterances, and in particular song in its most highly developed forms, offer highly significant material for the study of the general principles of pattern perception. We present a number of examples to support this conclusion, drawn from various sources but particularly from new material of our own, such as that supplied by the blackbird (*Turdus merula*). By no means all the examples are as thoroughly worked out as is desirable, but they do at least seem to hold out real promise for further study.

It is generally assumed that Gestalt psychology, which initiated the development of modern work on this topic, started with Wertheimer's classic paper in 1912. His work showed that the psychological and physiological concepts of the time were inadequate to account for his experimental results on visual perception. The full impact of this work was delayed by the first World War and not until publications by two associates of Wertheimer, namely Kurt Koffka (1924) and Wolfgang Koehler (1925), began to have their effect, were the far reaching implications of the new work understood.

To turn to the present day, in contrast to the 1940s and 1950s, one finds few references to the term 'Gestalt' in *Psychological Abstracts*. The reason for this is that there is no longer an 'orthodox' psychology and a 'Gestalt' psychology – rather field theories (although still debated) – have become an accepted part of the subject. To quote just one modern textbook (Chaplin & Krawiec, 1974), 'Since these earlier more or less classic experiments Gestalt Psychology has made its influence felt in ever widening circles... Indeed it is impossible today to find a chapter on perception in any general or experimental text-book of psychology that does not show the influence of this school'.

The word 'Gestalt' implies in ordinary German usage 'form' or 'shape'; but in psychological literature the word 'configuration' is usually employed as the English equivalent. But quite early on (1925) the concept of 'wholeness', which is also implicit in some of the colloquial German uses of the term, was introduced by Koehler as an absolutely essential feature of 'Gestalten'. Gibson extends this by saying that 'a Gestalt is a separate whole whose characteristics are determined by its configuration'. From this it follows that the units are percepts, not separate sensa or sense-data; and whether or not the elements first perceived are the parts or the whole, the meaning and relevance of the parts depend on the configuration of the whole. But no matter how intensive the analysis, the data are still percepts in that they are relational not absolute. Thus, strictly speaking, 'sensation' has disappeared, and we are left merely with perceptions of varying degrees of complexity. This idea was admirably summed up by Gibson in 1952 who said, 'The perceptual impression is the primary one, immediate and independent; the sensory impression is the secondary one, obtainable only by analysing the perception' – a view now supported by the physiological data on the visual system.

The Gestalt theorists, particularly in their early years, enunciated a multiplicity of 'laws' and 'factors', by no means all of which were ultimately found to be either helpful or necessary. Some were found very difficult to define in such a manner as to mean the same to all who used them; for the mere fact of giving names to perceptual experiences did not guarantee that such experiences were really identical for all people placed in the same situation (see Broadbent, 1961).

What then are the Gestalt laws that are relevant and important for us now? (1) The prime law is claimed to be that of '*Prägnanz*', according to which 'psychological organisation will always be as good as the controlling circumstances permit'. Here it should be explained that a 'good Gestalt' refers *inter alia* to such characteristics as regularity, symmetry, inclusiveness, unity and conciseness. Thus, in effect, the Law of Pregnance suggests the direction of events. Psychological or perceptual organisation tends to move in one general direction rather than in another direction, always towards the state of Pregnance, towards the 'good Gestalt'. A diagram which shows an interrupted circle or triangle, or one made up of a number of dots, but with the contour left with gaps in it, tends to be seen as closed. A figure which is not entirely symmetrical may nevertheless seem symmetrical if the observer does not examine it too closely; and rapid and successive stimulation of three triangularly located spots on the skin produces the impression of a circle. In each instance mentioned the subject organises the stimulus situation into 'an experience as good as controlling circumstances will permit. They are examples of the Law of Pregnance as it applies to pure perception' (Katz, 1951, p. 41).

Following on this primary Law of Pregnance we have (2) the Law of *Similarity*, which is a principle determining the formation of groups of perceptions, such as groups of lines or dots. When more than one kind of element is present, those which are similar tend to be seen as groups. This applies to items which are similar in form or in colour, or which show similar transitions, e.g. are alike in the steps separating them. A series of experiments with nonsense syllables, two-digit numbers and nonsense two-dimensional figures carried out by Koehler (1941) showed quite conclusively that similar (homogeneous) pairs were much more readily learned than dissimilar (heterogeneous) ones. The former illustrates the law of similarity. This idea has been much developed and enlarged by Restle (1970) who finds universal organising tendencies in sequential transitions.

Next comes the Law of *Proximity* (3). Perceptual groups are favoured according to the nearness of the parts. Thus if several parallel lines are spaced unevenly on a page, those nearer together will tend to form groups against a background of empty space. Patterning through proximity also holds within audition, as within the grouping of successive clicks: here the proximity is, of course, a temporal one.

The Law of *Closure* (4). Other things being equal, lines which enclose a surface tend to be seen as a unit; so that closed areas are more stable than unclosed ones and therefore more readily form figures in perception. In a problematic situation the whole is seen as incomplete and a tension, according to the law of pregnance, is set up towards completion. This strain to complete is an aid to learning, and to achieve closure is 'satisfying'.

Finally we have (5) the Law of *Good Continuation* or of *Common Movement*. This states that organisation in perception tends to occur in such a manner that a straight line appears to continue as a straight line, a part circle as a circle, and so on, even though many other kinds of perceptual structuring would be possible. It also states that elements are grouped in perception when they move simultaneously and in a similar manner. For instance if we have two projectors each projecting groups of dots on to the same part of a screen so that they form one haphazard collection; then, when one projector is moved the dots projected by it combine immediately and are set apart from the other, motionless, set of dots; and as long as motion continues the two groups do not merge. If, however, both projectors are moved at the same time in a different manner, for instance, one along a straight path and the other along a curved one, the two groups of dots are seen simultaneously in the course of separate movements, with the projection screen as a stationary point of reference (Katz, 1951, p. 27).

A glance at the literature will show that at least 90% of the original outburst of experimental work on Gestalten dealt with visual patterns and only by implication with other forms of perception. But there was an earlier worker, an Austrian, Christian von Ehrenfels, a professor in Vienna and later in Prague, who anticipated the Gestalt psychology with what he called 'Gestalt form-quality' and who, being an excellent musician, was particularly concerned with sound patterns. In 1890 he pointed out that melodic transposition means that a melody exists independently of the tones which constitute it. Again he remarked that even Ernst Mach, who was the apostle of pure sensation, when he talked about the direct perception of a sound-Gestalt, was 'obviously speaking of perception in a sense that differed from the ordinary'. For though the ordinary person thinks of the perception of spatial wholes as immediate, he seldom or never perceives melodies in the same way.

From facts such as these it might have been thought that von Ehrenfels would have been regarded as the real founder of Gestalt psychology. But he was primarily a logician who expanded his theories into a form which somewhat obscured their general significance for psychology. The unfortunate result was that the Gestalt characteristics of hearing were relatively neglected. This was regrettable because von Ehrenfels' work, from the point of view of hearing, was original and far in advance of its time, and it is this that has led us to consider it worthwhile to look again at the sound patterns of birds in the light of Gestalt views in general and of von Ehrenfels' work in particular (von Ehrenfels, 1890).

SOME ACOUSTIC EXAMPLES OF GESTALT PRINCIPLES

Very few songs strike us as made up of random clusters of notes. Some Emberizidae (e.g. the corn bunting, *Emberiza schoeniclus* and the grasshopper sparrow, *Ammodramus savannarus*) sound that way to the human ear; but when slowed down and analysed are shown to be highly organised (Thorpe & Lade, 1961).

In considering the analysis of sound patterns it is convenient to take the Gestalt principles in an order more or less the reverse of that used by those authors who are concerned primarily with visual perception. We will therefore start with

Proximity

If a human subject hears a series of metronome beats separated by short intervals it is practically impossible to think of each beat separately. Two, or several, will always combine to form a rhythmic series. The rhythmic effect becomes more distinct if all the beats are not presented at equal intervals, but in such a manner that two or three always follow each other with the same separation and are separated from the next group by a longer pause. Similarly, patterns based upon stepwise pitch progressions are thereby rendered easier to learn than patterns based on leaps. Thus clearly the law of proximity applies to acoustic forms.

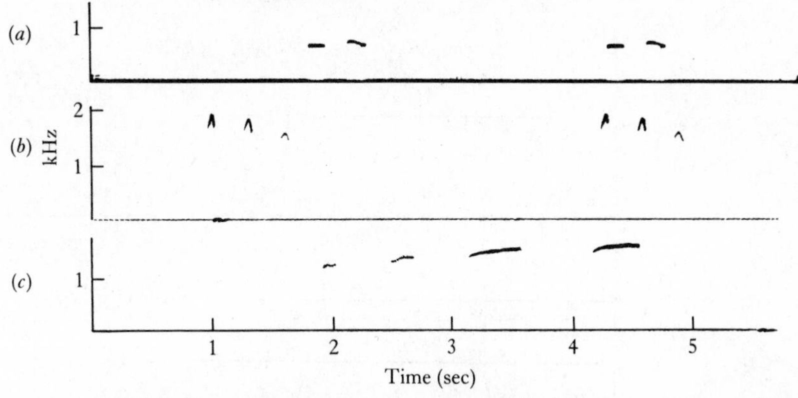

Fig. 1. Sound spectrograms (narrow band) of the songs of three African cuckoo species of the genus *Cuculus*, all of which are obligatory nest parasites, to illustrate pitch proximity (PP), temporal similarity (TS), common movement in pitch (CMP), progressive change in intensity (PCI) and progressive change in the duration of the units (PCD) (from recordings by Dr Claude Chappuis). (*a*) *Cuculus gularis*. PP and TS. C sharp 554, D sharp 622. (*b*) *Cuculus solitarius*. PP, TS, CMP and PCI. (*c*) *Cuculus cafer*. PP, TS, CMP, PCI, and PCD. D 1174, E flat 1244, F 1396, F sharp 1476. *Note*. The notes and frequencies are correct to the nearest semitone in the equally tempered scale.

Similarity

Here again the concept originally applied to visual figures is found to apply to hearing. Thus if, in a series of metronome beats at equal intervals, two loud beats are consistently followed by two soft ones, the soft beats will be grouped with soft, and the loud with loud. They will thus be heard in subrhythms of the total rhythmic pattern (Katz, 1951). Here again, similarity in tonal quality enables the listener readily to perform feats of acoustic pattern recognition in a complex aural environment.

Such findings suggest that all perception of rhythmic processes, in whatever context they occur, are in the first place based upon this tendency to group percepts according to these two primary features of proximity and similarity.

Through the kindness primarily of Dr Claude Chappuis we have been able to analyse the vocalisations of sixteen species and/or subspecies of cuckoos of the genera *Cuculus*, (four species, three subspecies), *Cercococcyx* (three species), *Chrysococcyx* (three species), and *Clamator* (three species). The results of this study are summarised in Figs. 1–3. It will be seen that all examples shown (though not those of the genus *Clamator*, the songs of which for the greater part, lack clear pitch differentiation) provide clear examples of *pitch proximity* and all except two of *temporal similarity*. This is to say that

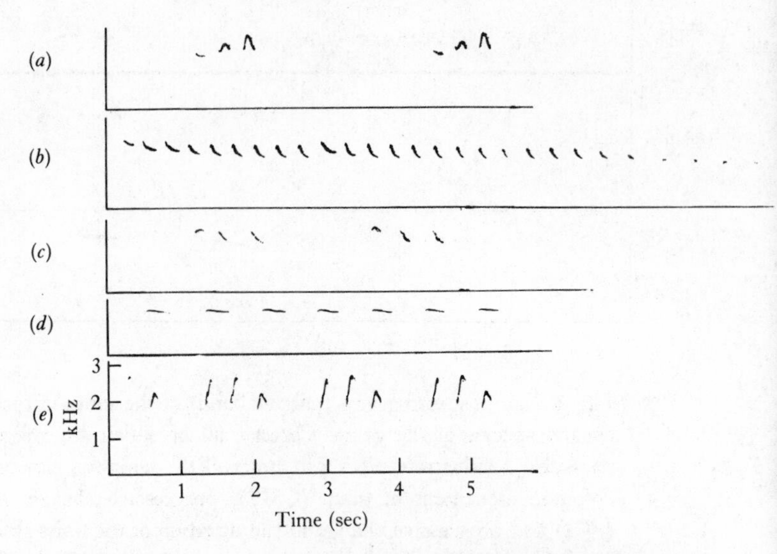

Fig. 2. Sound spectrograms (narrow band) of the songs of three African cuckoo species of the genus *Cercococcyx*, all of which are obligatory nest parasites. (*a, b, c, d* from recordings by Dr Claude Chappuis, *e* from a recording by Mr Stuart Keith.) Indications and note as in Fig. 1. (*a*) *Cercococcyx mechowi* PP, TS, CMP, and PCI. (*b*) *C. mechowi* PP, TS, CMP, PCI and PCD. (*c*) *C. olivinus* PP, TS, CMP, PCI and PCD. (*d*) *C. olivinus* PP and TS. (*e*) *C. montanus* PP and TS.

the songs of each species comprise notes gradually ascending and descending in pitch; and most species have songs showing gradual increase or decrease in duration and intensity. This is of special interest in that all these species are obligatory nest parasites and show, in the pattern and structure of their songs, no evidence of influence by the vocalisations of their hosts. It would seem to follow that the patterns cannot be learnt individually but must be basically the expression of genetic coding. Since the differences in song are not such as to be plausibly attributable to selection resulting from general ecological or environmental differences, we can only assume them to be the result of constraints imposed by the powers, preferences and limits of the auditory perception of the species concerned. Presumably selection has operated (against the background of the species' perceptual abilities) to assist specific discrimination by the hearer.

Rhythm plays an enormous part in natural communication systems of many different types and is undoubtedly an important basis for the perception of different sound patterns. It is a glimpse of the obvious to say that all life is rhythmic and it is equally obvious that rhythm is one of the most basic

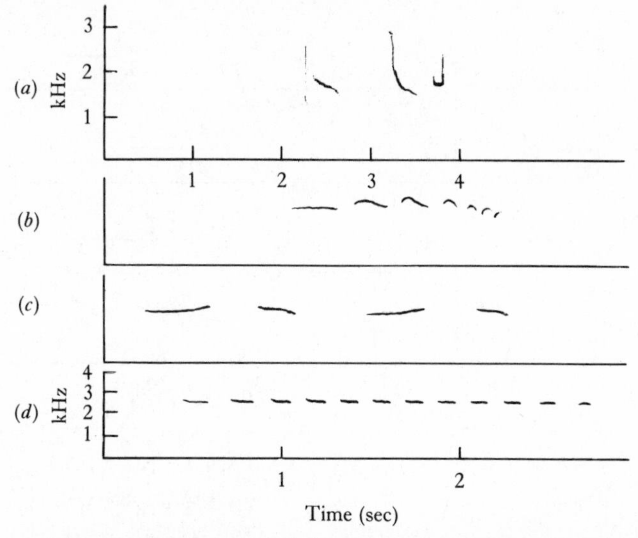

Fig. 3. Sound spectrograms (narrow band) of the songs of four African cuckoo species of the genus *Chrysococcyx*, all of which are obligatory nest parasites (from recordings by Dr Claude Chappuis). Indications and note as in Fig. 1. (*a*) *Chrysococcyx cupreus*. All portamento type units. (*b*) *Chrysococcyx caprius*. PP, TS, CMP, PCI and PCD E 2637, F sharp 2951, F sharp 2951, F sharp 2951, E 2637, E 2637, E flat 2488. (*c*) *Chrysococcyx klaasi*. PP and TS. Units 1 and 3 ascend D 2349 to E 2637. Units 2 and 4 descend E 2637 to D 2349. (*d*) *Chrysococcyx flavigularis*. PP, TS, CMP and PCD. E 2637 followed by very gradual descent in pitch to near 2500.

features of systems of communication by sound. Rhythm is in fact the basis of form in bird songs and call notes, and a sense of rhythm must be essential to the perception of bird songs – whether by the birds themselves or by man (Hall-Craggs, 1969). And it was Schopenhauer (1859) who suggested that 'Rhythm is in time what symmetry is in space and equilibrium is in matter'.

But man has an extraordinary ability to stretch or compress in time both action patterns (e.g. playing a musical instrument or writing) and perceptual recognition of patterns (e.g. a tune which is compressed or stretched without altering any other feature). Birds however seem to show little or no evidence of this particular ability.

There are of course other types of similarity which are used for grouping and distinguishing between acoustic stimuli. In the vertebrates, where the ear is capable of distinguishing between sounds of relatively pure tonal quality and those which contain a wide spectrum of frequencies, the 'tonal quality' of a sound is likely to be a significant feature: harsh sounds and pure sounds are likely to have different and contrasted meanings. A well-known

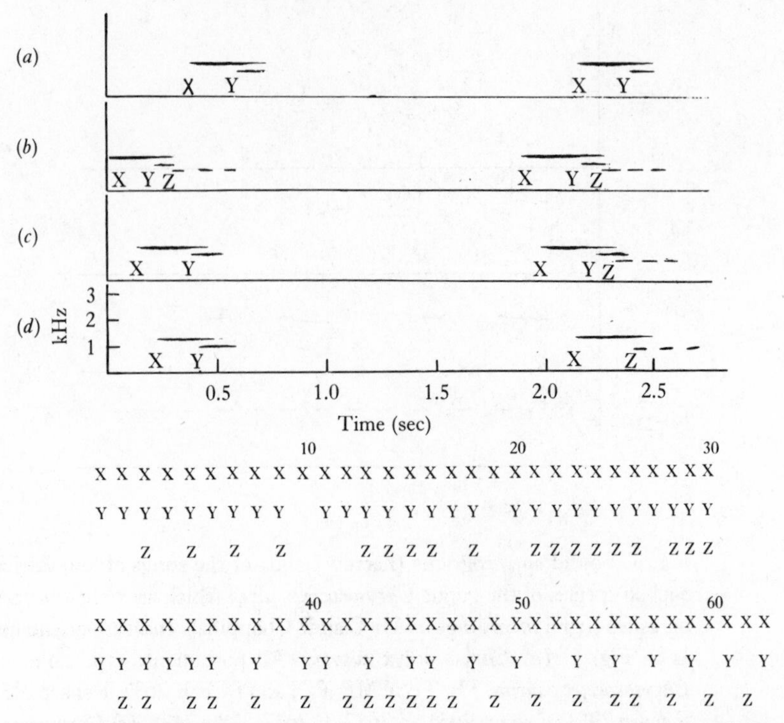

Fig. 4. An example of the development of a song form by three individuals of the tropical Boubou shrike (*Laniarius aethiopicus major*) singing together or in alternation for 74 consecutive bouts, lasting 2 min 58 sec. (Recorded by W. H. Thorpe at Meadow Point, Lake Nakuru, Kenya; 17 March 1964.) 'X' and 'Y' are presumed to be male and female. Sex of 'Z' not known. (*a*) Initial XY duet. (*b*) XYZ trio. (*c*) Duet and trio alternating. (*d*) Y and Z sing in alternation with X. Bouts 42 to 63 constituted the climax of this series which continued for a further 8 bouts after which an XY and a solo X resulted in irregular pauses which terminated the series at about 74.

example of this is provided by the differing types of call notes shown by many small passerine species – as in the Turdidae.

An aspect of the ability of birds to recognise and reproduce melodic lines is the remarkable degree to which they can perform 'acoustic abstractions' of the songs of an alien species, which contain all the important elements defining the song pattern imitated. Moreover, many birds are known to be adept at reproducing immediately the exact series of overtones in a complex utterance, thus giving a remarkably correct reproduction of the timbre of the original.

If to the ability to respond to relationships between *frequency* and *amplitude*

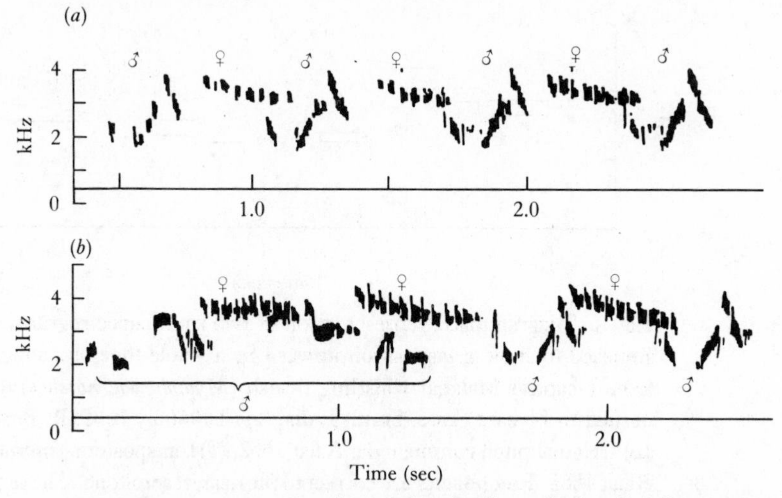

Fig. 5. Duet pattern of male and female *Cossypha heuglini* showing the maintenance of two independent melodic lines (from Thorpe, 1972, p. 168).

changes we add precise perception of *duration* of both the individual notes and of the overall pattern of phrases and of songs; and if we have (as we obviously do in varying degrees) recognition of the interaction of temporal relations with one another to produce rhythm – then we have considerably reinforced the evidence for the 'Gestalt' qualities of such utterances.

We shall be dealing mainly with the production of songs and other extended sequences of notes. But of course the above characteristics are also necessarily involved in the imitation of the human voice as well as in brief snatches of a rival's voice. This can often happen almost instantaneously; one well-attested case is provided by the Indian hill mynah 'Igwog' trained by Mrs Yvette Bower and analysed in detail by Klatt & Stefanski (1974).

Common movement

The auditory analogue of this concept is the ability to follow a train of changing pitches, as in a scale or simple melody, and to distinguish it from another simultaneous train moving at a different speed or in a different direction. Thus it confers the power to perceive a tune against an 'accompaniment' (cf. the cocktail party problem) or as distinct from another tune proceeding at the same time (polyphony). It is quite clear (Thorpe, 1972) that many birds have this ability in at least some degree (e.g. the polyphonic duetting of *Laniarius aethiopicus* and *Cossypha heuglini*: see Figs. 4 and 5). It is also evident in a number of instances of vocal imitation, when new

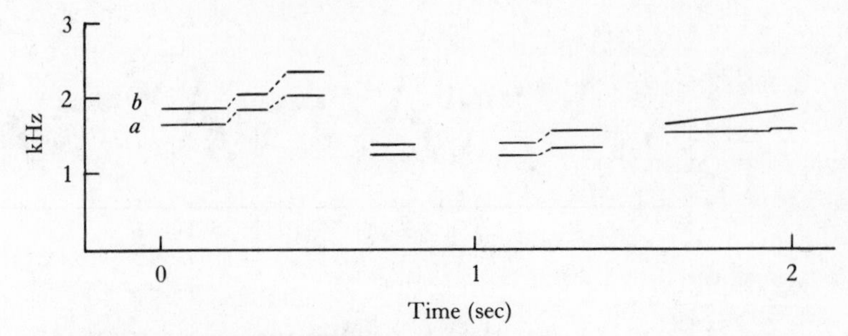

Fig. 6. Diagrammatic representation of two sound spectrograms superimposed to show transposition upward by a whole tone of a song phase from a captive Malabar whistling thrush (*Myiophoneus horsfieldii*). (Recorded in Mysore State, South India, April 1966 by B. C. R. Bertram.) (*a*) Original pitch commencing A flat 1662, (*b*) transposition commencing B flat 1865. Frequencies are correct to the nearest semitone. The temporal patterning is maintained but is less precisely matched than as shown in the simplified illustration.

sequences are incorporated into the existing repertoire of sound patterns. It is even known for a single individual bird to produce complex polyphonic patterning, e.g. the veery (*Hylocichla fuscescens*) (Nelson, 1973).

This topic of the recognition of tune or melody brings us back to von Ehrenfels, who was the first to point out that a tune cannot under any circumstances be comprehended as a mere sum of its constituent notes but must possess a particular form quality. This led him to his second point, equally new, it seems, to the psychologists of that time, that a tune is essentially transposable to a higher or lower pitch. The result of this may, of course, be that the original tune has not one single note in common with the transposed tune. Yet the tune itself is retained. It is the form quality that remains the same when such shifting takes places.

Examples of the immediate ability to transpose without apparent practice, have long been known in birds (Stadler, 1934). The finest example we have of this was obtained for us by Dr Brian Bertram in India and concerns the Malabar whistling thrush (*Myiophoneus horsfieldii*, see Fig. 6). But many more could be cited, e.g. *Cisticola nigriloris* (Thorpe, 1972), the bullfinch (*Pyrrhula pyrrhula*) (J. Nicolai, 1969, personal communication), the blackbird (*Turdus merula*) (Tretzel, 1967) and Swainson's thrush (*Hylocichla ustulata swainsoni*) (Nelson, 1973).

Innumerable songs show, of course, the 'musical characteristics of *accelerando*, *ritardando*, *crescendo* and *diminuendo*. The significant point to remember is that they all, in some degree, indicate a coordination of the whole vocalisation as a single unit. One clear example showing a striking *crescendo* with a certain degree of *accelerando* in *Cossypha heuglini* (the white-browed robin chat) will be found in Thorpe (1972, pp. 165–170 and figs. 9: 2–9: 5).

Besides the characteristics of organisation just mentioned, many songs strike the human ear as 'musical'. The vast majority of these are songs which include or comprise clearly pitched notes and which have a length and pattern in some respects characteristic of the species and/or the individual. But in many of these examples the evidence is strong that it is the overall relationships, not the particular notes, which constitute the signal-bearing essence of the song for the recipient.

Closure and Pregnance

These latter examples bring more and more strongly to the fore the problem of the nature of human music and the question whether there is any firm evidence for a similar quality in the sound production of birds or other animals. And it is here that the two remaining Gestalt categories become relevant – namely Closure and Pregnance. For our purposes these are conveniently dealt with together, and they are both essential to any concepts by which the 'musical' or 'good music' is to be judged. All the adjectives and epithets used by 'Gestalt' authors in explaining and elucidating these two categories will be found applicable to music and the musical tradition. Indeed if we agree with Helmholtz (1863, 1954) that music is primarily 'melodic motion' and that 'melodic motion is change of pitch in time' we can say that the more a piece of music complies with the definition of a good Gestalt as a form or whole, the characteristics of which are determined by its configuration, and where the meaning and relevance of the parts depend upon the whole, the more sure we can be that we are dealing with a worthy, beautiful and significant composition. A piece of music is a hierarchy of, or perhaps better, a concentric series of Gestalten; of forms within forms, each displaying all the characteristics of the classical Gestalt formulation.

THE DETECTION OF 'ERRORS' IN BEHAVIOUR PATTERNS AS EVIDENCE FOR GESTALT ORGANISATION/PERCEPTION

It is evident that species such as the blackbird, which manipulate the large quantities of material that comprise their song repertoires, deal in *Gestalten.* In the formation and extension of song-phrases, single units of sound are not used (one instance only has been recorded, Hall-Craggs, 1962); figures of at least three units with distinct temporal and tonal patterning and lasting at least 0.5 sec are the minima. Such figures are combined to form phrases which, in turn are combined to form sentences (previously called 'compound phrases'). The process is progressive during the whole course of the song season and the experienced bird manipulates increasingly longer sound forms. In the following pages we shall describe a case of persistent breakdown in the performance of one such extended song form but, before doing so, a short digression is necessary to suggest that the Gestalt principles of proximity and common movement may operate in the early stages of song acquisition by blackbirds.

Recordings were made of a small population in south Oxfordshire for a period of six years. When tracing the history of the song-phrases, it became evident that those most frequently transmitted from bird to bird were based upon small pitch intervals (proximity) and, in particular, small intervals constituting short scale-like passages, both ascending and descending (common movement). It is interesting to note that these features also rather closely resemble the organising tendencies described by Restle (1970) as a result of his experiments with human subjects learning serial patterns. Such subjects do not impose the order by sequential storage as a computer does but, instead, generate the order from a system of 'rules'. The universal organising tendencies that he has so far discovered concern what he calls 'runs' and 'trills', that is, stepwise movement between adjacent events and oscillation between any two events. The 'runs' are analogous to proximity and common movement; in musical terms they are analogous to the conjunct (i.e. stepwise) movement specified for the writing of melodic lines, especially in a contrapuntal context wherein one line accompanies one or more contrasting lines. 'Trills' conform to the principle of similarity and resemble scale passages in that they are one of the simplest musical strategies to learn. Learning of 'runs' by blackbirds is common; 'trills' are learned when the pitch interval is small.

These relatively simple rules govern unit relationships but do not apply to higher-level pattern organisation. However, at the higher organisational levels of the blackbird song repertoire – the combination of figures to form phrases – there is a rather widely applied 'rule': that certain sections shall be retained as introductory figures (prefixes), others are concluding figures (suffixes) and, in the longer phrases, some material may regularly fulfil the function of bridge passages, linking prefix to suffix. These prefixes, suffixes and bridge passages will almost always serve the same function in spite of their recombination in the formulation of 'new' phrases. It appears, however, that this rule has to be learned for, early in the song season, the function of prefix and suffix is not always established although it is usually rapidly acquired.

As a result of such organisational tendencies, and when phrases and sentences are established and maintained, it is a fairly simple matter to detect 'errors' in performance. The concept of 'error' supports the notion of Gestalt perception. Temporal or tonal displacement of a sound or sounds resulting in faulty relationship to the antecedent or subsequent sounds, if recognised as such by the bird, indicates the perception of relations. In the repertoires of regularly recorded birds, such errors are obvious, not merely by their statistical rarity but by the bird's response to its own fault.

Errors have been found in phrase performance of all recorded birds. Generally these fall into three types: (1) A false start to a phrase after which the bird stops and begins again. Two reasons are suggested. First, the bird has been distracted by hearing a rendering of one of its own phrases given by a neighbouring bird; in this case the subject is likely to reply immediately with its own version of the phrase, and consequently the normal interphrase timing may be disrupted. The second reason derives from a fault in the acoustic structure of the initial unit or units; this is perceived by the bird and corrected by repetition. (2) A procedure similar to (1) may occur mid-phrase resulting in hesitation followed by normal continuation or repetition of the last unit sung. (3) The bird may fail to complete a phrase correctly in that the normal ending of the phrase is replaced by a jumble of sounds which has not occurred previously and does not occur again, although a slightly similar series of sounds (often of an 'irritable' nature) may occur under similar circumstances. It must be emphasised that such 'jumbles' of sounds lack organisation and are distinct from the process of phrase development. Confused phrase endings are also quite distinct from the unfinished phrases that often occur when a bird is disturbed by external events.

There is one recorded case in blackbird song which does not fall into any of these categories but which is relevant to Gestalt theory, and in particular to the concept of 'closure'. In this instance – following a cessation of song for eight days during the ninth and tenth weeks of the song season – the bird, S, lost its ability to sing a sentence which had been fully developed and sung in its entirety only seven days prior to the cessation. When song was resumed, S repeatedly stopped half way through this sentence and became subject to a form of 'stuttering' comprising repetition, up to four times, of the same three units. The stuttering was usually followed by an attempt to complete the phrase in its original form; in this S was unsuccessful but ultimately achieved a compromise and was able to complete the sentence in the normal manner but with a modified central section.

S was recorded daily during the song season 1959 except during periods of song cessation and when weather made recording impossible. Its full territorial song commenced on 17 February and ended 17 May, thus confirming the finding that, at least in the particular area of south Oxfordshire – beech wood and garden habitat – the duration of individual blackbird song seasons is usually in the region of three calendar months. At the beginning of the season

Table 1. *The development of a song sentence from four originally discrete sections of a blackbird's repertoire*

Date	Figure A	Phrase B	Phrase C	Figure D
17 and 18 Feb.	As suffix to 2 other figures Not retained	In isolation	In isolation	—
21 Feb.	As prefix to B Retained	With A preceding Retained	In isolation	In isolation
	A, B established			
22 Feb.	A, B continued		With D following Retained	As suffix to C Retained
			C, D established	
11 April	A, B, C, D established			

this bird sang an assortment of 18 figures and phrases, most of them only short figures, as defined at the beginning of this section.

In this discussion we are concerned with only two of these figures A and D, and two phrases, B and C. These four originally discrete sections of the total repertoire were eventually combined to form the uninterrupted sentence A, B, C, D. Table 1 shows the development of this sentence.

On 11 April the bird first omitted the pause between the two phrases and the sentence A, B, C, D was established. In the five recordings made from 11 to 17 April inclusive, the sequence A, B pause C, D was recorded seven times and the sentence A, B, C, D 20 times. No recording was made on 18 April and the bird ceased singing (apart from brief snatches of song) from 19 to 25 April inclusive.

Upon resumption of song on 26 April, S seemed unable to sing the sentence A, B, C, D. The opening section A, B was sung normally but S stopped after the third unit of phrase C and frequently repeated these three units, without pause, from one to four times. These three units, henceforth referred to as X^n are regarded as a form of stuttering because, as a result of the bird's subsequent treatment of this material, it was obvious that it was making some effort to regain the original sentence and, of course, such repetition of figures within a phrase is alien to blackbird song syntax. In the repertoires of seven regularly recorded birds, no phrase incorporated more than one repetition of a figure and, when such repetition occurred, it was of a stable and predictable nature.

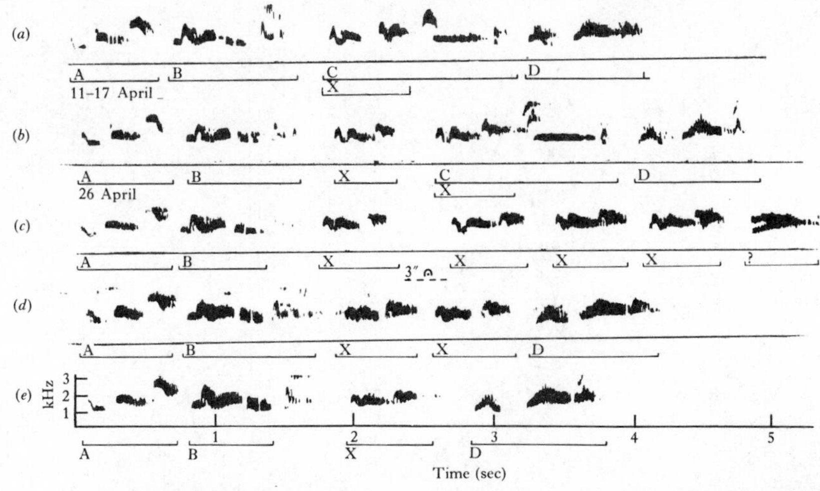

Fig. 7. Sound spectrograms showing how an individual blackbird (*Turdus merula*) modified a 'sentence' after a song cessation of eight days. (Recordings by J. Hall-Craggs, Oxfordshire, 1959.) When the bird resumed singing on April 26 it was apparently unable to sing phrase C without repeating the section X. (*a*) The original sentence. (*b*) The sentence with one repetition of X. (*c*) The sentence with three repetitions of X followed by a raucous sound in place of the original ending. (*d*) The attempt to complete the sentence by singing the original suffix D following a repetition of X in place of the original ending to C. (*e*) The final form A, B, X, D adopted to replace A, B, C, D. In this way the bird has eliminated its 'stuttering' over the figure X and achieved an apparently satisfactory substitute sentence of appropriate length and 'closure'. *Note.* Figure X comprises three units. These are clear in (*b*) but appear to merge in the other illustrations. This blurring is caused by sound reflection resulting from the bird's abnormal habit of singing from deep cover in a wood of mature beech trees.

Supposing this to have been a form of stuttering, the bird gave every evidence of trying to overcome the impediment and complete the sentence A, B, C, D as sung prior to the cessation. In this attempt five different processes emerged (Fig. 7). Where n = from 1 to 4 repetitions:

(1) A, B, X^n Stop.
(2) A, B, X^n Noise. (The noise was an unusual 'terminal decoration' not used in other contexts.)
(3) A, B, X^n, C, D.
(4) A, B, X, D.
(5) A, B, X, followed without pause by some different figure or phrase from the repertoire.

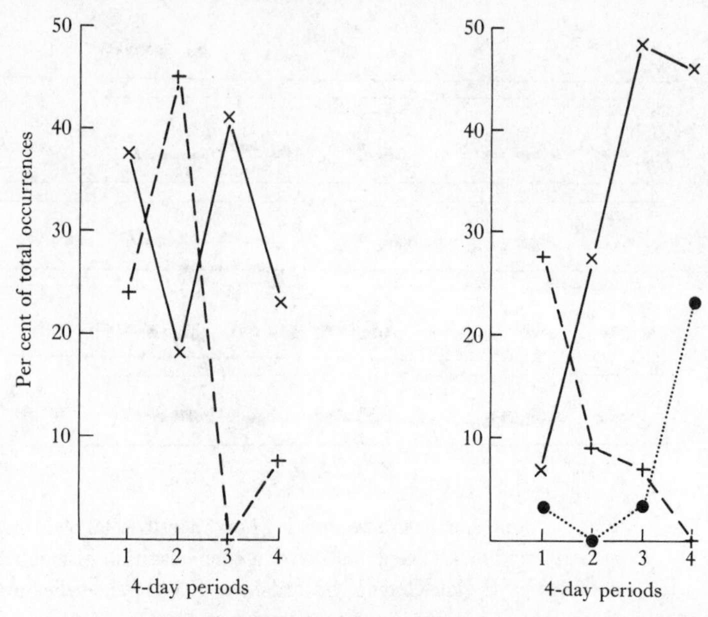

Fig. 8. Graph illustrating the gradual increase in frequency of occurrence of the replacement form A, B, X, D (see Fig. 7) balanced by decrease in the incidence of A, B, X^n and A, B, X^n, C, D. A, B, X^n ? (where ? signifies an unusual raucous noise) fluctuates. During the last four-day period, when A, B, X, D has become established, experimentation with substitute suffixes for D takes place. *Left*: A, B, X^n (+– – –+) and A, B, X^n, ? (×——×). *Right*: A, B, X^n, C, D (+– – –+); A, B, X, D (×——×); and A, B, X, *new* (●···●).

Between the resumption of song on 26 April and the end of the period of adjustment to a substitute sentence, 16 daily recordings were made. During this period 95 performances of the sentence in one or other of the forms listed above were recorded. The graphs, Fig. 8, show the distribution of these forms expressed as a percentage of the whole (ordinate) and broken down into four-day periods (abscissa). The salient points are:

(1) The original sentence A, B, C, D was never regained.

(2) Apart from the fluctuating A, B, X^n?, each form containing the stuttering X^n is reduced in frequency of occurrence towards the end of the total period.

(3) The replacement form A, B, X, D gains steadily in frequency of occurrence and shows only a slight drop during the last four-day period. By this time the stuttering was decreasing, and S had begun to experiment with new concluding figures and phrases; a process common in blackbird song under normal conditions. It therefore appears that the bird

strove to attain a 'correct' ending to the sentence, i.e. the appropriate suffix, in Gestalt terms to achieve closure, and to eliminate the stuttering. J. Fentress (personal communication) described an interesting analogue to this behaviour. In the ontogeny of movement sequences in some of his experimental mice, animals not infrequently interpolate some bizarre movement within a normal sequence of behavioural elements. As in the blackbird's song behaviour, these alien movements may be repeated, they may terminate a sequence and they may become permanently incorporated in the normal sequence.

The data presented here result from recording, indexing and sound spectrographic analysis; nevertheless it was clear at the time of recording and evident by ear alone that the bird was in trouble. Field notes made while the bird was singing contain such allusions as 'Bird getting in a muddle. Lack of practice?' and 'Sings "X" 3 or 4 times in succession'. If, indeed, this was a case of stuttering, assignment of a cause remains conjectural, but three factors may have contributed to it. In the first place, S was literally out of practice; all blackbirds have a slight and temporary reversion in song following such an extended period of cessation, but this rarely lasts for more than a day or two. Secondly, a neighbouring blackbird sang figure X as a prefix to one of its phrases but, as is usual in this species, its rendering was not identical with that of S. Thus, hearing a modified form of the figure for eight days when not itself singing may well have influenced or confused S; in fact the structure of X as sung by S after the cessation tends more towards the form as sung by the neighbour than to its own rendering of the three units when sung as the introductory section of the original phase C. Thirdly, S was an abnormally nervous bird. It habitually sang from deep cover (in itself an abnormality), took fright at any movement or change of circumstances in its territory and, unlike other recorded blackbirds, never showed any sign of habituation to the presence, much less to the movement, of a human while it was singing.

The stimulus for presenting this abnormal behavioural sequence after a time lapse of 16 years was due to the theoretical points raised by Dawkins (this volume) on correctly rounded-off serial patterns and the possibility of attaching meaning to the concepts of 'correctness' and 'incorrectness' in animal behaviour. The example cited fits well into Dawkins' model where A and B are two particular behaviour patterns and the sequence AXB is common, although X is not a particular pattern but any one of a set: 'the performance of A opens a sequence which demands to be closed by the occurrence of B regardless of what X happens to be'. Moreover, X may stand for not one behaviour pattern but a series of unfixed length. In the case of the bird S, however, it is assumed that the variable X is an unintended or erroneous behavioural element for the reason already stated: repetition of this kind is alien to the syntax of blackbird song. Another reason for the assumption is suggested by the lack of smooth transition from the sentence A, B, C, D to A, B, X, D. When a Blackbird introduces a new combination of figures or phrases there is usually a period when both the old and new forms coexist, the old being gradually phased out as the incidence of the new increases (Hall-Craggs, 1962). But with S the sentence A, B, C, D ceased abruptly.

Dawkins points out that an AXB pattern of this type would not be detected by Markov chain analysis, and it is a disturbing thought that a similar process may be operative at a higher level in the repertoire hierarchy, that is, in sequences of phrases and sentences separated by the normal pauses; a situation less readily detected either by ear or by extended indexing. There is even some rather sketchy evidence that such may be the case. For example, it sometimes happens that three disconnected phrases A, B and C may, for a time retain such a relationship despite variation of intervening phrases. A more intriguing situation is also suggested by extensive lateral indexing. In this, A.B.C. may form initially a sequence of adjacent but disconnected phrases. If B is then indexed in a central column there may be found roughly a stepwise lateral displacement of A and C with time, e.g. A.X.B.X.C, A.X.X.B.X.X.C, A.X.X.X.B.X.X.X.C and so on where each X is different.

The problems encountered in decoding such a process are threefold: (1) whether the progression is determined by the number of occurrences of the sequence in any particular form or (2) whether it is determined by the number of hours or days during which a particular presentation may occur. (3) To obtain proof and, thereby, a complete picture of the process, it would be necessary to record the total output of an individual throughout an entire song season. So far only daily samples have been recorded and these mainly confined to the dawn song when it is easier to predict where and when the bird will be singing. Thus although there is no more than a suggestion that the process occurs, the indications are strong enough to merit further investigation.

SUMMARY

A brief review of the principles of Gestalt psychology, as applied to visual forms and perception, leads to consideration of how far these principles might also be applied to temporal forms and auditory perception. The vocalisations of birds provide significant material for the study on account of their wide range – from patterns which are presumed to be genetically encoded and not subsequently modified by learning (e.g. some of the songs of the nest parasitic cuckoos) to those which appear to be entirely learned (e.g. some turdine repertoires). Simple Gestalt principles are found to be operating in the former, while the latter exhibit the higher organisational features of 'closure' and 'pregnance'. Melodic transposition is thought to be the ultimate criterion for the auditory perception of 'wholes' and this is found in the songs of a number of species. A description of error detection and correction by the blackbird is offered as further evidence for auditory Gestalt perception in birds.

REFERENCES

Broadbent, D. E. (1961). *Current Problems in Animal Behaviour*, ed. W. H. Thorpe & O. L. Zangwill, p. 249. Cambridge University Press: London.

Chaplin, J. P. & Krawiec, T. S. (1974). *Systems and Theories of Psychology*, 3rd edn. Rinehart, Holt.

Ehrenfels, C. von (1890). Ueber 'Gestaltqualitäten'. *Vierteljahrsschrift für wissenschaftlich Philosophie*, **14**, 249–292.

Gibson, J. J. (1952). *The Perception of the Visual World*. Allen & Unwin: New York & London.

Hall-Craggs, J. (1962). The development of song in the blackbird, *Turdus merula*. *Ibis*, **104**, 277–300.

Hall-Craggs, J. (1969). The aesthetic content of bird song. In *Bird Vocalizations*, ed. R. A. Hinde. Cambridge University Press: London.

Helmholtz, H. (1863, 1954). *On the Sensations of Tone*. Dover Publications: New York.

Katz, D. (1951). *Gestalt Psychology its Nature and Significance*. Methuen: London.

Klatt, D. H. & Stefanski, R. A. (1974). How does a mynah bird imitate human speech? *Journal of the acoustical Society of America*, **55**, 822–832.

Koehler, W. (1925). *The Mentality of Apes*. Kegan Paul: London.

Koehler, W. (1941). On the nature of associations. *Proceedings of the American philosophical Society*, **84**, 489–502.

Koffka, K. (1924). *Growth of the Mind*. Kegan Paul: London.

Nelson, K. (1973). Does the holistic study of behaviour have a future? In *Perspectives in Ethology*, ed. P. P. G. Bateson & P. H. Klopfer. Plenum Press: New York & London.

Restle, F. (1970). Theory of serial pattern learning: structural trees. *Psychological Reviews*, **77**, 481–495.

Schopenhauer, A. (1859, 1966). *The World as Will and Representation*. Dover: New York.

Stadler, H. (1934). Die Vogel kann transponieren. *Ornithologische Monatschrift*, **59**, 1–9.

Thorpe, W. H. (1972). Duetting and antiphonal song in birds. *Behaviour Monograph*, **18**. E. J. Brill: Leiden.

Thorpe, W. H. & Lade, B. I. (1961). The songs of some families of the order Passeriformes. *Ibis*, **103*a***, 231–259.

Tretzel, E. (1967). Imitation und Transposition menschlicher Pfiffe durch Amseln. *Zeitschrift für Tierpsychologie*, **24**, 137–161.

Wertheimer, M. (1912). Experimentelle Studien über das Sehen von Bewegung. *Zeitschrift für Psychologie*, **61**, 161–265.

EDITORIAL: 2

Although the first four authors in this section appear to take rather different stances, their contributions complement each other. Dawkins is concerned with the overall organisation of the animal's behavioural repertoire, and argues that the hierarchical principle is likely to be of general explanatory value. McFarland is concerned with decisions determining the particular way in which an animal behaves – decisions which may of course be incorporated within (or between) hierarchical systems – and attempts to derive principles from the relative costs and benefits to the individual of its various activities. Andrew's problem is similar, but whilst his approach might appear to be more straightforward in that he seeks to explain decisions in terms of preceding events, his application of human attention theory to this problem is quite new. Finally, Fentress's paper emphasises the difficulty of defining the basic units with which each of the others has been concerned, pointing out that, when an animal is in one state, a system of behaviour may be clearly recognised and defined but, as the animal's state changes, the boundaries between that system and others may change or disappear.

Certain features common to the approaches of these four authors deserve comment. First, all seek to understand the ways in which changes in an organism's activities are determined *at the behavioural level*. This should not be taken to mean that attempts to find neural answers to motivational problems have been abandoned – a glance at the journals would show that to be far from the truth. But it is the case that the task has proved much harder than was originally hoped.

The first four authors differ somewhat in their attitudes on this issue. Andrew and Fentress both seek non-physiological principles, but are not averse – especially Fentress – to using physiological evidence to support their cases. McFarland argues that, however much we know about the factors that influence motivational state, we shall not be able to predict behaviour unless we understand the processes whereby decisions between different types of behaviour are made, and such understanding can be gained only by consideration of the adaptive consequences of behaviour. Dawkins takes a more extreme view, holding that even if we could obtain complete neurophysiological explanations, they would be indigestible and inappropriate.

We must emphasise again here that we have not attempted to obtain a complete coverage of all the growing points in ethology. The omission of chapters making a strong case for the neural analysis of behaviour does not imply that we believe the subject is no longer interesting. A complete understanding of the mechanisms of behaviour will require an analysis of the hardware as well as the software, and principles of neural functioning are at least as accessible as principles concerned with the inter-relations of behavioural elements. Furthermore, knowledge of the hardware may often place constraints on the kinds of software explanations that are offered. At the moment, however, it is easier to see the possibilities of exciting new principles emerging from behavioural analyses than it is from a neural approach.

Another issue calling for emphasis is the extent to which all four authors, especially McFarland, rest their cases on functional considerations. This is, of course, in the best ethological tradition – Tinbergen (e.g. 1951) has often emphasised that causal analysis can be assisted by considerations of the job the mechanism is carrying out. But it is important to be clear about the relations between these approaches. Both involve lines of enquiry which are themselves legitimate and can be self-sufficient. The issue is the extent to which they are interfertile.

A third point concerns the extent to which each of these contributors is outward-looking. What we have in mind here is the contrast with some of the learning theories of the forties and fifties. Modelled on classical physics, they nevertheless attempted to be self-contained. Compare them with the first four contributions in this section. McFarland forms links with optimality and control theory; Dawkins with epistemology, control theory and artificial intelligence; Fentress with neurophysiology and Andrew with signal-detection and attention theory. The aim is not to produce an ingrown ethological theory, but to seek theories of behaviour using whatever tools are available.

The chapter by Thorpe & Hall-Craggs is in line with the four concerned with motivation, in as much as they are interested in principles stated at the behavioural level – though recent advances in visual physiology clearly indicate that an understanding of Gestalt principles in terms of neural functioning may soon be possible. They also place emphasis on functional considerations, and in using Gestalt theory they are re-establishing links that have largely withered since Lorenz's writings in the thirties.

REFERENCE

Tinbergen, N. (1951). *The Study of Instinct.* Clarendon Press: Oxford.

PART B

Function and evolution

EDITORIAL: 3

To a biologist observing behaviour, it is as natural to ask 'What is this behaviour for?', meaning how does it contribute to survival or reproduction, as it is to ask about the immediate causal factors that contributed to its occurrence. Questions about function lend themselves notoriously to armchair speculation: Thayer's (1909) suggestion that flamingoes are cryptic against a red sunset is a classic example (Tinbergen, 1965). However, it will already be apparent from McFarland's chapter that hard-headed approaches to the study of function are possible. Within ethology, questions of function have been studied in two main ways: comparative studies of related species in which inter-species variations in a character are related to environmental variables (see e.g. Crook, 1964; Lack, 1968); and detailed field studies involving either field experiments or the assessment of relations between variability in a trait and subsequent survival or reproductive success (e.g. Tinbergen *et al.*, 1962; Patterson, 1965; Tinbergen, Impekoven & Franck, 1967).

Three of the papers in this section have as their primary objective the explanation of observed behaviour, or differences in behaviour, in functional terms. They differ, however, in their scope. Clutton-Brock & Harvey take as their arena the social behaviour of primates, discussing the adaptive consequences on individuals of characters ranging from female sexual swellings to group size. Marler compares one part of the behavioural repertoire, vocalisations, of two species, seeking to account for the differences in terms of functional considerations related to their habitat and social structure. Bertram focusses on the reproductive system of the lion, and also assesses the extent to which its detailed characteristics can be understood in terms of kin selection – that is, through the consequences of behaviour on the perpetuation of the behaving individual's own genes, whether through his own offspring or related individuals.

In the course of evolution, functions can change. This is an issue that has engaged the attention of many ethologists in recent decades – movements which now play a role in communication are thought to be derived from ones previously used in quite different contexts (Tinbergen, 1959). Considerations of function must be applied not only to particular movements and behavioural traits, but also to behavioural capacities. Humphrey's chapter takes up this issue with respect to the group of capacities generally referred to as 'intelligence', pointing out that they are now used in many ways of little relevance to survival or reproduction, and asking in what context intelligence can originally have evolved.

REFERENCES

Crook, J. H. (1964). The evolution of social organisation and visual communication in the weaver birds (Ploceinae). *Behaviour Supplement*, **10**, 1–178.

Lack, D. (1968). *Ecological Adaptations for Breeding in Birds*. Methuen: London.

Patterson, I. J. (1965). Timing and spacing of broods in the black-headed gull, *Larus ridibundus*. *Ibis*, **107**, 433–459.

Thayer, C. H. (1909). *Concealing-coloration in the Animal Kingdom*. Macmillan: New York.

Tinbergen, N. (1959). Derived activities: their causation, biological significance, origin and emancipation during evolution. *Quarterly Review of Biology*, **27**, 1–32.

Tinbergen, N. (1965). Behaviour and natural selection. In *Ideas in Modern Biology*, Proceedings of the 16th International Zoology Conference, vol. **6**, ed. J. A. Moore. Natural History Press: New York.

Tinbergen, N., Broekhuysen, G. J., Feekes, F., Houghton, J. C. W., Kruuk, H. & Szulc, E. (1962). Egg shell removal by the black-headed gull, *Larus ridibundus*, L.; a behaviour component of camouflage. *Behaviour*, **19**, 74–117.

Tinbergen, N., Impekoven, M. & Franck, D. (1967). An experiment on spacing out as a defence against predation. *Behaviour*, **28**, 307–321.

6

Evolutionary rules and primate societies

T. H. CLUTTON-BROCK AND P. H. HARVEY

A primary aim of primate socio-biology (or socio-ecology) is to explain variation in social behaviour in terms of biological function. To do this, it is necessary to consider the possible consequences of differences in behaviour (Hinde, 1969, p. 159). Some of these may be selectively advantageous, some neutral and some disadvantageous. Throughout this paper we refer to those in the first category as 'functions'.

Clearly differences in behaviour are not always functional. Ecological changes may occur which obviate the advantages of particular differences (Fisher, 1958; VanValen, 1973). Phylogenetic variation may cause species to adapt to similar environmental situations in different ways (Gartlan, 1973; Clutton-Brock, 1974). Extreme environmental pressures may produce non-functional differences between populations of the same species: the pathological responses shown by many species maintained at unnatural population densities provide one clear example (Christian, 1963; von Holst, 1974). In this paper we have tried to focus on consistent differences in behaviour which are unlikely to have these complications.

To investigate the functional significance of inter-specific differences in social behaviour, workers have traditionally attempted to relate gross behavioural variation (including variation in group size, group sex ratio and range size) to gross ecological differences (Crook & Gartlan, 1966; Crook, 1970; Denham, 1971; Eisenberg, Muckenhirn & Rudran, 1972; Jolly, 1972). More recently, research has concentrated on the behavioural correlates of more detailed ecological differences (Altmann, 1974; Clutton-Brock, 1974, 1975; Sussman, 1974; Hladik, 1976). However, functional aspects of more detailed behavioural differences and, in particular, of social relationships within groups, have still been largely ignored. Attempts to relate these directly to ecological variation are unlikely to be illuminating. This is because much intra-group behaviour is adapted to the social environment (Kummer, 1970) rather than to the external environment. As a result, it is usually less easy to interpret the functional significance of behavioural differences at this level. For example, while it is relatively easy to guess why species living in stable environments often live in small groups and are territorial (Brown, 1964; E. O. Wilson, 1975), it is less obvious why, among primates, paternal care should be commonest in the same species (see p. 198). Before we can understand the relations between environmental factors and social structures, we need to know the usual functions of different behavioural traits.

Social behaviour can be analysed at a variety of levels (Hinde, this volume). A proper understanding of functional aspects of social behaviour has been retarded by attempts to investigate the adaptive significance of different social systems. This approach has led some authors to suggest explanations of differences in social behaviour which rely on group benefit rather than individual advantage. The point has recently been somewhat overstressed by Alexander (1974, p. 327): '...with few exceptions, essentially every effort to analyze or interpret primate social organization assumes that adaptations exist that assist groups and not individuals...'.

There is, however, no evidence that selection operates between groups. Attempts to explain the evolution of traits which benefit the group but decrease the fitness of the individuals carrying them face a major theoretical problem: such altruistic traits will be eliminated by selection operating between members of the same group. Even if situations occur where an altruistic genotype drifts to fixation within particular groups, selfish mutants or migrants will be at a selective advantage and the situation will be evolutionarily unstable. While it is probable that group selection will have some slight effect where populations are divided into small, isolated groups (Maynard Smith, 1964; Gilpin, 1975) the degree of isolation required for an altruistic genotype to spread through a population is so great as to be improbable in primate societies (Maynard Smith, 1976).*

In our experience, all behavioural traits suggested as examples of group selection can be adequately explained as the product of selection operating on individuals. The primary step in building an evolutionary theory of social behaviour must be investigation of the advantages and disadvantages incurred by individuals who behave in different ways in social encounters. One important implication of this approach is that traits may evolve which increase their carriers' fitness to the detriment of the reproductive rate of the species (see Hamilton, 1971*b*).

The rest of this paper attempts to explain variation in different aspects of

* A more recent model of group selection (D. S. Wilson, 1975) has been claimed by some (May, 1975; E. O. Wilson, 1975) to have wider generality than previous models. This seems unlikely to be the case since the conditions under which the model could operate are highly specialised (Maynard Smith, 1976).

social behaviour in terms of selection operating on individuals. When considering the advantages and disadvantages of different actions to individuals, we shall use the convenient shorthand of discussing what they ought to do to increase their fitness. This usage never implies any form of conscious action.

We are well aware that, in many cases, the data do not fit our hypotheses exactly and that alternative explanations of the variation are possible. Exceptions to any evolutionary rule must be expected, both because different behavioural strategies may evolve in similar environmental situations (see above) and because similar traits may have different functions. It is not within the scope of this paper to investigate such complexity or to attempt to explain all exceptions within the framework of the hypotheses we suggest. Rather, our aim is to outline possible approaches which may help to generate fruitful questions.

BENEFICIENT BEHAVIOUR

Social actions can be classified into two functional categories: beneficient and disruptive (or maleficient). The first enhance the recipient's fitness while the second diminish it. This section discusses the evolutionary mechanisms which may have produced actions of the first kind.

Parental behaviour

The most obvious form of beneficient behaviour is that directed by parents at their offspring. Clearly, it is usually to an individual's advantage to protect and assist its offspring if this increases the latter's chance of survival. The extent to which parental behaviour is likely to evolve will be constrained by its costs in terms of future reproductive output, a topic which has been extensively discussed by E. O. Wilson (1975, pp. 63–105).

In all mammalian societies studied so far, as well as in most birds, females invest more time and energy in caring for their offspring than do males. This situation has evolved because males can produce more young per unit time than females (Trivers, 1972). Since each infant represents a smaller part of a male's potential reproductive output than a female's, males should be likely to expend less time and energy on rearing individual offspring. This is the case in most primates (Mitchell, 1969; Mitchell & Brandt, 1972).

One prediction of this theory is that, in societies where individual males have the opportunity to breed with a large number of females, extensive paternal care is least likely to evolve. Conversely, in monogamous species (where the reproductive potential of males is effectively the same as that of females) paternal care is more likely to occur. In the majority of monogamous bird species, the male assists the female to rear their young (Lack, 1968). Among primates, too, the most developed examples of paternal care occur in monogamous species. In marmosets (*Callithrix jaccus*), tamarins (*Saguinus geoffroyi*), night monkeys (*Aotus trivirgatus*), titis (*Callicebus moloch*), and siamangs (*Symphalangus syndactylus*) the male carries the infant for much of the day (Moynihan, 1964; Epple, 1967; Chivers, 1974; Mason, 1974). Such behaviour is not recorded in polygynous species, though males may carry infants occasionally and frequently protect them (Spencer-Booth, 1970).

Pair bonds

While it is possible to explain many differences in paternal behaviour as adaptations to different breeding systems, it is far less easy to understand the distribution of breeding systems themselves. Polygyny is likely to evolve in situations where females are able to rear young successfully without additional help (Lack, 1968). Here, males will maximise their fitness by breeding with several females instead of helping a single mate to rear their young. If the advantage to the male is sufficiently great, polygyny may even evolve in cases where it is to the female's disadvantage to breed polygynously (Hamilton, 1971*b*). For example, by seizing and defending a territory which can support several females, a male effectively forces them to breed polygynously – a situation in which antagonistic relationships between females would be likely to occur. The extent to which polygyny will occur in such situations will be limited both by the relative disadvantages to females of breeding in this way and by the costs to the male of acquiring and defending such a large territory.

One way in which males may minimise such disadvantages is to form breeding cooperatives with related individuals (e.g. Zahavi, 1974; Wrangham, 1975). This would allow them to increase the area to which each has access. Consider a group of male territories, imposed on a population of smaller female territories, in which one male settled next to a related individual. The advantage to each of defending their joint boundary would be negatively correlated with their degree of blood relationship. In such a situation, it might well be advantageous to both individuals to neglect these boundaries if this allowed them to extend those with unrelated individuals, thus increasing the number of females to which each had access. From such a situation, it is not difficult to derive a breeding cooperative involving several related males. Given an approximately linear relationship between the number of males and the *length* of the boundary which they could defend, individuals in larger groups would have access to a relatively larger *area* (and thus to a relatively greater number of females) than individuals in smaller groups. Moreover, the group's ability to defend the territory against contestants would also be likely to increase with group size in a non-linear fashion.

In monogamous species, males presumably maximise their reproductive success by assisting their mates to rear joint offspring (Lack, 1968; Goss-Custard, Dunbar & Aldrich-Blake, 1972) rather than by breeding with a greater number of females (Trivers, 1972; Geist, 1974). E. O. Wilson (1975) outlines three situations in which monogamy is likely to develop:

(1) where a scarce or valuable resource has to be defended against conspecific competitors,
(2) where the physical environment is so harsh that the female cannot provide food for the young without assistance,
(3) where the male's assistance allows the date of breeding to be advanced.

One might therefore expect to find monogamy in species where male assistance is most necessary (see above). In birds (Lack, 1968) it is more common among insectivorous and carnivorous species (where food is generally difficult to collect) than among frugivorous or graminivorous ones. And in marmosets, monogamy is associated with the production of twins (though why twinning itself should have evolved is obscure).

One problem with this approach is the absence of direct paternal care in a number of monogamous species. In several monogamous birds where the male does not help to feed the young, he may provide indirect assistance by guarding the nest site or feeding the female (Lack, 1968, p. 150). Perhaps in monogamous primates in which the same situation occurs, the male's contribution lies in the defence of a feeding territory (see Goss-Custard *et al.*, 1972). Certainly in both *Indri indri* and *Hylobates lar* the male plays a larger part in territorial defence than the female (J. Pollock, personal communication; Ellefson, 1970). A similar explanation has been proposed for the function of monogamy in ungulate societies (Geist, 1974).

This argument still fails to provide a satisfactory explanation of the distribution of monogamy among primates. Geist (1974) argues that in species living in relatively stable environments where population density is closely constrained by resource availability, competition will be most intense and male assistance will be most advantageous. This may explain why monogamy tends to occur in forest-dwelling species (e.g. the Hylobatidae, *Indri* and *Callicebus*) but does not account for its occurrence in only some of these species. Detailed ecological comparisons of monogamous and polygynous species occupying similar habitats are clearly needed. The fact that typically monogamous primates apparently rarely or never show polygyny suggests that monogamy must have considerable advantages in those situations where it occurs.

Even where monogamy has evolved, males should be unfaithful to their mates when opportunity arises (Trivers, 1972), for it would be to a male's advantage to father offspring which would be reared by other couples. Females would gain little from preventing such infidelity though they should stop their mates spending any time or energy helping to rear such 'illegitimate' offspring. In contrast, males should enforce rigid fidelity on their mates, as it would be disadvantageous for them to rear the offspring of a competitor. In mountain bluebirds (*Sialia currucoides*) (D. P. Barash, in Kolata, 1975) males returning to their territories to find their mates consorting with strange males may drive both the strange male and their own mate out of the territory. After the female lays, males may eject such strangers but do not attack their mates. Thus it is surprising that females in some monogamous primates (see Mason, 1974) occasionally mate with males of another group, apparently with little interference.

Beneficience directed at other relatives

In many vertebrate societies, as well as in invertebrates, individuals behave beneficiently towards animals other than their own offspring. Hamilton (1963, 1964*a*, *b*) has pointed out that individuals may increase their genetic contribution to subsequent generations by beneficient acts directed at relatives sharing a large proportion of their genotype. The degree of beneficience evolving in this way will depend on three variables: the degree of relationship, the cost (in terms of detriment to individual fitness) of assistance to the benefactor, and the benefit to the recipient (Hamilton, 1964*b*). (The mathematical basis of this theory, usually referred to as 'kin selection' (Maynard Smith, 1964), has recently been elaborated (Orlove, 1975; West-Eberhard, 1975).) Kin selection provides the best explanation so far of the evolution of sterile worker castes in the social Hymenoptera (Hamilton, 1972; West-Eberhard, 1975). It may also be important in explaining such phenomena as cooperative mating (Watts & Stokes, 1971; Maynard Smith & Ridpath, 1972) and of cooperative rearing (Skutch, 1961; Lack, 1968; Kleiman & Eisenberg, 1973) in birds and carnivores.

In primates, kin selection may help to explain a variety of cooperative relationships (Alexander, 1974; E. O. Wilson, 1975).

(*a*) In many species, adult females show pronounced interest in the infants of other females, holding them for prolonged periods and grooming or nursing them (Spencer-Booth, 1970; Klopfer, 1974). 'Aunting' may assist the mother in rearing the young, providing a possible substitute in the event of her death (e.g. Bernstein, 1969; E. O. Wilson, 1975), and may permit the 'aunt' to become acquainted with the process of infant rearing (Horwich & Manski, 1975). Some evidence suggests that the latter function is likely to be more important than the former in some species. In macaques (*Macaca* spp.), aunting behaviour is more often shown by nulliparous females. And in many species, females show interest in infants but are prevented from handling them by their mothers. 'Aunts' are frequently previous offspring or sisters of the infant's mother (Spencer-Booth, 1970).

(*b*) There is widespread evidence of cooperation between related females. In troops of Japanese macaques (*Macaca fuscata*) and rhesus macaques (*Macaca mulatta*), the rank of daughters is related to the rank of their mothers (Kawamura, 1958; Kawai, 1958; Yamada, 1963; Sade, 1967). This appears to be the case partly because mothers or other relatives may intervene on their side in aggressive interactions (Sade, 1967).

(*c*) In several cercopithecoid species, adolescent males leaving their natal group often join troops which contain related animals (Sade, 1968; Boelkins & Wilson, 1972). Strange individuals attempting to join established groups are frequently attacked (Southwick, Siddiqi, Farooqui & Pal, 1974; Lindburg, 1969; Stolz & Saayman, 1970).

(*d*) Within groups, adult males may assist each other in competitive interactions (e.g. Bramblett, 1970; Saayman, 1971). In several studies, males which support each other have proved to be related (Koyama, 1967, 1970). This may also explain why, in some species, males often change groups together with one or more sibs (Drickamer & Vessey, 1973).

(*e*) In langurs (*Presbytis* spp.) and gorillas (*Gorilla gorilla*) mature males may sometimes tolerate the presence of their mature male offspring (Yoshiba, 1968; A. H. Harcourt, personal communication). And in hamadryas baboons (*Papio hamadryas*) (Kummer, 1968) young males may enter existing harems, adopting submissive behaviour, and gradually acquire sexual access to females as the leader ages and spends more time in non-sexual activities. One explanation is that it is to the advantage of the ageing male to share sexual access to his females with a close relative during his declining years, if this enhances his ability to defend the group and diminishes the chance that the group will be occupied by an unrelated animal after his death. Such a mechanism might be more likely to evolve in species, like hamadryas baboons or gorillas, where females frequently leave their natal group and where the chance of inbreeding is reduced.

(*f*) Proximity and grooming are usually more pronounced between relatives than between unrelated animals (Yamada, 1963; Sade, 1965; van Lawick-Goodall, 1968; Missakian, 1974).

This last example illustrates an important point. Associations between social relationships and degree of kinship need not necessarily indicate that kin selection has been involved. For example, one inevitable effect of extended parent/offspring relationships is that individuals are likely to spend more time close to relatives than to unrelated individuals, and thus may be expected to interact more with the former. One way of distinguishing true kin selection is to compare social relationships in groups which differ in the degree of relatedness between members.

Conflicts between benefactors and recipients

Even in cooperative relationships, conflicts of interest will occur between benefactors and recipients. In any relationship, it will be to the benefactor's advantage to set limits to the assistance it provides and to the recipient's advantage to persuade the benefactor to overstep them. This may lead to the occurrence of agonistic behaviour in otherwise cooperative relationships. One example of such a conflict of interests, associated with agonistic behaviour, is provided by the weaning process (Trivers, 1974).

At some stage in the growth of the infant, it will be to its mother's advantage to terminate assistance and to conserve her resources for subsequent offspring. Unless other factors are involved, it will be in the infant's interest to extract further support from her. This may result in a period of agonistic interaction between mother and child* (e.g. Hinde & Spencer-Booth, 1967, 1971). It may be that the frequent attacks made by infants on their mother's subsequent consorts as in orangs (*Pongo pygmaeus*) and chimps (*Pan troglodytes*) (MacKinnon, 1974; van Lawick-Goodall, 1968) can be explained in similar terms. By such behaviour, infants could occasionally prevent successful insemination, thus prolonging the period of maternal support, though evidence on this point is lacking. However, this is not an adequate explanation of all cases of the phenomenon for, in some species, unrelated infants are as likely to attack the consort as those of the copulating female (Gouzoules, 1974).

Beneficience directed at non-relatives

An alternative evolutionary pathway for the development of cooperative behaviour is provided by reciprocal altruism (Trivers, 1971). Cooperative

* In a recent paper, Alexander (1974) provides a model which appears to preclude the possibility of parent/offspring conflict. It is argued that if a selfish mutant arises in one member of a sib group, and reduces its parent's inclusive fitness, it is unlikely to spread (because the same genes will be operative in the adult offspring as in the parent). Taken to its logical conclusion, this suggests that offspring should always act in the interests of their parents.

This may be so in asexual organisms where parent and offspring are genetically identical, but such a mutant may well spread in a sexually reproducing population where parents and offspring have only half their genes in common.

relationships between unrelated individuals may develop where a beneficient action adds more to the recipient's fitness than it detracts from the benefactor's and is likely to be returned in the future. Reciprocal altruism is distinguished from symbiosis (cases where both participants gain immediately in fitness from an interaction) by the fact that it involves a time lag between the beneficient act and its reward.

A variety of social situations fulfil the criteria for the evolution of reciprocal altruism. For example, in supportive coalitions between breeding males (see above) the advantage gained by the recipient may be high and cost to the benefactor low, while the possibilities for repayment may also be high. This may also be the case in situations where individuals combine to attack potential predators (e.g. Harlow, 1969; Yoshiba, 1968).

It is not known how important this mechanism is in the evolution of cooperative relationships among primates. Reciprocal altruism is usually difficult to identify since it is most likely to occur between relatives, when its effects will be indistinguishable from those of kin selection. However, the latter can be excluded in cases of cooperation between members of different species (Trivers, 1971). Certain polyspecific associations (e.g. in *Cercopithecus* spp.; Gartlan & Struhsaker, 1972) could have evolved by reciprocal altruism. Wide-ranging species may profit from the detailed knowledge of resource distribution acquired by those with smaller ranges while the latter gain additional shelter from predators (see Hamilton, 1971*a*), though members of both species suffer the disadvantages of increased feeding competition.

It is important to note that many interactions cited as examples of reciprocal altruism can be more simply explained as cases of symbiosis. For example, in polyspecific associations involving species whose sensory modalities are differently developed (such as baboons or vervets and bushbuck (*Tragelaphus seriptus*) or impala (*Aepyceros melampus*) – Altmann & Altmann, 1970; Elder & Elder, 1970) both species probably gain immediately from the association and it is not necessary to postulate a delayed reward.

DISRUPTIVE BEHAVIOUR

Disruptive or selfish actions diminish the fitness of the recipient. Interactions involving disruption are likely to occur where two or more individuals compete for limited resources and can, for our purposes, be regarded as one form of ecological competition (see Nicholson, 1955; Brown, 1964; Miller, 1967; Geist, 1974). Perhaps because disruption poses a less obvious problem to evolutionary theory than beneficience, it has attracted less attention.

Aggression

However aggression is defined, it usually falls within the category of disruptive behaviour. The functions of aggression in animal societies are frequently viewed in terms which rely on group selection. For example, Lorenz (1966, p. 22) argues that it fulfils a 'species preserving function' (see also Scott, 1962, 1974; Washburn & Hamburg, 1968). Similar examples are widespread in recent literature on primate behaviour. Sorenson (1974, p. 14, quoting George, 1966) argues that, to have survival value, aggression must 'bring order into the social circle'. And in the same volume, Nagel & Kummer (1974, p. 175) suggest that its function is to 'maintain social structure and to relate it to ecological resources'.

There is no need to view the function of aggression in this way. The sifaka (*Propithecus verreauxi*) which fights to gain access to females (Richard, 1974*b*) or the mangabey (*Cercocebus albigena*) which successfully disputes a clumped food source by threatening its neighbour (Chalmers, 1968) is likely to increase its own fitness. Behavioural traits which enable individuals to do this can spread through the population even if they diminish average reproductive success (see Hamilton, 1971*b*). For example, in several primate species, males may disrupt each other's reproductive activities. In hanuman langurs (*Presbytis entellus*), males which occupy a troop after successfully ejecting the previous male(s) may attack and kill infants (Sugiyama, 1965*a*, *b*; Mohnot, 1971; Hrdy, 1974). Similar behaviour may occur in black-and-white colobus (*Colobus polykomos*) and hamadryas baboons (J. F. Oates, personal communication; Kummer, Gotz & Angst, 1974). By infanticide, such males remove potential competitors of their own offspring and ensure that the females of the group come quickly into oestrus. In addition, males of some species attack other males who are copulating (Gouzoules, 1974) and sexual behaviour in subordinates may be inhibited by the presence of a dominant animal (Perachio, Alexander & Marr, 1973). Gouzoules (1974) has documented a number of cases of harassment in primate societies. His suggestion that the function of harassment may be to direct the copulating male's aggression away from his consort is difficult to state in terms of individual advantage. It seems more likely that, by harassing copulating animals, males may increase their own chances of fertilising the females, though there is no available evidence on this point.

This approach to aggressive relationships requires us to explain why animals are not more aggressive. It is evident that aggression incurs costs as well as benefits (Tinbergen, 1951; Hutchinson & MacArthur, 1959; Ripley, 1961; Geist, 1974; E. O. Wilson, 1975). Theoretical models of contest strategies (Maynard Smith & Price, 1973; Parker, 1974; Gale & Eaves, 1975) show that neither the most persistent fighting strategy nor the most savage one is necessarily most advantageous because such strategies are likely to escalate costs to both contestants. This does not explain why, in contests where both animals have committed themselves and one has won, the winner does not kill the loser to avoid further competition. One possibility is that, even here, an attempt to kill or severely wound by the winner might lead to a violent counter-attack by the loser (Geist, 1974). For example, when rhesus monkeys have been defeated they may passively accept incisor bites but will counter-attack viciously if the winners use their canines (Bernstein & Gordon, 1974). In such situations, the cost of escalating the contest may exceed the advantages of removing a potential competitor. Alternatively, such advantages may fail to outweigh the time and energy which would be needed to kill the rival. Finally, in those species where winners are inhibited from killing losers this could be because they might be closely related and the act would diminish the winner's inclusive fitness.

We must not expect the relationship between temporary variation in resource availability and the frequency (or intensity) of aggression to be a simple one. Indeed, there is already some evidence that it is not. In several species, aggressive interactions are common during periods of intermediate food availability. When food is abundant, the frequency of aggression drops (Kruuk, 1972; E. O. Wilson, 1975) presumably because the benefit of winning contests does not justify the expenditure of time, energy or risk to further reproductive potential involved in fighting. At periods of very low food-availability, the frequency of aggression is also reduced (e.g. Hall, 1963; Southwick, 1967; Loy, 1970), presumably because time/energy budgets are so finely adjusted that all forms of expenditure must be minimised. Similar rules are likely to apply to all species though we should expect to find inter-specific differences in the form of the relationship.

Sexual dimorphism and aggression

The theory of sexual selection states that the sex with lower reproductive potential will be competed for by the other sex (Fisher, 1958; Trivers, 1972). In most animals, the reproductive potential of males is considerably higher than that of females and their reproductive success usually varies more widely (Trivers, 1972). Consequently, we should expect selection to favour greater development, in males, of traits which enhance successful competition for males. Many sex differences in body size, weapon development (e.g. antlers, horns, canines) and pelage (Goss-Custard *et al.*, 1972; Schaller, 1972; Geist, 1974) can thus be explained in terms of individual advantage and explanations relying on group advantage (e.g. DeVore & Washburn, 1963; Coelho, 1974) are unnecessary. Aggressiveness can be seen as a functionally similar trait since it may assist a male to compete successfully for females. In the great majority of primate species, males appear to be more aggressive than females (see Crook, 1972; Chalmers, 1973; Holloway, 1974) and are generally dominant to them. (There is no obvious explanation why in some species, including ring-tailed lemurs (*Lemur catta*) (Jolly, 1966), *Indri* (J. Pollock, personal communication) and talapoins (*Miopithecus talapoin*) (Wolfheim & Rowell, 1972) females are dominant to males.)

Convincing evidence that this view of the evolution of male characteristics is correct comes from studies of polyandrous bird species where females compete for access to males. Here, one would predict that females would possess most of the traits which usually typify males: they should be larger and more aggressive, initiating courtship and taking primary responsibility for defence of the territory. This is the case in most of these species (Jenni, 1974).

The same approach helps us to understand the evolution of certain inter-specific differences (Trivers, 1972; Alexander, 1974). We should expect to find the most marked sex differences in those societies where breeding competition is most intense. Thus, sex differences should be greatest in strongly polygamous societies and least in truly monogamous ones. By and large, this is true for primates as well as for other mammalian groups (e.g. ungulates; Jarman, 1974). Species such as patas monkeys (*Erythrocebus patas*) or gelada baboons (*Theropithecus gelada*) which show the most unequal socionomic sex ratios* are characterised by the greatest sex differences in body size. In contrast, monogamous species such as gibbons or titis show the least. We would also predict that differences in aggressiveness between males and females should be most marked in the most polygynous societies, although suitable comparative data are rare (Nagel & Kummer, 1974). For the same reasons, males of strongly polygamous species should be more likely to fight dangerously and to escalate conflicts than males of monogamous species.

The situation is probably further complicated by differences in aggressiveness between species which are not the product of sexual selection. For example, differences in aggressiveness between marmot species appear to control differences in the timing and dispersal in the young and may have developed for this reason (Barash, 1974). Similarly, inter-specific differences in life span may affect aggressiveness: the longer the life span, the less willing individuals should be to risk reproductive potential for immediate access to resources (Geist, 1974). Consequently, aggression may be commoner and more intense in short-lived species than in animals with long life spans.

* i.e. the ratio of adult males to adult females in reproductive groups.

Finally, the nature of food supplies may be important. Where they are evenly dispersed, highly transient or super-abundant, the advantages of aggression may be minimal (Brown, 1964; Geist, 1974).

Age differences in aggression

Age differences are likely to affect the advantages of aggression to the individual (Parker, 1974). Two processes will be involved. First, as an individual matures the chances of it successfully winning disputes are likely to increase initially and then decrease as it passes its prime. Secondly, older individuals should be prepared to invest more in these aggressive encounters than younger ones because their reproductive potential is lower (Trivers, 1972). Age changes in aggressiveness in primate societies have not yet been well documented. However, there is evidence that juveniles and adolescents relatively rarely show aggression (e.g. Chalmers, 1973) compared with adults. In addition, detailed studies of hunting forays in chimpanzees (Wrangham, 1975) suggests that older individuals may be more prepared to contest access to supplies of meat.

Kinship and aggression

Kin relationships, too, may modify aggression (see Hamilton, 1971*b*). One might predict that the amount of tolerance extended to different relatives would correlate with their degree of genetic similarity. This appears to be the case among Japanese macaques where genealogical relationships extending over several generations are known in a number of troops. To examine the extent of tolerance between different relatives, Yamada (1963) scattered wheat over ten areas of approximately one square metre each and recorded the frequency with which animals fed together at the same area. Co-feeding was limited by the tendency of dominant animals to threaten subordinates feeding at the same site. Consequently the frequency with which individuals co-fed with different subordinates provided an approximate measure of the degree of tolerance extended to them (though there is an obvious danger that co-feeding frequency was also affected by the likelihood of subordinates approaching and feeding with dominant relatives). From Yamada's data, it is possible to calculate the average frequency with which (*a*) mothers tolerated children, (*b*) sibs tolerated other sibs, (*c*) grandmothers tolerated grandchildren, (*d*) aunts tolerated nieces and nephews, (*e*) mature females tolerated unrelated individuals. (In each case the first mentioned was dominant to the second.)

Table 1. *Average frequency of feeding tolerance (see text) shown by Japanese macaques to subordinate individuals compared with their degree of genealogical relationship (data from Yamada, 1963)*

Dominant	Mother	Sib	Grandmother	Aunt	Mature female
Subordinate	Child	Sib	Grandchild	Niece or nephew	Unrelated
Degree of relationship	0.5	0.25–0.5	0.25	0.125–0.25	0
Average frequency of observations where co-feeding occurred	18.3	12.6	2.7	3.3	1.1
Number of pairs of individuals involved	16	14	4	6	154

Table 1 compares the average frequency with which different relatives tolerated each other with their estimated degree of genealogical relationship. As can be seen, there is a close association between the two measures.

Kin selection may help to explain the function of unsolicited aggression (see p. 209) in animal societies. It is not uncommon for dominant individuals to threaten or attack animals which are neither initially close to them nor are competing for any obvious resources (Hall & Mayer, 1967; Richard, 1974*a*, *b*). By doing this, they may possibly be depressing the latters' status and thus increasing their own relative rank or that of their offspring. Similarly, females may attempt to prevent other females from breeding. In several species of Old World monkeys, females harass copulating couples (see above) while, in wild dogs, breeding females may even kill litters born to other group members (van Lawick & van Lawick-Goodall, 1970).

Inter-specific differences in the genealogical composition of social groups may help to explain some differences in the distribution of aggressive encounters. For example one would predict that in monogamous societies or in species living in single male troops where sibs have 50% of their genotype in common, they would be more tolerant of each other than in societies where promiscuous mating occurs (where they are likely to share only 25%). However, the relevant evidence is lacking in mammals.

Parental manipulation

In general, we should not expect parents to be aggressive to their own offspring (see above). However, in some situations, this may increase their inclusive fitness and it may even be to the advantage of the parent to kill one or more of its offspring (Alexander, 1974; Trivers, 1974). For example, some birds preferentially feed older offspring (e.g. Lockie, 1955) with the probable result that the youngest fail to survive and may even be eaten by their sibs when food supplies are short. And in groups of Japanese and rhesus macaques, female offspring of the same mother tend to rank immediately below her, their rank within the matriline being inversely related to their age (Yamada, 1963; Koyama, 1967; Sade, 1967). This may partly occur as a result of the mother's support for her younger offspring in agonistic interactions (see Koyama, 1967, 1970). By doing this, she may help to protect her younger daughters from harmful competition with her older offspring, thereby increasing her inclusive fitness. In this case, we should predict that orphans would be out-ranked by their older sibs, though the available evidence for this is ambiguous (Sade, 1967; Missakian, 1972).

Spiteful behaviour

So far, we have only discussed disruptive actions which are to the benefit of the initiator. In certain cases, disruptive behaviour may evolve which either does not benefit the individual's inclusive fitness or even diminishes it (Hamilton, 1970, 1971*b*). These are cases where an individual 'spitefully' disrupts other members of the population thus decreasing their fitness relative to its own. Hamilton (1970) argues that such behaviour is unlikely to evolve for three reasons: that all spiteful actions incur appreciable costs; that animals will not be able to differentiate between individuals which are of less than average relatedness to them and those which are more closely related, and may therefore damage their own kin; and that the trait is only likely to spread in small populations which it may help to extinguish. But these conditions are met in higher mammals. In species where marked dominance hierarchies exist, disruptive actions directed by dominants at subordinates may cost the dominant little. Also there is evidence that individuals may be able to distinguish their degree of relatedness to other animals quite nicely (see above). Finally, if one supposes that spiteful aggression is only shown in certain contexts (for example at high population densities) there is no reason to think that it would extinguish the lineage in which it arose. While it therefore seems possible that spite may evolve, it will be extremely difficult to distinguish from selfish disruption. For example, even where apparently unsolicited aggression occurs, it is possible that it increases the initiator's inclusive fitness (see above).

One possible example of spite might be called 'punishment'. Where a dominant individual's access to resources is jeopardised by the behaviour of a subordinate, it may be to the former's advantage to punish the latter, reducing its fitness enough to make further attempts unprofitable. Hyenas

(*Crocuta crocuta*) that intrude into their neighbours' ranges may be attacked and killed even after they have submitted and are attempting to escape (Kruuk, 1972). By occasionally killing such intruders, clans must reduce the potential advantages of poaching as well as removing competitors. And in Japanese macaques adults may punish individuals which attack other troop members (Kawamura, 1967). Punishment is only likely to occur between individuals where there is an initial asymmetry in contesting ability, so that the costs of punishing are relatively low. It is particularly likely to occur where the punishment is experienced or observed by several individuals, thus increasing the benefits of punishing. The reverse of punishment, where a dominant rewards a subordinate for behaviour which increases the former's fitness, constitutes one form of reciprocal altruism and has already been discussed (see p. 202).

Similar behaviour may occur which does not involve transactions in fitness. Among most higher animals, the behaviour of individuals in social interactions is probably extensively conditioned by the consequences they experience as a result of each others' actions (see Hinde & Stevenson-Hinde, this volume). We should expect the reactions which different individuals evoke to be adapted (so that social actions received by an individual which increases his/her fitness will be positively reinforcing and those that diminish it will be negatively reinforcing). However, it is difficult to believe that reactions could be equally well adapted to all contexts and it is possible that individuals may learn to take advantage of each others' learning systems. For example, if B behaves towards A in such a way as to increase A's fitness, A's most advantageous manoeuvre might be to reward B in 'artificial currency', trading a reinforcement which does not increase, or even detracts from, the latter's fitness. Similarly, it may be less costly to punish subordinates with behaviour which they find uncomfortable than actually to diminish their fitness. Clearly, this will not be a stable situation, for selection is likely to increase individuals' abilities to distinguish between 'real' and 'artificial' rewards or punishments. However, we might expect to find animals attempting to pull off such confidence tricks, particularly in social contexts which are highly variable. As Humphrey (this volume) has pointed out, selection operating through transactions of this kind would be likely to favour the evolution of intelligence.

Sexual swellings

The second part of this section considered the adaptive significance of sexual dimorphism in a number of traits which are more marked in males than females. It is clearly relevant to consider briefly the functional significance of traits which are more developed in the female. 'In my "Descent of Man"' wrote Darwin (1876) 'no case interested and perplexed me so much as the brightly coloured hinder ends and adjoining parts of certain monkeys.'

Pronounced cyclical swellings of the perineal or para-callosal area show a strange distribution throughout the primates. They occur in a few prosimians (including mouse lemurs, tarsiers (*Tarsius spectrum*) and at least some species of *Lemur*); they are widespread in certain genera of Old World monkeys (including *Papio, Macaca* and *Cercocebus*) while in others they occur in a proportion of species (e.g. *Colobus*) or are absent altogether (e.g. *Cercopithecus, Presbytis*); among the apes, they occur in chimpanzees, but not in gibbons, siamangs, gorillas or orangs (Napier & Napier, 1967; Wickler, 1967). In many species, there is evidence that swellings are most pronounced during oestrous periods (Wickler, 1967; Rowell, 1972) and that they are associated with increased frequency of sexual approaches by males (Saayman, 1970, 1971; Tutin & McGrew, 1973*a*, *b*). While it is extremely likely that swellings serve to signal the females' reproductive state, this still fails to account for their distribution across species.

Two alternative hypotheses exist which may account for the distribution. Swellings may help to ensure that the female attracts the attention of at least one male during the course of oestrous periods. In this case, one might expect to find swellings in species living in habitats where visibility is reduced, where individuals are widely separated or where there is a heavily biased socionomic sex ratio in favour of females. Alternatively, swellings may help to attract several males, enabling the female to choose (either actively or passively) between them. In this case, one would expect to find sexual swellings in species where females have access to several potential mating partners. More specifically they should occur in species which are typically found in multi-male troops, but not in those which usually live in single-male groups.

There is little evidence to support the first explanation. Indeed, Darwin himself (1876) pointed out that sexual swellings tend to occur in open-country species where visibility is good. Though most of the Old World monkeys do live in groups with an uneven sex ratio, there is no obvious association between the occurrence of sexual swellings and socionomic sex ratios which are more than usually imbalanced.

In contrast, there is a close association between cyclical swellings and multi-male troops. The three genera where swellings are commonest (*Papio, Cercocebus* and *Macaca*) typically live in multi-male groups. Within the genus *Colobus*, female sexual swellings occur in red colobus (*Colobus badius*) and in olive colobus (*Colobus verus*) (Hill, 1952; Kuhn, 1972) which are found in large groups (Clutton-Brock, 1974; Hill, 1952) but not in black-and-white colobus which usually live in smaller, single-male groups (Marler, 1969). Among the guenons and related species, they occur in talapoins, which also live in large troops (Gautier-Hion, 1970) but not in *Cercopithecus* spp. (Rowell, 1972) which mostly live in one-male troops (Struhsaker, 1969). Similarly, pronounced sexual swellings occur in female chimpanzees but not in gibbons, orangs or gorillas which live in smaller groups (Napier & Napier, 1967). Although some of the prosimian species which possess swellings (including tarsiers and lemurs) do not live in permanent social groups, their social systems are probably of a kind which permits female access to several males.

One possibility is that, by attracting several males, the female increases her chance of being mated by a relatively high-ranking male: in a recent study of sifakas, Richard (1974*b*) found that females would only copulate with males who were dominant in the context of the breeding season. Receptive female gorillas and baboons direct their sexual advances at the dominant male(s) (Hess, 1973; Saayman, 1970) and are less likely to tolerate the attention of subordinates. This may either confer genetic advantages on the female's offspring or, in species where a female's rank is affected by the rank of her consort, improve her social position in the group (see below).

There are several problems with this point of view. A number of species which typically live in multi-male groups do not show cyclical swellings. In most of these cases, swellings do not occur in any species belonging to the same taxonomic group. For example, they are apparently absent both in New World monkeys and in Asiatic colobines. Since we should expect evolution to produce different adaptations in different phyla, exceptions at this level should not surprise us (Clutton-Brock, 1974). The case of the gelada shows that species can evolve different ways of signalling cyclical sexual changes. Here, oestrus is associated with the development of wart-like blisters on the female's chest patch and in the paracallosal area. Although both some ano-genital swelling and changes in colouration of the sexual skin occur, these are associated with the female's age and general reproductive state (cycling versus non-cycling) rather than with oestrus (Dunbar & Dunbar, 1974).

The distribution of cyclical swellings within the genus *Macaca* poses a more important problem. Here some species (e.g. *M. sylvana*, *M. nemestrina* and *M. silenus*) show conspicuous cyclical swellings while others (including *M. mulatta*, *M. speciosa*, *M. radiata* and *M. fuscata*) show colour changes in the ano-genital area but little or no swelling (Hill, 1966; Napier & Napier, 1967). Both groups of species typically live in multi-male troops and the absence of swellings in some is hard to understand. However, all the species which do not show cyclical swellings show colour changes (Napier & Napier, 1967) and it is possible that they have evolved alternative ways of attracting males, perhaps employing olfactory or behavioural signals to a greater extent (see Michael & Keverne, 1968, 1970).

Another problem is presented by *Papio hamadryas* where marked sexual swellings occur in a species which lives in one-male breeding units. There is no obvious explanation of this since not only do females seldom or never mate with adults other than their unit male but they join their groups before starting to cycle (Kummer, 1968).

In addition to changes associated with the oestrous cycle, females of several species show swellings or colour changes associated with particular stages of the life cycle. For example, in juvenile rhesus macaques ano-genital swellings occur which are not found in adult females (Rowell, 1963) and in geladas the patch of sexual skin on the chest swells around puberty (Dunbar & Dunbar, 1974). Changes in ano-genital or para-callosal colouring associated with the latter stages of pregnancy occur in a number of species, including baboons, rhesus macaques, white-cheeked mangabeys, vervets (*Cercopithecus aethiops*), talapoins and geladas (Rowell, 1972; Dunbar & Dunbar, 1974). While it may be advantageous to females to signal their reproductive status clearly it is not known why these changes should occur in some species but not in others.

The presence of cyclical changes in females' genitalia has led to the evolution of analogous structures in the males of a variety of species, including hamadryas baboons and vervets (Wickler, 1967). Wickler argues that mimicking the genitalia of an oestrous female helps subordinate males to appease dominant animals in conflict situations. This argument is strengthened by two recent studies. First, young male red colobus and olive colobus possess a perineal organ which closely mimics female sexual swellings and which becomes progressively less obvious as the animal grows older (Kuhn, 1972). It is during the same period that, in red colobus at least, young males are regularly attacked by mature males (T. Clutton-Brock, personal observation), perhaps in an attempt to make them leave the troop. Secondly, in spotted hyenas, where the female is larger and more dominant than the male, females mimic male genitalia, possessing an elaborate pseudo-penis and pseudo-scrotum which is used in greeting and appeasement signals (Kruuk, 1972).

The distribution of male mimicry of female characteristics is puzzling (Crook, 1972). Accepting Wickler's hypothesis, one might expect it to occur in species living in multi-male troops, yet it reaches its most elaborate form in hamadryas baboons and is not found in talapoins (Wickler, 1967). There does not appear to be any obvious explanation.

One final point concerning the evolution of mimicry needs to be mentioned. It is commonly argued that the mimic 'exploits' the reaction of the animal which it is appeasing. While this may be the case, it is difficult to believe that the latter really cannot distinguish the mimic from its model. And, in so far as being 'exploited' is disadvantageous, selection would presumably have favoured the ability to distinguish between the two. Although the original mimics may have exploited other animal's reactions, presumably the mimicry has now become part of a ritualised appeasement signal and the dominant animals are well able to distinguish it. This suggests that they should only respond to it in so far as this is to their own advantage.

FUNCTIONAL ASPECTS OF SOCIAL STRUCTURE

Having considered the functions of cooperative and disruptive actions, it is now possible to examine the functional significance of variation in social structure in similar terms. This section considers four aspects of social structure in primate societies.

Territoriality

The adaptive significance of territoriality in birds and primates has been extensively reviewed (Hinde, 1956; Tinbergen, 1957; Carpenter, 1958; Bates, 1970; Maynard Smith, 1974*a*). However, explanations seldom differentiate between advantages to different members of the same group. For example, it is argued that territoriality may (1) minimise disturbance by conspecifics (2) ensure a reliable food supply for the breeding unit (3) reduce predation by dispersing the population (see Hinde, 1956; Tinbergen, 1957).

While it is possible that, in some species, territoriality may benefit all members of the group equally, in other cases it may only benefit particular individuals. In an originally monogamous species, males which oust neighbouring males and defend large territories may increase their fitness by gaining breeding access to several females (Hamilton, 1971*b*). In such a situation, territoriality is likely to spread through the population even if it is not to the advantage of the females (see p. 198). This interpretation may explain why, in most territorial primates, it is the male who is largely responsible for defence (though this situation could have arisen by other evolutionary routes).

Group sex ratio

It is commonly argued that the adaptive significance of socionomic sex ratios which are biased towards females is that they result in more efficient use of resources than more equal ratios (Crook & Gartlan, 1966; Gartlan & Brain, 1968; Kummer, 1971). In the way the argument is usually stated, it relies on group selection though this need not be the case if it is assumed that

(i) females eject males to avoid food competition,

(ii) dominant males eject subordinate males to prevent them competing for food with their females and young (Geist, 1974).

Both functions are feasible, though there is little evidence for the first. However, in societies where adult males eject young or adolescent males (Sugiyama, 1965*a*, *b*; Poirier, 1969), it is more likely that they do so because this helps to ensure exclusive breeding access to the group females (Goss-Custard *et al.*, 1972).

In fact, it is harder to understand why dominant males should ever tolerate the presence of reproducing subordinates since the disadvantages of sharing breeding access would be considerable (Maynard Smith & Ridpath, 1972). The presence of additional males may benefit all other group members by increasing the efficiency of predator detection or cooperative defence (Crook, 1970; Altmann & Altmann, 1970). Alternatively, their presence may help the dominant to defend his position against attempts to displace him by extra-troop males (see Jay, 1965; Sugiyama, 1965*a*, *b*). The disadvantages of tolerance to the dominant animal may be minimised by allowing only related males to remain (see Bertram, this volume) and by maintaining access to ovulating females. However, the latter argument provides no explanation of why multi-male troops have developed in some species but not in others. This remains one of the most striking enigmas in primate social behaviour.

Social dominance

The concept of social dominance has been extensively employed since it was first described (Schjelderup-Ebbe, 1922). Recently, several reviews have rightly condemned its uncritical use and have stressed that relationships between individuals cannot be described adequately in terms of linear hierarchies (Rowell, 1974*a*; Syme, 1974). Since at least one reviewer (Gartlan, 1968) has argued that dominance hierarchies may represent a pathological response to abnormal environments, it is necessary to justify our acceptance of the concept before discussing its function.

Dominance has been defined in many ways (Rowell, 1974*a*). In functional terms, its central statement is that particular individuals in social groups have regular priority of access to resources (or of avoidance where noxious stimuli are involved) in competitive situations. If it is to be a useful explanatory concept or 'intervening variable' (Hinde, 1974) the access of individuals to different resources and in different contexts should be correlated (Richards, 1974). In its most developed form, dominance rank is 'transitive' (i.e. if male A has priority over male B he will also have precedence over all individuals over which B takes precedence) and is related to behaviour in agonistic and affiliative interactions as well as in asocial contexts (Bartlett & Meier, 1971; Richards, 1972).

Gartlan (1968) offers two important criticisms of dominance. First, he points out that access to different resources may be poorly correlated (see also Syme, 1974). In particular, there is evidence that the frequency with which individuals copulate is often poorly correlated with their rank in other situations (Kummer, 1957; Southwick, Beg & Siddiqi, 1965; Jay, 1965; Jolly, 1966; Bygott, 1974). Secondly, he stresses that linear hierarchies occur in caged groups or in populations where density is unnaturally high and are usually associated with frequent aggression. The inference here is that hierarchies seldom occur in natural populations under normal conditions.

Neither of these two points requires us to reject the concept though both emphasize its complexity. In fact, evidence from a wide variety of species shows that access to many different resources is correlated in many species, though not in all (Bernstein, 1969, 1970; Richards, 1974; Syme, 1974), and that agonistic dominance is generally associated with breeding success (see refs. in Wynne-Edwards, 1962; Loy, 1971; Suarez & Ackerman, 1971; E. O. Wilson, 1975). In addition, some degree of hierarchical ranking is present in most social species (Carpenter, 1954; Washburn, Jay & Lancaster, 1965; Mazur, 1973) and in field studies where no hierarchy is reported (e.g. Carpenter, 1934; Rowell, 1966, 1967; Neville, 1972) it is usually difficult to be certain that this is not a product of inadequate sampling. However, there is evidence that the predictive value of ranking is lower in some species than in others (see below). The significance of these differences becomes clearer when one considers the adaptive significance of dominance.

Three common explanations of the function of dominance are that it reduces aggression (Collias, 1953; Scott, 1962; Poirier, 1974), that it restricts the access of the most 'expendable' individuals to limiting resources (Chance & Jolly, 1970; Kummer, 1971) and that it helps to regulate population density (Wynne-Edwards, 1962). Such arguments rely on group selection, and have arisen because a hierarchy has been regarded as a behavioural trait, whereas it is no more than 'the statistical consequence of a compromise made by each individual in its competition for food, mates and other resources. *Each compromise is adaptive but not the statistical summation*' (Williams, 1966, p. 218).

The evolution of hierarchies is not difficult to explain. Individuals will benefit by gaining access to food and mates in competitive situations. Whether it will be to their advantage to contest access will depend on the potential benefit of acquiring the resource and on the potential costs of the interaction. Where possible benefits are high and costs low, individuals are likely to contest access and where benefits are low and costs high they will not do so (see Parker, 1974). Hierarchies occur because competitive ability inevitably varies between individuals and because less successful animals learn not to contest access to encounters where they are unlikely to win thus saving time and energy. They will be transitive if individuals' abilities to compete successfully are independent of the identity of their opponents (though not, obviously, of the competitive ability of the latter). Animals may initiate dominance interactions outside the immediate context of competitive encounters (e.g. Struhsaker, 1967*a*, *b*; Saayman, 1972; Bygott, 1974) if this enhances their ability to compete successfully in future.

Thus dominance may be analogous to territoriality (Wilson, 1971). Both traits enable some individuals to obtain a larger-than-average share of available resources by disrupting competitors. The switch from territoriality at low population density to dominance behaviour at high density in many species (e.g. Archer, 1970), suggests that the two traits may be causally as well as functionally related.

Rowell (1974*a*) proposes a different view of dominance, stressing two points. First, since the majority of approach-retreat interactions are initiated by subordinates, we should regard the hierarchy as being maintained by the subordinates rather than the dominants. Secondly, ranking may develop through observational learning in competitive situations. Thus if two strange monkeys are repeatedly presented with a raisin, which one of them consistently snatches, the other will learn that it is unlikely to get the reward and will eventually give up trying.

Neither point constitutes a serious criticism of the view of dominance suggested here. Even if the subordinates initiate the majority of interactions, they do not necessarily control the relationship. For example, monkey B may retreat from monkey A in 99 cases in the absence of any action or display on A's part. On the hundredth occasion he fails to retreat, and is attacked and chased away by A. As a result he is likely to initiate subsequent retreats without waiting for A to respond. For every retreat initiated by A, the observer will score a far greater number which are initiated by B, yet it is A's behaviour which largely controls the relationship.

Moreover, while observational learning may well be involved in the ontogeny of dominance hierarchies, it is only likely to be important in cases where it provides a reliable indicator of the observing animal's chances of acquiring the resource. Thus if the raisin, in Rowell's example, was an oestrus female which, by chance, immediately submitted to the blandishments of the first monkey, we should not expect the second to fail to contest such situations in future. Natural selection would be most unlikely to encourage such a fatalistic attitude.

Both costs and benefits of contesting another individual's priority in competitive dyadic encounters (as well as the costs and benefits of submitting) will be affected by the same factors that influence the advantages and disadvantages of aggressive behaviour generally (see p. 204). If our view of dominance is correct, it should be possible to understand many of the currently unexplained complexities of hierarchies (which so often confuse attempts to generalise about dominance, e.g. Gartlan, 1964) in terms of variation in the costs and benefits of contesting and submitting (see Maynard Smith, 1974*b*; Parker, 1974).

This appears to be the case. The following subsections discuss the possible explanation of variation in interaction patterns associated with four different variables.

Absolute value of resource. Where the values of individual resource units are extremely low, it may not be advantageous to contest access to them. This may help to explain the apparent absence of hierarchical food access in some herbivorous and folivorous animals (see Geist, 1974). Conversely, when the value of resources is extremely high, contesting will almost always be advantageous. In a number of species which live in individual territories or harem groups, males which do not gain a harem or a territory usually fail to breed. (Kummer, 1968; Watson & Moss, 1972). In several such species, males usually fail to form stable hierarchies when caged together (e.g. Zuckerman, 1932; Scott, 1966; von Holst, 1974).

Inter-individual differences in resource value. Where the costs or benefits of exploiting different resources vary between individuals, individual access to different resources should also be expected to vary. Cases where access to different kinds of food are not closely correlated (e.g. Harding, 1973; Dunbar & Crook, 1975; Wrangham, 1975) may be a product of differences in dietetic needs and hence in food values: rhesus monkeys raised on low-protein diets outrank individuals raised on high-protein diets on food access but not on avoidance competition (Wise & Zimmerman, 1973). It is, however, very difficult to explain the high frequency of copulations achieved by subordinates in many cases (e.g. Bygott, 1974; Enomoto, 1974) and the tolerance of breeding activity in subordinates by dominant animals belonging to the same group (e.g. Hanby, Robertson & Phoenix, 1971; Saayman, 1971). It is clearly important that studies should investigate the number of successful fertilisations achieved by high and low ranked individuals: in a troop of Japanese macaques Hanby *et al.* found that although some of the most sexually active males were low in rank, dominant animals ejaculated more frequently. Where tolerance of 'genuine' breeding actually exists, the usual degree of relatedness between dominants and subordinates must be determined.

Changes in resource value. In situations where either costs or benefits to individuals of acquiring resources vary through time, we should expect their willingness to contest access to do so too (Parker, 1974). Where individuals have recently held a particular kind of resource, the advantage of maintaining access to it is likely to diminish. This may help to explain (i) why the longer initially dominant individuals have fed at a food source, the more likely they are to be displaced or to distribute a part of their food to other individuals when challenged (see Wrangham, 1975); (ii) why deprivation, which is likely to affect individuals differentially, may produce changes in rank (Boelkins, 1967; Castell & Heinrich, 1971).

Alternatively, previous access to resources may increase their value (Parker, 1974). In experiments with caged hamadryas baboons, Kummer *et al.* (1974) introduced a strange female to two strange males. In such cases, the two males would contest access to the female, and the winner then formed a social bond with her. However, if one of the males was allowed prior access to the female and the other was able to observe the two interacting before being introduced into the same enclosure, he 'respected' the bond and did not attempt to acquire the female even if he was dominant to the first male. Where the second male was much higher in dominance rank than the first, his inhibition sometimes disappeared and he attacked the first male. One possible explanation is that, in a natural situation, a female is likely to escape from an individual who defeats her established male (perhaps because the former may kill her infant – see p. 209). In this case, the value of the female to the holding male is greater than to the attacker and the latter might be expected to contest the situation only when he could do so at low cost. This inhibition may be absent in males which greatly out-rank the holders because the rank difference enables them to acquire the female at little potential cost. It is significant to note that when Kummer presented food (a resource whose value would not differ between the holder and the contestant in the same way as a female) to the animals in the same way, males were not inhibited from taking it away from its previous possessor.

Changes in resource value may be responsible for the appearance of dominance hierarchies only at certain times of year. For example, dominance rank is only apparent in male squirrel monkeys (*Saimiri sciureus*) during the mating season (Baldwin, 1968, 1971). Changes in rank are also common at the onset of mating seasons, as in *Propithecus verreauxi* (Richard, 1974*a*). Young males may be particularly likely to challenge at this time of year because the difference in fighting ability between them and older animals will have steadily decreased over the past year as they have grown. The onset of the breeding season will produce a sudden increase in the benefits of dominance and may consequently make challenging advantageous. We would predict that seasonal changes in rank stability would be most marked in species showing a well defined mating season.

Age changes, too, may affect the value of resources (Trivers, 1972) and might be expected to influence the distribution of contests. In many long-lived primates, males show a relatively short period of reproductive activity (Kummer, 1968; Rowell, 1974*a*). When individuals reach this period of their life, they should become more likely to contest encounters owing to the greater benefits that are possible. As individuals age, both fighting ability and potential loss in encounters will decline (see p. 207). Situations may occur where it is to an old male's advantage to contest access to valuable resources but not to do so for less valuable ones. This may help to explain cases where old males who have fallen in rank are still responsible for a large proportion of copulations (e.g. Hall & DeVore, 1965; Saayman, 1971). However, this phenomenon may also occur if females tend to select males with whom they have mated previously.

Predictability of contest. Aggressive interactions and unstable dominance relationships will be more likely to occur in situations where neither of the contestants can judge the outcome of the encounter (Parker, 1974; Maynard Smith, 1974*b*). This may help to explain why aggressive interactions are relatively common (i) in newly formed groups (e.g. Bernstein, 1969), (ii) between members of established groups and strangers (e.g. gorillas, A. H. Harcourt, personal communication), (iii) between members of the same group who have not met for a while (e.g. Vessey, 1971; Bygott, 1974), (iv) between rank neighbours within established groups (Mazur, 1973) and (v) in species living in very large groups where the same individuals are unlikely to encounter each other frequently.

Group dynamics

In natural populations of primates the composition of groups* is continuously changing. Not only may individuals move between groups but groups themselves may join or divide. Such processes have been described in detail in Japanese and rhesus macaques (Furuya, 1968, 1969, 1973; Drickamer & Vessey, 1972). Group dynamics are often regarded as homeostatic mechanisms operating, at least in part, to the advantage of the species (e.g. Koyama, 1970). Recently, theoretical approaches have shown how dispersal processes may evolve through selection operating at the individual level (see E. O. Wilson, 1975). The most useful approach is to consider the advantages of leaving the group to particular individuals. Below, we consider three questions concerning group leaving:

* In this section we are using Crook's (1970) notation and only refer to stable, long-lasting social units as 'groups'. Temporary aggregations have already been considered elsewhere (Cohen, 1971).

(1) *What are the usual advantages of leaving?* In many animal species, there is evidence of high mortality in emigrants (see E. O. Wilson, 1975). Since this is probably the case in primates too, we should expect the advantages of group leaving to be considerable if they are to outweigh those of remaining.

In many cases, leaving is a functional response to increased population density. Animals which leave presumably gain advantages in terms of food availability, decreased breeding disturbance and reduced susceptibility to disease. Such advantages probably account for instances where groups split (e.g. Furuya, 1968, 1969, 1973) and one of them adopts a new range.

In some species, individuals leave groups as a result of aggression directed at them by dominant members of the group (see below). Adult males are often intolerant of the proximity of adolescent males (e.g. *Presbytis* spp., Sugiyama, 1965*a*, *b*; Poirier, 1969) and adolescents may leave their natal troops to avoid persecution, either becoming solitary or forming all-male groups.

Finally, there is growing evidence that individuals may leave their natal troops to avoid inbreeding (Itani, 1972). Patterns of dispersal occur which are difficult to account for on either of the two hypotheses described above. For example, in *Macaca mulatta* and in *Papio anubis* adolescent males regularly transfer between groups, apparently of their own choice (Sade, 1968; Drickamer & Vessey, 1973; Packer, 1975). That the functional significance of such dispersal is the avoidance of inbreeding is supported by the fact that females often prefer to mate with strange males. In *Papio anubis*, females present more frequently to males which have transferred than to those born into their troop (C. Packer, personal communication) and similar preferences for strange males have been reported in rodents as well as in some invertebrates (E. O. Wilson, 1975). Lastly, there is now good evidence that inbreeding can severely depress the fitness of the offspring, both from studies of rodents (Hill, 1974) and of humans (*Homo sapiens*) (Schull & Neel, 1965; E. Seemanova, quoted in E. O. Wilson, 1975).

(2) *Why is it generally males rather than females which leave?* Since breeding competition between males is likely to be more intense than between females (see p. 205) males are more likely to persecute and eject each other than are females. However, this provides a satisfactory explanation of why males leave only in those species (see above) where persecution occurs.

In others, such as rhesus macaques and olive baboons (*Papio anubis*), where males leave of their own choice, the explanation is not obvious. Four factors are presumably involved: (i) the benefits of moving to a new troop, (ii) the costs of moving to a new troop, (iii) the benefits of staying in the same troop,

(iv) the costs of staying in the same troop. Because of their greater reproductive potential (see p. 205) the possible benefits of moving are likely to be greater to males than to females. Not only does moving give them the chance of finding a troop where there are only a few ageing males which can be easily dominated (thus allowing them to increase their reproductive success considerably) but it enables them to take advantage of the females' preference for mating with 'strange' males (see above). On the other hand, the costs of inbreeding to females would be likely to be greater than to males since the latter's investment in offspring is considerably smaller (see p. 197). This would probably be the case even though males will reduce their inclusive fitness by mating with relatives (assuming that this reduces the latter's fitness). Finally, while the relative costs of moving are likely to vary between the sexes, they would depend on the ease with which 'strange' males and females could join new groups, which probably varies between species.

In baboons and macaques, where males can evidently move quite easily between troops, the most likely explanation is that the advantages of transferring to males are considerable. Consequently, they may outweigh the disadvantages of leaving at a lower level of inbreeding than that at which it would be advantageous for females to leave. Once the males had evolved dispersal mechanisms, thus reducing the danger of inbreeding, the females would be unlikely to do so.

Recent studies of chimpanzees (Nishida & Kawanaka, 1972; Wrangham, 1975; A. Pusey, personal communication) suggest that in this species females are more likely to move between communities than males. Wrangham's interpretation of chimpanzee social structure is that females and their immediate dependents occupy small, relatively well-defined ranges, while males live in kin groups occupying, and perhaps even defending, large ranges covering those of a number of females. He suggests that small range size and group size may be imposed on females by energetic constraints resulting from lactation and pregnancy. Males may maximise reproductive success (see p. 198) by forming kin groups which jointly defend large ranges, overlapping those of a number of females to which they share breeding access. Such kin groups are likely to be hostile to strangers (see Bygott, 1974) and consequently the costs of transferring will be high and the benefits low. This may increase the level of inbreeding at which it is advantageous to males to transfer to a point where it exceeds that for females, with the result that females instead of males have evolved dispersal mechanisms. This theory predicts that female dispersal is most likely to occur in species where related males form breeding cooperatives. There are few such species for which data are available. Consequently, it is significant that at least one of them, the Arabian babbler, shows female dispersal (Zahavi, 1974).

(3) *Why do some individuals of particular age/sex categories leave while others remain?* In a number of mammals where social dynamics have been studied, including lions (Schaller, 1972), Japanese macaques (Yamada, 1963) and rhesus macaques (Drickamer & Vessey, 1973) some individuals of particular age/sex categories leave their natal group while others do not. To understand why this may be, it is worth considering the simplified situation in a matriarchal group. In most social mammals, groups consist of one or more founding females, their daughters, grand-daughters and pre-adolescent male descendants (Eisenberg, 1966, 1971). Consider a situation where

(i) each group is founded by a matriarch, all offspring remain in the group, and females (including the matriarch) start breeding in their third year, subsequently producing one daughter per year;
(ii) groups are territorial, group living is advantageous and emigration from the group territory reduces breeding success;
(iii) individuals within the group are ranked in a dominance hierarchy according to their age, and animals of the same age have equal rank;
(iv) the frequency with which dominant animals direct aggression at subordinates depends on their degree of relatedness in the female line.

In such a group, it is possible to calculate the probable degree of relatedness (in the female line) between all pairs of group members (see Fig. 1). The number of threats which each individual receives will depend on the total amount of unrelated genotype vested in individuals which are dominant to it (see Fig. 2). Hence it is possible to predict the relative number of threats which each individual born into the group is likely to receive.

As group size starts to grow and food availability declines, the number of aggressive interactions will increase. At some stage, the frequency with which the most-threatened individual is disrupted will outweigh the advantages of remaining in the group and she will leave. Since emigrants are likely to be young and vulnerable to predators, it may be to their mothers' advantage to accompany them if this increases the latters' inclusive fitness. In such cases, we should predict that the probability of a particular mother leaving the group will increase with the number of offspring she has produced.

Many variations may be played on the same theme. For example, in several societies, daughters' ranks are inversely related to their order of birth (see p. 208). In such situations, older daughters would incur more threats and be more likely to leave than the younger ones. It is also possible that particular individuals or coalitions might indulge in manoeuvres (and counter-manoeuvres) to drive other animals from the group. For example, in situations where the matriarch's oldest grandchildren will outrank her youngest daughters, it might be to her advantage to ensure that the former leave the

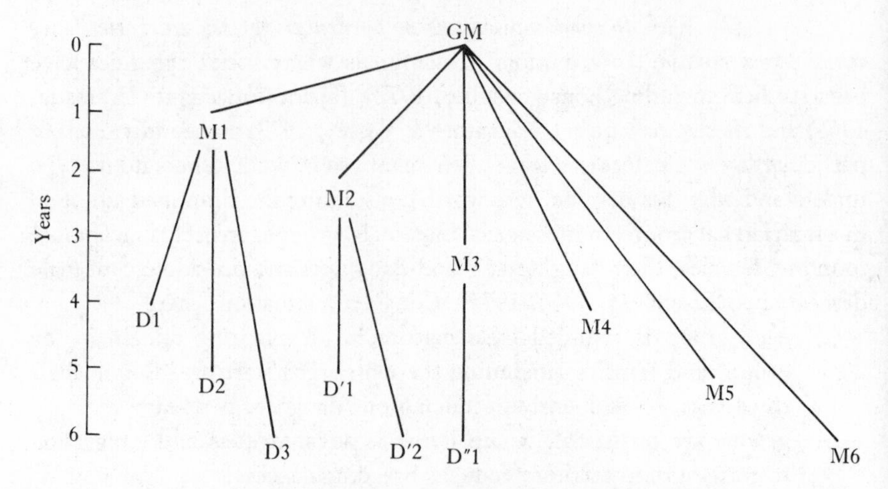

Fig. 1. Year of birth and matrilineal relationships in a model matriarchal society. GM = grandmother, M = mother, D = daughter.

group before the latter are born. In such cases, one might expect to find that a disproportionate number of threats would be given by the matriarch to her oldest grandchildren and, perhaps, to their mother too. This underlines the point that both in simple groups and in more complex societies, individuals which appear to initiate changes in group membership are not necessarily those who are responsible for them.

In societies where the male is permanently present in the group and outranks all other group members, further complexities may arise. If, as in langurs, individual males do not hold groups long enough for incest to occur, it should be to their advantage to ensure that their offspring remain in the group and that unrelated individuals leave it. This may bring them into conflict with the older females who should attempt to eject the most recently born group members (see above). Cases where dominant males intervene in interactions between females (see Rowell, 1974*b*), taking the side of the younger female may be the result of such conflicts of interest. Alternatively, where the male holds the group long enough for incest to occur, he may attempt to eject his daughters, and to retain unrelated females. In such cases, it might be to his advantage to tolerate the presence of another, unrelated breeding male.

In multi-male troops the potential complexities are so great that it is difficult to formulate predictions. Studies of group fission in Japanese macaques (see above) show that, when troops divide, the lowest ranking matrilines tend to form a splinter troop which may occupy a new range – a situation

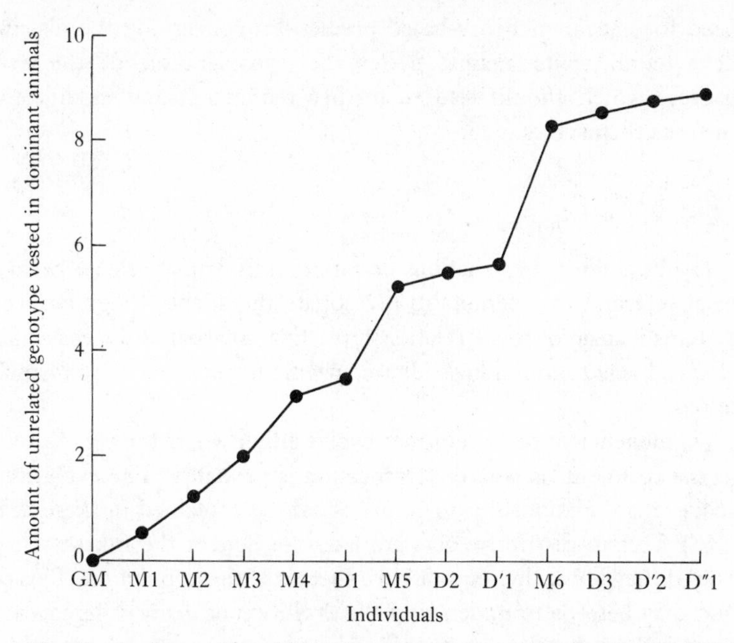

Fig. 2. Amount of unrelated genotype (see text) vested in animals dominant to each individual in Fig. 1 during the sixth year (for abbreviations see Fig. 1).

closely analogous to that predicted in simple matriarchal groups by our model. Though cases of group fission are often associated with changes in the rank of one or more central males (Furuya, 1969), the behavioural mechanisms underlying them are unclear and it is not known to what extent aggression between females is involved.

These models are clearly over simple though most of their assumptions are plausible. However, they offer a predictive approach to studying group dynamics which may be useful in the future. In the almost total absence of relevant data, it is not useful to extend speculation further. We can only hope that the development of long-term field studies of particular populations will eventually provide the relevant information.

In conclusion, we wish to stress two points. First, that consideration of all aspects of social structure can be broken down so that it is possible to consider advantages at the individual level. In particular, functional arguments concerning grouping tendencies (e.g. Cody, 1971; Hamilton, 1971*a*; Clutton-Brock, 1974) and social roles (Gartlan, 1968) might benefit from this approach. Secondly, that functional interpretations of social behaviour, including those outlined in this paper, must be accepted with reserve while they rely only on the evidence of a few examples. There is now a growing need for checking theory-based predictions concerning the distributions of behavioural traits against systematic consideration of the evidence – a process which should lead to further refinement and modification of the theories themselves.

SUMMARY

(1) Functional explanations of variation in primate social behaviour have been retarded by attempts to investigate the adaptive significance of social systems instead of social relationships. Examination of the costs and benefits of social acts to individuals allows specific hypotheses to be formulated and tested.

(2) Beneficient behaviour may evolve either where the benefactor is related to the recipient or where reciprocation is probable. The evolution of many cooperative relationships in primates can be explained in these terms.

(3) Certain disruptive acts are likely to benefit the fitness of individuals even if they lower the reproductive success of the population. This consideration may help us to understand the evolution of many differences in aggressiveness both between and within primate species.

(4) Differences in social structure are commonly explained in terms which rely on group advantage. Territoriality, group sex ratios, social dominance and group dynamics are interpreted in terms of individual competition for resources.

We are particularly grateful to Pat Bateson and Robert Hinde both for organising the conference and for the encouragement which they have provided for field work on social behaviour over the last decade. We should also like to thank John Maynard Smith and Richard Andrew for their encouragement and support at Sussex. Finally, we are grateful to the following for help, advice, encouragement or criticism: Dr S. Altmann, Drs R. and M. Dawkins, Bob Gibson, P. J. Greenwood, Dr A. M. Harvey, Dr P. J. Hogarth, G. Mace, Dr P. R. Marler, C. Packer, Dr L. Partridge, J. Pollock, Dr T. E. Rowell and Dr P. J. B. Slater.

REFERENCES

Alexander, R. D. (1974). The evolution of social behavior. *Annual Review of Ecology and Systematics*, **5**, 325–383.

Altmann, S. A. (1974). Baboons, space, time and energy. *American Zoologist*, **14**, 221–248.

Altmann, S. A. & Altmann, J. (1970). *Baboon Ecology*. University of Chicago Press: Chicago.

Archer, J. (1970). Effects of population density on behaviour in rodents. In *Social Behaviour in Birds and Mammals*, ed. J. H. Crook, pp. 169–210. Academic Press: London

Baldwin, J. D. (1968). The social behaviour of adult male squirrel monkeys (*Saimiri sciureus*) in a seminatural environment. *Folia primatologica*, **9**, 281–314.

Baldwin, J. D. (1971). The social organization of a semifree-ranging troop of squirrel monkeys (*Saimiri sciureus*). *Folia primatologica*, **14**, 23–50.

Barash, D. P. (1974). The evolution of marmot societies: a general theory. *Science, Washington*, **185**, 415–420.

Bartlett, D. P. & Meier, G. W. (1971). Dominance status and certain operants in a communal colony of rhesus macaques. *Primates*, **12**, 209–219.

Bates, B. C. (1970). Territorial behaviour in primates: a review of recent field studies. *Primates*, **11**, 271–284.

Bernstein, I. S. (1969). Stability of the status hierarchy in a pigtail monkey group (*Macaca nemestrina*). *Animal Behaviour*, **17**, 452–458.

Bernstein, I. S. (1970). Primate status hierarchies. In *Primate Behavior*, vol. **1**, ed. L. A. Rosenblum, pp. 71–109. Academic Press: New York.

Bernstein, I. S. & Gordon, T. P. (1974). The function of aggression in primate societies. *American Scientist*, **62**, 304–311.

Boelkins, R. C. (1967). Determination of dominance hierarchies in monkeys. *Psychoanalytic Science*, **7**, 317–318.

Boelkins, R. C. & Wilson, A. P. (1972). Intergroup social dynamics of the Cayo Santiago rhesus (*Macaca mulatta*) with special reference to changes in group membership by males. *Primates*, **13**, 125–140.

Bramblett, C. A. (1970). Coalitions among gelada baboons. *Primates*, **11**, 327–333.

Brown, J. L. (1964). The evolution of diversity in avian territorial systems. *Wilson Bulletin*, **6**, 160–169.

Bygott, D. (1974). Agonistic behaviour in wild male chimpanzees. Ph.D. thesis, University of Cambridge.

Carpenter, C. R. (1934). A field study of the behaviour and social relations of howling monkeys. *Comparative Psychology Monographs*, **10**, no. 2.

Carpenter, C. R. (1954). Tentative generalisations on the grouping behaviour of non-human primates. *Human Biology*, **26**, 269–276.

Carpenter, C. R. (1958). Territoriality: a review of concepts and problems. In *Behavior and Evolution*, ed. A. Roe & G. G. Simpson, pp. 224–250. Yale University Press: New Haven.

Castell, R. & Heinrich, B. (1971). Rank order in a captive female squirrel monkey colony. *Folia primatologica*, **14**, 182–189.

Chalmers, N. R. (1968). The social behaviour of free living mangabeys in Uganda. *Folia primatologica*, **8**, 263–281.

Chalmers, N. R. (1973). Differences in behaviour between some arboreal and terrestrial species of African monkeys. In *Comparative Ecology and Behaviour of Primates*, ed. R. P. Michael & J. H. Crook, pp. 69–100. Academic Press: London.

Chance, M. & Jolly, C. (1970). *Social Groups of Monkeys, Apes and Men*. Cape: London.

Charles-Dominique, P. & Martin, R. D. (1972). Behaviour and ecology of nocturnal prosimians. *Zietschrift für Tierpsychologie Supplement*, **9**.

Chivers, D. J. (1974). *The Siamang in Malaya*. Contributions to Primatology, **4**. Karger: Basel.

Christian, J. J. (1963). The pathology of overpopulation. *Military Medicine*, **128**, 571–603.

Clutton-Brock, T. H. (1974). Primate social organisation and ecology. *Nature, London*, **250**, 539–542.

Clutton-Brock, T. H. (1975). Feeding behaviour of red colobus and black-and-white colobus in East Africa. *Folia primatologica*, **23**, 165–207.

Cody, M. L. (1971). Finch flocks in the Mohave Desert. *Theoretical Population Biology*, **2**, 142–158.

Coelho, A. M. (1974). Socio-bioenergetics and sexual dimorphism in primates. *Primates*, **15**, 263–269.

Cohen, J. E. (1971). *Casual Groups of Monkeys and Men: Stochastic Models of Elemental Social Systems*. Harvard University Press: Cambridge, Mass.

Collias, N. E. (1953). Social behavior in animals. *Ecology*, **34**, 810–811.

Crook, J. H. (1970). The socio-ecology of primates. In *Social Behaviour in Birds and Mammals*, ed. J. H. Crook, pp. 103–166. Academic Press: London.

Crook, J. H. (1972). Sexual selection, dimorphism and social organisation in the primates. In *Sexual Selection and the Descent of Man 1871–1971*, ed. B. Campbell, pp. 231–281. Aldine-Atherton: Chicago.

Crook, J. H. & Gartlan, J. S. (1966). Evolution of primate societies. *Nature, London*, **210**, 1200–1203.

Darwin, C. (1876). Sexual selection in relation to monkeys. *Nature, London*, **15**, 18–19.

Denham, W. N. (1971). Energy relations and some basic properties of primate social organization. *American Anthropologist*, **73**, 77–95.

DeVore, I. & Washburn, S. (1963). Baboon ecology and human evolution. In *African Ecology and Human Evolution*, ed. F. C. Howell & F. Boulière, pp. 335–367. Wenner–Gren Foundation: New York.

Drickamer, L. C. & Vessey, S. H. (1973). Group changing in free-ranging male rhesus monkeys. *Primates*, **14**, 359–368.

Dunbar, R. I. M. & Crook, J. H. (1975). Aggression and dominance in the weaver bird, *Quelea quelea*. *Animal Behaviour*, **23**, 450–459.

Dunbar, R. I. M. & Dunbar, P. (1974). The reproductive cycle of the gelada baboon. *Animal Behaviour*, **22**, 203–210.

Eisenberg, J. F. (1966). *The Social Organisation of Mammals*. Handbüch der Zoologie, vol. **8**.

Eisenberg, J. F. (ed.) (1971). *Man and Beast: Comparative Social Behavior*. Smithsonian Institution Press: Washington, D.C.

Eisenberg, J. F., Muckenhirn, N. A. & Rudran, R. (1972). The relation between ecology and social structure in primates. *Science, Washington*, **176**, 863–874.

Elder, W. H. & Elder, N. L. (1970). Social groupings and primate associations of the bushbuck (*Tragelaphus scriptus*). *Mammalia*, **34**, 356–362.

Ellefson, J. O. (1970). Territorial behaviour in the common white-handed gibbon, *Hylobates lar* Linn. In *Primates, Studies in Adaptation and Variability*, ed. P. C. Jay, pp. 180–199. Holt, Rinehart & Winston: New York.

Enomoto, T. (1974). The sexual behaviour of Japanese monkeys. *Journal of Human Evolution*, **3**, 351–372.

Epple, G. von (1967). Vergleichende Untersuchungen über Sexual- und Social-verhaltan der Krallenhaffen (Hapalidae). *Folia primatologica*, **7**, 37–65.

Fisher, R. A. (1958). *The Genetical Theory of Natural Selection*. Dover: New York.

Furuya, Y. (1968). On the fission of troops of Japanese monkeys: 1. Five fissions and social changes between 1955 and 1966 in the Gagyusan troop. *Primates*, **9**, 323–350.

Furuya, Y. (1969). On the fission of troops of Japanese monkeys: 2. General view of troop fission of Japanese monkeys. *Primates*, **10**, 47–69.

Furuya, Y. (1973). Fissions in the Gagyusan colony of Japanese monkeys. In *Behavioral Regulators of Behavior in Primates*, ed. C. R. Carpenter, pp. 107–114. Bucknell University Press: New Jersey.

Gale, H. S. & Eaves, L. J. (1975). Logic of animal conflict. *Nature, London*, **254**, 463–464.

Gartlan, J. S. (1964). Dominance in East Africa monkeys. *Proceedings of the East African Academy*, **2**, 75–79.

Gartlan, J. S. (1968). Structure and function in primate society. *Folia primatologica*, **8**, 89–120.

Gartlan, J. S. (1973). Influences of phylogeny and ecology on variations in the group organization of primates. In *Symposia of the Fourth international Congress of Primatology*, ed. E. W. Menzel, vol. **1**, pp. 88–102. Karger: Basel.

Gartlan, J. S. & Brain, C. K. (1968). Ecology and social variability in *Cercopithecus aethiops* and *C. mitis*. In *Primates: Studies in Adaptation and Variability*, ed. P. C. Jay, pp. 253–292. Holt, Rinehart & Winston: New York.

Gartlan, J. S. & Struhsaker, T. T. (1972). Polyspecific associations and niche separation of rain-forest anthropoids in Cameroon, West Africa. *Journal of Zoology*, **168**, 221–226.

Gautier-Hion, A. (1970). L'organisation sociale d'une bande de talapoins (*Miopithecus talapoin*) dans le nord-est de Gabon. *Folia primatologica*, **12**, 116–141.

Geist, V. (1974). On the relationship of social evolution and ecology in ungulates. *American Zoologist*, **14**, 205–220.

George, J. (1966). Why do animals fight? *Audubon*, **68**, 18–20.

Gilbert, L. E. & Singer, M. E. (1973). Dispersal and gene flow in a butterfly species. *American Naturalist*, **107**, 58–72.

Gilpin, M. (1975). *Group Selection in Predator–Prey Communities*. Princeton University Press: Princeton.

Goss-Custard, J. D., Dunbar, R. I. M. & Aldrich-Blake, F. P. G. (1972). Survival, mating and rearing strategies in the evolution of primate social structure. *Folia primatologica*, **17**, 1–19.

Gouzoules, H. (1974). Harassment of sexual behaviour in the stumptail macaque, *Macaca arctoides*. *Folia primatologica*, **22**, 208–217.

Hall, K. R. L. (1963). Variations in the ecology of the chacma baboon. *Symposia of the zoological Society of London*, **10**, 1–28.

Hall, K. R. L. & DeVore, I. (1965). Baboon social behavior. In *Primate Behavior*, ed. I. DeVore, pp. 53–110. Holt, Rinehart & Winston: New York.

Hall, K. R. L. & Mayer, B. (1967). Social interaction in a group of captive patas monkeys, *Erythrocebus patas*. *Folia primatologica*, **5**, 213–236.

Hamilton, W. D. (1963). The evolution of altruistic behaviour. *American Naturalist*, **97**, 354–356.

Hamilton, W. D. (1964*a*). The genetical evolution of social behaviour. I. *Journal of theoretical Biology*, **7**, 1–16.

Hamilton, W. D. (1964*b*). The genetical evolution of social behaviour. II. *Journal of theoretical Biology*, **7**, 17–52.

Hamilton, W. D. (1970). Selfish and spiteful behaviour in an evolutionary model. *Nature, London*, **228**, 1218–1220.

Hamilton, W. D. (1971*a*). Geometry for the selfish herd. *Journal of theoretical Biology*, **31**, 295–311.

Hamilton, W. D. (1971*b*). Selection of selfish and altruistic behaviour in some extreme models. In *Man and Beast: Comparative Social Behavior*, ed. J. F. Eisenberg & W. S. Dillon, pp. 55–91. Smithsonian Institution Press: Washington, D.C.

Hamilton, W. D. (1972). Altruism and related phenomena, mainly in social insects. *Annual Review of Ecology and Systematics*, **3**, 193–232.

Hanby, J. P., Robertson, L. T. & Phoenix, C. H. (1971). The sexual behaviour of a confined troop of Japanese macaques (*Macaca fuscata*). *Folia primatologica*, **16**, 123–143.

Harlow, H. F. (1969). Age-mate or peer affectional systems. *Advances in the Study of Behavior*, **2**, 333–383.

Harding, R. S. O. (1973). Predation by a troop of olive baboons (*Papio anubis*). *American Journal of Physical Anthropology*, **38**, 589–591.

Hess, J. P. (1973). Some observations on the sexual behaviour of captive lowland gorillas. In *Comparative Ecology and Behaviour of Primates*, ed. R. P. Michael & J. H. Crook, pp. 507–581. Academic Press: London.

Hill, J. L. (1974). *Peromyscus*: effect of early pairing on reproduction, *Science, Washington*, **186**, 1042–1044.

Hill, W. C. O. (1952). The external and visceral anatomy of the olive colobus monkey (*Procolobus verus*). *Proceedings of the zoological Society of London*, **122**, 127–186.

Hill, W. C. O. (1966). *Primates*, vol. **6**, *Cercopithecoidea*. Edinburgh University Press: Edinburgh.

Hinde, R. A. (1956). The biological significance of the territories of birds. *Ibis*, **98**, 340–369.

Hinde, R. A. (ed.) (1969). *Bird Vocalizations*. Cambridge University Press: London.

Hinde, R. A. (1974). *Biological Bases of Human Social Behaviour*. McGraw-Hill: New York.

Hinde, R. A. & Spencer-Booth, Y. (1967). The behaviour of socially living rhesus monkeys in their first two and a half years. *Animal Behaviour*, **15**, 169–196.

Hinde, R. A. & Spencer-Booth, Y. (1971). Effects of brief separation from mother on rhesus monkeys. *Science, Washington*, **173**, 111–118.

Hladik, M. (1976). Ecology, diet and social patterning in Old and New World primates. In *Socio-ecology and Psychology of Primates*, ed. R. H. Tuttle. Moutton: The Hague, in press.

Holloway, R. L. (1974). *Primate Aggression, Territoriality and Xenophobia*. Academic Press: New York.

Holst, A. von (1974). Social stress in the tree-shrew: its causes and physiological and ethological consequences. In *Prosimian Biology*, ed. R. D. Martin, G. A. Doyle & A. C. Walker, pp. 389–411. Duckworth: London.

Horwich, R. H. & Manski, D. (1975). Maternal care and infant transfer in two species of *Colobus* monkeys. *Primates*, **16**, 49–73.

Hrdy, S. B. (1974). Male–male competition and infanticide among the langurs (*Presbytis entellus*) of Abu' Rajasthan. *Folia primatologica*, **22**, 19–58.

Hutchinson, G. E. & MacArthur, R. H. (1959). On the theoretical significance of aggressive neglect in interspecific competition. *American Naturalist*, **93**, 133–134.

Itani, J. (1972). A preliminary essay on the relationship between social organisation and incest avoidance in non-human primates. In *Primate Socialization*, ed. F. E. Poirier, pp. 165–171. Random House: New York.

Jarman, P. J. (1974). The social organisation of antelope in relation to their ecology. *Behaviour*, **48**, 215–267.

Jay, P. (1965). The common langur of North India. In *Primate Behavior*, ed. I. DeVore. Holt, Rinehart & Winston: New York.

Jenni, D. A. (1974). Evolution of polyandry in birds. *American Zoologist*, **14**, 129–144.

Jolly, A. (1966). *Lemur Behavior*. Chicago University Press: Chicago.

Jolly, A. (1972). *The Evolution of Primate Behavior*. Macmillan: New York.

Kawai, M. (1958). On the system of social ranks in a natural troop of Japanese monkeys. I. Basic and dependent rank. *Primates*, **1**, 111–130.

Kawamura, S. (1958). Matriarchal social ranks in the Minoo-B troop: a study of the rank system of Japanese monkeys. *Primates*, **2**, 181–252.

Kawamura, S. (1967). Aggression as studied in troops of Japanese monkeys. *Brain Function*, **5**, 195–223.

Kleiman, D. G. & Eisenberg, J. F. (1973). Comparisons of canid and felid social systems from an evolutionary perspective. *Animal Behaviour*, **21**, 637–659.

Klopfer, P. H. (1974). Mother–young relations in lemurs. In *Prosimian Biology*, ed. R. D. Martin, G. A. Doyle & A. C. Walter, pp. 271–292. Duckworth: London.

Kolata, G. B. (1975). Sociobiology (1): models of social behavior. *Science, Washington*, **187**, 50–51.

Koyama, N. (1967). On dominance rank and kinship of a wild Japanese monkey troop in Arashiyama. *Primates*, **8**, 189–216.

Koyama, N. (1970). Changes in dominance rank and division of a wild Japanese monkey troop in Arashiyama. *Primates*, **11**, 335–390.

Kruuk, H. (1972). *The Spotted Hyena*. Chicago University Press: Chicago.

Kuhn, H. J. (1972). On the perineal organ of male *Procolobus badius*. *Journal of Human Evolution*, **1**, 371–378.

Kummer, H. (1957). Soziales Verhatten einer Mantelpavian-Gruppe. *Beiheft Schweiz Zeitschrift für Psychologie*, **33**, 1–91.

Kummer, H. (1968). *Social Organisation of Hamadryas Baboons*. Chicago University Press: Chicago.

Kummer, H. (1970). Behavioral characteristics in primate taxonomy. In *Old World Monkeys*, ed. J. H. Napier & P. H. Napier, pp. 25–34. Academic Press: London.

Kummer, H. (1971). *Primate Societies: Group Techniques of Ecological Adaptation*. Aldine-Atherton: Chicago.

Kummer, H., Gotz, W. & Angst, W. (1974). Triadic differentiation: an inhibitory process protecting pair bonds in baboons. *Behaviour*, **49**, 62–87.

Lack, D. (1968). *Ecological Adaptations for Breeding in Birds*. Methuen: London.

Langford, J. B. (1963). Breeding behaviour of *Hapale jacchus* (common marmoset). *South African Journal of Science*, **59**, 229–230.

Lawick, H. van & Lawick-Goodall, J. van (1970). *Innocent Killers*. Collins: London.

Lawick-Goodall, J. van (1968). The behaviour of free-living chimpanzees in the Gombe Stream Reserve. *Animal Behaviour Monographs*, **1**, 161–311.

Lindburg, D. G. (1969). Rhesus monkeys: mating season mobility of adult males. *Science, Washington*, **166**, 1176–1178.

Lockie, J. D. (1955). The breeding and feeding of jackdaws and rooks. *Ibis*, **97**, 341–369.

Lorenz, K. (1966). *On Aggression*. Methuen: London.

Loy, J. (1970). Behavioural responses of free-ranging rhesus monkeys to food shortage. *American Journal of Physical Anthropology*, **33**, 263–271.

Loy, J. (1971). Estrous behaviour of free-ranging rhesus monkeys (*Macaca mulatta*). *Primates*, **12**, 1–31.

MacKinnon, J. (1974). The ecology and behaviour of wild orang-utans (*Pongo pygmaeus*). *Animal Behaviour*, **22**, 3–74.

Marler, P. (1969). *Colobus guereza*: territoriality and group composition. *Science, Washington*, **163**, 93–95.

Mason, W. A. (1974). Differential grouping patterns in two species of South American monkey. In *Ethology and Psychiatry*, ed. N. F. White, pp. 153–169. Toronto University Press: Toronto.

May, R. (1975). Group selection. *Nature, London*, **254**, 485.

Maynard Smith, J. (1964). Group selection and kin selection: a rejoinder. *Nature, London*, **201**, 1145–1147.

Maynard Smith, J. (1974*a*). *Models in Ecology*. Cambridge University Press: London.

Maynard Smith, J. (1974*b*). The theory of games and the evolution of animal conflict. *Journal of theoretical Biology*, **47**, 209–221.

Maynard Smith, J. (1976). Group selection. *Quarterly Review of Biology*, in press.

Maynard Smith, J. & Price, G. R. (1973). The logic of animal conflict. *Nature, London*, **246**, 15–18.

Maynard Smith, J. & Ridpath, M. G. (1972). Wife sharing in the Tasmanian native hen, *Tribonyx mortierii*: a case of kin selection? *American Naturalist*, **106**, 447–452.

Mazur, A. (1973). A cross-species comparison of status in small established groups. *American sociological Review*, **28**, 513–530.

Michael, R. P. & Keverne, R. B. (1968). Pheromones in the communication of sexual status in primates. *Nature, London*, **218**, 746–749.

Michael, R. P. & Keverne, E. B. (1970). Primate sex pheromones of vaginal origin. *Nature, London*, **225**, 84–85.

Miller, R. S. (1967). Pattern and process in competition. *Advances in ecological Research*, **4**, 1–74.

Missakian, E. A. (1972). Genealogical and cross-genealogical dominance relations in a group of free-ranging rhesus monkeys (*Macaca mulatta*) on Cayo Santiago. *Primates*, **13**, 169–180.

Missakian, E. A. (1974). Mother–offspring grooming relations in rhesus monkeys. *Archives of sexual Behaviour*, **3**, 135–141.

Mitchell, G. D. (1969). Paternalistic behaviour in primates. *Psychological Bulletin*, **71**, 399–417.

Mitchell, G. D. & Brandt, E. M. (1972). Paternal behavior in primates. In *Primate Socialization*, ed. F. E. Poirier, pp. 173–206. Random House: New York.

Mohnot, S. M. (1971). Some aspects of social changes and infant killing in the hanuman langur, *Presbytis entellus* (Primates: Cercopithecidae) in Western India. *Mammalia*, **35**, 175–198.

Moynihan, M. (1964). Some behavior patterns of platyrhine monkeys. I. The night monkey (*Aotus trivirgatus*). *Smithsonian miscellaneous Collections*, vol. **146**, no. 5.

Myers, J. H. & Krebs, C. J. (1971). Genetic behavioral and reproductive attributes of dispersing field voles *Microtus pennsylvanicus* and *Microtus ochrogaster*. *Ecological Monographs*, **41**, 53–78.

Nagel, U. & Kummer, H. (1974). Variation in cercopithecoid aggressive behavior. In *Primate Aggression, Territoriality and Xenophobia*, ed. R. L. Holloway, pp. 159–184. Academic Press: New York.

Napier, J. R. & Napier, P. J. (1967). *A Handbook of Living Primates*. Academic Press: London.

Neville, M. K. (1972). Social relations within troops of red howler monkeys (*Alouatta seniculus*). *Folia primatologica*, **18**, 47–77.

Nicholson, A. J. (1955). An outline of the dynamics of animal populations. *Australian Journal of Zoology*, **2**, 9–65.

Nishida, T. & Kawanaka, K. (1972). Inter unit-group relationships among wild chimpanzees in the Mahali Mountains. *Kyoto University African Studies*, **7**, 131–169.

Orlove, M. J. (1975). A model of kin selection not invoking coefficients of relationships. *Journal of theoretical Biology*, **49**, 289–310.

Packer, C. (1975). Male transfer in olive baboons. *Nature, London*, **225**, 219–220.

Parker, G. A. (1974). Assessment strategy and the evolution of fighting behaviour. *Journal of theoretical Biology*, **47**, 223–243.

Perachio, A. A., Alexander, M. & Marr, L. (1973). Hormonal and social factors affecting evoked sexual behaviour in rhesus monkeys. *American Journal of physical Anthropology*, **38**, 227–232.

Poirier, F. E. (1969). The Nilgiri langur troop: its composition, structure, function and change. *Folia primatologica*, **10**, 20–47.

Poirier, F. E. (1974). Colobine aggression: a review. In *Primate Aggression, Territoriality and Xenophobia*, ed. R. L. Holloway, pp. 123–158. Academic Press: New York.

Richard, A. (1974*a*). Intra-specific variation in the social organization and ecology of *Propithecus verreauxi*. *Folia primatologica*, **22**, 178–207.

Richard, A. (1974*b*). Patterns of mating in *Propithecus verreauxi*. In *Prosimian Biology*, ed. R. D. Martin, G. A. Doyle & A. C. Walker, pp. 49–74. Duckworth: London.

Richards, S. M. (1972). Tests for behavioural characteristics in rhesus monkeys. Ph.D. thesis, University of Cambridge.

Richards, S. M. (1974). The concept of dominance and methods of assessment. *Animal Behaviour*, **22**, 914–930.

Ripley, S. D. (1961). Aggressive neglect as a factor in inter-specific competitions in birds. *Auk*, **78**, 366–371.

Rowell, T. E. (1963). Behaviour and female reproductive cycles of rhesus macaques. *Journal of Reproduction and Fertility*, **6**, 193–203.

Towell, T. E. (1966). Forest living baboons in Uganda. *Journal of Zoology*, **147**, 344–364.

Rowell, T. E. (1967). A quantitative comparison of the behaviour of a wild and caged baboon troop. *Animal Behaviour*, **15**, 499–509.

Rowell, T. E. (1972). Female reproduction cycles and social behaviour in primates. *Advances in the Study of Behavior*, **4**, 69–105.

Rowell, T. E. (1974*a*). The concept of social dominance. *Behavioural Biology*, **11**, 131–154.

Rowell, T. E. (1974*b*). Contrasting male roles in different species of non-human primates. *Archives of sexual Behaviour*, **3**, 143–149.

Saayman, G. S. (1970). The menstrual cycle and sexual behaviour in a troop of free-ranging chacma baboons, *Papio ursinus*. *Folia primatologica*, **15**, 123–143.

Saayman, G. S. (1971). Behaviour of the adult males in a troop of free-ranging chacma baboons (*Papio ursinus*). *Folia primatologica*, **15**, 36–57.

Saayman, G. S. (1972). Aggressive behaviour in free-ranging chacma baboons (*Papio ursinus*). *Journal of Behavioural Science*, **1**, 77–83.

Sade, D. S. (1965). Some aspects of parent-offspring and sibling relations in a group of rhesus monkeys, with a discussion of grooming. *American Journal of physical Anthropology*, **23**, 1–18.

Sade, D. S. (1967). Determinants of dominance in a group of free-ranging rhesus monkeys. In *Social Communication among Primates*, ed. S. A. Altmann, pp. 99–114. Chicago University Press: Chicago.

Sade, D. S. (1968). Inhibition of son–mother mating among free-ranging rhesus monkeys. *Scientific Psychoanalyst*, **12**, 18–27.

Schaller, G. B. (1972). *The Serengeti Lion*. Chicago University Press: Chicago.

Schjelderup-Ebbe, Th. (1922). *Beiträge zur Biologie und Social-und Individual Psychologie bei Gallus domesticus*. Muller: Greifswald, Germany.

Schull, W. J. & Neel, J. V. (1965). *The Effects of Inbreeding in Japanese Children*. Harper & Row: New York.

Scott, J. P. (1962). Hostility and aggression in animals. In *Roots of Behavior*, ed. E. L. Bliss, pp. 167–178. Harper & Row: New York.

Scott, J. P. (1974). Agonistic behavior in mice and rats: a review. *American Zoologist*, **6**, 683–701.

Scott, J. P. (1974). Agonistic behaviour of primates: a comparative perspective. In *Primate Aggression, Territoriality and Xenophobia*, ed. R. L. Holloway, pp. 417–434. Academic Press: New York.

Skutch, A. F. (1961). Helpers among birds. *Condor*, **63**, 198–226.

Smythe, N. (1970). The adaptive value of the social organisation of the coati (*Nasua narica*). *Journal of Mammalogy*, **51**, 818–820.

Sorenson, M. W. (1974). A review of aggressive behaviour in the tree shrews. In *Primate Aggression Territoriality and Xenophobia*, ed. R. L. Holloway, pp. 13–30. Academic Press: New York.

Southwick, C. H. (1967). An experimental study of intragroup agonistic behaviour in rhesus monkeys (*Macaca mulatta*). *Behaviour*, **28**, 182–209.

Southwick, C. H., Beg, H. A. & Siddiqi, M. R. (1965). Rhesus monkeys in North India. In *Primate Behavior*, ed. I. DeVore, pp. 111–159. Holt, Rinehart & Winston: New York.

Southwick, C. H., Siddiqi, M. F., Farooqui, M. Y. & Pal, B. C. (1974). Xenophobia among free-ranging rhesus groups in India. In *Primate Aggression, Territoriality and Xenophobia*, ed. R. L. Holloway, pp. 185–210. Academic Press: New York.

Spencer-Booth, Y. (1970). The relationships between mammalian young and conspecifics other than mothers and peers: a review. *Advances in the Study of Behavior*, **3**, 120–194.

Stolz, L. P. & Saayman, G. S. (1970). Ecology and behaviour of baboons in the Northern Transvaal. *Annals of the Transvaal Museum*, **26**, 99–143.

Struhsaker, T. T. (1967*a*). Behavior of vervet monkeys (*Cercopithecus aethiops*). *University of California Publications in Zoology*, **82**, 1–64.

Struhsaker, T. T. (1967*b*). Social structure among vervet monkeys (*Cercopithecus aethiops*). *Behaviour*, **29**, 83–121.

Struhsaker, T. T. (1969). Correlates of ecology and social organisation among African cercopithecines. *Folia primatologica*, **11**, 80–118.

Suarez, B. & Ackerman, D. R. (1971). Social dominance and reproductive behaviour in male rhesus monkeys. *American Journal of physical Anthroplogy*, **35**, 219–222.

Sugiyama, Y. (1965*a*). Behavioural development and social structure in two troops of hanuman langurs (*Presbytis entellus*). *Primates*, **6**, 213–247.

Sugiyama, Y. (1965*b*). On the social change of hanuman langurs (*Presbytis entellus*) in their natural conditions. *Primates*, **6**, 381–418.

Sussman, R. W. (1974). Ecological distinctions in sympatric species of *Lemur*. In *Prosimian Biology*, ed. R. D. Martin, G. A. Doyle & A. C. Walker, pp. 75–108. Duckworth: London.

Syme, G. J. (1974). Competitive orders as measures of social dominance. *Animal Behaviour*, **22**, 931–940.

Tinbergen, N. (1951). *The Study of Instinct*. Oxford University Press: London.

Tinbergen, N. (1957). The functions of territory. *Bird Study*, **4**, 14–27.

Trivers, R. L. (1971). The evolution of reciprocal altruism. *Quarterly Review of Biology*, **46**, 35–57.

Trivers, R. L. (1972). Parental investment and sexual selection. In *Sexual Selection and the Descent of Man, 1871–1971*, ed. B. Campbell, pp. 136–179. Aldine–Atherton: Chicago.

Trivers, R. L. (1974). Parent–offspring conflict. *American Zoologist*, **14**, 249–264.

Tutin, C. E. G. & McGrew, W. C. (1973*a*). Sexual behaviour of group-living adolescent chimpanzees. *American Journal of physical Anthropology*, **38**, 195–200.

Tutin, C. E. G. & McGrew, W. C. (1973*b*). Chimpanzee copulatory behaviour. *Folia primatologica*, **19**, 237–256.

VanValen, L. (1973). A new evolutionary law. *Evolutionary Theory*, **1**, 1–30.

Vessey, S. H. (1971). Free-ranging rhesus monkeys: behavioural effects of removal, separation and reintroduction of group members. *Behaviour*, **40**, 216–227.

Washburn, S. L. & Hamburg, D. A. (1968). Aggressive behavior in Old World monkeys and apes. In *Primates: Studies in Adaptation and Variability*, ed. P. C. Jay, pp. 458–478. Holt, Rinehart & Winston: New York.

Washburn, S. L., Jay, P. C. & Lancaster, J. B. (1965). Field studies of old world monkeys and apes. *Science, Washington*, **150**, 1541–1547.

Watson, A. & Moss, R. (1972). A current model of population dynamics in red grouse. *Proceedings of the 15th international Ornithology Conference*, pp. 134–149. E. J. Brill: Netherlands.

Watts, C. R. & Stokes, A. W. (1971). The social order of turkeys. *Scientific American*, **224**, 112–118.

West-Eberhard, M. J. (1975). The evolution of social behaviour by kin selection. *Quarterly Review of Biology*, **50**, 1–34.

Wickler, W. (1967). Socio-sexual signals and their intra-specific initiation among primates. In *Primate Ethology*, ed. D. Morris, pp. 69–147. Weidenfeld & Nicolson: London.

Williams, G. C. (1966). *Adaptation and Natural Selection*. Princeton University Press: Princeton.

Wilson, D. S. (1974). A theory of group selection. *Proceedings of the National Academy of Science, USA*, **72**, 143–146.

Wilson, E. O. (1971). Competitive and aggressive behaviour. In *Man and Beast: Comparative Social Behavior*, ed. J. F. Eisenberg & W. S. Dillon, pp. 182–217. Smithsonian Instituton Press: Washington, D.C.

Wilson, E. O. (1975). *Sociobiology*. Harvard University Press: Cambridge, Mass.

Wise, L. A. & Zimmerman, R. R. (1973). The effect of protein deprivation on dominance measured by shock avoidance competition and food competition. *Behavioural Biology*, **9**, 317–329.

Wolfheim, J. H. & Rowell, T. E. (1972). Communication among captive talapoin monkeys. *Folia primatologica*, **18**, 224–255.

Wrangham, R. W. (1975). The behavioural ecology of chimpanzees in Gombe National Park, Tanzania. Ph.D. thesis, University of Cambridge.

Wynne-Edwards, V. C. (1962). *Animal Dispersion in Relation to Social Behaviour*. Oliver & Boyd: Edinburgh.

Yamada, M. (1963). A study of blood-relationship in the natural society of the Japanese macaque. *Primates*, **4**, 43–65.

Yoshiba, K. (1968). Local and intertroop variability in ecology and social behavior of common Indian langurs. In *Primates: Studies in Adaptation and Variability*, ed. P. C. Jay, pp. 217–242. Holt, Rinehart & Winston: New York.

Zahavi, A. (1974). Communal nesting by the Arabian babbler. *Ibis*, **116**, 84–87.

Zuckerman, S. (1932). *The Social Life of Monkeys and Apes*. Routledge: London.

7

Social organization, communication and graded signals: the chimpanzee and the gorilla

PETER MARLER

The communication system of a species is a basic component in its social design and must contribute to the organization of its societies. Thus to understand communication systems and how they operate is a step in explaining how the diversity of animal social systems is maintained. Conversely, the social system of a species provides the background against which one may hope to interpret the function that its communication signals serve. One approach is to select a pair of species with contrasting social systems, to define their signal repertoires and how they are used, preferably in a quantitative fashion, and look for correlations between differences in social organization and differences in the nature, patterns and frequency of signal usage.

The chimpanzee (*Pan troglodytes*) and the gorilla (*Gorilla gorilla*) differ in their social organization. Information on them is now sufficient to begin to appraise these differences and those in their vocal behaviour in a search for possible relationships between them. The following is a first attempt at a quantitative comparison of the frequency of usage of vocal signals in these two primate species, with some theoretical reflections on their emphasis on graded vocal sounds.

The data derive primarily from published accounts by Schaller (1963) and Fossey (1972, 1974) on gorilla vocal behaviour and social organization and by van Lawick-Goodall (1968*a*, 1971) and Marler & Hobbet (1975) on the chimpanzee. This was supplemented by additional data on chimpanzee vocalizations at Gombe National Park gathered by the author between 28 June and 28 August 1967, subsequently analysed in collaboration with Linda Hobbet. In addition I have been given generous access to unpublished data on chimpanzees by Bygott (1974) and Wrangham (1975) and on gorillas by Harcourt (*in litt.*) and Stewart (*in litt.*) Quantitative data on call use derive only from Fossey (1972) and from Marler & Hobbet (unpublished). These are sometimes referred to in what follows as the 'Fossey data' and the 'Marler data' for reasons that will be evident from the context.

VOCALIZATIONS OF THE CHIMPANZEE

The present discussion derives from several accounts of vocal behaviour of the chimpanzee (Reynolds & Reynolds, 1965; van Hooff, 1962, 1967, 1971), and especially from the verbal description of vocalizations of chimpanzees in the Gombe National Park in Tanzania by van Lawick-Goodall (1968*a*, *b*, 1971) augmented with data from films and sound recordings of the Gombe population (Marler, 1969; Marler & Hobbet, 1975; Marler & Tenaza, 1976).

The original catalogue of vocalizations prepared by van Lawick-Goodall (1968*a*) included 24 classes. For purposes of comparison with the gorilla, this list has been reduced to a basic repertoire of 13, for the following reasons. Chimpanzee vocalizations are highly graded and this will be discussed as a separate issue later. This list of 24 includes several calls that are closely related, morphologically graded variations on a common theme (e.g. pant-hoots and pant-shrieks; barks and shrieks). These have been merged into single categories. For purposes of a primary classification I have also merged sounds distinguished mainly by context wherever our acoustical analysis failed to reveal reliable characters for distinguishing them (e.g. hoo and hoo-whimpering and whimper, scream calls and screaming, panting and copulatory panting). In fact most sounds are given in several contexts, the classification of which is a distinct and complex operation, requiring more information than is yet available. Sounds distinguished by the age and sex of the vocalizer were also grouped, although acoustical differences were sometimes detectable (e.g. screaming and infant screaming). Two rarely used sounds were not recorded (soft grunt, groan), and, since to the ear each resembles another category (grunting, rough grunting), they are not treated separately. New names have been coined when acoustical analysis suggested a more appropriate term, or to avoid defining sounds by their accompanying posture or facial expression (e.g. bobbing pants, changed to pant-grunting; cf. also mixed names used by van Hooff (1971) etc.).

The 13 basic vocal categories are listed on the left side of Table 1 in order of their frequency of use by one particular study population, as indicated in a moment. The quantitative data in this section derive from 23 hours of tape-recorded vocal behaviour from which a total of 2656 vocalizations uttered by individuals were characterized by ear and by sound spectrographic analysis. Recordings of unidentified individuals and of calls that could not be characterized because of synchronous calling of several animals were rejected. One event was noted for each consecutive sequence of the same type. Observations totalling 600 hours were distributed through the day, all made at the feeding area or close by. Distortions introduced by the aggregation of animals at the provisioning area, where all present data were gathered, and by sampling

Calls with similar acoustical morphology are placed opposite. ( ) = synonyms with van Lawick-Goodall. [ ] = calls designated separately by van Lawick-Goodall, here lumped with the call above, either because they seem acoustically indistinguishable, because they intergrade without clear separation, or because they are different renditions of the same basic call by male or female, or by particular age-classes. Calls are listed in order of frequency of usage, with the most frequent first. These rankings derive from the raw data, gathered as indicated in the text, and are not corrected for differing proportions of sex and age classes in the study populations. Corrections were applied for the rankings in Table 3.

Chimpanzee		Gorilla		
Vocalization (in order of use)	Circumstances (from van Lawick-Goodall, 1968*a*)	Circumstances (from Fossey, 1972)	Equivalent vocalization*	
1. PANT-HOOT [pant-shriek and roar]	Hearing distant group; rejoining group, meat-eating in nest at night; general arousal	Inter-group encounters with aggressive component	HOOT SERIES	3
2. PANT-GRUNT (bobbing pants) [pant-shriek]	Subordinate approaching or being approached by dominant	Mild threat within group	PANT SERIES	12
3. LAUGHTER	Playing, especially being tickled	Social play, tickling	CHUCKLES	15
4. SQUEAK	Being threatened, submission, close to dominant	Infant separated, in difficulty	CRIES	8
5. SCREAM [infant scream, screaming]	Fleeing attack, submission, when lost, while attacking dominant; copulating female	Aggressive disputes within group; copulating female	SCREAM	5
6. WHIMPER [hoo-whimper, quiet hoo]	Begging, infant–parent separation, strange sound or object	Danger of injury or abandonment (see also CRIES)	WHINE	13
7. BARK (soft bark)	Vigorous threat	Alerting to mild alarm; group movement initiation	HOOT BARK	2
		Very mild alarm or curiosity	QUESTION BARK	7
		Very mild alarm or curiosity	HICCUP BARK	10
8. WAA BARK	Threat to other, often dominant, at distance	See WRAAGH		
9. ROUGH GRUNT (grunts) [barks, shrieks, groans]	Approaching and eating preferred food	Feeding; group contentment	BELCH	6
10. PANT [copulatory pants]	Copulating male, grooming, meeting another as prelude to kissing etc.	(COPULATORY PANTING Harcourt *in litt.*)		?
11. GRUNT [soft grunt]	Feeding, mild general arousal, social excitement	Mild aggression in moving group (cf chimp cough – Harcourt)	PIG GRUNT	4
12. COUGH	Mild, confident threat to subordinate chimpanzee, baboon	See PIG GRUNT		
13. WRAAA	Detection of human or other predator, also dead chimpanzee; may be threat component	Sudden alarming situation; loud noise; unexpected contact with buffalo, with aggressive elements	WRAAGH	1
No Equivalent		May be anomalous; ailing animal	WHINNIE	14
No Equivalent		Mild aggression in stationary group	GROWL	11
No Equivalent		Strong aggression of silverback ♂ to predator or other group	ROAR	9

* Numbers indicate order of use in Fossey sample.

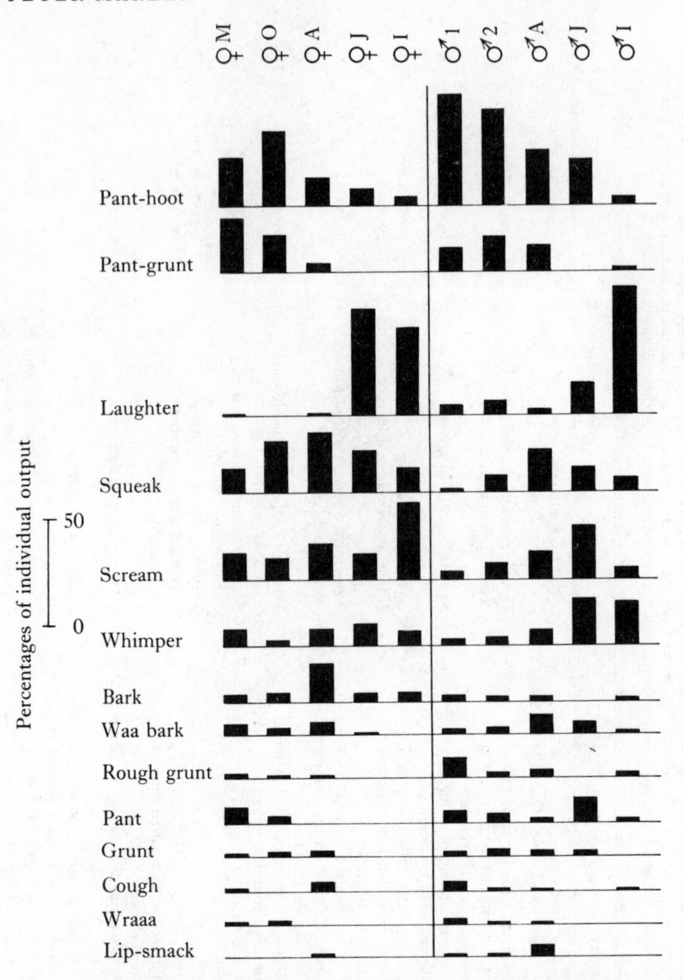

Fig. 1. Histograms of the frequency of use in a sample of tape-recordings of chimpanzees at Gombe Stream of 13 vocal and one non-vocal sound expressed as percentages of the average output of a representative member of that sex and age classes. ♀M – adult female with infant; ♀0 – adult female without infant; ♀A – adolescent female; ♀J – juvenile female; ♀I – infant female; ♂1 – older adult male; ♂2 – younger adult male; ♂A – adolescent male; ♂J – juvenile male; ♂I – infant male.

biases, will be discussed elsewhere. Here the quantitative evidence will be considered only in broadest outline.

The vocal output of a typical member of each age–sex class was estimated by dividing the frequencies with which each vocal type was recorded by the number of individuals in each class in the study population at Gombe at the time (1967). These ranged from one (juvenile male) to nine (mother and infant). These estimates were then converted into percentages of the totals

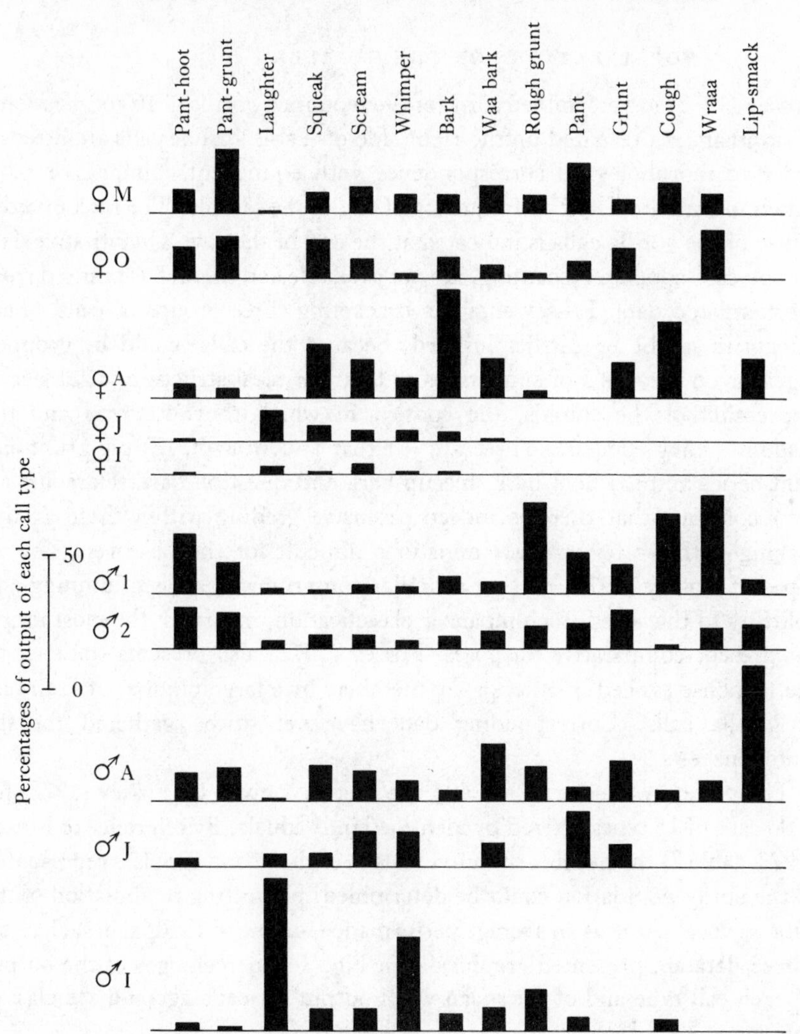

Fig. 2. Histograms of the frequency of use by chimpanzees of 14 sounds expressed as percentages of the output of that call type by a representative member of each sex and age classes. Abbreviations as in Fig. 1.

for each class of sounds and of the total output for each class of individuals. The latter, presented as histograms (Figs. 1 and 2), facilitate inspection of changes in vocal usage with age and show how use of each call type was distributed with sex and age in the sample gathered. Data on chimpanzee vocalizations and their physical structure are available elsewhere (Marler & Tenaza, 1976; see also van Hoof, 1971). All are illustrated on the sound track of the film *Vocalizations of Wild Chimpanzees* (Marler & van Lawick-Goodall, 1971). A resumé of the circumstances in which each is uttered is presented in Table 1.

VOCALIZATIONS OF THE GORILLA

Fossey (1972) has assembled a list for the mountain gorilla of 16 vocalizations. A summary is presented on the right side of Table 1. The calls are listed in order of morphological correspondence with equivalent chimpanzee calls, which in turn are listed by frequency of use in the sample. The rank in order of use of the gorilla calls is indicated at the end of the row. Quantitative data on one call, copulatory panting, are not available and this call is omitted from the tabulated data. Fossey suggests bracketing three groups of calls where categories might be further lumped, because the calls 'could be grouped together on the basis of similarities in their physical structure, a subjective impression of the sounds, the context in which they occurred and the responses they elicited'. These are (1) roar and wraagh, (2) pig grunt and pant series and (3) hoot bark, hiccup bark and question bark. Harcourt (*in litt.*) confirms that there is indeed extensive grading within these groups making within-category discriminations difficult for the observer. The 16 types of Fossey (1972) present a similar compromise between lumping and splitting to that used in chimpanzee classification, making it the most useful for present comparative purposes. Fossey (1972) also presents statistics on the response evoked in other group members by a large number of examples of gorilla calls. Corresponding data have yet to be gathered for the chimpanzee.

The pattern of usage by sex and age classes is given by Fossey (1972) for 1700 calls of 15 types uttered by identified individuals. By reference to Fossey (1972, table 2) the number of individuals in each sex and age class represented in the study population could be determined, permitting retabulation of the data on vocal usage as an average performance for one individual in each class. These data are presented graphically in Fig. 3 as percentages of the output of each call type and of the mean vocal output for each age and sex class of animals.

Fossey divides the vocalizations into seven functional groupings: aggressive calls (3), mild alarm calls (2), fear and alarm calls (2), distress calls (2), group coordination vocalizations (3), calls for inter-group communication (1), and finally miscellaneous calls (3). However, this type of functional classification involves many ambiguities and nothing equivalent has been presented for the chimpanzee. Instead I have tried to summarize van Lawick-Goodall's views on the circumstances in which each vocal type is typically given in Table 1.

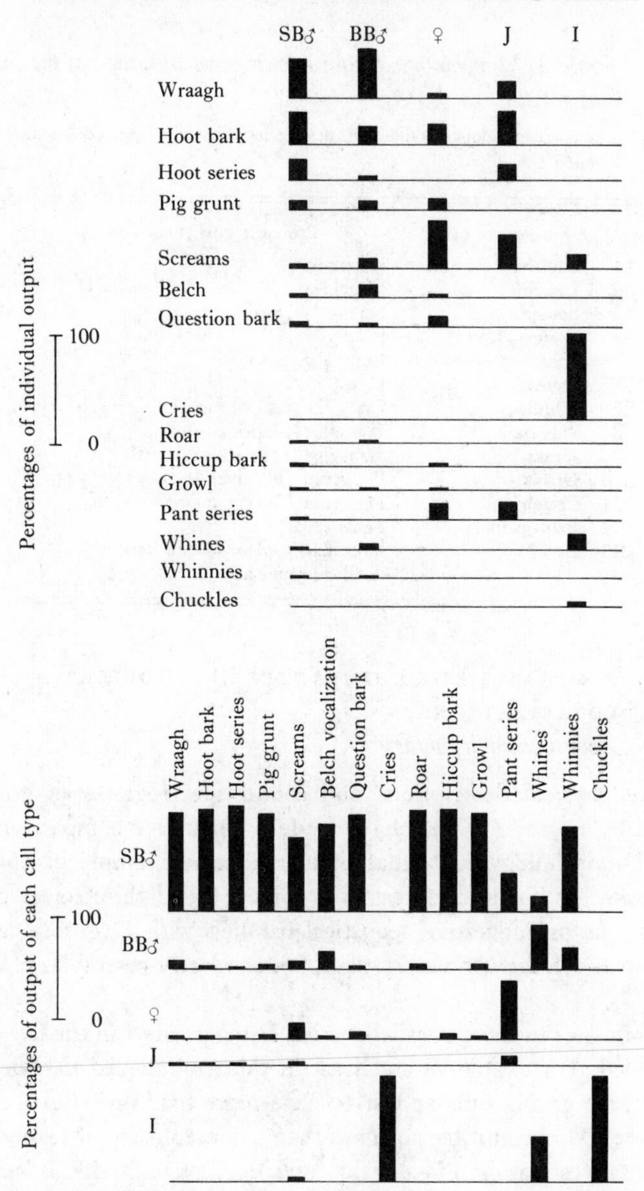

Fig. 3. Histograms of the frequency of use by gorillas of 15 sounds expressed as percentages of the output by an average member of that sex and age class, and of the output by an average individual of that call type. (Data adapted from Fossey, 1972.) SB – silverback; BB – blackback.

Table 2. *Morphological equivalents in vocalization of the chimpanzee and gorilla*

The correspondence is deemed most reliable in pairs labelled A and least in those labelled C.

	Chimpanzee call	Probable gorilla equivalent
A1	Pant-hoot	Hoot series
A2	Laughter	Chuckles
A3	Scream	Scream
A4	Rough grunt	Belch
A5	Pant	Copulatory pant
B6	Squeak	Cries (1)
B7	Whimper	Cries (2) (also whine ??)
B8	Waa bark	Wraagh (1) (short form)
B9	Wraaa	Wraagh (2) (long form)
B10	Grunt	Pig grunt (1) (given in train)
B11	Cough	Pig grunt (2) (given singly)
C12	Pant-grunt?	Pant series?
C13	Bark?	Hoot bark? (Also hiccup bark and question bark?)

COMPARISON OF CHIMPANZEE AND GORILLA VOCALIZATIONS

Acoustical morphology

Using the physical descriptions and sound spectrograms of gorilla calls presented by Fossey (1972) I have made a qualitative comparison of their acoustical morphology with that of our extensive sample of chimpanzee vocalizations. As Table 2 indicates, to my eye all 13 chimpanzee calls have either obvious or suggestive acoustical parallels with 11 gorilla calls. This estimate includes the copulatory pant, which closely resembles chimpanzee panting.

My confidence in these parallels varies, being greatest in the five comparisons labelled A in Table 2, and least in the two labelled C. For those in group B three gorilla calls appear to have more than one equivalent in the chimpanzee. This could be no more than a consequence of taxonomic bias in different investigators. I favour this interpretation in B6–B9. It is clear from Fossey's descriptions that the category of gorilla cries includes both of those we have designated in the chimpanzee as the squeak and the whimper. Similarly with the gorilla wraagh, Fossey's sound spectrograms illustrate both short and long forms, the former resembling the chimpanzee waa bark, the latter the chimpanzee wraaa. Intriguingly, she presents these as individual differences rather than as different forms that may be uttered by a single individual, as occurs in the chimpanzee.

I feel less confident about comparisons with the gorilla pig grunt. By their nature, grunting sounds are difficult to characterize acoustically. However, the available evidence suggests that the chimpanzee grunt may correspond to the pig grunt as given in the rapid sequence by gorillas, while the chimpanzee cough may correspond to the version of the pig grunt given singly.

Most speculative of the comparisons given in Table 2 are the last two. Fossey's description of the gorilla pant series is brief, and no sound spectrograms are illustrated. However, the description suggests a possible parallel with chimpanzee pant-grunting, the latter perhaps being more voiced, especially on the exhalation phase.

Both the chimpanzee and the gorilla have barks. Although they seem to be similar in quality, they differ in temporal organization. While the chimpanzee bark is monosyllabic, the three gorilla barks listed by Fossey, the hoot bark, question bark, and hiccup bark, may consist of one to three syllables. However, these three seem closely related both in structure and context, and Harcourt (*in litt.*) did not distinguish between hoot and hiccup barks in his subsequent field studies of gorilla behaviour.

Finally, it is not clear whether the chimpanzee has an equivalent of the gorilla whine. Fossey was unable to record this sound, but to judge from her description, it might be a more drawn-out version of the chimpanzee whimper, which could also be described as 'puppy-like'. It has tentatively been included in Table 2 as a whimper equivalent.

We are left with three gorilla sounds which have no feasible equivalent in the chimpanzee. One of these is the whinnie, apparently an idiosyncracy of one animal, and this will not be considered further. The two remaining gorilla calls, the roar and the growl seem to have no chimpanzee counterpart. Fossey's description of the roar is fairly complete, and I feel confident that the chimpanzee has no corresponding call. Although the growl was not tape-recorded by her, the verbal description is of a quite distinctive, though soft sound. Probably this too represents a significant species difference in vocal behaviour.

Perhaps the most unexpected conclusion to emerge from this comparison of vocalizations of the chimpanzee and the gorilla is the extent of correspondence between them. Although a few of the comparisons are still speculative, it seems likely that every item in the chimpanzee vocal repertoire has a fairly close morphological equivalent in the gorilla, and this with two species assigned by taxonomists to different genera. Similarly with the gorilla, it seems likely that further research will confirm that only two of the 16 vocalizations recorded lack a morphological equivalent in the chimpanzee (omitting the idiosyncratic whinnie). It is intriguing that both of these unshared calls, the roar and the growl, are used by gorillas in aggressive situations.

Frequency of use

The relative ease with which morphological equivalents can be discerned in chimpanzee and gorilla vocalization in turn facilitates comparisons of the frequency of use. In contemplating such a comparison, many problems of sampling arise. Ideally, one would record data longitudinally from several individuals representing each age class, sex, and major social role, such as high-ranking, or low-ranking, mothers with or without an infant, and so on.

The observations analysed here were not gathered with such foresight, and there are sampling problems, some serious, to be discussed in detail elsewhere. Nevertheless the data offer a unique first opportunity to compare quantitatively the signalling behaviour of two closely related primates.

The publication of Fossey's gorilla data on the frequency of calls obtained in an early stage of her field studies on the gorilla prompted a parallel analysis of data gathered as extensive tape-recordings during a 10 week period with the Gombe Stream chimpanzee population. Both studies were subject to some special conditions. Fossey's subjects were relatively unhabituated during the period when her quantitative data on vocalizations were gathered. As will be noted later, there is evidence of subsequent changes in the frequency of vocal use as the same population became more accustomed to the presence of human observers. By contrast, the Gombe Stream chimpanzees were totally habituated to human proximity during the study period. However, they were subject to heavy provisioning with bananas at this time, and all sound recordings and observations were made in close proximity to the feeding area. Provisioning caused unusual patterns of aggregation, as manifest, for example, in the relatively high frequency of aggressive interactions (Wrangham, 1974). Recordings were often made of animals that were relaxing after a good meal, with possible consequences for signalling behaviour. Also certain vocalizations, such as laughter, were favoured in sound recording. After reviewing the data, an attempt will be made to reinterpret them in the light of the special conditions.

Certain equivalent calls rank similarly in the chimpanzee and gorilla with regard to frequency of use, falling at least in the same quartile of ranking (Table 3). With regard to overall rankings, chimpanzee pant-hoots and gorilla hoot series rank similarly (1 and 3). The same is true of screams, ranking fifth in both cases. At the opposite extreme are certain call pairs with very different rankings. Thus laughter ranks third for the Marler chimpanzee data, while gorilla chuckles rank fifteenth in the Fossey sample. While the chimpanzee wraaa and the waa bark rank thirteenth and eighth in the chimpanzee, the gorilla wraagh ranks first in the Fossey data. There is also a considerable discrepancy between the chimpanzee grunt and cough (eleventh

Table 3. *The ranking of morphologically similar calls of the chimpanzee (1–14, includes lip-smacking) and gorilla (1–15) according to frequency of use by sex and age classes after correction for varying proportions of sex and age classes in the study populations (gorilla data from Fossey, 1972). In brackets are Harcourt's rankings after habituation*

C	*Chimpanzee*	Overall	♂1	♂2	♀	J	I
G	*Gorilla*	Overall	SB♂	BB♂	♀	J	I
C	Pant-hoot	1	1	1	1	3	5
G	Hoot series	3 (5)	3	6	10	3	—
C	Laughter	3	7	5	13	5	1
G	Chuckles	15 (2)	—	—	—	—	4
C	Squeak and whimper	4 & 6	9 & 13	4 & 7	3 & 5	1 & 4	4 & 2
G	Cries	8 (6)	(15)	(15)	(15)	(15)	1
C	Scream	5	5	3	4	2	3
G	Screams	5 (8)	5	4	1	1/2	2
C	Rough grunt	9	3	10	10	12	6
G	Belch	6 (1)	6	3	7	(15)	(15)
C	Grunt and cough	11 & 12	11 & 9	8 & 14	9 & 11	10 & 12	(14) & 11
G	Pig grunt	4 (3)	4	5	4	(15)	(15)
C	Wraaa & waa bark	13 & 8	10 & 12	13 & 9	11 & 7	14 & 7	14 & 8
G	Wraagh	1 (12)	1	1	6	5	(15)

SB = silverback; BB = blackback.

and twelfth) and the gorilla pig grunt (fourth). Finally, there are some equivalent pairs which rank differently in the two species, but to a less extreme degree. These are the chimpanzee squeak and whimper (fourth and sixth) and the gorilla cries (eighth) and finally the chimpanzee rough grunt (ninth) and the gorilla belch (sixth). Some of these differences are obviously interpretable from the differing circumstances in which the Marler chimpanzee and Fossey gorilla samples were gathered. Others may reflect more basic differences between the two species.

As already noted, Fossey's study population of gorillas was less well habituated than the chimpanzee population at the time of her quantitative study of vocalizations. She notes that while 'alarm calls' (wraaghs) were the most frequent vocalizations at that time, it is probable that with later data from the same groups the alarm calls would have been relatively less frequent, and the belch or other group coordination vocalizations might be the most frequent for nearly all age and sex classes (Fossey, 1972, p. 40). Subsequent research, yet to be analysed in quantitative terms, has confirmed this prediction.

Working with the same study population as Fossey, but now better habituated, Harcourt (*in litt.*) suggests that the rank order of usage is now more like the following: (1) belch, (2) chuckles, (3) pig grunt, (4) hoot bark or hiccup bark (not distinguished), (5) hoot series, (6) whine and cries, (7) question barks, (8) screams, (9) pant series, (10) copulatory pants (not quantified in Fossey's sample), (11) growl, (12) wraagh, (13) roar, and (14) whinnie. Harcourt expresses less confidence in the relative rankings of (4)–(14) than in that of the first three, and notes that chuckles might rank first on a duration measure though third on an onset measure. Where appropriate, these new estimated rankings have been included in brackets in Table 3.

While quantitative data are desirable, these already hint at some dramatic changes, presumably attributable primarily to habituation of the gorilla study groups. The change in estimated rank of the gorilla wraagh from the first to the twelfth is as Fossey had predicted. The frequent calling in the early study was presumably triggered by the observer. If Harcourt's data are confirmed quantitatively, then this becomes another case of close resemblance between chimpanzee and gorilla. The estimated decline in rank of gorilla cries from 8 to 6 in frequency of use is perhaps another manifestation of the habituation process, as is the decline in ranking of screams from fifth to eighth. The increased ranking of chuckles as estimated by Harcourt to second from fifteenth is perhaps a manifestation of the other side of the habituation coin, namely that certain activities would have been inhibited during observation of nervous animals, notably play behaviour. This presumably became relatively more common as habituation proceeded.

Perhaps the most intriguing of these changes in relative frequency of use of gorilla calls is the estimated rank increase of the belch from sixth to first, bringing the gorilla data into even more marked contrast with that for the chimpanzee, where the equivalent call, 'rough grunting', ranked ninth in the overall Marler sample. The situations that Fossey (1972) describes for gorilla belching include not only feeding, but also sunning, grooming and play, with calling of one individual often evoking belching in several others. This is certainly a broader array of contexts than those in which chimpanzee rough grunting occurs, and it seems likely that this is a significant difference between the two species. The gorilla belch vocalization is now under intensive study by A. H. Harcourt and K. Stewart.

One consequence of the provisioning of the chimpanzee population during gathering of the Marler sample is the presence of satiated animals, possibly favouring the occurrence of play (Wrangham, *in litt.*). Gathering of animals in the provisioning area may have made age-mates more readily available for activities such as play. Furthermore the presence of the Flo family in the camp area, an unusually large and coherent family group, may also have favoured the occurrence of play around the camp. Thus chimpanzee laughter, a subject of special study, is surely over-represented in the Marler sample. It now seems conceivable that it may actually be less frequent than in the gorilla in normally dispersed chimpanzee populations. Again quantitative data are needed.

We have noted the probable increase in aggressive activities in the Gombe Stream study population consequent upon provisioning (Wrangham, 1974), and it is possible that the high frequency of use of pant-hooting is to some extent attributable to the high level of general arousal maintained in the camp area. Again quantitative data are needed to explore this possibility.

Although there is a clear and intriguing contrast in the ranking of the chimpanzee rough grunt and the gorilla belch in frequency of use, perhaps the most striking conclusion to be drawn from the data is again the surprising degree of correspondence between the two species, especially as manifest in the revised estimate from Harcourt. However, it must be borne in mind that we are still dealing only with overall rate of use. As the next section demonstrates, there are species differences in use in relation to sex and age classes in the two species.

Sexual and individual differences in acoustic morphology

Sexual differences in vocal morphology have an important bearing on social organization. As can be seen in Figs. 1 and 2, the basic categories of all chimpanzee vocalizations are uttered by both sexes in the study sample. In the gorilla, on the other hand, there are three adult male vocalizations lacking from the female repertoire according to Fossey (1972). These are the roar, the whine and the whinnie. Schaller (1963) did record the whine from adult females. Fossey (1972) and Harcourt (*in litt.*) regard the whinnie as a somewhat unusual case associated mainly with one sick adult male, obviously calling for further study. However, there can be no ambiguity about the restriction of the roar to adult male gorillas. Females utter nothing similar.

In sounds used by both sexes there is also the possibility of within-category differences in the morphology of male and female renditions. In addition to quantitative differences in such parameters as pitch, duration and spectral organization, which have yet to be studied, there is one illustration of a sexual difference in the more general patterning of a vocalization, in chimpanzee pant-hooting (Marler & Hobbet, 1975).

The pant-hooting of adult female chimpanzees differs consistently from that of adult and adolescent males in the absence of a 'climax' section (Marler & Hobbet, 1975). The average duration of female pant-hooting sequences studied was greater than that of male sequences, though the ranges overlap greatly. The pitch of the first harmonic of female loud calls – their equivalent to a climax – tends to be deeper than that of the climax calls of most males, their duration is shorter, and their shape tends to be more arched. From what

Table 4. *A tally of those measures in which the pant-hooting of pairs of individual chimpanzees showed significant differences* ($p < 0.05$, two-tailed t-test)

Female 'loud' calls are equivalent to the male pant-hooting 'climax'. (A) Duration of complete sequences. (B) Peak frequency of climax of loud exhalation-calls. (C) Duration of climax or loud exhalation-calls. (D) Interval between climax or loud exhalation-calls (from Marler & Hobbet, 1975).

	Mike	Humphrey	Charlie	Faben	Figan	Nova	Flo
Mike		BCD	B D	B	C	ABCD	BCD
Humphrey			ABCD	ABCD	AB D	A CD	A C
Charlie				D	CD	BC	BCD
Faben					C	BCD	B
Figan						BCD	BC
Nova							D

we know of ape hearing (Stebbins, 1971), there seems little doubt that these differences between the pant-hooting of adult male and female chimpanzees are audible to them, providing a potential basis for sexual identification at a distance.

Individuality in the pant-hooting of seven chimpanzees at Gombe National Park was analysed statistically by calculating the significance of differences between all possible pairs (Marler & Hobbet, 1975). The four measures compared were (*a*) duration of pant-hooting sequences, (*b*) peak frequency of climax exhalation-calls, (*c*) duration of climax exhalation-calls and (*d*) interval between climax exhalation-calls. All pairs differed significantly in at least one measure, with a mean of 2.3 significantly different measures per pair analysed, out of four possible (Table 4).

One's ear confirms the impression from analysis that there are ample cues available for a listening chimpanzee to establish the sex and probably individual identity of a distant pant-hooting animal. There are also intriguing hints of a relationship between the duration of pant-hooting and age in males. The sequences of two older males studied were significantly shorter than those of younger males. This difference also correlates with dominance rank, such that higher ranking animals had shorter pant-hooting sequences. The sequences of the two females studied, longer than in males, conform to this relationship. More study is needed to establish its generality. There is probably individuality in other chimpanzee vocalization, though much less striking and consistent than that of pant-hooting.

Individual differences have been noted in gorilla vocalizations by Fossey (1972) who reports an ability to distinguish individual males by some calls and chest-beating. It will be interesting to see whether further analysis

Table 5. *A comparison of the frequency of gorilla and chimpanzee vocal behaviour by sex and age classes (gorilla data after Fossey, 1972)*

	Gorilla					Chimpanzee				
	Ad♂	Ad♀	Juv.	Inf.	Total	Ad♂	Ad♀	Juv.	Inf.	Total
A. Number of individuals in study population	31	36	20	19	106	15	12	10	7	44
B. Number of vocalizations of known individuals	1583	77	7	33	1700	831	633	506	343	2313
C. Percentage of total vocal output	93	5	0.5	2	—	36	27	22	15	—
D. Average vocal output per class member (B/A)	51.1	2.1	0.35	1.7	16.0	55.4	52.7	50.6	49	52.57
E. Percentage each class member makes of total (D)	92	4	0.6	3	—	27	25	24	24	—

Ad. = adult, Juv. = juvenile, Inf. = infant.

confirms Harcourt's (*in litt.*) impression that the individuality of gorilla sounds is less striking than that of chimpanzee pant-hooting, which was readily detected by van Lawick-Goodall (1968*a*, 1971) by ear, without the aid of sound spectrograms.

Sexual differences in vocal usage

Sexual differences in vocal behaviour may occur not only in call morphology but in the frequency of use of equivalent categories of vocalization. The quantitative Fossey & Marler data on chimpanzee and gorilla calling contain many such asymmetries. Most striking is the overwhelming domination of gorilla vocalization by adult males. This applies to all but one of the 14 vocalizations used by gorillas (Fossey, 1972, and Fig. 3). Thus only one vocalization, pant series, is used more by adult females. Even here adult males contribute a quarter of the total and the overall numbers are small, this being one of the least frequently used vocalizations. There is no record in Fossey's data of a gorilla vocalization unique to adult females.

The difference between gorilla and chimpanzee has been made clearer by retabulating the Marler chimpanzee data in classes that correspond more closely to those used by Fossey. Table 5 also shows the number of individuals in each sex and age classes in the study population, permitting calculation of the average number of vocalizations uttered by each class member, given directly (D) and as a percentage (E). A striking difference emerges.

Whereas vocal behaviour of the chimpanzee is evenly distributed throughout all classes of individuals, adult male gorillas contribute more than 90% of all vocal behaviour recorded. This is true both of the overall figures and as reduced to the output of an average class member. Fossey's data also show that there is a further asymmetry in the vocal behaviour of adult males, silverback males making a much greater contribution than blackback males.

As the gorilla population became habituated to observers frequency of the wraagh declined drastically, as already noted. However, there is evidence of a compensatory increase in use of other calls. Thus Harcourt (*in litt.*) estimates, on the basis of data from one group, rates of belch vocalizing for individual adults of 4.5/hour (silverback), 1.5/hour (blackback ♂) and 0.5/hour (♀, mean of five animals). This estimate excludes periods of belch chorussing. For the group of seven adults this gives an average rate of 8.5/hour, much higher than in the Fossey sample, with less well-habituated subjects. In spite of these changes the domination of vocal output by silverback males persists in the habituated gorilla population (Harcourt, *in litt.*).

The domination of vocal output by adult males, so striking in the gorilla, is absent in the chimpanzee. Although there are interesting differences in the details of vocal use between male and female chimpanzees in the Marler sample, the data are most notable for the extent of sharing between the sexes of all vocal categories that they suggest. There is no call type unique to one sex, and they contribute roughly equal proportions to the overall frequency of vocal behaviour. This implies similarity of their social roles, but not of course identity. For example difference in adult sexual roles in the chimpanzee is implied by consistent differences in male and female renditions of at least one call. As already indicated, although both male and female chimpanzees engage in pant-hooting, there are differences in morphology that are striking enough to permit a human observer to determine the sex of the vocalizer (Marler & Hobbet, 1975).

Differences in vocal usage by sex and age

One approach to comparing the extent of shared use of a vocalization by different sex and age classes is to employ an information theory index of diversity (e.g. Baylis, 1974). Applying the formula indicated in Table 6, to the quantitative Fossey & Marler data, as converted to the output of average animals, $H_i$max is the value if all sex and age classes contributed equally to the use of a given vocalization. The extent of departure of the realized values of H_i from the maximum reflects the degree of inequality of sex and age class contributions to that call type.

As can be seen from Table 6, the values for many chimpanzee vocalizations are close to the maximum. H_i is greater than half of the maximum value for

Table 6. *Diversity of usage of chimpanzee and gorilla sounds by age and sex classes, as measures by the diversity index of Baylis (1974)*

$H_i = -\Sigma_i \log_2 P_i$ and $H_i\max = \log_2 a$, where a is the number of event categories and P_i is the calculated relative frequency with which either the ith sex/age class uttered that vocalization or the ith vocalization was uttered by that age–sex class. The ratio of $H_i/H_i\max$ indicates the degree to which the diversity of usage approaches the maximum possible. Items are listed in order of increasing diversity.

Chimpanzee vocalization types	H_i	Gorilla vocalization types	H_i	Age and sex classes ($H_i\max = 8.77$)	
($H_i\max = 2.32$)		($H_i\max = 2.32$)		Chimpanzees	H_i
Waa bark	2.22	Whines	1.46	♂ 1	2.64
Scream	2.22	Screams	1.34	♂ 2	2.78
Bark	2.20	Pant series	1.32	♀	2.89
Pant	2.13	Belch vocalization	0.73	Juvenile	3.28
Squeak	2.03	Whinnies	0.72	Infant	2.10
Grunt	1.97	Question bark	0.53	($H_i\max = 3.81$)	
Pant-hoot	1.95	Pig grunts	0.44		
Whimper	1.92	Growl	0.44	Gorillas	H_i
Pant-grunt	1.92	Wraagh	0.37	♂ Blackback	2.49
Rough grunt	1.77	Hoot bark	0.32	♂ Silverback	2.59
Cough	1.62	Hiccup bark	0.26	♀	2.55
Wraaa	1.45	Roar	0.24	Juvenile	2.24
Lip-smack	1.45	Hoot series	0.12	Infant	1.18
Laughter	1.17	Cries	—	($H_i\max = 3.91$)	
		Chuckles	—		

all 14 chimpanzee sounds, with laughter least diverse. The situation in the gorilla is very different, with H_i less than half of $H_i\max$ in 11 of 14 sounds, an obvious reflection of the lower diversity of usage of most gorilla sounds. The usage diversity of the great majority of chimpanzee sounds exceeds that of even the most diversely used gorilla vocalization.

In the gorilla, the domination of vocal behaviour by adult males results almost entirely from silverbacks, blackbacked males differing much less from adult females than silverbacks do in the frequency of use of different call types. Thus there are many behavioural markers signifying the age class that silverbacked males represent.

The Fossey gorilla data do not permit further characterization of intra-sexual age classes, but they reveal two calls unique to infants and one used more frequently by them than by any other group. Similarly in the chimpanzee, the vocalization of laughter is dominated by infants and juveniles of both sexes, though especially by males. Other strong differences in age-class usage of vocal types can be seen in Figs. 1 and 2. In the particular sample gathered, adolescent females tend to dominate the bark vocalization, older adult males the wraaa and rough grunting, infant males dominate whimpering, and older adult females dominate pant-grunting. It will be most interesting to compare vocal usage with sex and age classes in non-provisioned, wider-ranging chimpanzees, and in well-habituated gorillas. In the latter, for example, Harcourt (*in litt.*) reports that silverback males rarely scream, and that adult females scream most often, in the context of intra-group squabbles.

The preponderance of use of a call by one age class may be merely statistical, often insufficient to serve as a reliable marker of that age class to other individuals. In particular, there is no drastic distinction between older and younger adult male chimpanzees comparable to the difference between blackback and silverback gorillas. The pattern of vocal usage by older and younger adult male chimpanzees does differ (Fig. 1), however, in ways that may again correlate with more subtle changes in social role with age. Thus, in this sample, mature males used the aggressive cough more often than younger ones, as well as the wraaa call, with both alarm and aggressive connotations. The same applies to rough grunting, given at food which, since they are higher in rank than younger males, they tend to control.

Differences in overall rates of vocalization

Since the quantitative data on chimpanzee and gorilla vocal behaviour were gathered in different ways a definitive comparison of rates of vocalization is not possible. Fossey (1972) records a total of 1700 vocalizations from identified individuals during 2255 hours of study. A rough estimate of the rate of vocalizing in an average gorilla group is once per hour, assuming that Fossey usually had just one group under observation most of the time. In making this calculation I am assuming that each vocal event noted represents a single call and not a sequence. As already mentioned, Fossey's subjects were poorly habituated for much of the period that data on vocalizations were gathered. The frequent use of the wraagh is probably explicable in this way. The wraagh represents almost a third of all calls noted from identified individuals (547: 1700). Thus a reduction in the rate of vocalizing by habituated gorillas might be expected. However, there could be a compensatory increase in the rate of use of other calls as Harcourt (*in litt.*) has indeed suggested. He estimates rates of vocalization for one well-habituated group to be as high as 10–20+ calls per hour, much greater than indicated by the Fossey data, perhaps as a result of habituation to human proximity.

In the chimpanzee study 2656 vocalizations were recorded and identified from known individuals during 23 hours of tape recordings. From these figures we derive a rate of about 115 vocalizations per hour. However, there are many problems with this estimate. Tape-recordings were made not on a random or systematic schedule, but when behaviour was occurring, thus leading to an overestimation of the rate of chimpanzee calling. Furthermore, the aggregation of animals in the provisioning area is likely to have increased rates of social interaction and thus of vocalization. On the other hand, underestimation results from the fact that many vocalizations were excluded from this estimate, because of failure to identify the vocalizer or vocal type, and because the chimpanzee data actually comprise vocal sequences in many cases rather than single calls, such as Fossey presumably recorded for the gorilla. The 23 hours of chimpanzee recordings were obtained during 600 hours of observation. Many unnoted vocalizations also occurred during the 577 hours in which no tape-recordings were made.

Trying to balance the trends toward over- and underestimation, I estimate that the rate of vocalizing by an average provisioned chimpanzee group in the feeding area is somewhere between 10 and 100 calls per hour. It would probably be lower in free-ranging animals. Thus these estimates of frequency of vocalizing, unsatisfactory though they are, suggest rather similar rates of vocalizing in chimpanzees and gorillas living under similar conditions.

Comparisons of temperament in the two species emphasize the stolid nature of the gorilla (Schaller, 1963) and data on vocal behaviour lend some support to this view. Consider for example the use of distress calls – the chimpanzee whimper and gorilla cries. Whereas the latter are completely confined to infant gorillas, no doubt more timid and emotional than adults, the corresponding chimpanzee calls are used by all ages and both sexes, even though infants still predominate as users of the whimper. This might be taken as a sign of greater emotionality and expressiveness of chimpanzees of all ages as compared with the gorilla. However, it may be noted that screams, which rank fifth in frequency of general usage for both species, are uttered by gorillas of all ages and both sexes. In the chimpanzee, although all classes of individuals engage in screaming, it is less frequent in males than in females, and is notably infrequent in the mature, higher ranking males. By contrast silverback male gorillas were the most frequent to scream in Fossey's study, although this seems to have changed in favour of adult females with habituation (Harcourt, *in litt.*).

Comparisons of temperament in the two species should not overlook the remarkable vocal range of silverback males. If vocal output is indeed to be used as an index of temperament, then silverback males can hardly be viewed as any less expressive and emotional than a chimpanzee, however phlegmatic other classes of individuals may appear. One wonders whether the restraint of the latter might not be correlated with the relative exuberance and assertiveness of the silverback male and whether groups lacking a silverback might behave differently.

Generalizations on chimpanzee and gorilla vocal behaviour

Thus far the comparison of vocal behaviour of the chimpanzee and the gorilla suggests the following points. The numbers of vocal categories used by field observers to describe the behaviour are similar – 13 for the chimpanzee and 16 for the gorilla. Leaving aside for the moment questions about the validity of these categories, visual inspection of sound spectrograms suggests roughly similar vocal repertoires for the two species. There seem to be morphological similarities between 11 gorilla calls and all 13 chimpanzee calls. The circumstances of use of these shared calls also have much in common, suggesting homologies in both form and function.

While there are similarities in chimpanzee and gorilla vocal repertoires, striking differences occur in their patterns of usage. Chimpanzee vocal behaviour is distributed through all age classes and both sexes. There are no vocal types unique to one sex. By contrast male gorillas dominate the vocal behaviour of the species, and most of the burden falls on one age class – the silverback male. While quantitative aspects of gorilla vocal usage seem to change with increasing habituation of the study group, domination by silverback males still prevails (Harcourt, *in litt.*). One vocalization, the roar, can confidently be judged as unique to adult male gorillas. There is probably no consistent difference in frequency of vocalization in the two species. There is a strong element of individuality in at least one chimpanzee vocalization. There are hints that although there is individuality in some gorilla vocalizations, it may be less prominent than in chimpanzee pant-hooting.

SOCIAL ORGANIZATION OF THE CHIMPANZEE AND GORILLA

In seeking to relate vocal behaviour to social organization, a first problem is to define the basic social units. This requires knowledge of how individuals are distributed in space, their patterns of contact and separation, and the ways in which they interact with one another, either competitively, so limiting one another's access to resources, or cooperatively, by mutual aid in reproduction, resource exploitation, and the avoidance of threats to life and health. Defining social groupings and distinguishing competitive from cooperative interactions are less easy than they might appear.

Attempting to summarize a complex situation, there are some solitary adult male gorillas, but most animals live in a social group. The groups averaged 15 members in the nine troops studied by Fossey (1972), usually with one but up to three fully mature silverback males per group. The balance is made up of blackbacked males, adult females, and their offspring. The overlap between the extensive home ranges of gorilla groups is considerable.

While Schaller (1963) and Fossey (1974) state that their gorilla study populations were not territorial, each group had an area of exclusive use, apparently as a consequence of an admixture of mutual avoidance and aggressive repulsion of neighbouring groups (Schaller, 1963; Fossey, 1972, 1974).

The pattern of social organization in the chimpanzee is harder to define. Rather than durable groupings that move synchronously as coherent units, the tendency is for individuals and smaller groups to separate and coalesce in variable combinations that vary in composition from day to day. Chimpanzees were most commonly seen by van Lawick-Goodall (1968*a*, 1971) in groups of two to six animals. More recent data suggest that average group sizes are even smaller than this. Thus Wrangham (*in litt.*) working with the same chimpanzee population found mean group sizes of 1.1 in one season, 2.0 in another, from samples in which at least one member was an adult male. The subgroupings of chimpanzees in the Gombe Stream population are constantly changing in composition, apart from the durable subgroups consisting of mothers and their immature offspring. In striking contrast with the gorilla, peaceful subgroupings of chimpanzees commonly include more than one fully adult male at a time.

If it is indeed possible to define a larger group on the basis of predominantly peaceful and cooperative interactions of subgroupings when they meet (there is in fact a good deal of aggression when reunions occur) this grouping is probably larger than in the gorilla, perhaps of the order of 50 animals. The evidence, still somewhat equivocal, suggests that such a 'community' at the Gombe National Park, has neighbouring 'communities' to both north and south, with both of which there is some peaceful intermingling.

However, there is evidence that high-ranking adult males exclude themselves from this intermingling (van Lawick-Goodall, 1971; Wrangham, 1975), perhaps with a degree of mutual avoidance (Nishida & Kawanaka, 1972), so that one may begin to think of territoriality as existing in male chimpanzees, if not in females.

Whereas recent work suggests that the several adult males who are members of a group may each have somewhat different sectors of the home range where activities are concentrated – so called 'core areas' – they do also share a common overall living space (Wrangham, 1975). Females seem to have even better-defined 'core-areas' than males, although they are highly mobile during oestrus. While a given female's core area may overlap with those of neighbouring females, contacts with more distant females, even though they are within the same male group home range, may be rarer. It even seems feasible, according to Wrangham (1975) that females seen in contact with more than one community actually have 'core-areas' that fall in a boundary area. Thus the chimpanzee community may be a less coherent unit than had been thought previously.

Attempting to compare and contrast the patterns of social organization in chimpanzee and gorilla in terms that might relate to their systems of vocal communication, they seem to have more in common in patterns of between-group than within-group relationships. Most recent research suggests that there is territoriality in both species, and that adult males take a prime role in territorial defence in both the chimpanzee and the gorilla. In the former the territory is maintained by a group of males, while in the latter the onus tends to fall on one silverback male. The two species also differ in the closeness of identification of adult females with a particular group and its home range and territorial commitments. However, in neither species is there evidence as yet of a female role in territorial maintenance. Thus while inter-group relationships in chimpanzee and gorilla do differ in a number of significant respects they also have major features in common.

With regard to within-group relationships, the contrast between the two species is more striking. Groups of gorillas are relatively compact and coherent. Individuals may get out of visual contact with companions feeding in dense vegetation, and vocalizations, especially belching are used to maintain contact in this circumstance (Harcourt, *in litt.*). The belching vocalization is in fact a dominant component in the vocal behaviour of habituated gorillas. The distances involved are small, and mechanisms for reestablishing contact seems relatively direct and uncomplicated, at least by comparison with the chimpanzee.

Chimpanzee group members are now viewed as being even more dispersed much of the time than had originally been thought. Some adult males and females spend as much as 80–95% of their time alone (Wrangham, *in litt.*). To communicate with other group members, they must often signal vocally over great distances, with companions out of sight to them. During their long periods of isolation adults of either sex may encounter predators. In rejoining group members they must engage in the signal exchanges required for re-establishment of relationships with fellow group members after long periods of visual if not vocal separation. The latter are occasions of high arousal, and may involve a variety of extended vocal and visual signalling.

In summary, patterns of chimpanzee and gorilla social organization differ most strikingly in the organization of within-group relationships. The chimpanzee has a large, dispersed social group, containing adult males, with members recombining from day to day in different subgroupings of adult males, females and young. They also spend much time alone. Group members are often separated by long distances. The gorilla has smaller, more coherent social groups, usually with only one fully adult male. Within-group vocal signalling is over much shorter ranges. The compactness of the group is such that they confront such exigencies as predator detection, dissemination of alarm and defence as a group rather than on an individual basis. This difference in within-group organization provides some basis for speculating about the social significance of differences in their vocal behaviour.

SOCIAL DESIGN AND VOCAL COMMUNICATION

Given these differences in social organization in the two species, can they be correlated with any of the differences in vocal behaviour? If so, can a case be made for the relationship being a causal one?

The basic gorilla social group is small and compact. While group members do depart and rejoin, it seems clear from published and unpublished accounts that the rate of within-group reunions, and the level of social arousal associated with them, is much lower than in the chimpanzee. Given a compact social group of relatively stable social composition, it is possible for one sex and age class to assume responsibility for many of the communicative decisions required for the maintenance of within-group social organization, and defence against predators, as well as the maintenance of relations with other social groups and solitary individuals. An argument such as this makes the domination of gorilla social behaviour by silverback males to some extent intelligible. However, although gorilla group composition and coherence *permits* the silverback male to assume the main prerogative of vocal signalling, it is by no means clear that this is a *required* consequence.

There seems little doubt that a significant and perhaps major proportion of chimpanzee vocal behaviour is occasioned by renewed contacts between within-group social subunits. The highly dispersed, fluid nature of chimpanzee society favours, in all sex and age groups other than dependent infants, the assumption of competence to cope with major environmental and social contingencies on an individual basis. The obvious exception here is territorial defence, which seems to be emerging as a solely male responsibility, as in the gorilla. The major vocalization involved in territorial exchanges, pant-hooting, also serves other functions incorporating long-distance vocal signalling, and this perhaps explains its presence in the female repertoire as well. Moreover, we have noted that the female form of this vocalization is discriminable from that of adult males. The other acoustical signal involved in between-community interactions, drumming, is restricted to adult males.

There is ample evidence that chimpanzee subgroupings signal vocally to one another over distances of hundreds of metres, something that probably never occurs within a gorilla group (Harcourt, *in litt.*). An obvious case is meat-eating where loud sustained vocalizations attract distant group members to the event (van Lawick-Goodall, 1968*a*, 1971; Teleki, 1973). Coordination of movements and other behaviours within the community is presumably modulated by these exchanges, in which all animals other than infants may participate on occasion, presumably to the benefit of the group as well as to themselves. This pattern of within-group dispersal and coalescence must favour the assumption by all individuals of an ability to cope with a wide variety of communicative interactions. The remarkable spread of the use of all vocalizations throughout the chimpanzee community membership is perhaps interpretable in these terms, in striking contrast to the domination of gorilla vocal behaviour by one sex and age class, the silverback males.

We have noted the prominence of individuality in one long-range vocalization of the chimpanzee (Marler & Hobbet, 1975). Pant-hooting is a major vehicle for the maintenance of social contact between dispersed group members in a chimpanzee community. Individuality is likely to be of value to them in both within-group and between-group discriminations.

By contrast, Harcourt (*in litt.*) states that the gorilla hoot series, the equivalent of the chimpanzee pant-hoot, is most frequent during inter-group encounters or in interactions between a lone silverback male and the group. Again it is usually given when neither participant is visible to the other. However, a silverback male gorilla, especially if he is the only silverback in a group, knows immediately that another animal giving hoot series is a stranger. Also, there are often extraneous clues available to the gorilla which help in identification of a stranger, such as the area in the range where the interaction is taking place.

It is hard to imagine, however, that there would be *no* additional benefit accruing to individuality of hoot series. Further research may reveal that the difference between gorilla and chimpanzee here, if any, is more one of degree than kind, with gorilla call individuality at a moderate level, complemented by the accessibility of other cues, especially within the group, where many individual discriminations will be possible on the basis of age-correlated variations in behaviour and appearance.

We can also speculate about the adaptive significance of similarities that exist in the vocalizations of the chimpanzee and the gorilla. In both, the onus of territorial maintenance seems to fall primarily on mature adult males. In both species, the vocalization most commonly employed in inter-group encounters, pant series and pant-hooting, are given more frequently by the age class responsible for territorial maintenance, mature males, even though the asymmetry in usage is much less striking in the chimpanzee than in the gorilla. The similarity of the non-vocal sounds used in such encounters, chest-beating and drumming can result only from convergence. It remains to be determined whether such mechanical sounds have a unique advantage in this context, such as better carrying-power than vocalizations, or whether interactions between the two species at some stage of their history might have favoured inter-specific territoriality, and thus convergence on similar signals for inter-group spacing.

With regard to the apparent degree of correspondence in the morphology of vocalizations of the two species, a majority of the repertoire must function primarily for within-group signalling. If the details of vocal behaviour are shaped by adaptive social function with any degree of precision, and we should not forget that this is by no means a proven assumption, then we might infer that the major environmental and social contingencies met in within-group behaviour of the two species are similar in nature, although differing in the distribution of responsibilities through the group. Some of the quantitative differences in ranking of use of equivalent vocalizations may also be adaptive. The need to maintain greater group coherence in the gorilla perhaps explains the more frequent use and broader array of contexts of gorilla belching, as compared with the less frequently used and more restricted chimpanzee rough grunting. 'Belching' often seems to serve foraging gorillas as a 'keeping in contact' signal (Harcourt, *in litt.*). Chimpanzees have no call that functions as a general, close-range, contact signal.

It is interesting to note that several equivalent vocal pairs in the two species rank rather similarly in frequency of usage in spite of the differences in their social organization.

These then are the kinds of correlations between social organization and vocal signalling that throw some light on the behavioural similarities and differences in the chimpanzee and the gorilla. While the data are still imperfect, we can begin to see more clearly what additional observations are desirable. Data on the intensity and range of reception of vocalizations with different functions are needed. Above all, new approaches should be sought to characterize the *functions* of different vocalizations, so that more subtle interspecies comparisons of the proportions of a signal repertoire devoted to different kinds of adaptive tasks may be possible.

We have noted that apart from the gorilla roar, the only other gorilla call that definitely seems to lack a counterpart in the chimpanzee repertoire is the growl. If an aggressive function can indeed be assigned to this call, can we infer that there is a need for a close-range, within-group, aggressive vocalization in the gorilla which is absent in the chimpanzee, or met in a different way? Within-group aggression is often a noisier and more highly aroused interaction in the chimpanzee, including elaborate and highly ritualized aggressive displays. Playback of recorded vocalizations applied recently to primates in the field by Waser (1974, 1975, 1976) in studies of vocal communication in mangabeys (*Cercocebus albigena*) is one approach to such functional questions.

The difficulties in understanding communication in free-living primates are not only practical but logical. It is hard to understand a communicative system without participating in it oneself in a rather complete way. In the process of disentangling the relationships between social organization and signalling behaviour there is a prospect of learning more about the general principles underlying animal communication, undoubtedly more complex and subtle than is often supposed. In particular, the graded nature of many chimpanzee and gorilla vocalizations, commented on in this review and by all investigators of these two species, presents many challenges for future research.

DISCRETE AND GRADED SIGNALS AND THEIR SIGNIFICANCE

Since Rowell and Hinde described the remarkable development of graded sounds in the vocal repertoire of the rhesus macaque (*Macaca mulatta*) (Rowell, 1962; Rowell & Hinde, 1962), it has become clear that higher primates exhibit an unusual emphasis upon graded signals in their vocal behaviour. This is not to say that discrete sound signals are not to be found in primates. On the contrary, there are several well-analysed examples. The vocalizations of the squirrel monkey (*Saimiri sciureus*) are discrete for the most part, although showing different degrees of within-category variation (Winter, Ploog & Latta, 1966; Winter, 1969). Vocalizations of the vervet monkey are discretely separate from one another for the most part (Struhsaker, 1967), and the same is true of some sounds of two other species of *Cercopithecus* monkeys, the blue monkey (*C. mitis*) and the red-tailed monkey (*C. ascanius*) (Marler, 1973). The signals still show within-category variation in form, as well as in amplitude, completeness and frequency of delivery. Some blue monkey calls are more variable than others, and the differences probably have communicative significance. Thus discretely organized signals still show variation. Some species illustrate an intermediate condition, such as the black-and-white colobus (*Colobus polykomos*) (Marler, 1972). The situation in species such as the rhesus macaque differs in the extent to which there are few if any truly discrete items in the vocal repertoire. All morphological types tend to grade with one or more other items in the repertoire.

While the difference between vocal repertoires that are predominantly discretely organized, and those that are graded, is one of degree, it is a striking fact that a number of higher primates fall at one extreme on this continuum. In addition to the rhesus macaque, we now know that the stumptail macaque (*Macaca speciosa*) and the Japanese macaque (*M. fuscata*) have both extensively graded vocal repertoires (Bertrand, 1969; Chevalier-Skolnikoff, 1974; Green, 1975). Highly graded vocal repertoires have also been described in the red colobus (*Colobus badius*) and the talapoin monkey (*Miopithecus talapoin*), differing from the other cases of highly graded vocal repertoires mentioned thus far in living in rainforest habitats rather than the forest edge and savanna (Marler, 1970; Gautier, 1974; Struhsaker, 1975). Both the gorilla and the chimpanzee manifest a trend towards highly graded vocalizations (Marler, 1969; van Hooff, 1971; Fossey, 1972; Marler & Tenaza, 1976).

Some hints as to the functional significance of signal grading are suggested by reviewing the major features of social organization of those primate species in which the nature of the vocal repertoire has been well documented. The closest correlation with the dichotomy between discretely organized and graded vocal repertoires occurs with territoriality. Whether living in forest or savanna, non-territorial primates seem more prone to develop graded vocalizations. The major exceptions to this rule at present are the squirrel monkey, thought to be non-territorial, yet with a significant number of discrete vocalizations in its repertoire (Winter *et al.*, 1966) and the chimpanzee (see p. 259). There is also a weaker correlation with the nature of group membership, with a tendency for emphasis upon graded sounds in those species with a multi-male type of group organization.

The speculation has been offered that territorial organization, in the maintenance of which adult males assume special responsibilities, favours the evolution of loud, far-crying signals to serve, among other things, as inter-group signals. By contrast, signals normally functioning over shorter ranges are perhaps more prone to be graded. It may be significant that close-range sounds frequently function in concert with visual signals, often also highly graded. Thus effective transmission distance may be one of the prime correlates in the evolution of a discrete type of vocal organization. Long-distance sound signalling, with no support from other sensory modalities, is perhaps served best by relatively invariant signals that are discretely distinct from one another, with no intergradations.

Signals functioning over shorter transmissions distances may be less constrained towards discreteness and invariance, both because reception will be less hindered by noise in the environment, and because of the possibility of redundant visual signalling, aiding in the accurate and reliable reception of subtle signal gradations. This is a shift of emphasis one might expect to find in non-territorial species, with less of their signal repertoire devoted to inter-group signalling necessarily involving longer distances. Any potential that graded signals may have for the communication of more refined information than can be conveyed by discrete signals is likely to be exploited in such circumstances.

THE FUNCTIONAL SIGNIFICANCE OF SIGNAL GRADING

What interpretation is to be placed on the extensive intergradation of vocal categories found in such animals as macaques, the chimpanzee and the gorilla? The weakest hypothesis would be that it represents disorderly, erratic variation resulting from a loose relationship between vocal morphology and the physiological determinants of other ongoing behaviours. This is clearly not the case. Several observers of graded vocal behaviour in primates have noted correlated variations in associated activities. For example van Lawick-Goodall (1968*a*) discerned correlations between the type of vocalization and ongoing motivational states, as inferred from circumstances of production and the accompanying behaviour. Precision of vocal control can obviously be achieved by the chimpanzee. The most frequent vocalization, pant-hooting, is one of the more stereotyped chimpanzee vocalizations, in which consistent individual differences are maintained in the face of involvement in a graded relationship of other vocalizations such as screaming (van Lawick-Goodall, 1968*a*; Marler & Hobbet, 1975). Similarly Gautier (1974) has shown that variations within the graded repertoire of the talapoin monkey along one or another acoustical dimension are by no means random, but match other characteristics of the situation.

Perhaps the most elaborate attempt to map continuous vocal gradations against the accompanying context of the vocalizing animal was conducted by Green (1975) on the Japanese macaque. During a 14 month field study, extensive recordings of vocalizations were noted along with assessments of the social situation in which each was produced. Sound spectrographic analyses were subjected to a systematic empirical taxonomy based on physical properties. Ten classes were diagnosed by various distinctive features some relating to temporal pattern of delivery, some to the mechanism of phonation, others to more or less arbitrary acoustical features.

One of the groups will serve as an illustration, the class II 'coos'. These were divided into seven types according to the features shown in Fig. 4. In a separate analysis the circumstances of each were assessed and classified, along the lines illustrated. The two sets of information were then put together to see whether the subdivisions into which the sounds had been classified bore either orderly or random relationships to the circumstances of production.

As may be seen from Fig. 4, the relationship was by no means random. Instead each variant was highly correlated with one situation or cluster of related circumstances. Thus the relationship between signal grading and circumstances of production in this group of Japanese macaque calls is highly orderly, and has the potential of conveying subtle and complex information about the circumstances of sound production and the identity and state of

Coo type	Distinguishing criteria			
Name	Midpoint pitch	Position of highest peak	Duration	Other features
Double	⩽510 Hz	N.A.	N.A.	Two overlapping harmonic series
Long low	⩽510 Hz	N.A.	⩾0.20 sec	N.A.
Short low	⩽590 Hz	≠1	⩽0.19 sec	N.A.
Smooth early high	⩾520 Hz	<2/3	N.A.	No dip
Dip early high	⩾520 Hz	<2/3	N.A.	Dip
Dip late high	⩾520 Hz	⩾2/3	N.A.	Dip
Smooth late high	⩾520 Hz	⩾2/3	N.A.	No dip

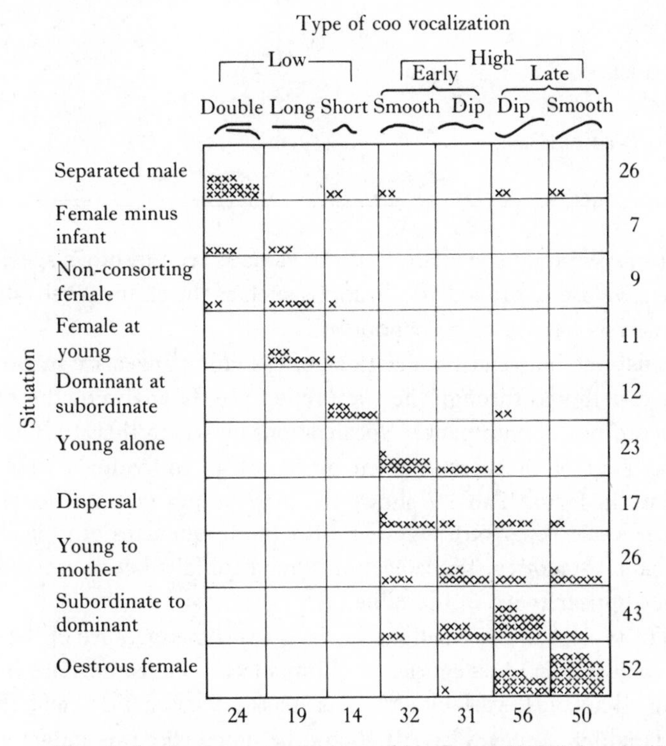

Fig. 4. At the top are criteria used by Green (1975) for classifying 'coo' sounds of Japanese macaques into seven categories. N.A. = not applied for separation of types. Below is a tabulation of the situations in which 226 occurrences of coo-calls were observed in the field in Japan.

Table 7. *A sample of 343 intermediate chimpanzee calls classified by the categories they fall between (A – two entries for each intermediate call) as compared with the number of typical examples of that category recorded (B). The ratio of B/A is one measure of the variability of each category. They are listed in order of decreasing variability*

	Call category	A (Intermediates)	B (Typical)	B/A
1	Waa bark (most variable)	92	84	0.9
2	Scream	173	216	1.25
3	Bark	54	95	1.76
4	Grunt	25	43	1.79
5	Squeak	146	264	1.81
6	Wraaa	7	18	2.57
7	Whimper	54	175	3.24
8	Pant-grunt	64	285	4.45
9	Pant	13	81	6.23
10	Rough grunt	7	83	11.86
11	Pant-hoot	52	648	12.0
12	Cough	0	33	33
13	Laughter (least variable)	0	271	271
	Totals	343(×2) = 686	2313	6.7

the vocalizer. We may speculate that the same is true in other species that make extensive use of graded vocalizations such as the chimpanzee, although such an analysis has yet to be conducted.

Intergradations between vocalizations of the chimpanzee are by no means uniformly distributed through the vocal repertoire. In analysing the 23 hours of tape-recordings of chimpanzee vocalizations on which this study is based, a note was kept of those calls given by identified individuals which were intermediate in form. Table 7 shows the distribution of transitional forms through the vocal repertoire together with the frequencies of typical forms of those same categories. Since each intermediate falls between two typical forms, they appear twice in the table.

A ratio of typical to intermediate forms gives some measure of the degree to which each of the 13 categories of chimpanzee vocalizations are involved in grading. The most variable, the waa bark, is listed first, and the least variable, laughter, appears last. It should be noted that this judgement has no relationship to the degree of within-category variation, referring only to variability manifest as gradation with another category. Laughter for example is exceedingly variable in form, but it was never noted as grading into any other item in the vocabulary.

THE SIGNIFICANCE OF VARIATION IN CHIMPANZEE VOCALIZATIONS

Variation from one rendition of a call type to another may take several forms. As already noted, a category that is discretely separate from other types may exhibit within-category variation. Such variation might be accidental and random, or it might be orderly or even highly organized. As already noted, chimpanzee laughter is perhaps the most discrete item in the vocal repertoire, yet it varies considerably in ways likely to have communicative significance.

In addition to this within-category variation, there is the graded variation between categories such that they become connected by intermediate forms. Variation of this type, illustrated by graded chimpanzee vocalizations, raises many questions. For one thing, it renders uncertain the significance of the original categories, recognized by field workers, and accepted as meaningful in the main body of this paper. Even though observers confronted with such a graded vocal system feel reasonably confident in discerning categories, these judgements may be based on the more frequent usage of modal forms than of intermediates. Table 7 shows that intermediate forms are often rarer than the typical categories.

There are at least two different forms of delivery of graded vocalizations, either 'adjacent' or 'separate'. In the former case, an individual produces a string of continuous variants with the morphology changing from type A to type B in a series of small consecutive steps rather than a single jump. In such a series the differences between adjacent pairs may be slight. One wonders how meaningful single steps are to the animals. However, if they are given in a string with only brief intervals separating them, each might provide a frame of reference for the next, still available in short-term memory, so that directions and rates of change of vocal morphology in the string might be more readily detectable. This might be how some vocal gradations assume communicative significance.

One might predict a correlation between degrees of vocal grading and the tendency of calls to be uttered in rapid strings. This occurs with a number of those in the upper half of Table 7, categories especially involved in between-category grading. It is true also that some of those lower in the table tend to be delivered at longer intervals such as the rough grunt and the cough. Pant-hooting sequences also tend to be spaced out from one another. However, laughter is given in loud, rapid, varying strings, yet it is the least involved in between-category gradation of all. Additional data are needed on the temporal pattern of delivery calls more or less involved in grading.

In addition to grading of 'adjacent' utterances, variation may also occur in vocalizations that are rendered separately. Thus an observer recording vocalizations given singly at different times may, upon analysis, find them to form a graded series when compared with each other. Given the difficulty in identifying a single signal on a highly graded continuum, it is hard to imagine how graded sounds given in 'separate' fashion could be used effectively in communication, unless additional cues were provided by the signalling animal. It may be that when two animals are communicating at close range, visual signals provided simultaneously with, before or following a graded signal provide the additional information needed for accurate identification by a potential respondent.

However, all such speculations must lie in abeyance until we discover how a potential respondent in fact perceives variations along a continuously graded signal dimension. Once we determine whether they are perceived categorically or in a continuous mode, speculations about the significance they may have to another member of the species will become more meaningful. The issue is well illustrated by recent research on the perception of speech, which may have a bearing on the interpretation of vocal communication in other primate species.

SIGNAL GRADATIONS IN SPEECH AND THEIR PERCEPTION

As one listens to the flow of normal speech, the different phonemes from which words are constructed sound discretely separate from one another. It is thus something of a surprise to learn from descriptive analyses of the acoustical structure of speech, exploring the variability of speech sounds as an ethologist studies fixed action patterns, that many phonemic categories are in fact graded as they occur in normal speech. The best data concern voiced and voiceless stop consonants, their essential characteristics being among the easiest to discern in sound spectrograms of speech. Examples in English are – b d g – which are voiced stop consonants while – p t k – are so-called voiceless stop consonants.

The graded nature of the morphology of related phonemic categories is well illustrated by the consonant–vowel combinations [pa] and [ba] (Lisker & Abramson, 1964, 1970; Abramson & Lisker, 1970). The essential difference between the two consonants lies in the interval between the first release of air, in this case through the lips, and the onset of laryngeal voicing. The interval is brief or non-existent in [ba], whereas in [pa] it is about 40 msec. Histograms of the distribution of this so-called 'voice onset time' in speech reveals the remarkable fact that for many samples the distributions for these two phonemes are not discretely separate. Rather they form immediately adjacent categories, patterned in many cases as a bimodal distribution in which the trough fails to reach the baseline. It is in this sense that the two sounds grade into one another. Although the trough indicates a boundary, the voice-onset times for [pa] and [ba] overlap. A similar condition has been found in several other speech sounds which have a close acoustical relationship, although functioning as distinct phonemic categories.

Aspects of this work have great potential significance to ethologists. For example, Abramson & Lisker have plotted the voice-onset time boundaries at which speakers tend to separate such sounds as [pa] and [ba] not only in English, but also in ten other languages. Although languages differ somewhat in where the criterion for phonemic distinction is drawn, the most remarkable outcome from this comparative survey is the repeated recurrence of certain distinctions with roughly the same value. Thus the voice-onset time boundary between [pa] and [ba] sounds recurs at about 25–30 msec in one language after another with sufficient regularity that one can begin thinking of such a distinction as a speech universal.

We have become accustomed to thinking of universals as being present in the deep structure of language, as revealed by comparative structural linguistics (Chomsky, 1967) and by psychological studies of early sentence construction by children (e.g. McNeil, 1966; Brown, 1973). It is novel to encounter the idea that universals may also exist in the superficial acoustical aspects of word structure. Widely shared species-specific behavioural traits inevitably raise ontogenetic questions and the possibility of genetic determinants. There is a growing body of evidence that infants are predisposed to observe some of these same categorical boundaries between speech sounds (Eimas, Siqueland, Jusczyk & Vigorito, 1971; Eimas, 1976) – predispositions that a child might employ in early perception and development of speech in much the same way that auditory templates have been postulated to guide early song development in birds (Marler, 1975, 1976).

Our present concern is with the grading of speech sounds, reminiscent of the gradations we have described in such animal sounds as the vocalizations of the chimpanzee. Studies of speech perception show that even though the sounds may be graded, they are nevertheless perceived categorically.

EVIDENCE FOR THE CATEGORICAL PERCEPTION OF GRADED SPEECH SOUNDS

Just as ethological analyses of the structure and function of animal signals have benefited from the use of synthetic stimuli, so the understanding of speech perception has advanced dramatically with the employment of synthetic speech. Beginning with the Haskins Pattern Playback developed in the early fifties (Cooper, Liberman & Borst, 1951) and now employing computer techniques, it has become possible, for example, to synthesize series of [pa] and [ba] sounds with the voice-onset time as a variable. Thus a continuum of sounds can readily be created, each differing from its adjacent neighbour by a 10 msec increment in voice-onset time, the extremes ranging, say, from +50 msec to −50 msec. An English-speaking subject asked to identify a selection of such sounds, divides the continuum into two categories. All sounds with a voice-onset time longer than about 25 msec are identified as [pa] and those on the other side of this boundary as [ba]. This is of course what we do in listening to normal speech, making allowance for variation in voice-onset times, contextual and otherwise, within as well as between categories.

In tests of the discriminability of different speech sounds, as opposed to studies of how they are classified, the so-called ABX is commonly used. One sound is presented first, then its neighbour in the series, and finally one or the other as a third. The subject has to decide whether the third stimulus (X – actually A or B) is identical with the first (A) or the second (B). This approach led to the remarkable discovery that it is difficult for an untrained listener even to hear small variations in such parameters as voice-onset time when they occur within a normal phonemic category. On the other hand, pairs of sounds separated by a similar degree of difference but standing across the phonemic boundary are perceived as different with a high degree of accuracy (Liberman, Harris, Kinney & Lane, 1961).

This so-called categorical mode of perception (Studdert-Kennedy, Liberman, Harris & Cooper, 1970) seems especially adapted for speech sounds. Parallel experiments with sounds differing from each other in essentially the same way as the speech stimuli but not identified by subjects as speech, reveal no increase in distinguishability at stimulus values in the region corresponding to the location of the phoneme boundary implying a continuous, non-categorical mode of perception. Notably, the distinguishability of control stimuli is poorer than that of speech sounds that stand across an inter-category boundary, suggesting that the distinctiveness of the latter is especially emphasized. Intriguingly within-category speech variants are responded to differently if the criterion is reaction time (Pisoni & Tash, 1974) showing that at one level of sensory processing they are distinguishable though not at another.

The relevance of these findings on the perception of speech sounds to the significance of graded vocalizations in higher primates is obvious. Sounds of animals such as the gorilla and the chimpanzee share with speech sounds the characteristic of continuous grading in some of their acoustical morphology. With no other information about the sounds of speech than their graded nature, one might readily misinterpret the way in which speech sounds are heard by another person. The knowledge that such speech sound continua are processed categorically molds all of one's further interpretations about how the system operates. Until we discover whether higher primates such as the chimpanzee process the vocal continua that they produce in continuous or categorical perceptual fashion, we can hardly hope to understand their communicative significance.

SOME FINAL SPECULATIONS ON THE PERCEPTION OF GRADED SIGNALS BY PRIMATES

Vocal signalling systems that make extensive use of continuous acoustical gradations are not unique to primates, but higher primates have elaborated such systems to an unusual degree. Nevertheless, discretely organized signals are present in forest-living primates (e.g. Marler, 1973), so that both extremes of organization are represented. Although there is no physiological information to go on, it seems likely that discrete signals would be matched on the receptor side by perceptual mechanisms that would tend to be categorical. Indeed, such a notion is to some extent implicit in the ethological concept of innate release mechanisms and their matching releasers. Special features of such categorical perception would be a tendency to impose boundaries on stimulus series, although the stimuli need not exhibit discontinuities, amplified sensitivity to stimulus variations around boundaries, and reduced sensitivity to within-category variations (Studdert-Kennedy *et al.*, 1970).

An unusual emphasis upon graded signals invites speculation about how their corresponding perceptual mechanisms operate in species that use them. If, as one suspects, the sounds serve especially to communicate variations of mood, permitting companions to adjust and orient their behaviour to the continuously varying prospects of future action and reaction of the vocalizer, it would seem appropriate to postulate continuous sensory processing of such signals rather than the segmentalization that categorical perception implies. Thus extensive employment of a continuous mode of perceptual processing of signals may have been something of an innovation especially characteristic of non-territorial, group-living primates.

What methods are available to discover how in fact primates such as macaques and chimpanzees perceive their own vocal signals? One approach would be to condition them to respond differently to sounds in their own vocal repertoire and then present to them a series of intermediates to see where they delineate a boundary. If anything equivalent to the categorical boundaries for speech exist in the primates, attempts to move the boundary by changing the training stimuli in one direction or the other along the stimulus continuum should meet with resistance.

Another possible approach combines study of average evoked potentials with an habituation technique. Several studies with human subjects have established that categorical processing of speech sounds can be detected by an evoked potential technique (e.g. Wood, Goff & Day, 1971; Dorman, 1974). The method relies on reference to the amplitude of certain parts of the evoked potential wave form as an index of the attention-catching property of a stimulus. After habituating the response by repeated applications of the stimulus, a second stimulus is applied. The extent of revival of the criterion response is used as an index of the degree to which the pre- and post-habituation stimuli resemble one another perceptually. Application of this habituation technique, similar in principle to that employed by Eimas *et al.* (1971) to demonstrate categorical perception of speech sounds in human infants, to the perception of speech sounds by adults, reveals boundaries that coincide with those indicated by other techniques. Such a method may prove applicable to the perception of vocal sounds by primates, especially if synthetic stimuli can be used, with all parameters under full control. Preparations are being made to test this possibility with macaques.

The method also has the added advantage that it reveals lateralization of cerebral processing of speech sounds in man (e.g. Morrell & Salamy, 1971; Matsumiya, Tagliasco, Lombroso & Goodglass, 1972), and the possibility of asymmetries in the processing of vocal sounds by other primates should be explored. The consistent bi-lateral asymmetry discovered in skulls of the mountain gorilla is intriguing from this viewpoint (Groves & Humphrey, 1973). The differential emphasis on modes of stimulus processing in the right and left hemispheres of our brain is interpreted by some as emphasizing categorical processing in the dominant speech hemisphere and continuous processing in the subordinate hemisphere (e.g. Kimura, 1961, 1964; Cohn, 1971). I have speculated elsewhere that lateralization might actually have originated to permit unusual specialization of the continuous perceptual mode in our pre-human ancestors, for exploiting communicative advantages that graded vocal signals provide (Marler, 1975).

The intermingling of categorical and non-categorical processing of sound signals in pre-human primates may have called for some radical changes in the physiology of perception. A shift from continuous perceptual processing of vocal sounds to the categorical speech perception of man would in a sense have been a step back to an earlier condition, if it is correct to assume that innate release mechanisms, widely distributed in animals, process stimuli in categorical fashion. However, the application in this case may have been a novel one, to the processing of signals that are not necessarily discretely separate, but are at least sometimes organized in a continuously graded fashion. The concomitant retention of continuous processing of non-speech sounds, song, and the paralinguistic aspects of speech, which we think of as a primitive trait for man, but was perhaps an advanced one for other primates, may have created physiological problems resolved by the assignment of prime responsibility of these two perceptual modes to different cerebral hemispheres.

Whatever truth there is in these speculations, the importance of discovering how non-human primates process conspecific vocal stimuli is obvious, especially those species that include extensive graded vocal continua. Any evidence on cerebral dominance will be a case of serendipity. Meanwhile one may at least entertain the hypothesis that cerebral dominance originated in advanced non-human primates to facilitate the emergence of what was for them the relatively novel accomplishment of continuous processing of an extensive part of their vocal repertoire. This ability, originally specialized for the processing of communication signals, might then have been retained by us, only to become subordinate once more to the categorical speech processing of the dominant hemisphere.

If we can discover how graded vocal sounds are in fact processed by primates, it will then be appropriate to re-examine the ethology of vocal communication under natural conditions, thus to learn more of what they are actually signalling about with the graded sounds that are such a distinctive attribute of their vocal behaviour. To the extent that graded sounds seem especially addressed to subtler communicative events taking place within the social group, where cooperative interactions are perhaps most likely to be found, there is a prospect of achieving a better understanding of those aspects of social communication that receive special emphasis in primates as compared to other animals. This in turn will be a step towards a better appreciation of how an elaborate social organization is sustained by the system of social communication that its members employ.

SUMMARY

(1) A sample of 2656 tape-recorded vocalizations of individually-known chimpanzees is compared with data of Fossey (1972) on an equivalent sample of mountain gorilla vocalizations, in an attempt to relate vocal behaviour to social organization in these two species.

(2) Lumped vocal categories are estimated at 13 for the chimpanzee and 16 for the gorilla (Fossey). Qualitative inspection of sound spectrograms and acoustical descriptions indicates extensive correspondence between chimpanzee and gorilla vocalizations.

(3) Quantitative comparison of output of equivalent calls is complicated by different observation conditions known to affect vocal output – sampling of habituated and provisioned chimpanzees (Marler) and unhabituated gorillas (Fossey). More recent gorilla data (Harcourt) reveal changes in vocal emphasis after habituation.

(4) Chimpanzee and gorilla differ in patterns of usage of apparently equi-

valent calls, such as chimpanzee rough grunting and gorilla belching. The former ranks low in frequency of usage, the latter high, in habituated animals. Both occur as 'food calls', but gorilla belching is also a generalized 'contact call'. The contrast correlates with maintenance of the sustained coherence of the gorilla social group, which is lacking in the chimpanzee.

(5) Production of chimpanzee vocalizations is distributed through all sex and age classes. By contrast, three gorilla vocalizations are restricted to adult males, which also dominate production of all but one call.

(6) Strong individuality and male–female differences are demonstrated in a long-distance chimpanzee vocalization, pant-hooting.

(7) The two species differ in social organization, especially in relationships within the group. Differences in vocal behaviour seem to be correlated. Members of a chimpanzee group spend much time alone and in small subgroups. Within-group reunions are frequent and occasion high arousal and much signal exchange. All sex and age classes except infants must therefore cope with many environmental and social situations, correlating with dispersion of chimpanzee vocal usage through all sex and age classes. The strong individuality and loudness of pant-hooting may aid in maintaining organization of the dispersed chimpanzee group and in between-group relations.

(8) Members of the compact gorilla group face environmental and between-group social situations together. Within-group separations are typically brief and short-range, and highly ritualized reunions are rare. Unlike the situation in the chimpanzee this makes it possible for one sex and age class to make many communicative decisions for the group as a whole, and perhaps permits the qualitative and quantitative domination of gorilla vocal behaviour by silverback males. The causal significance of such correlations remains to be determined.

(9) Some chimpanzee vocalizations are extensively graded in acoustical morphology, others are more discrete, as revealed by analysis of 343 intermediate calls. Our understanding of the significance of graded primate vocalizations is hindered by ignorance of the perceptual processing of such vocal stimulus continua. Human 'categorical' perceptual processing of graded sounds such as occur in normal speech is reviewed, providing a possible model for analysis of vocal perception in higher non-human primates, which make especially frequent use of graded vocal continua in their social behaviour.

I am indebted to Drs Pat Bateson and Robert Hinde for criticisms of the manuscript. I have especially benefitted from advice of Drs A. H. Harcourt and R. Wrangham, who commented extensively on an early draft and gave generous access to unpublished material. Linda Hobbet and Peter Skaller analysed the chimpanzee recordings and tabulated data. Drs Jeffrey Baylis, Robert Dooling, Steven Green and Stephen Zoloth criticized the manuscript and advised on statistical matters. Baron Hugo van Lawick and Dr Jane van Lawick-Goodall invited the study and aided in many ways. Dr Patrick McGinnis, Alice Ford, Dr Patti Moehlman and Dr Richard Zigmond generously made field notes available. The research was supported by grants from NSF (GB 33102) and NIMH (MH 14651).

REFERENCES

Abramson, A. S. & Lisker, L. (1970). Discriminability along with voicing continuum: Cross-language tests. In *Proceedings of the Sixth International Congress of Phonetic Sciences, Prague 1967*, pp. 569–573. Academia: Prague.

Baylis, J. (1974). A quantitative, comparative study of courtship in two sympatric species of the genus, *Cichlasoma* (Teleostei, Cichlidae). Ph.D. thesis, University of California.

Bertrand, M. (1969). The behavioural repertoire of the stumptail macaque. *Bibliotheca Primatologica*, **11**, 1–273.

Brown, J. D. (1973). *A First Language, the Early Stages*. Harvard University Press: Cambridge, Mass.

Bygott, J. D. (1974). Agonistic behaviour and social relationships among adult male chimpanzees. Ph.D. thesis, University of Cambridge.

Chevalier-Skolnikoff, S. (1974). The ontogeny of communication in the stumptail macaque (*Macaca arctoides*). In *Contributions to Primatology*, vol. **2**, pp. 1–174. Karger: Basel.

Chomsky, N. (1967). Appendix A. The formal nature of language. In *Biological Foundations of Language*, ed. E. H. Lenneberg, pp. 379–442. Wiley: New York.

Cohn, R. (1971). Differential cerebral processing of noise and verbal stimuli. *Science, Washington*, **172**, 599–601.

Cooper, F. S., Liberman, A. M. & Borst, J. M. (1951). The interconversion of audible and visible patterns as a basis for research in the perception of speech. *Proceedings of the national Academy of Sciences, USA*, **37**, 318–325.

Dorman, M. F. (1974). Auditory evoked potential correlates of speech sound discrimination. *Perception and Psychophysics*, **15**, 215–220.

Eimas, P. D. (1976). Speech perception in early infancy. In *Infant Perception*, ed. L. B. Cohen & P. Salapatek, pp. 193–231. Academic Press: New York.

Eimas, P. D., Siqueland, E. R., Jusczyk, P. & Vigorito, J. (1971). *Science, Washington*, **171**, 303–306.

Fossey, D. (1972). Vocalizations of the mountain gorilla (*Gorilla gorilla beringei*). *Animal Behaviour*, **20**, 36–53.

Fossey, D. (1974). Observations on the home range of one group of mountain gorillas (*Gorilla gorilla beringei*). *Animal Behaviour*, **22**, 568–581.

Gautier, J.-P. (1974). Field and laboratory studies of the vocalizations of talapoin monkeys (*Miopithecus talapoin*). *Behaviour*, **60**, 209–273.

Green, S. (1975). Communication by a graded vocal system in Japanese monkeys. In *Primate Behavior, vol.* **4**, ed. L. A. Rosenblum, pp. 1–102. Academic Press: New York.

Groves, C. P. & Humphrey, N. K. (1973). Asymmetry in gorilla skulls: evidence of lateralized brain function? *Nature, London*, **244**, 53–54.

Hooff, J. A. R. A. M. van (1962). Facial expressions in higher primates. *Symposium of the zoological Society of London*, **8**, 97–125.

Hooff, J. A. R. A. M. van (1967). The facial displays of the Catarrhine monkeys and apes. In *Primate Ethology*, ed. D. Morris, pp. 7–68. Weidenfeld & Nicolson: London.

Hooff, J. A. R. A. M. van (1971). A structural analysis of the social behaviour of a semi-captive group of chimpanzees. In *Aspecten van het social Gedrag en de Communicatie bij Humane en Hogere Niet-humane Primaten*, pp. 11–127. Diss: Utrecht.

Kimura, D. (1961). Cerebral dominance and the perception of verbal stimuli. *Canadian Journal of Psychology*, **15**, 166–171.

Kimura, D. (1964). Right–left differences in the perception of melodies. *Quarterly Journal of experimental Psychology*, **16**, 355–358.

Lawick-Goodall, J. van (1968*a*). A preliminary report on expressive movements and communication in the Gombe Stream chimpanzees. In *Primates. Studies in Adaptation and Variability*, ed. P. C. Jay, pp. 313–374. Holt, Rinehart & Winston: New York.

Lawick-Goodall, J. van (1968*b*). The behaviour of free-living chimpanzees in the Gombe Stream Reserve. *Animal Behaviour Monograph*, **1**, 161–311.

Lawick-Goodall, J. van (1971). *In the Shadow of Man.* Collins: London.

Liberman, A. M., Harris, K. S., Kinney, J. & Lane, H. (1961). The discrimination of relative onset time of the components of certain speech and non-speech patterns. *Journal of experimental Psychology*, **61**, 379–388.

Lisker, L. & Abramson, A. S. (1964). A cross-language study of voicing of initial stops. Acoustical measurements. *Word*, **20**, 384–422.

Lisker, L. & Abramson, A. S. (1970). The voicing dimension: Some experiments in comparative phonetics. In *Proceedings of the 6th international Congress of Phonetic Sciences, Prague 1967*, pp. 563–567. Academia: Prague.

Marler, P. (1969). Vocalizations of wild chimpanzees, an introduction. In *Proceedings of the 2nd international Congress of Primatology, Atlanta 1968*, vol. **1**, pp. 94–100. Karger: Basel.

Marler, P. (1970). Vocalizations of East African monkeys. I. Red colobus. *Folia Primatologica*, **13**, 81–89.

Marler, P. (1972). Vocalizations of East African monkeys. II. Black and white colobus. *Behaviour*, **42**, 175–197.

Marler, P. (1973). A comparison of vocalizations of red-tailed monkeys and blue monkeys: *Cercopithecus ascanius* and *C. mitis* in Uganda. *Zeitschrift für Tierpsychologie*, **33**, 223–247.

Marler, P. (1975). On the origin of speech from animal sounds. In *The Role of Speech in Language*, ed. J. F. Kavanagh & J. E. Cutting, pp. 11–37. MIT Press: Cambridge, Mass.

Marler, P. (1976). Species-specific behavior: A role for sensory templates in vocal development. In *Simpler Networks*, ed. J. Fentress. Sinauer Assoc.: New York, in press.

Marler, P. & Hobbet, L. (1975). Individuality in a long-range vocalization of wild chimpanzees. *Zeitschrift für Tierpsychologie*, **38**, 97–109.

Marler, P. & Lawick-Goodall, J. van (1971). *Vocalizations of Wild Chimpanzees* (sound film). Rockefeller University Film Service: New York.

Marler, P. & Tenaza, R. (1976). Signalling behavior of wild apes with special reference to vocalization. In *How Animals Communicate*, ed. T. Sebeok. Indiana University Press: Bloomington, in press.

Matsumiya, U., Tagliasco, V., Lombroso, C. T. & Goodglass, H. (1972). Auditory evoked responses: meaningfullness of stimuli and inter-hemispheric asymmetry. *Science, Washington*, **175**, 790–792.

McNeill, D. (1966). Developmental psycholinguistics. In *The Genesis of Language*, ed. F. Smith & G. A. Miller, pp. 15–84. MIT Press: Cambridge, Mass.

Morrell, L. K. & Salamy, J. G. (1971). Hemispheric asymmetry of electro-cortical responses to speech stimuli. *Science, Washington*, **174**, 164–166.

Nishida, T. & Kawanaka, K. (1972). Inter-unit-group relations among wild chimpanzees in the Mahali Mountains. *Kyoto University African Studies*, **7**, 131–169.

Pisoni, D. B. & Tash, J. (1974). Reaction times to comparisons within and across phonetic categories. *Perception and Psychophysics*, **15**, 285–290.

Reynolds, V. & Reynolds, F. (1965). Chimpanzees of the Budongo Forest. In *Primate Behavior*, ed. I. DeVore, pp. 368–424. Holt, Rhinehart & Winston: New York.

Rowell, T. E. (1962). Agonistic noises of the rhesus monkey. *Symposium of the zoological Society of London*, **8**, 91–96.

Rowell, T. E. & Hinde, R. A. (1962). Vocal communication by the rhesus monkey (*Macaca mulatta*). *Proceedings of the zoological Society of London*, **138**, 279–294.

Schaller, G. B. (1963). *The Mountain Gorilla.* University of Chicago Press: Chicago.

Stebbins, W. C. (1971). Hearing. In *Behavior of Nonhuman Primates*, vol. **3**, ed. A. M. Schrier & F. Stollnitz, pp. 159–192. Academic Press: New York.

Struhsaker, T. (1967). Auditory communication among vervet monkeys (*Cercopithecus aethiops*). In *Social Communication among Primates*, ed. S. A. Altmann, pp. 281–324. University of Chicago Press: Chicago.

Struhsaker, T. (1975). *Behavior and Ecology of Red Colobus Monkeys.* University of Chicago Press: Chicago.

Studdert-Kennedy, M., Liberman, A. M., Harris, K. S. & Cooper, F. S. (1970). Motor theory of speech perception: a reply to Lane's critical review. *Psychological Review*, **77**, 234–249.

Teleki, G. (1973). *The Predatory Behavior of Wild Chimpanzees.* Bucknell University Press: Lewisburg.

Waser, Peter. (1974). Intergroup interaction in a forest monkey: the mangabey *Cercocebus albigena.* Ph.D. thesis, Rockefeller University.

Waser, P. (1975). Experimental playbacks show vocal mediation of inter-group avoidance in monkeys. *Nature, London*, **255**, 56–68.

Waser, P. (1976). Vocal control of intergroup spacing in a forest monkey. *Behaviour*, in press.

Winter, P. (1969). The variability of peep and twit calls in captive squirrel monkeys (*Saimiri sciureus*). *Folia primatologica*, **10**, 204–215.

Winter, P., Ploog, D. & Latta, J. (1966). Vocal repertoire of the squirrel monkey (*Saimiri sciureus*). Its analysis and significance. *Experimental Brain Research*, **1**, 359–384.

Wood, C. C., Goff, W. R. & Day, R. S. (1971). Auditory evoked potentials during speech perception. *Science, Washington*, **173**, 1248–1251.

Wrangham, R. W. (1974). Artificial feeding of chimpanzees and baboons in their natural habitat. *Animal Behaviour*, **22**, 83–93.

Wrangham, R. W. (1975). Behavioural ecology of chimpanzees in Gombe National Park, Tanzania. Ph.D. thesis, University of Cambridge.

8

Kin selection in lions and in evolution

B. C. R. BERTRAM

Students of ecology in the field have generally been interested in the way in which a species maintains itself, or in the reproductive success of that species in a broad sense. Students of behaviour in the field, on the other hand, have tended to concentrate more on the behaviour of individuals, or on the relationships between individuals. A result of the relatively recent development of behavioural ecology has been a useful merging of these two approaches, with emphasis being placed on the reproductive success of individuals. It is becoming increasingly clear that there is a great deal of variation in reproductive success among different individuals within a species. It is also being increasingly recognised that, for the larger vertebrates at least, it is these differences in reproductive success between individuals which determine both the direction in which the species is evolving, and the speed at which it is changing. The environment, especially other potentially competing species, sets broad limits within which the species is constrained, but there is a considerable amount of latitude within those limits. Evolutionary change within the limits is caused by selection resulting from differences between conspecifics in the number of offspring they leave; field work is showing that such differences are considerable.

Field work in behavioural ecology has also included studies of the same individuals over long periods of time. For the more social bird and mammal species, these studies have indicated that animals within social groups are often genetically related to one another. This has been demonstrated in primates (Sade, 1967; Missakian, 1973; Clutton-Brock & Harvey, this volume); in carnivores (Mech, 1970; Kruuk, 1972; Schaller, 1972; Bertram, 1973, 1975); in ungulates (Klingel, 1965; Lincoln & Guiness, 1973); in elephants (Douglas-Hamilton, 1973); and in birds (Watts & Stokes, 1971; Brown, 1972; Ridpath, 1972; Zahavi, 1974). Doubtless there are many other instances.

Hamilton in 1964 introduced the concept of kin selection, pointing out that replicas of an individual's genes were likely to be present not only in that animal's own offspring but also in its close relatives. In effect this was a modification of the way in which natural selection was presumed to work: by the selection of genes rather than by the selection of the individuals carrying those genes. Thus an animal's genes are passed on both via that animal's own young and via its relatives' young. Genes which cause their carrier animal to produce more replicas of themselves increase in frequency in the population, or in other words are selected for. Consequently one would expect selection for genes which cause relatives to be given favoured treatment. Such favours could take the form of cooperating, of refraining from competing or of helping to rear relatives' offspring. But not all relatives can be favoured maximally, partly because the total quantity of these favours is finite (for example food which can be found), and partly because favouring relatives may reduce an animal's own future reproductive potential. Therefore the genes selected for would be expected to be those which cause such favours to be distributed most to those individuals most closely related (generally own offspring), and to a decreasing extent to more distant relatives.

This kin selection pressure is one of the many selective pressures operating on the individual in any species. It will clearly be strongest in those social species where closely related individuals are in close proximity to one another, and where opportunities for cooperation are greatest. In the absence, as yet, of very long-term records of the genealogy of individuals in social groups in the wild, it is rarely known empirically how closely related those individuals are to one another. But given certain assumptions, it is possible to calculate how closely related they are on average, as the following examples indicate.

AVERAGE DEGREES OF RELATEDNESS

The *degree of relatedness* (r) of two individuals is the probability that a gene in one is a replica of a gene in the other, by virtue of descent from a shared ancestor. Thus between a parent and its offspring (in diploid species) $r = \frac{1}{2}$, because through meiotic division a parent contributes half of its genes to each of its gametes. By the same process, it can be shown (Hamilton, 1971*a*) that half-siblings are related by ¼, full siblings by ½, grandparent to grandchild by ¼, full cousins by ⅛, and so on.

A simple bird case

Imagine a bird species which is regularly polyandrous, with two equal males both of whom mate with a single female; her offspring are therefore fathered by either male. We want to calculate how closely related to one another the offspring of such trios would be on average. We will take three possible cases:

(*a*) The adult males are not related to one another.

Any two offspring have an equal chance of being either full siblings (so related by $\frac{1}{2}$) or else half-siblings (so related by $\frac{1}{4}$). The average relatedness among any two offspring is therefore $(\frac{1}{2}+\frac{1}{4})/2$, i.e. $\frac{3}{8}$, $= 0.38$.

(*b*) The adult males are full brothers.

Any two offspring have a 50% chance of being full siblings, in which case they are related by $\frac{1}{2}$. Also they have a 50% chance of being half-siblings with the same mother but different full-sibling fathers; in this latter case they are related by $\frac{1}{4}$ via their mothers plus $\frac{1}{4}\times\frac{1}{2}$ via their fathers. The mean relatedness among any two offspring is thus $\frac{1}{2}[\frac{1}{2}+(\frac{1}{4}+\frac{1}{8})]$, which equals $\frac{7}{16}$, $= 0.44$.

(*c*) The adult males are the offspring of a typical trio, which is how the species has been reproducing for generations. Let the degree of relatedness between them be z.

Any two offspring, as above, have a 50% chance of being full sibs and therefore related by $\frac{1}{2}$. They also have a 50% chance of having the same mother, but different fathers related to one another by z; in this case they are related by $\frac{1}{4}$ via their mother, plus $\frac{1}{4}\times z$ via their fathers, i.e. by $\frac{1}{4}(z+1)$. The average relatedness among any two offspring is thus $\frac{1}{2}[\frac{1}{2}+\frac{1}{4}(z+1)]$. But if we assume that the reproductive system has been stable for generations, any two male offspring will bear the same average relatedness to one another as their fathers do (i.e. z). Therefore $z = \frac{1}{2}[\frac{1}{2}+\frac{1}{4}(z+1)]$, which equals $\frac{3}{7}$, $= 0.43$.

Clearly, additional refinements could be included, such as one male getting more than 50% of the successful copulations, or the female occasionally mating with unrelated strangers. And clearly too, they will not make a great deal of difference to the average degree of relatedness among offspring.

This example has been cited to demonstrate the principles of calculating average degrees of relatedness in a relatively simple case. It is in fact almost exactly the situation found in the Tasmanian native hen (*Tribonyx mortierii*) (Maynard Smith & Ridpath, 1972; Ridpath, 1972).

Lion case

The lion (*Panthera leo*) social system is a great deal more complicated, but sufficient data are now available for the average degrees of relatedness among certain individuals to be calculated. Although the details of the method refer only to lions, the general principles of the method are applicable to a wide variety of social species.

We first make a number of statements about the lion social system, in the form of assumptions (nos. 1 to 11 below) concerning the reproduction of a 'typical' pride from the data of Bertram (1973, 1975) and Schaller (1972). For this typical case, the average relatedness among males and among females is calculated, thus demonstrating the principle of the method simply. There is of course considerable variation in pride size, structure and reproductive performance; I show below that this variation has relatively little effect on the degree of relatedness found.

Assumptions. The representative lion pride and its reproduction is illustrated in Fig. 1.

(1) A pride has two adult males and seven adult females.
(2) Four of these seven females give birth at about the same time, and rear their cubs together.
(3) The litter size is three cubs.
(4) The fathering of litters is shared equally among the adult males.
(5) Three female cubs when subadult remain in the pride.
(6) They replace three adult females which died or left during that period, and thus the pride size remains constant.
(7) The synchronising of births, survival of cubs, recruitment or departure of female subadults, and death of adults are all independent of relatedness within the pride.
(8) All male subadults are expelled from the pride.
(9) The adult males grew up together in another similar pride; they left it before maturity and stayed together; they have taken over, and are reproducing in, a pride which is not the one in which they were born.
(10) The adult males do not retain tenure of a pride for long enough to father more than one batch of young female recruitment to the pride.
(11) Prides are stable and last for generations, reproducing in this way.

As a result of this system, the adult males are related to one another (by a quantity x, to be determined), and the adult females are related to one another (by a quantity y); but the males are not related to the females with whom they are reproducing, and therefore no inbreeding occurs.

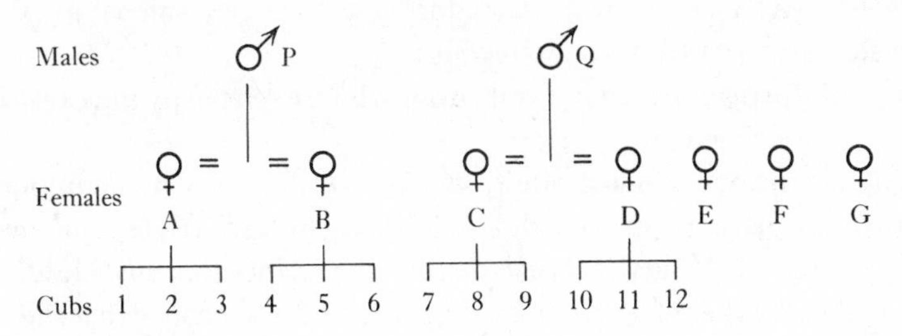

Fig. 1. Diagram of typical case of pride's reproduction, to show possible pairings among cubs.
Pairing of cubs with
Same father–same mother are
1–2, 1–3, 2–3, 4–5, 4–6, 5–6, 7–8,... (total 12).
Same father–different mother are
1–4, 1–5, 1–6, 2–4, 2–5, 2–6, 3–4, 3–5, 3–6, 7–10,... (total 18).
Different father–different mother are
1–7, 1–8, 1–9, 1–10, 1–11, 1–12, 2–7, 2–8,... (total 36).
Total number of pairings is 66.

Calculations. Two premises are made. The first is that because males stay together, therefore the relatedness (x) between adult males is the same as the average relatedness among any two cubs growing up in the pride together. But this relatedness (x, therefore) among the cubs can also be calculated in another way. It is the mean of the following:

(i) 12 pairings of full-sibling cubs (i.e. same father and same mother; see Fig. 1); these are related by ¼ (via their father) plus ¼ (via their mother).

(ii) 18 pairings of half-sibling cubs, related by ¼ (via their same father) plus $\frac{1}{4}y$ (via their mothers who are related to one another by y).

(iii) 36 pairings of cubs which are related by $\frac{1}{4}x$ (via their related fathers) plus $\frac{1}{4}y$ (via their related mothers).

Thus of the total of 66 pairings among the 12 cubs, the mean relatedness

$$x = \frac{(\frac{1}{4}+\frac{1}{4})12+(\frac{1}{4}+\frac{1}{4}y)18+(\frac{1}{4}x+\frac{1}{4}y)36}{66}$$

which simplifies to

$$38x-9y = 7. \tag{1}$$

The second premise is that because the system is stable and repeating over the generations, the mean relatedness among females in a pride remains on average constant with time. Thus y stays the same after replacement of three adult females by subadults. The pride now has four adult and three subadult females in it. The new mean relatedness (still y) among all pairs of these seven females is the mean of the following:

(i) 3 pairings of subadult–subadult, who are related by x because they were cubs together.

(ii) 6 pairings of adult–adult, who are related by y by definition.

(iii) 12 pairings of subadult–adult; these are related on average by $(1/7)1/2+(6/7)1/2y$ (because there is a 1/7 chance that any adult female was the mother of a particular subadult, and a 6/7 chance that she was not).

Thus of the total of 21 pairings among the 7 females, the mean relatedness, which still equals

$$y = \frac{(x)3+(y)6+\left(\frac{1}{14}+\frac{6y}{14}\right)12}{21} \tag{2}$$

which simplifies to

$$7x-23y = -2.$$

Solving equations 1 and 2 gives the relatedness between adult males $x = 0.22$, and the relatedness between females $y = 0.15$.

The effects of variation. To investigate the effects of the variation between prides, it is necessary to put the equations 1 and 2 into general form, and then to apply the same approach. Let a be the number of adult females in the pride; let b be the number of young females recruited into the pride as replacements for the same number of adults; let c be the number of females giving birth at one time; let d be the litter size; and let e be the number of adult males. We now recalculate equations 1 and 2. It can be shown in the same way as before, that the mean relatedness x among cubs also equals

$$\frac{\left(\frac{1}{2}\right)\left[\binom{d}{2}c\right]+(1/4+1/4y)\left[\binom{c/e}{2}d^2e\right]+(1/4x+1/4y)\left[\binom{e}{2}\frac{d^2c^2}{e^2}\right]}{\binom{cd}{2}}$$

which simplifies to

$$x\left(3cd+\frac{cd}{e}-4\right)+y(1-c)d = d+\frac{cd}{e}-2. \tag{1a}$$

Similarly the mean relatedness y among females in the reconstructed pride also equals

$$\frac{x\left[\binom{b}{2}\right]+y\left[\binom{a-b}{2}\right]+\left[\frac{1}{2a}+\left(\frac{a-1}{a}\right)\frac{y}{2}\right](a-b)b}{\binom{a}{2}}$$

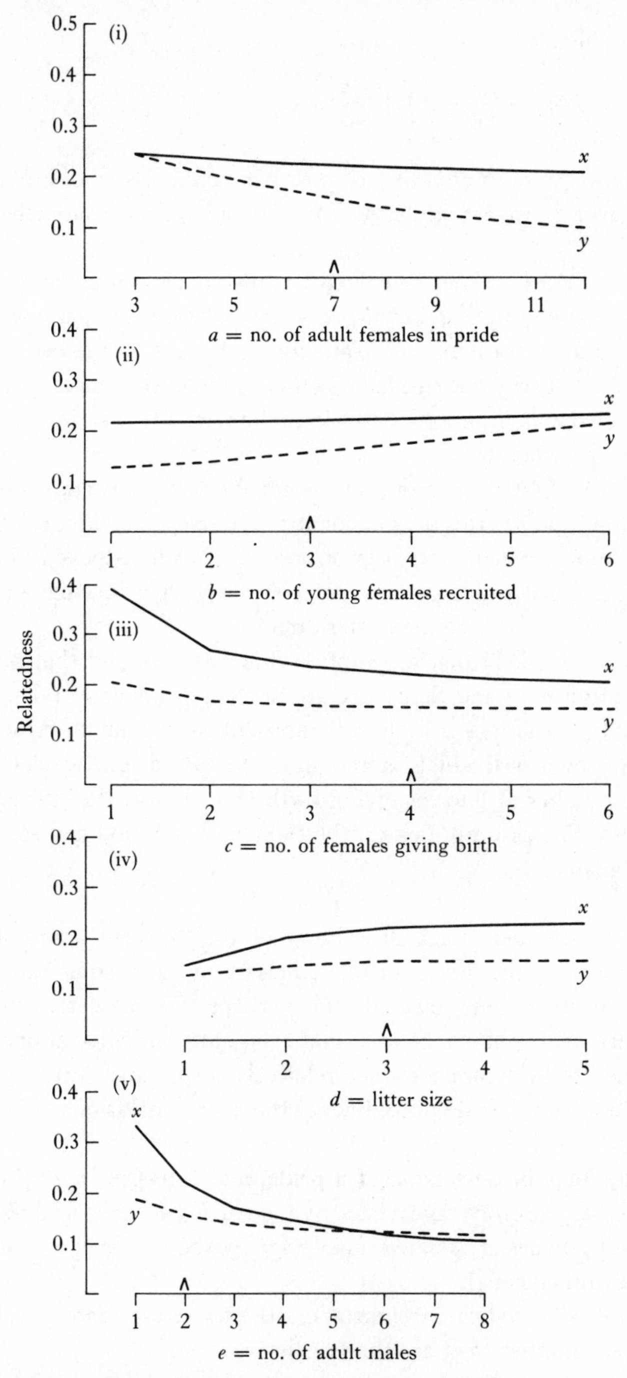

Fig. 2. Relatedness among males (x, solid line) and among females (y, broken line) in lion prides, with parameters a to e varying. ∧ indicates typical case.

which simplifies to

$$x(b-1)b+\left(\frac{b^2}{a}-ab\right)=-\frac{b}{a}(a-b). \tag{2a}$$

(For integers, $\binom{n}{2}$ is the number of possible combinations of 2 items which can be drawn from a population of n items; it is numerically equal to $\frac{1}{2}n(n-1)$.)

Equations 1*a* and 2*a* can be solved for any values of the parameters a to e. Fig. 2 (i–v) shows that varying any one of these five parameters while keeping the others constant has relatively little effect on the overall degrees of relatedness. x and y rise and fall together. The consequences of departures from most of the assumptions 1 to 11 can be tested by varying one or more of these five parameters.

For instance, if fatherhood of litters is not shared equally among both adult males (assumption 4), this is equivalent to reducing the number of males to between 2 and 1, which results in an increase in relatedness in both sexes (Fig. 2, v). But males in larger groups (3 or more) reproduce more effectively (Bertram, 1975), which counters this effect.

Similarly, it is likely that assumption 8 is not met, and that births, cub survival, recruitment and departure are not independent of relationhips in the pride. For example a high cub mortality may tend to remove entire litters rather than individuals at random; but this is equivalent to a reduction in the number of females giving birth (Fig. 2, iii), the effects of which can be seen. The assumptions can be tightened and modified as more data become available.

Conclusions. Lack of validity of the different assumptions has different effects on the degree of relatedness x and y; in some cases they are raised and in other cases lowered, and perhaps to some extent they cancel one another out. But it can be seen that even quite large deviations from the case assumed to be typical produce relatively small effects on these degrees of relatedness, and they do not influence the main conclusions. These are the following:

(i) Male lions in possession of a pride, and cubs growing up in a pride together, are on average related by about 0.22 (i.e. almost half-siblings).

(ii) Adult females are related on average by about 0.15 (i.e. a little closer than full cousins).

(iii) Under all practical circumstances the males are more closely related to one another than are the females.

It must be stressed that the degrees of relatedness calculated are mean figures, made up of a range of differing degrees of relatedness. For example, some animals will be full siblings, related by 0.5. The next most closely related will be related by 0.38; they will be lions who shared a father and whose mothers were full sisters. And so on. Thus the values of relatedness are not normally distributed; instead there is a range of degrees of relatedness, more or less discrete at the higher values, and scattered asymmetrically and discontinuously about a mean of 0.22 or 0.15. This is assuming that the males are totally unrelated to the females, which is unrealistic in practice. However, if they were, for example, their third cousins they would be related to the females by less than 0.01, which is negligible compared with the much closer degrees of relatedness within each sex or peer group of offspring.

Applicability

There are a number of cases of animals cooperating with related individuals. Refraining from competition with them is another form of cooperation which may be much commoner although less easily recognisable. At least 60 bird species are known to nest cooperatively (Skutch, 1961; Harrison, 1969; Brown, 1972; Fry, 1972). In a variety of mammal species also, individuals other than the parents have been shown to assist in feeding or protecting the young (Kühme, 1965; Mech, 1970; Schaller, 1972; Rood, 1974; Hrdy, 1976). In many of these cases the individuals are likely to be related to one another. In some cases there are now sufficient data to produce a rough general model of the reproduction of the species, and so to calculate how closely related they are, using methods similar to the one outlined in this paper. To do so would be a valuable first step in quantifying the relative costs and benefits of cooperative behaviour towards other group members, and so understanding the evolution of many aspects of their social behaviour. Examples where such a method would be applicable are the Tasmanian native hen (Maynard Smith & Ridpath, 1972; Ridpath, 1972), Arabian babblers (*Turdoides squamiceps*) (Zahavi, 1974), turkeys (*Meleagris gallopavo*) (Watts & Stokes, 1971), wolves (*Canis lupus*) (Mech, 1970; E. Zimen, personal communication), wild dogs (*Lycaon pictus*) (van Lawick & van Lawick-Goodall, 1971; Schaller, 1972), red foxes (*Vulpes vulpes*) (D. W. Macdonald, personal communication), banded mongooses (*Mungos mungo*) (J. P. Rood, personal communication), elephants (*Loxodonta africana*) (Douglas-Hamilton, 1973), and a large number of primate species (see e.g. Clutton-Brock & Harvey, this volume).

KIN SELECTION PRESSURES IN LIONS

The lion social system contains elements of both cooperation and competition among individuals. Lions cooperate in hunting, in driving out intruders, and particularly in rearing their young. They compete to some extent for limited resources: for food, for oestrous females, and for a place in the pride. One of the many factors influencing the compromise between competition and cooperation in any such situation is likely to be the degree of relatedness among those individuals, through the operation of kin selection. We will consider three aspects of this cooperation where a kin selection pressure is likely to have operated in its evolution, and will consider the other selective pressures which are also likely to have been involved.

Communal suckling

Adult females in a lion pride generally suckle one another's cubs, if those cubs are of similar age or younger (Schaller, 1972, p. 149; Bertram, 1975). The suckling is not necessarily indiscriminate, and individuals vary in the extent to which they will allow others' offspring to suckle from them.

The degree of relatedness between a female and the offspring of another female in the pride is $\frac{1}{2} \times 0.15$, or about one-seventh of that between her and her own offspring. If the benefit to the other cub exceeds seven times the cost to her own cub, such communal suckling should evolve through kin selection alone. One can conceive of cases where this disproportion could occur (for example as a result of the deaths of particular females or cubs, or extreme hunger), but it would appear to be unlikely in many cases. Nonetheless, kin selection provides a selective pressure contributing towards communal suckling.

However, it is not the only selective pressure operating. Trivers (1971) showed that reciprocal altruism could be selected for in animals of long-lived species able to detect any failure by companions to reciprocate. It seems reasonable to suppose that a lioness could detect and discriminate against the young of another female who 'cheated' by not allowing communal suckling from her. In a sense, this is a case of a delayed benefit to the individual who feeds another's young: her own offspring will in turn be fed by others later.

There appear to be other advantages to the individual. Cubs born synchronously and reared communally were found to be more likely to survive than cubs born asynchronously (Bertram, 1975). It is possible that the presence of other cubs themselves is beneficial to the survival of a female's own cubs (for example by making their mothers more likely to stay together and to hunt cooperatively and therefore more efficiently). When a cub is reared to subadulthood, it benefits in several ways from the presence of companions. And in particular, males have a considerably longer and more effective reproductive life if they have companions (Bertram, 1975). Thus it is to the advantage of a lioness for her own cubs' companions to survive, and therefore communal suckling could be selected for even in the absence of kin selection.

Male tolerance towards cubs at kills

When food is in short supply, there is a large amount of squabbling and competition for it at kills. Larger animals rob smaller ones of small carcasses or pieces of food, or prevent them from feeding by threatening or attacking them if they come near. The exception which has occasioned frequent surprise is that males more than females will often allow small cubs to feed from a piece of food in their possession (Schaller, 1972, p. 152).

A male approached by a cub is on average related to that cub by 0.31 (i.e. there is a 50% chance that it is his own offspring and so related to him by ½, and a 50% chance that it is his companion's offspring and so related to him by ½×0.22). It is most unlikely that a male can distinguish which are his own offspring, for a variety of reasons: females come into heat at very irregular intervals, they mate very frequently and in many cases with more than one male in succession, and most oestrous periods do not result in cubs.

A female approached by a cub is related to that cub by either ½ (if it is her own), or ½×0.15 (if it is not hers). If we assume that she is one of the four females who gave birth, her average relatedness to the cub is 0.18 (there is a 25% chance that it is her own offspring, and a 75% chance that it is a companion's). It is likely that a female generally can recognise her own offspring, although it may be difficult in practice for her to discriminate among cubs during competition among a number of hungry animals at a kill. In the apparent absence of such discrimination, a male is likely to be considerably more closely related than a female, to a cub which tries to feed from his portion of food. Thus the benefit to a male is greater than to a female from any improved cub survival which results from this tolerance at kills.

One must not forget that there may well be other selective pressures contributing towards this difference in behavour. For example, males have a larger food intake than females do (Schaller, 1972, p. 270). It is possible that the cost to a male of the loss of a small fraction of its food supply is less serious than it would be to a female. (Note that here, as in the previous example, 'cost' and 'benefit' to an animal are defined in terms of that animal's future reproductive success; they are therefore virtually unmeasurable at present.) In addition, a male has a much shorter effective reproductive life than a female does (Bertram, 1975). Therefore the cubs fathered by him that are alive at any time represent a much greater proportion of his lifetime's reproductive output than is the case for a female, and so those cubs are more valuable to males than they are to females.

Competition among males for oestrous females

There is no dominance order among pride males. The first to encounter an oestrous female stays with her and mates with her. He is temporarily dominant while with her, and rarely allows other males to come within 10 metres. Other males make no attempt to do so, but often wait nearby. A female may change males during her oestrous period (which usually lasts two to four days and during which mating takes place at intervals averaging about 15 minutes in length). The lack of a dominance order among males, and the insignificant amount of agonistic behaviour, imply a surprisingly low level of competition among the males in lions as compared with other species.

A male is related on average by 0.22 to his companion, whose offspring thus carry a proportion of his own genes. His own offspring would carry four to five times as many, but the fact that some of his genes are transmitted to the next generation via his related companion must reduce the selective pressure for competition with that companion for matings.

It is interesting to compare the reproductive success of different male lions in terms of offspring produced, with their success as measured by genes in common with the next generation. Table 1 shows for a group of six companion males the relative proportions of offspring presumed to have been fathered by each male; this is on the basis of the number of times each male was seen in possession of an oestrous female. (I appreciate that there are pitfalls in assuming a correlation between the frequencies of copulation and of paternity, but this problem is irrelevant to the present argument.) On the basis of copulations, the most successful male produced 3½ times as many offspring as the least successful. But males are also 'reproducing by proxy' via their companions. One half of a male's gene complement is in one of his own offspring, and one-ninth (i.e. $\frac{1}{2} \times 0.22$) in each cub of one of his companions. Table 1 shows the relative proportions of each male's genes in the next generation. It can be seen that in terms of genes transmitted, the most successful male did only 1½ times as well as the least successful. It is this latter difference (which is still appreciable, but much less than 3½ times) which will determine how strong is the selective pressure for a male to compete with his related companion for an oestrous female. Thus kin selection will be a factor responsible for the low level of competition observed.

However, it is not the only factor. The large number of matings in each oestrous period (about 300), the low probability of cubs being produced as a result (about 20% of oestrous periods result in births), a mean litter size of two to three cubs, and the high cub mortality (75–80%) all mean that on average a very large number of matings are required for each cub reared in the next generation. The probable benefit from a copulation is thus extremely

Table 1. *Relative reproductive success of six companion male lions*

Male lion number	1	3	4	5	6	8	Total
Number of times seen with an oestrous female	10	9	4	14	9	6	52
Hence % of oestrous females seen with him = presumed proportion of offspring fathered by him. The best does 3½ times as well as the worst	19.2	17.3	7.7	26.9	17.3	11.6	100
Gene replicas in own offspring (i.e. line above multiplied by ½)	9.6	8.7	3.9	13.5	8.7	5.8	50.2
% of offspring fathered by his companions:	80.8	82.7	92.3	73.1	82.7	88.4	
Gene replicas in offspring of companions (i.e. line above multiplied by ½×0.22)	9.0	9.2	10.3	8.1	9.2	9.8	55.6
Total of his gene replicas	18.6	17.9	14.2	21.6	17.9	15.6	105.8
Hence % of his gene replicas in next generation (therefore the best does 1½ times as well as the worst)	17.6	16.9	13.4	20.4	16.9	14.8	100.0

small, which again reduces the selective pressure on a male to compete for it.

The costs of competing, on the other hand, are probably very high. Serious injuries can be inflicted during fights, and such injuries can be fatal. Even if not injured himself, the winner of a fight which leaves him without a male companion is likely to be expelled much sooner by a new group of males (Bertram, 1975). When driven out, his reproductive period is effectively terminated; also his offspring in the pride from which he has been expelled suffer an increase in mortality at the takeover of the pride by the new males (Bertram, 1975).

THE STRENGTH OF KIN SELECTION PRESSURES

I have dealt at length with the lion example above because it illustrates well the great array of selective pressures operating on animals. Kin selection is only one of these pressures. It has not proved possible so far to demonstrate conclusively that kin selection does operate in the wild, because there are likely always to be a large number of other selective pressures, some operating in the same direction and others pushing in different directions. But it is not reasonable to say that because kin selection cannot be demonstrated to operate it can therefore be ignored. (By analogy, the fact that a lump of iron falls to the ground under the influence of gravity does not mean that there are no other forces acting upon it, such as air resistance, magnetic, and inertial forces, all influencing its descent.) Kin selection pressures provide the most probable explanation for the evolution of non-reproductive castes in social insects (Hamilton, 1964; Wilson, 1971). It has not been shown on theoretical grounds that genes cannot be selected for regardless of their carrier animals, and therefore Hamilton's (1964) logical modification of the mechanism by which natural selection operates should be accepted. It is surprising that it has not received wider acceptance already.

In the examples above, I have considered only the average degrees of relatedness between individuals interacting with one another. In theory, there should be selection for any gene which enabled its carrier animal to be better at recognising its degree of relatedness to any other individual; the animal could then discriminate precisely the extent to which it should cooperate (or compete) with any other animal according to how close a relative it was. However, there are a number of complicating factors, at different levels, preventing great precision.

First, there are considerable practical problems involved in the recognition of relatives. Older siblings, for example, can observe directly that younger siblings have come from the same mother; but younger animals cannot distinguish older relatives except by the behaviour of those older siblings or of other individuals.

There can be considerable doubts over paternity, in species where females mate with more than one male. It is difficult to see how the offspring of a female could determine whether they had the same or a different father. Similarly, it is difficult to see how offspring of different females could determine whether they were fathered by the same or by different males.

In the case of two animals which were the offspring of totally different parents, there would seem to be no way in which the offspring could determine whether either parent of one was related to any extent to either parent of the other.

In addition, this is a case where there may be conflict between mother and offspring. It is less to the advantage of the mother for her offspring to be able to recognise their father, because they may then expend resources in cooperating with their relatives on the father's side; there will be kin selection pressures on them to do so. In the absence of inbreeding, the mother is not related to the father's relatives and has no genetic interest in their wellbeing; it is in her interests for her offspring to be altruistic only to her own relatives, not to his. Thus countering any selection on offspring for recognition of relatives, there will be selection on their mother for making difficult the recognition of their father's relatives. Selection pressure on the father will be similar but in the opposite direction, and generally will involve greater practical problems, depending on the social organisation of the species.

Further, altruistic behaviour by her offspring towards her relatives increases the mother's inclusive fitness more than that of the offspring; this is another source of potential conflict between parent and offspring (Trivers, 1974). It would be to the mother's advantage to mislead her offspring into treating her relatives as closer relatives than they really are.

Other individuals too, relatives or not, would usually benefit by misleading a potental altruist into treating them as though they were more closely related to him than they really are. Selection for this ability to deceive would make more difficult the task of determining precisely the degree of relatedness of another individual.

Thus for these practical and evolutionary reasons, there is likely to be a considerable range of uncertainty as to the relatedness of any other individual. The degree of relatedness between two animals can be considered as being composed of two portions, one detectable and the other undetectable. The detectable portion of their relatedness comes from such observed facts as a shared mother. The undetectable portion comes from uncertainties over paternity, adoptions, and relatedness between parents and ancestors.

The relative sizes of the detectable and undetectable portions of the relatedness between two animals will depend on a variety of factors, particularly on the rigidity of the social system, the level of perceptive ability of the species, and the degree of cultural transmission of information about relatedness. Between two individuals, the relative sizes of the two portions are not necessarily the same for each animal; for example, older animals are in a position to possess more knowledge of the parentage of younger animals than vice versa.

However, in the course of evolution, the strength of the kin selection pressure depends on the total degree of relatedness, regardless of the relative proportions which the animals are able to detect. For instance, in the lion case, one might expect to find greater cooperation between brothers than the average within the male group, and other types of sub-alliance within the pride. It is not known to science, nor presumably to lions, whether all the cubs in a lion litter produced by a female who mated with more than one male were necessarily fathered by the same or by different males. Litter mates are therefore related by 0.25 (the detectable portion) plus an undetectable portion of between 0 and 0.25 depending on the identities and relatedness of their fathers. Two offspring of different mothers may well have had the same father, and their mothers may well have been related to one another; in consequence, although the degree of relatedness between two such animals is probably composed almost entirely of the undetectable portion, it may sometimes be as great as in the case of the two littermates above. With these large amounts of uncertainty, a lion cannot determine how close its various relatives are. On average, a littermate will be more closely related than other animals, and therefore one would expect that he would receive favoured treatment; but since another peer will on average be appreciably related too, the selective pressure to aid one rather than the other will be small.

A further complication arises in considering the strength of kin selection pressures. The degree of relatedness is not the only factor to be taken into account when behaving altruistically towards another animal. The expected future reproductive success of the relative being helped is also important, and the extent to which that can be improved by cooperation. Clearly it is of little benefit to an individual to give food to a relative who is about to die anyway, or to a relative who has enough food. While the degree of relatedness to another animal may be difficult to determine, it may be relatively easy to determine that animal's age, and to some extent its needs.

SELECTIVE PRESSURES, FUNCTIONS AND ADAPTIVENESS

The assumption used to be widespread that every characteristic of an animal species, and every feature of its social system had evolved for 'the good of the species' as a whole. The biologist would try to think of a possible way in which a particular feature benefitted the species, and the most plausible one was put forward as the explanation. There are several drawbacks to such an approach.

First, it was not clear what was most to the benefit of the species. To have the greatest possible number of individuals? Or to have animals of the highest possible quality? Or to have animals of as diverse genotypes and therefore greatest adaptability as possible?

Second, different observers might well produce a different explanation according to their subjective expectations. And if the characteristic or feature of the species had been the opposite of that observed, a different explanation might have been satisfactorily found for that too. This subjectiveness is due to the impossibility of testing what were the causes of events which took place in the past; all one can do is to suggest a cause which now would produce the same result. In view of what has been said about the great variety of selective pressures acting on animals, to select a single one as *the* cause seems extremely dubious.

Third, it is nowadays realised that characters may be selected for without necessarily being beneficial to the species as a whole. Hamilton's (1971*a*) model, quoted by Clutton-Brock & Harvey (this volume), showed that a gene responsible for selfish behaviour could be selected for, regardless of its effect on the species, which evolves in the direction of those individuals within it which leave most descendants.

One of the clearest and most striking instances of this is shown by the lion.

I have found that male lions taking over a pride are liable to kill cubs they find there (Bertram, 1975); by doing so they make the mothers of those cubs give birth sooner to those males' own offspring, which are then also more likely to survive. The new males thus increase their own reproductive output at the expense of the parents of those cubs. Such behaviour does not benefit the species, but the genes responsible for it can be selected for because they make their carriers leave more descendants. In theory, if there were a second lion species living in the same ecological niche and differing from the first only in the lack of cub-killing behaviour, then this second lion species should outbreed the first and eventually eliminate it. However, in practice one does not find such situations, at least in the larger vertebrates. And if one did, it would be only a matter of time before a cub-killing gene appeared as a mutation in the second species and spready rapidly because of the reproductive advantages it would give to the individuals who carried it. Similar cases of infanticide have been reported in several primate studies (summarised by Hrdy, 1974, and Clutton-Brock & Harvey, this volume). Infanticide is clearly not adaptive to both participants in the relationship.

It has been suggested (Hamilton, 1971*b*) that much gregarious behaviour, such as herding or shoaling, may have been selected for in the same way. The reduced probability of predation on an individual which gets closer to another makes him leave more offspring on the average, and therefore propagate the genes contributing towards such gregariousness, even if, as could sometimes be the case, this helped the predator to catch more of that prey species.

Instances of behaviour which benefits the individual and is detrimental to the species are not easy to find, but are I think likely to be more common than is generally supposed. Much territorial behaviour might come into this category, and much of the aggressive behaviour of dominant animals in social groups.

The problem of the evolution of a character or of a behaviour pattern can be approached from either of two directions.

(1) The shorthand approach often taken is to say 'This animal shows a certain behaviour: what is its function?'. A function here is a beneficial consequence of the behaviour (Hinde, 1969, p. 159). This is an approach most useful at the species level.

(2) The approach taken throughout this paper has been via selective pressures. One starts by saying 'This animal shows a certain behaviour: what selective pressures are acting on it to make it show this behaviour to a greater or lesser extent?' In my view this second approach is the more useful when considering selection at an intra-specific level, and particularly when dealing with complex relationships between individuals. It is easier to see how kin selection can be incorporated into answers to questions phrased in terms of

selective pressures rather than in terms of functions; also it helps to draw attention to the diversity of consequences of a particular behaviour, on all of which natural selection will act.

For example, for a lioness to allow rather than prevent her companions' cubs from suckling from her has a great variety of possible consequences: some related cubs may get more milk and perhaps therefore are more likely to survive; her own cubs may get less milk and perhaps therefore are less likely to survive; her own cubs if they do survive may be more likely to have companions and all may therefore do better reproductively; exchange of odour among animals may make her own cubs able to suckle from other females; she may have to delay having her next litter; she may therefore look after her present cubs for better or for longer. Some of these selection pressures will be stronger than others, and they will act in different directions, but all play a part in determining through natural selection the compromise degree of communal suckling in lions. To label one or two of the beneficial consequences as functions may lead one to neglect the other consequences.

INDIVIDUAL SELECTION

It is impossible to test what selective pressures in the past caused the evolution of a particular character. However, it is possible to compare the reproductive performance of individuals which possess that character to a greater or lesser extent. Then if a difference is found, the direction which evolution will take can be predicted, and the way in which the selective pressures are operating can be investigated. This is the approach which is increasingly being taken in evolutionary considerations. However, five points need especially to be borne in mind.

(*a*) We are interested primarily in the frequencies of genes in future generations. Looking at the number of young produced by particular individuals is only a very first stage. We should remember that not all offspring are equal. Young of high-ranking mothers may rank higher than the offspring of lower-ranking mothers (Sade, 1967), and this may enhance their reproductive prospects. Larger young may have better survival chances and better competitive ability than smaller young, and the relative advantages of producing many small young or fewer larger young may change in different circumstances (Lack, 1968). Young which are driven out have a small but finite chance of founding new colonies; expulsion of offspring to seek new breeding places elsewhere may be a good average strategy for an adult, even if many of those offspring perish as a result. When, in a relatively few years, probably, it becomes possible to determine parentage from examination of blood proteins or by other means, and when long-term studies of vertebrate groups have been continued for much longer, then the reproductive success of individuals will be able to be assessed over a few generations.

(*b*) If we can detect a difference in reproductive output during a field study, this implies that the selective pressures are strong ones; weak ones will not be detectable.

(*c*) We must remember that individuals which differ in the characteristic which we are investigating probably differ also in other characteristics which may be much less obvious. For example, in comparing the reproductive performance of peak time versus late breeders, one should bear in mind that the late breeding may be by individuals which are in some way inferior to the majority. In some cases, differences between individuals could be produced by manipulation by the observer, so overcoming this problem.

(*d*) We must remember that there may be physical or physiological constraints on animal's reproductive performance. In colonially nesting birds, for example, those at the centre may do better than those at the edge, but there is a physical limit to how many centre nests there can be (Kruuk, 1964). Tits nesting earlier in the season do better than those nesting later, but there are physiological and food-supply limits to how early they can nest (Perrins, 1970). Larger animals may compete better, but may need more food, or may wear themselves out sooner.

(*e*) The operation of kin selection through the reproduction of relatives should not be forgotten.

Eventually it will be possible to understand, and hopefully to measure, the many different selective pressures acting on individuals which have been kept under study: at present the kin selection pressure is the only one which can be quantified. Then, deciding how and why the species evolved to its present day form, behaviour and social system will be slightly less guesswork than it now is.

SUMMARY

Animals in social groups are often genetically related to one another. Given certain information on their social system and reproduction, it is possible to calculate how closely related they are likely to be.

Such a calculation is carried out for lions (*Panthera leo* L.). The adult males in a typical pride are on average about as closely related as half-siblings, and the females as full cousins. It is shown that variation among prides has little effect on these average degrees of relatedness.

Since these related lions in a pride cooperate with one another, a kin selection pressure is likely to be operating. Its importance is discussed in connection with communal suckling, males' tolerance towards cubs, and the lack of competition for oestrous females. It is shown that individual selective pressures are also operating in these cases.

One would expect selective pressure for enhancing animals' ability to distinguish precisely their degree of relatedness to other individuals; some of the problems hindering this are outlined.

It is emphasised that evolutionary change occurs in response to the net effect of a great array of selective pressures, acting on individuals and exerting their 'pressure' in different directions; kin selection pressure is only one of them.

It is a pleasure to thank the Sub-Department of Animal Behaviour at Madingley for offering me a base at intervals over the past 10 years. I am most grateful to 13 colleagues, many of them at Madingley, for help, criticism and comments on parts of this paper at various stages.

The information on lions, which provides the backbone of the paper, was collected during four years in the Serengeti, by kind permission of the Director of Tanzania National Parks and of the Serengeti Research Institute; this is SRI Publication No. 171. I wish to thank the Natural Environment Research Council, The Royal Society, and the African Wildlife Leadership Foundation for financial support of my work there.

REFERENCES

Bertram, B. C. R. (1973). Lion population regulation. *East African Wildlife Journal*, **11**, 215–225.

Bertram, B. C. R. (1975). Social factors influencing reproduction in wild lions. *Journal of Zoology*, **177**, 463–482.

Brown, J. L. (1972). Communal feeding of nestlings in the Mexican jay (*Aphelocoma ultramarina*): interflock comparisons. *Animal Behaviour*, **20**, 395–403.

Douglas-Hamilton, I. (1973). On the ecology and behaviour of the Lake Manyara elephants. *East African Wildlife Journal*, **11**, 401–403.

Fry, C. H. (1972). The social organisation of bee-eaters (Meropidae) and co-operative breeding in hot-climate birds. *Ibis*, **114**, 1–14.

Hamilton, W. D. (1964). The genetical evolution of social behaviour. I and II. *Journal of theoretical Biology*, **7**, 1–16 and 17–52.

Hamilton, W. D. (1971*a*). Selection of selfish and altruistic behavior in some extreme models. In *Man and Beast: Comparative Social Behavior*, ed. J. F. Eisenberg & W. S. Dillon, pp. 57–91. Smithsonian Institution Press: Washington, D.C.

Hamilton, W. D. (1971*b*). Geometry for the selfish herd. *Journal of theoretical Biology*, **31**, 295–311.

Harrison, C. J. O. (1969). Helpers at the nest in Australian birds. *Emu*, **69**, 30–40.

Hinde, R. A. (1969). (ed.) *Bird Vocalizations*. Cambridge University Press: London.

Hrdy, S. B. (1974). Male–male competition and infanticide among the langurs (*Presbytis entellus*) of Abu, Rajasthan. *Folia primatologica*, **22**, 19–58.

Hrdy, S. B. (1976). The care and exploitation of non-human primate infants by conspecifics other than the mother. *Advances in the Study of Behavior*, **6**, in press.

Klingel, H. (1965). Notes on the biology of the plains zebra, *Equus quagga boehmi* Matschie. *East African Wildlife Journal*, **3**, 86–88.

Kruuk, H, (1964). Predators and anti-predator behaviour of the black-headed gull (*Larus ridibundus* L.). *Behaviour Supplement*, **11**, 1–130.

Kruuk, H. (1972). *The Spotted Hyena: A Study of Predation and Social Behaviour.* University of Chicago Press: Chicago.

Kühme, W. (1965). Communal food distribution and division of labour in African hunting dogs. *Nature, London*, **205**, 443–444.

Lack, D. (1968). *Ecological Adaptations for Breeding in Birds.* Methuen: London.

Lawick, H. van & Lawick-Goodall, J. van (1971). *Innocent Killers.* Houghton Mifflin Company: Boston.

Lincoln, G. A. & Guiness, F. E. (1973). The sexual significance of the rut in red deer. *Journal of Reproduction and Fertility, Supplement*, **19**, 475–489.

Maynard Smith, J. & Ridpath, M. G. (1972). Wife-sharing in the Tasmanian native hen, *Tribonyx mortierii*: a case of kin selection? *American Naturalist*, **106**, 447–452.

Mech, L. D. (1970). *The Wolf: The Ecology and Behaviour of an Endangered Species.* Natural History Press: New York.

Missakian, E. A. (1973). Genealogical mating activity in free-ranging groups of rhesus monkeys (*Macaca mulatta*) on Cayo Santiago. *Behaviour*, **45**, 225–241.

Perrins, C. M. (1970). The timing of birds' breeding seasons. *Ibis*, **112**, 242–255.

Ridpath, M. G. (1972). The Tasmanian native hen. *Tribonyx mortierii.* II. The individual, the group, and the population. *CSIRO Wildlife Research*, **17**, 53–90.

Rood, J. P. (1974). Banded mongoose males guard young. *Nature, London*, **248**, 176.

Sade, D. S. (1967). Determinants of dominance in a group of free-ranging rhesus monkeys. In *Social Communication Among Primates*, ed. S. A. Altmann, pp. 99–114. University of Chicago Press: Chicago.

Schaller, G. B. (1972). *The Serengeti Lion: A Study of Predator–Prey Relations.* University of Chicago Press: Chicago.

Skutch, A. F. (1961). Helpers among birds. *Condor*, **63**, 198–226.

Trivers, R. L. (1971). The evolution of reciprocal altruism. *Quarterly Review of Biology*, **46**, 35–57.

Trivers, R. L. (1974). Parent–offspring conflict. *American Zoologist*, **14**, 249–264.

Watts, C. R. & Stokes, A. W. (1971). The social order of turkeys. *Scientific American*, **224**, 112–118.

Wilson, E. O. (1971). *Insect Societies.* Harvard University Press: Cambridge, Mass.

Zahavi, A. (1974). Communal nesting by the Arabian babbler: a case of individual selection. *Ibis*, **116**, 84–87.

9

The social function of intellect

N. K. HUMPHREY

Henry Ford, it is said, commissioned a survey of the car scrap yards of America to find out if there were parts of the Model T Ford which never failed. His inspectors came back with reports of almost every kind of breakdown: axles, brakes, pistons – all were liable to go wrong. But they drew attention to one notable exception, the *kingpins* of the scrapped cars invariably had years of life left in them. With ruthless logic Ford concluded that the kingpins on the Model T were too good for their job and ordered that in future they should be made to an inferior specification.

Nature is surely at least as careful an economist as Henry Ford. It is not her habit to tolerate needless extravagance in the animals on her production lines: superfluous capacity is trimmed back, new capacity added only as and when it is needed. We do not expect therefore to find that animals possess abilities which far exceed the calls that natural living makes on them. If someone were to argue – as I shall suggest they might argue – that some primate species (and mankind in particular) are much cleverer than they need be, we know that they are most likely to be wrong. But it is not clear why they would be wrong. This paper explores a possible answer. It is an answer which has meant for me a re-thinking of the function of intellect.

A re-thinking, or merely a first-thinking? I had not previously given much thought to the biological function of intellect, and my impression is that few others have done either. In the literature on animal intelligence there has been surprisingly little discussion of how intelligence contributes to biological fitness. Comparative psychologists have established that animals of one species perform better, for instance, on the Hebb–Williams maze than those of another, or that they are quicker to pick up learning sets or more successful on an 'insight' problem; there have been attempts to relate performance on particular kinds of tests to particular underlying cognitive skills; there has (recently) been debate on how the same skill is to be assessed with 'fairness' in animals of different species; but there has seldom been consideration given to why the animal, in its natural environment, should *need* such skill. What is the use of 'conditional oddity discrimination' to a monkey in the field (French, 1965)? What advantage is there to an anthropoid ape in being able to recognise its own reflection in a mirror (Gallup, 1970)? While it might indeed be 'odd for a biologist to make it his task to explain why horses can't learn mathematics' (Humphrey, 1973*a*), it would not be odd for him to ask why *people can*.

The absence of discussion on these issues may reflect the view that there is little to discuss. It is tempting, certainly, to adopt a broad definition of intelligence which makes it self-evidently functional. Take, for instance, Heim's (1970) definition of intelligence in man, 'the ability to grasp the essentials of a situation and respond appropriately': substitute 'adaptively' for 'appropriately' and the problem of the biological function of intellect is (tautologically) solved. But even those definitions which are not so manifestly circular tend nonetheless to embody value-laden words. When intelligence is defined as the 'ability' to do this or that, who dares question the biological advantage of being *able*? When reference is made to 'understanding' or 'skill at problem-solving' the terms themselves seem to quiver with adaptiveness. Every animal's world is, after all, full of things to be understood and problems to be solved. For sure, the world *is* full of problems – but what exactly are these problems, how do they differ from animal to animal and what particular advantage accrues to the individual who can solve them? These are not trivial questions.

Despite what has been said, we had better have a definition of intelligence, or the discussion is at risk of going adrift. The following formula provides at least some kind of anchor: 'An animal displays intelligence when he modifies his behaviour on the basis of valid inference from evidence'. The word 'valid' is meant to imply only that the inference is logically sound; it leaves open the question of how the animal benefits in consequence. This definition is admittedly wide, since it embraces everything from simple associative learning to syllogistic reasoning. Within the spectrum it seems fair to distinguish 'low-level' from 'high-level' intelligence. It requires, for instance, relatively low-level intelligence to infer that something is likely to happen merely because similar things have happened in comparable circumstances in the past; but it requires high-level intelligence to infer that something is likely to happen because it is entailed by a *novel* conjunction of events. The former is, I suspect, a comparatively elementary skill and widespread through the animal kingdom, but the latter is much more special, a mark of the 'creative' intellect which is characteristic especially of the higher primates. In what follows I shall be enquiring into the function chiefly of 'creative' intellect.

Now I am about to set up a straw man. But he is a man whose reflection I have seen in my own mirror, and I am inclined to treat him with respect. The opinion he holds is that the main role of creative intellect lies in *practical invention.* 'Invention' here is being used broadly to mean acts of intelligent discovery by which an animal comes up with new ways of doing things. Thus it includes not only, say, the fabrication of new tools or the putting of existing objects to new use but also the discovery of new behavioural strategies, new ways of using the resources of one's own body. But, wide as its scope may be, the talk is strictly of 'practical' invention, and in this context 'practical' has a restricted meaning. For the man in question sees the need for invention as arising only in relation to the external physical environment; he has not noticed – or has not thought it important – that many animals are *social* beings.

You will see, no doubt, that I have deliberately built my straw man with feet of clay. But let us nonetheless see where he stands. His idea of the intellectually challenging environment has been perfectly described by Daniel Defoe. It is the desert island of Robinson Crusoe – before the arrival of Man Friday. The island is a lonely, hostile environment, full of technological challenge, a world in which Crusoe depends for his survival on his skill in gathering food, finding shelter, conserving energy, avoiding danger. And he must work fast, in a truly inventive way, for he has no time to spare for learning simply by induction from experience. But was that the kind of world in which creative intellect evolved? I believe, for reasons I shall come to, that the real world was never like that, and yet that the real world of the higher primates may in fact be considerably *more* intellectually demanding. My view – and Defoe's, as I understand him – is that it was the arrival of Man Friday on the scene which really made things difficult for Crusoe. If Monday and Tuesday, Wednesday and Thursday had turned up as well then Crusoe would have had every need to keep his wits about him.

But the case for the importance of practical invention must be taken seriously. There can be no doubt that for some species in some contexts inventiveness does seem to have survival value. The 'subsistence technology' of chimpanzees (Goodall, 1964; Teleki, 1974) and even more that of 'natural' man (Sahlins, 1974) involves many tricks of technique which appear *prima facie* to be products of creative intellect. And what is true for these anthropoids must surely be true at least in part for other species. Animals who are quick to realise new techniques (in hunting, searching, navigating or whatever) would seem bound to gain in terms of fitness. Why, then, should one dispute that there have been selective pressures operating to bring about the evolution of intelligence in relation to practical affairs? I do not of course dispute the general principle; what I question is how much this principle *alone* explains.

How clever does a man or monkey need to be before the returns on superior intellect become vanishingly small? If, despite appearances, the important practical problems of living actually demand only relatively low-level intelligence for their solution, then there would be grounds for supposing that high-level creative intelligence is wasted. Even Einstein could not get better than 100% at O-level. Can we really explain the evolution of the higher intellectual faculties of primates on the basis of success or failure in their 'practical exams'?

My answer is no, for the following reason: even in those species which have the most advanced technologies the exams are largely tests of knowledge rather than imaginative reasoning. The evidence from field studies of chimpanzees all points to the fact that subsistence techniques are hardly if ever the product of premeditated invention; they are arrived at instead either by trial-and-error learning or by imitation of others. Indeed it is hard to imagine how many of the techniques could in principle be arrived at otherwise. Teleki (1974) concluded on the basis of his own attempts at 'termiting' that there was no way of predicting *a priori* what would be the most effective kind of probe to stick into a termite hill, or how best to twiddle it or, for that matter, where to stick it. He had to learn inductively by trial-and-error or, better, by mimicking the behaviour of Leakey, an old and experienced chimpanzee. Thus the chimpanzees' art would seem to be no more an invention than is the uncapping of milk-bottles by tits. And even where a technique could in principle be invented by deductive reasoning there are generally no grounds for supposing that it has been. Termiting by human beings is a case in point. In northern Zaire, people beat with sticks on the top of termite mounds to encourage the termites to come to the surface. The technique works because the stick-beating makes a noise like falling rain. It is just possible that someone once upon a time noticed the effect of falling rain, noticed the resemblance between the sound of rain and the beating of sticks, and put two and two together. But I doubt if that is how it happened; serendipity seems a much more likely explanation. Moreover, whatever the origin of the technique, there is certainly no reason to invoke inventiveness on the part of present-day practitioners, for these days it is culturally transmitted. My guess is that most of the practical problems that face higher primates can, as in the case of termiting, be dealt with by learned strategies without recourse to creative intelligence.

Paradoxically, I would suggest that subsistence technology, rather than requiring intelligence, may actually become a substitute for it. Provided the *social* structure of the species is such as to allow individuals to acquire subsistence techniques by simple associative learning, then there is little need for individual creativity. Thus the chimpanzees at Gombe, with their superior technological culture, may in fact have *less* need than the neighbouring baboons to be individually inventive. Indeed there might seem on the face of it to be a negative correlation between the intellectual capacity of a species and the need for intellectual output. The great apes, demonstrably the most intellectually gifted of all animals, seem on the whole to lead comparatively undemanding lives, less demanding not only than those of lower primates but also of many non-primate species. During two months I spent watching gorillas in the Virunga mountains I could not help being struck by the fact that of all the animals in the forest the gorillas seemed to lead much the simplest existence – food abundant and easy to harvest (provided they *knew* where to find it), few if any predators (provided they *knew* how to avoid them) . . . little to do in fact (and little done) but eat, sleep and play. And the same is arguably true for natural man. Studies of contemporary Bushmen suggest that the life of hunting and gathering, typical of early man, was probably a remarkably easy one. The 'affluent savage' (Sahlins, 1974) seems to have established a *modus vivendi* in which, for a period of perhaps 10 million years, he could afford to be not only physically but intellectually lazy.

We are faced thus with a conundrum. It has been repeatedly demonstrated in the artificial situations of the psychological laboratory that anthropoid apes possess impressive powers of creative reasoning, yet these feats of intelligence seem simply not to have any parallels in the behaviour of the same animals in their natural environment. I have yet to hear of any example from the field of a chimpanzee (or for that matter a Bushman) using his full capacity for inferential reasoning in the solution of a biologically relevant practical problem. Someone may retort that if an ethologist had kept watch on Einstein through a pair of field glasses he might well have come to the conclusion that Einstein too had a hum-drum mind. But that is just the point: Einstein, like the chimpanzees, displayed his genius at rare times in 'artificial' situations – he did not use it, for he did not *need* to use it, in the common world of practical affairs.

Why then do the higher primates need to be as clever as they are and, in particular, that much cleverer than other species? What – if it exists – is the natural equivalent of the laboratory test of intelligence? The answer has, I believe, been ripening on the tree of the preceding discussion. I have suggested that the life of the great apes and man may not require much in the way of practical invention, but it does depend critically on the possession of wide factual knowledge of practical technique and the nature of the habitat. Such knowledge can only be acquired in the context of a *social* community – a community which provides both a medium for the cultural transmission of information and a protective environment in which individual learning can occur. I propose that the chief role of creative intellect is to hold society together.

In what follows I shall try to explain this proposal, to justify it, and to examine some of its surprising implications.

To me, as a Cambridge-taught psychologist, the proposal is in fact a rather strange one. Experimental psychologists in Britain have tended to regard social psychology as a poor country cousin of their subject – gauche, undisciplined and slightly absurd. Let me recount how I came to a different way of thinking, since this personal history will lead directly in to what I want to say. Some years ago I made a discovery which brought home to me dramatically the fact that, even for an experimental psychologist, *a cage* is a bad place in which to keep a monkey. I was studying the recovery of vision in a rhesus monkey, Helen, from whom the visual cortex had been surgically removed (Humphrey, 1974). In the first four years I'd worked with her Helen had regained a considerable amount of visually guided behaviour, but she still showed no sign whatever of three-dimensional spatial vision. During all this time she had, however, been kept within the confines of a small laboratory cage. When, at length, five years after the operation, she was released from her cage and taken for walks in the open field at Madingley her sight suddenly burgeoned and within a few weeks she had recovered almost perfect spatial vision. The limits on her recovery had been imposed directly by the limited environment in which she had been living. Since that time, in working with laboratory monkeys I have been mindful of the possible damage that may have been done to them by their impoverished living conditions. I have looked anxiously through the wire mesh of the cages at Madingley, not only at my own monkeys but at Robert Hinde's. Now, Hinde's monkeys are rather better-off than mine. They live in social groups of eight or nine animals in relatively large cages. But these cages are almost empty of objects, there is nothing to manipulate, nothing to explore; once a day the concrete floor is hosed down, food pellets are thrown in and that is about it. So I looked – and seeing this barren environment, thought of the stultifying effect it must have on the monkey's intellect. And then one day I looked again and saw a half-weaned infant pestering its mother, two adolescents engaged in a mock battle, an old male grooming a female whilst another female tried to sidle up to him, and I suddenly saw the scene with new eyes: forget about the absence of *objects*, these monkeys had *each other* to manipulate and to explore. There could be no risk of their dying an intellectual death when the social environment provided such obvious opportunity for participating in a running dialectical debate. Compared to the solitary existence of my own monkeys, the set-up in Hinde's social groups came close to resembling a simian School of Athens.

Several of the other contributors to this book consider the dialectics of social interaction, and do so with much more authority than I can. None of them,

I think, would claim that scientific study of the subject is yet far advanced. Much of the best published literature is in fact genuinely 'literature' – Aesop and Dickens make, in their own way, as important contributions as Laing, Goffman or Argyle. But one generalisation can I think be made with certainty: the life of social animals is highly problematical. In a complex society, such as those we know exist in higher primates, there are benefits to be gained for each individual member both from preserving the overall structure of the group and at the same time from exploiting and out-manoeuvring others within it (see later). Thus social primates are required by the very nature of the system they create and maintain to be calculating beings; they must be able to calculate the consequences of their own behaviour, to calculate the likely behaviour of others, to calculate the balance of advantage and loss – and all this in a context where the evidence on which their calculations are based is ephemeral, ambiguous and liable to change, not least as a consequence of their own actions. In such a situation, 'social skill' goes hand in hand with intellect, and here at last the intellectual faculties required are of the highest order. The game of social plot and counter-plot cannot be played merely on the basis of accumulated knowledge, any more than can a game of chess.

Like chess, a social interaction is typically a *trans*action between social partners. One animal may, for instance, wish by his own behaviour to change the behaviour of another; but since the second animal is himself reactive and intelligent the interaction soon becomes a two-way argument where each 'player' must be ready to change his tactics – and maybe his goals – as the game proceeds. Thus, over and above the cognitive skills which are required merely to perceive the current state of play (and they may be considerable), the social gamesman like the chess player must be capable of a special sort of forward planning. Given that each move in the game may call forth several alternative responses from the other player this forward planning will take the form of a decision tree, having its root in the current situation and growing branches corresponding to the moves considered in looking ahead from there at different possibilities. It asks for a level of intelligence which is, I submit, unparalleled in any other sphere of living. There may be, of course, strong and weak players* – yet, as master or novice, we and most other members of complex primate societies have been in this game since we were babies.

* 'Weak players grow short bushy trees, looking a short way ahead at a mass of poorly differentiated possibilities; strong players prune the tree much more efficiently and... construct long thin trees, looking much deeper into a few critical variations. This pruning is the heart of the problem... Which branches are critical, and which are redundant and can safely be cut off?' – from an article in the *New Scientist* (vol. **66**, p. 119, 1975) on the first World Computer Chess Championship. It may be that the acquisition of social skill involves the learning of standard 'gambits' and 'defences' – relatively stereotyped patterns of interaction – which allow transactions to proceed quickly and smoothly from one critical decision point to another.

But what makes a society 'complex' in the first place? There have probably been selective pressures of two rather different kinds, one from without, the other from within society. I suggested above that one of the chief functions of society is to act as it were as a 'polytechnic school' for the teaching of subsistence technology. The social system serves the purpose in two ways: (i) by allowing a period of prolonged dependence during which young animals, spared the need to fend for themselves, are free to experiment and explore; and (ii) by bringing the young into contact with older, more experienced members of the community from whom they can learn by imitation (and perhaps, in some cases, from more formal 'lessons'). Now, to the extent that this kind of education has adaptive consequences, there will be selective pressures both to prolong the period of untrammelled infantile dependency (to increase the 'school leaving age') and to retain older animals within the community (to increase the number of experienced 'teachers'). But the resulting mix of old and young, caretakers and dependents, sisters, cousins, aunts and grandparents not only calls for considerable social responsibility but also has potentially disruptive social consequences. The presence of dependents (young, injured or infirm) clearly calls at all times for a measure of tolerance and unselfish sharing. But in so far as biologically important resources may be scarce (as subsistence materials must sometimes be, and sexual partners will be commonly) there is a limit to which tolerance can go. Squabbles are bound to occur about access to these scarce resources and different individuals will have different interests in participating in, promoting or putting a stop to such squabbles. In the last resort every individual should give priority to the survival of his own genes, and following the theoretical analysis outlined by Hamilton and Trivers (see Bertram and Clutton-Brock & Harvey, this volume) we may predict considerable conflicts of interest among the members of any community which spans more than a single generation; the greater the number of generations present the more complex the picture becomes. Thus the stage is set within the 'collegiate community' for considerable political strife. To do well for oneself whilst remaining within the terms of the social contract on which the fitness of the whole community ultimately depends calls for remarkable reasonableness (in both literal and colloquial senses of the word). It is no accident therefore that men, who of all primates show the longest period of dependence (nearly 30 years in the case of Bushmen!), the most complex kinship structures, and the widest overlap of generations within society, should be more intelligent than chimpanzees, and chimpanzees for the same reasons more intelligent than cercopithecids.

Once a society has reached a certain level of complexity, then new internal pressures must arise which act to increase its complexity still further. For, in a society of the kind outlined, an animal's intellectual 'adversaries' are members of his own breeding community. If intellectual prowess is correlated with social success, and if social success means high biological fitness, then any heritable trait which increases the ability of an individual to outwit his fellows will soon spread through the gene pool. And in these circumstances there can be no going back: an evolutionary 'ratchet' has been set up, acting like a self-winding watch to increase the general intellectual standing of the species. In principle the process might be expected to continue until either the physiological mainspring of intelligence is full-wound or else intelligence itself becomes a burden. The latter seems most likely to be the limiting factor; there must surely come a point where the time required to resolve a social 'argument' becomes insupportable.

The question of the time given up to unproductive social activity is an important one. The members of my model collegiate community – even if they have not evolved a run-away intellect – are bound to spend a considerable part of their lives in caretaking and social politics. It follows that they must inevitably have less time to spare for basic subsistence activities. If the social system is to be of any net biological benefit the improvement in subsistence techniques which it makes possible must more than compensate for the lost time. To put the matter baldly: if an animal spends all morning in non-productive socialising, he must be at least twice as efficient a producer in the afternoon. We might therefore expect that the evolution of a social system capable of supporting advanced technology should only happen under conditions where improvements in technique can substantially increase the return on labour. This may not always be the case. To take an extreme example, the open sea is probably an environment where technical knowledge can bring little benefit and thus complex societies – and high intelligence – are contra-indicated (dolphins and whales provide, maybe, a remarkable and unexplained exception). Even at Gombe the net advantage of having a complex social system may in fact be marginal; the chimpanzees at Gombe share several of the local food resources with baboons, and it would be instructive to know how far the advantage that chimpanzees have over baboons in terms of technical skill is eroded by the relatively large amount of time they give up to social intercourse. It may be that what the chimpanzees gain on the swings of technical proficiency they lose on the roundabouts of extravagant socialising.* As it is, in a year of poor harvest the chimpanzees in fact become much

* MacFarland (see his discussion of 'optimisation' in this volume) might like to draw an isocline linking points of 'equal net productivity' in a space defined by the two axes, 'technical skill' and 'time given over to social activity'. It is, of course, intrinsic to my argument that these axes are not independent, since I am suggesting that social activity is a prerequisite of technical skill. However, the same is probably true of his own illustrative example (p. 62), since a university lecturer's teaching ability is almost certainly not independent of his research ability.

less sociable (Wrangham, 1975); my guess is that they simply cannot spare the time (cf. Gibb, 1956; Baldwin & Baldwin, 1972). The ancestors of man, however, when they moved into the savanna, discovered an environment where technical knowledge began to pay new and continuing dividends. It was in that environment that the pressures to give children an even better schooling created a social system of unprecedented complexity – and with it unprecedented challenge to intelligence.

The outcome has been the gifting of members of the human species with remarkable powers of social foresight and understanding. This social intelligence, developed initially to cope with local problems of inter-personal relationships, has in time found expression in the institutional creations of the 'savage mind' – the highly rational structures of kinship, totemism, myth and religion with characterise primitive societies (Lévi-Strauss, 1962). And it is, I believe, essentially the same intelligence which has created the systems of philosophical and scientific thought which have flowered in advanced civilisations in the last four thousand years. Yet civilisation has been too short lived to have had any important evolutionary consequences; the 'environment of adaptiveness' (Bowlby, 1969) of human intelligence remains the *social* milieu.

If man's intellect is thus suited primarily to thinking about people and their institutions, how does it fare with *non-social* problems? To end this paper I want to raise the question of 'constraints' on human reasoning, such as might result if there is a predisposition among men to try to fit non-social material into a social mould (cf. Hinde & Stevenson-Hinde, 1973).

When a man sets out to solve a social problem he may reasonably have certain expectations about what he is getting in to. First, he should know that the situation confronting him is unlikely to remain stable. Any social transaction is by its nature a developing process and the development is bound to have a degree of indeterminacy to it. Neither of the social agents involved in the transaction can be certain of the future behaviour of the other; as in Alice's game of croquet with the Queen of Hearts, both balls and hoops are always on the move. Someone embarking on such a transaction must therefore be prepared for the problem itself to alter as a consequence of his attempt to solve it – in the very act of interpreting the social world he changes it. Like Alice he may well be tempted to complain 'You've no idea how confusing it is, all the things being alive'; that is not the way the game is played at Hurlingham – and that is not the way that non-social material typically behaves. But, secondly, he should know that the development *will* have a certain logic to it. In Alice's croquet game there was real confusion, everyone played at once without waiting for turns and there were no rules; but in a social transaction there are, if not strict rules, at least definite constraints on what is allowed and definite conventions about how a particular action by one of the transactors should be answered by the other. My earlier analogy with the chess game was perhaps a more appropriate one; in social behaviour there is a kind of turn-taking, there are limits on what actions are allowable, and at least in some circumstances there are conventional, often highly elaborated, sequences of exchange.

Even the chess analogy, however, misses a crucial feature of social interaction. For while the good chess player is essentially selfish, playing only to win, the selfishness of social animals is typically tempered by what, for want of a better term, I would call *sympathy*. By sympathy I mean a tendency on the part of one social partner to identify himself with the other and so to make the other's goals to some extent his own. The role of sympathy in the biology of social relationships has yet to be thought through in detail, but it is probable that sympathy and the 'morality' which stems from it (Waddington, 1960) is a biologically adaptive feature of the social behaviour of both men and other animals – and consequently a major constraint on 'social thinking' wherever it is applied. Thus our man setting out to apply his intelligence to solve a social problem may expect to be involved in a fluid, transactional exchange with a sympathetic human partner. To the extent that the thinking appropriate to such a situation represents the customary mode of human thought, men may be expected to behave inappropriately in contexts where a transaction cannot in principle take place: if they treat inanimate entities as 'people' they are sure to make mistakes.

There are many examples of fallacious reasoning which would fit such an interpretation. The most obvious cases are those where men do in fact openly resort to animistic thinking about natural phenomena. Thus primitive – and not so primitive – peoples commonly attempt to *bargain* with nature, through prayer, through sacrifice or through ritual persuasion. In doing so they are explicitly adopting a social model, expecting nature to participate in a transaction. But nature will not transact with men; she goes her own way regardless – while her would-be interlocutors feel grateful or feel slighted as the case befits. Transactional thinking may not always be so openly acknowledged, but it often lies just below the surface in other cases of 'illogical' behaviour. Thus the gambler at the roulette table, who continues to bet on the red square precisely because he has already lost on red repeatedly, is behaving as though he expects the behaviour of the roulette wheel to respond eventually to his persistent overtures; he does not – as he would be wise to do – conclude that the odds are unalterably set against him. Likewise, the man in Wason's experiments on abstract reasoning, who, when he is given the task of discovering a mathematical rule typically tries to substitute *his own* rule for the predetermined one (Wason & Johnson-Laird, 1972), is acting as though he expects the problem itself to change in response to his trial solutions. The comment of one of Wason's subjects is revealing: 'Rules are relative. If you were the subject, and I were the experimenter, then I would be right'. In general, I would suggest, a transactional approach leads men to refuse to accept the intransigence of facts – whether the facts are physical events, mathematical axioms or scientific laws; there will always be the temptation to assume that the facts will respond like living beings to social pressures. Men expect to argue *with* problems rather than being limited to arguing *about* them.

There are times, however, when such a 'mistaken' approach to natural phenomena can be unexpectedly creative. While it may be the case that no amount of social pleading will change the weather or, for that matter, transmute base metals into gold, there are things in nature with which a kind of social intercourse is possible. It is not strictly true that nature will not transact with men. If we mean by a transaction essentially a developing relationship founded on mutual give and take, then several of the relationships which men enter into with the non-human things around them may be considered to have transactional qualities. The cultivation of plants provides a clear and interesting example: the care which a gardener gives to his plants (watering, fertilising, hoeing, pruning etc.) is attuned to the plants' emerging properties, which properties are in turn a function of the gardener's behaviour. True, plants will not respond to ordinary social pressures (though men *do* talk to them), but the way in which they give to and receive from a gardener bears, I suggest, a close structural similarity to a simple social relationship. If Trevarthen (1974) can speak of 'conversations' between a mother and a two-month old baby, so too might we speak of a conversation between a gardener and his roses or a farmer and his corn. And the same can be argued for men's interactions with certain wholly inanimate materials. The relationship of a potter to his clay, a smelter to his ore or a cook to his soup are all relationships of fluid mutual exchange, again proto-social in character.

It is not just that transactional thinking is typical of man, transactions are something which people actively seek out and will force on nature wherever they are able. In the Doll Museum in Edinburgh there is a case full of bones clothed in scraps of rag – moving reminders of the desire of human children to conjure up social relationships with even the most unpromising material. Through a long history, men have, I believe, explored the transactional possibilities of countless of the things in their environment and sometimes, Pygmalion-like, the things have come alive. Thus many of mankind's most prized technological discoveries, from agriculture to chemistry, may have had their origin not in the deliberate application of practical intelligence but in the fortunate misapplication of social intelligence. 'Once Nature had set up men's minds the way she has, certain 'unintended' consequences followed – and we are in several ways the beneficiaries' (Humphrey, 1973*b*).

The rise of classical scientific method has in large measure depended on human thinkers disciplining themselves to abjure transactional, socio-magical styles of reasoning. But scientific method has come to the fore only in the last few hundred years of mankind's history, and in our own times there are everywhere signs of a return to more magical systems of interpretation. In dealing with the non-social world the former method is undoubtedly the more immediately appropriate; but the latter is perhaps more natural to man. Transactional thinking may indeed be irrepressible: within the most disciplined Jekyll is concealed a transactional Hyde. Charles Dodgson the mathematician shared his pen amicably enough with Lewis Carroll the inventor of Wonderland but the split is often neither so comfortable nor so complete. Newton is revealed in his private papers as a Rosicrucian mystic, and his intellectual descendants continue to this day to apply strange double-standards to their thinking – witness the way in which certain British physicists took up the cause of Uri Geller, the man who, by wishing it, could bend a metal spoon (e.g. Taylor, 1975). In the long view of science, there is, I suspect, good reason to approve this kind of inconsistency. For while 'normal science' (in Kuhn's sense of the words) has little if any room for social thinking, 'revolutionary science' may more often than we realise derive its inspiration from a vision of a socially transacting universe. Particle physics has already followed Alice down the rabbit hole into a world peopled by 'families' of elementary particles endowed with 'strangeness' and 'charm'. *Vide*, for example, the following report: 'The particles searched for at SPEAR were the *cousins* of the psis made from one *charm* quark and one *uncharmed* antiquark. This contrasts with the *siblings* of the psis . . .' (*New Scientist*, vol. **67**, p. 252, 1975, my italics). Who knows where such 'sociophysics' may eventually lead?

The ideology of classical science has had a huge but in many ways narrowing influence on ideas about the nature of 'intelligent' behaviour. But no matter what the high priests, from Bacon to Popper, have had to say about how people ought to think, they have never come near to describing how people *do* think. In so far as an idealised view of scientific method has been the dominant influence on mankind's recent intellectual history, biologists should be the first to follow Henry Ford in dismissing recent history as 'bunk'. Evolutionary history, however, is a different matter. The formative years for human intellect were the years when man lived as a social savage on the plains of Africa. Even now, as Browne wrote in *Religio medici*, 'All Africa and her prodigies are within us'.

Postscript

My attention has been drawn to a paper by Jolly (1966) on 'Lemur social behaviour and primate intelligence' which anticipates at several points the argument developed here. I have not attempted to re-write my own paper in a way that would do justice to Jolly's ideas; I hope that people who are intrigued by the relation between social behaviour and intelligence will refer directly to her original and interesting discussion.

In relation to both Jolly's paper and my own the question arises how can the hypotheses be tested. My central thesis clearly demands that there should be a positive correlation across species between 'social complexity' and 'individual intelligence'. Does such a correlation hold? It is not hard to find confirmatory examples; nor is it hard to find excuses for rejecting examples which are seemingly contrary – e.g. wolves (high social complexity without the requisite intelligence?) or orang-utans (high intelligence without the requisite social complexity?). But the trouble is that too much of the evidence is of an anecdotal kind: we simply do not have agreed definitions or agreed ways of measuring either of the relevant parameters. What, I think, is urgently needed is a laboratory test of 'social skill' – a test which ought, if I am right, to double as a test of 'high-level intelligence'. The essential feature of such a test would be that it places the subject in a transactional situation where he can achieve a desired goal only by adapting his strategy to conditions which are continually changing as a consequence partly but not wholly of his own behaviour. The 'social partner' in the test need not be animate (though my guess is that the subject would regard it in an 'animistic' way); possibly it could be a kind of 'social robot', a mechanical device which is programmed on-line from a computer to behave in a pseudo-social way.

SUMMARY

I argue that the higher intellectual faculties of primates have evolved as an adaptation to the complexities of social living. For better or worse, styles of thinking which are primarily suited to social problem-solving colour the behaviour of man and other primates even towards the inanimate world.

REFERENCES

Baldwin, J. D. & Baldwin, J. (1972). The ecology and behavior of squirrel monkeys (*Saimiri oerstedi*) in a natural forest in Western Panama. *Folia primatologica*, **18**, 161–184.

Bowlby, J. (1969). *Attachment and Loss*, Vol. 1. Hogarth: London.

French, G. M. (1965). Associative problems. In *Behavior of Non-Human Primates*, ed. A. M. Schrier, H. F. Harlow & F. Stollnitz. Academic Press: London.

Gallup, G. G. (1970). Chimpanzees: Self-recognition. *Science, Washington*, **167**, 86–87.

Gibb, J. (1956). Food, feeding habits and territory of the rock pipit *Anthus spinoletta*. *Ibis*, **98**, 506–530.

Goodall, J. (1964). Tool using and aimed throwing in a community of free-living chimpanzees. *Nature, London*, **201**, 1264–1266.

Heim, A. W. (1970). *The Appraisal of Intelligence*. Methuen: London.

Hinde, R. A. & Stevenson-Hinde, J. (1973). *Constraints on Learning: Limitations and Predispositions*. Academic Press: London & New York.

Humphrey, N. K. (1973*a*). Predispositions to learn. In *Constraints on Learning: Limitations and Predispositions*, ed. R. A. Hinde & J. Stevenson-Hinde. Academic Press: London & New York.

Humphrey, N. K. (1973*b*). The illusion of beauty. *Perception*, **2**, 429–439.

Humphrey, N. K. (1974). Vision in a monkey without striate cortex: a case study. *Perception*, **3**, 241–255.

Jolly, A, (1966). Lemur social behavior and primate intelligence. *Science, Washington*, **153**, 501–506.

Lévi-Strauss, C. (1962). *The Savage Mind*. Weidenfeld & Nicholson: London.

Sahlins, M. (1974). *Stone Age Economics*. Tavistock Publications: London.

Taylor, J. (1975). *Superminds*. Macmillan: London.

Teleki, G. (1974). Chimpanzee subsistence technology: materials and skills. *Journal of Human Evolution*, **3**, 575–594.

Trevarthen, C. (1974). Conversations with a two-month old. *New Scientist*, **62**, 230–235.

Waddington, C. H. (1960). *The Ethical Animal*. Allen & Unwin: London.

Wason, P. C. & Johnson-Laird, P. N. (1972). *Psychology of Reasoning*. Batsford: London.

Wrangham, R. W. (1975). The behavioural ecology of chimpanzees in Gombe National Park, Tanzania. Ph.D. thesis, University of Cambridge.

EDITORIAL: 4

Although the chapters in this section contain no examples of the experimental approach to the study of function pioneered by Tinbergen (e.g. Tinbergen *et al.*, 1962; Tinbergen, Impekoven & Franck, 1967), they provide further demonstration that the study of the consequences of behaviour is not only a respectable branch of behaviour study in its own right, but also one which has an important contribution to make to studies of causation. Studies of function are, of course, linked also to studies of how behaviour evolved and to studies of the role of behaviour in evolution. The first of these goes back to the very roots of ethology when behavioural characters were first used in taxonomy (e.g. Heinroth, 1911; Whitman, 1919; Lorenz, 1941). It reached maturity with the attempts to formulate principles describing the ritualisation of signal movements (e.g. Tinbergen, 1959; Blest, 1961) but only a few other areas have so far been tackled (e.g. maintenance activities, Baerends, 1966, and Jander, 1966; mouth breeding in cichlids, Oppenheimer & Barlow, 1968; predatory behaviour, Eisenberg & Leyhausen, 1972). The importance of behaviour in evolution has been fully recognised more recently, but the situation has changed dramatically in the last 25 years, with full recognition of the importance of habitat selection, food selection, mate selection and other aspects of social behaviour.

One point, though familiar, deserves emphasis because it arises in all these chapters. In discussion of function, behavioural characters cannot be considered in isolation. Characters of behaviour and structure form an adaptive complex such that changes in one may have ramifying consequences through the whole, and the significance of any one trait can only be understood with reference to others. It is in this spirit that Clutton-Brock & Harvey speculate about the behavioural differences amongst primates, Marler interprets the nature of chimpanzee vocalisations with respect to their social structure, Bertram examines the characteristics of lion reproductive behaviour, and Humphrey discusses the evolution of intelligence.

At this exciting stage in the study of the functions of behaviour, it is perhaps especially worthwhile to be precise about the use of the term. Any character of structure or behaviour may have diverse consequences, and those consequences may be advantageous, neutral or disadvantageous for gene survival. If we ask 'What is it good for?', we are asking merely for a list of advantageous consequences. Such a list may tell us little about how and in what ways variation in the character in question affects gene propagation. If, however, we ask about function in the strong sense of 'Through what consequences does natural selection act to maintain this character?', we are asking a question more closely related to the dynamics of evolution, though harder to answer. For natural selection may pre-empt its operation on one consequence of a behaviour pattern by acting on another consequence of the same pattern (Hinde, 1956, 1975). For example, consider two possible beneficial outcomes of egg-shell removal in gulls – increased crypticity of the nest-site, and protection of the young from the sharp egg-shell. If selection through the former acted with the result that *all* gulls removed their egg-shells whilst the young were still wet and immobile and incapable of cutting themselves, it could not act through the latter consequence. Both would be beneficial consequences in that, if egg-shells were not removed, gulls would be worse off in both ways. However, only the former would be a function in the strong sense of a beneficial consequence through which natural selection acted. Such a view is, of course, in no way contradictory to the view that every characteristic must be the outcome of competition between diverse selective forces. But it does contrast with the view that natural selection acts through every beneficial consequence: this latter view is implicit in, for instance, McFarland's (p. 61) statement that the removal of the sharp egg-shell 'may also exert a small selective pressure', and in the definition of function given by Clutton-Brock & Harvey. The difference between the two views revolves, of course, around the frequency of certain events – for instance, whether or not a gull occasionally delays egg-shell removal, and whether or not its chicks incur injury thereby. But the point that a beneficial consequence could be essential for survival but trivial from the point of view of natural selection is, we feel, worth emphasising.

Evolutionary thinking has been galvanised by Hamilton's (1964) argument that natural selection may promote behaviour disadvantageous to the actor if that behaviour increases the chance of kin surviving. Bertram in particular emphasised that the effect of natural selection is to favour genes rather than individuals. The importance of this insight should not, however, distract attention from the detailed mechanics of evolution. Natural selection acts on phenotypic characters. The problems of how those characters develop from the genotype is therefore of crucial significance to those seeking to explain behaviour in functional terms. Problems involved in the study of behavioural development are considered in the next section.

REFERENCES

Baerends, G. P. (1966). Uber einen möglichen Einfluss von Triebkonflikten auf die Evolution von Verhaltensweisen ohne Mitteilungsfunktion. *Zeitschrift für Tierpsychologie*, **23**, 385–394.

Blest, A. D. (1961). The concept of ritualization. In *Current Problems in Animal Behaviour*, ed. W. H. Thorpe & O. Zangwill. Cambridge University Press: London.

Eisenberg, J. F. & Leyhausen, P. (1972). The phylogenesis of predatory behavior in mammals. *Zeitschrift für Tierpsychologie*, **30**, 59–93.

Hamilton, W. D. (1964). The genetical evolution of social behaviour. I. *Journal of theoretical Biology*, **7**, 1–16.

Heinroth, O. (1911). Beiträge zur Biologie, namentlich Ethologie und Psychologie der Anatiden. *Verhandlungen 5 International Kongresses Ornithologie*, pp. 589–702.

Hinde, R. A. (1956). The biological significance of the territories of birds. *Ibis*, **98**, 340–369.

Hinde, R. A. (1975). The concept of function. *Function and Evolution of Behaviour*, ed. G. Baerends, C. Beer & A. Manning, pp. 3–15. Clarendon Press: Oxford.

Jander, U. (1966). Untersuchungen zur Stammesgeschichte von Putzbewegungen von Tracheaten. *Zeitschrift für Tierpsychologie*, **23**, 799–844.

Lorenz, K. (1941). Vergleichende Bewegungsstudien an Anatinen. *Journal für Ornithologie, Supplement*, **89**, 194–294.

Oppenheimer, R. W. & Barlow, G. W. (1968). Dynamics of parental behaviour in the black-chinned mouthbreeder, *Tilapia melanotheron*. *Zeitschrift für Tierpsychologie*, **25**, 889–914.

Tinbergen, N. (1959). Derived activities: their causation, biological significance, origin and emancipation during evolution. *Quarterly Review of Biology*, **27**, 1–32.

Tinbergen, N., Broekhuysen, G. K., Feekes, F., Houghton, J. C. W., Kruuk, H. & Szulc, E. (1962). Egg shell removal by the black-headed gull, *Larus ridibundus*, L.; a behaviour component of camouflage. *Behaviour*, **19**, 74–117.

Tinbergen, N., Impekoven, M. & Franck, D. (1967). An experiment on spacing out as a defence against predation. *Behaviour*, **28**, 307–321.

Whitman, C. O. (1919). The behavior of pigeons. *Publications of the Carnegie Institute*, **257**, 1–161.

PART C

Development

EDITORIAL: 5

Lorenz regarded behaviour as involving the intercalation of 'learned' and 'instinctive' elements. Since his approach to development was primarily that of a naturalist (e.g. Lorenz, 1965), he was particularly interested in how an animal's behaviour came to be adapted to the environment in which it lived. He therefore recognised only two 'sources of information' through which that adaptedness was acquired – natural selection in phylogeny, and learning in ontogeny. Adaptedness in the 'instinctive' elements was determined by the genes, adaptedness in the learned elements by individual experience. Lorenz's approach enabled him largely to bypass the issues of prime concern to psychologists and embryologists – the processes of development as an organism continuously interacts with its environment.

A major change in thinking about development came when the ideas of Kuo and Schneirla were assembled by Lehrman (1953) in his famous critique of Lorenz's theory. One major issue raised by Lehrman was that 'experience' may affect development in many ways that would not come within any generally accepted definition of learning (see also Hinde, 1968; Lehrman, 1970). The effects of experience may be either specific to a particular aspect of behaviour, or quite general. Similarly, the effects of genes vary from the specific to the general. But interaction between genetic and experiential factors both having general effects can lead to highly distinctive differences between individuals (Bateson, 1976). A hypothetical example would be a *Drosophila* mutant whose reduced rate of courtship was known to be caused by the general effects of a single gene on its visual system. If this gene is most likely to express itself when *Drosophila* are reared at a certain temperature, and temperature of rearing has an influence on other patterns of behaviour, then this distinctive pattern of courtship would indeed arise from the interaction of determinants with non-specific effects.

This is swampy ground and the problems of behavioural development can easily be lost in a morass of postulated interactions between the developing animal and its environment. One firm plank for research in this area has been to operate on the principle of uncovering sources of differences in behaviour, ignoring how specific those influences on development may be. So, if animals which are known to differ genetically are reared in similar environments, then any differences in their behaviour must ultimately have genetic origins. Similarly differences in genetically identical animals reared in different environments must be attributed ultimately to the environmental conditions. Of course, genetically based differences may be mediated through the external environment and environmentally based differences may rely on the induction of particular genes. Identification of the sources of differences is only the first step in unravelling the developmental interaction. Once the source of a difference is identified, further analysis is necessary to understand how it produces its effects.

Work on the processes involved in development, as distinct from the sources of differences, has undoubtedly been stimulated by the mounting interest in the internal control of learning. The issues were already recognised in earlier ethological work. Lorenz (1935), for instance, described differences in the imprintability of different precocial birds (see also Lorenz, 1965). Another illustration was Thorpe's (1958) demonstration that, although the song of chaffinches (*Fringilla coelebs*) is influenced by early experience, they will only learn a limited range of songs. The view that what a species learns is determined by species-characteristic constraints and predispositions contrasts with the optimistic hope of learning theorists in the thirties and forties that it would be possible to formulate simple laws of learning valid for all species and all situations. It was not until the experimental sophistication of learning theorists was coupled with a knowledge of the natural behaviour of a wide range of species that it began to be possible to move towards a productive synthesis (Seligman & Hager, 1972; Hinde & Stevenson-Hinde, 1973).

This theme is picked up by Manning in his chapter. He is concerned primarily with the means by which genetic factors influence behaviour, though he also has important things to say about how genetic studies can facilitate behavioural analysis. Rosenblatt takes the particular case of altricial mammals to illustrate the subtle interplay between the organism and its environment at each stage – and it is implicit that the environment is partly imposed, and partly selected by the animal.

Simpson, focussing on primates, discusses aspects of the complexity of early learning – and the complexity of studying it. Finally, Bateson first considers the usefulness of postulating control mechanisms that ensure that development finishes up in more or less the same place despite vicissitudes on the way. He then considers how such control mechanisms could be tuned by the environment so that the animal adapts itself to the particular conditions in which it lives.

EDITORIAL 5

REFERENCES

Bateson, P. P. G. (1976). Specificity and the origins of behavior. *Advances in the Study of Behavior*, **6**, 1–20.

Hinde, R. A. (1968). Dichotomies in the study of development. In *Genetic and Environmental Influences on Behaviour*, ed. J. M. Thoday & A. S. Parkes. Oliver & Boyd: London.

Hinde, R. A. & Stevenson-Hinde, J. (ed.) (1973). *Constraints on Learning: Limitations and Predispositions*. Academic Press: New York & London.

Lehrman, D. S. (1953). A critique of Konrad Lorenz's theory of instinctive behaviour. *Quarterly Review of Biology*, **28**, 337–363.

Lehrman, D. S. (1970). Semantic and conceptual issues in the nature–nurture problem. In *Development and Evolution of Behavior*, ed. L. R. Aronson, E. Tobach, D. S. Lehrman & J. S. Rosenblatt. Freeman: San Francisco.

Lorenz, K. (1935). Der Kumpan in der Umwelt des Vogels. *Journal für Ornithologie*, **83**, 137–203, 289–413.

Lorenz, K. (1965). *Evolution and Modification of Behavior*. University of Chicago Press: Chicago.

Seligman, M. E. P. & Hager, J. J. (ed.) (1972). *Biological Boundaries of Learning*. Appleton-Century-Crofts: New York.

Thorpe, W. H. (1958). The learning of song patterns by birds, with especial reference to the song of chaffinch, *Fringilla coelebs*. *Ibis*, **100**, 535–570.

10

The place of genetics in the study of behaviour

AUBREY MANNING

Studies in behaviour genetics have covered a wide field: motivation, development, sensory capacities, intelligence, learning, evolution, neuromorphology and neurochemistry have all been approached using genetic techniques, and there are probably others. Whilst it is at present impossible to construct any unities one must accept that many such studies have as their common aim one of the most fundamental problems in biology: how is behavioural potential encoded in genetic terms and expressed in the course of development?

The relative enormity of this problem is often matched by its inaccessibility. It cannot be claimed that there is any agreed view of the way forward and much of the work has frankly to be opportunistic – seizing on some favourable material or a useful new analytical technique to gain a limited objective. Consequently, behaviour genetics often presents a confusing picture of numerous disjointed studies, with workers proceeding on their separate paths.

Perhaps I am asking too much of a relatively new field. How many other fields have a really clear view of the way ahead? Nor is it really reasonable to expect everyone to keep their eye on, or even to agree upon, the long-term aim of behaviour genetics. There are many fascinating and important problems to be tackled along the way, and we can record some healthy achievements. Certainly behaviour geneticists have managed finally to convince all but a conservative rump of psychologists, that genotype is a variable to be contended with in the development of behaviour. To do so they sometimes had to operate in an almost evangelical style and records of this phase in the field are to be found in various papers in Hirsch (1967*b*) and Spuhler (1967).

It is quite difficult to select the best way to review progress and potential in the field. Genetic analysis of any kind requires that we have access to variation in the trait we are examining. This can be achieved by using contrasting inbred lines, selected strains, diverse natural populations or even species; we can also induce mutants which are screened for their behavioural effects. However, a classification based on the type of material used does not adequately bring out the nature of the problems in behaviour genetics. Elsewhere (Manning, 1975), I have described three approaches based loosely on their strategies. The first – 'parsing the phenotype' (Ginsburg's (1958) term) – has used genetics to study behavioural organization (although it provides some genetical insights also). The second uses the techniques of quantitative genetics to study the genetic architecture of continuously varying traits, and the third concentrates on the nature (both behavioural and more purely physiological) of gene action. There is considerable overlap and many studies have dual aims. Here I shall concentrate on the kinds of behavioural problem to which genetics has made a contribution.

CAUSAL ASSOCIATIONS

The regular association of two behavioural events often suggests that one causes the other or that the two are linked by a common causal factor. It is not always easy to test such an hypothesis directly but genetic analysis sometimes provides a rapid and elegant answer. Goy & Jackway (1962) found particular associations of behavioural components in the sexual behaviour of female guinea pigs from two inbred lines. Following ovariectomy and identical levels of hormone replacement, one line showed high duration of maximum levels of lordosis combined with a high frequency of mounts by the female, whilst in the other line both measures had low values. An association of this kind could arise from differences in responsiveness to circulating oestrogens, high in one case, low in the other, with the two components representing different aspects of this same response. However, the F_2 hybrid generation showed a complete range of combinations of these elements and this hypothesis must therefore be rejected. In fact a detailed examination of the timing of maximum lordosis and mounting through the cycle would suggest that they may be independent – the generation of new recombinant classes in the F_2 clinches the matter.

Genetic analysis also served to reject an hypothesis which linked learning ability with levels of acetylcholinesterase in the cerebral cortex of rats (Roderick, 1960). It also showed that the association of high playful aggressiveness and high timidity in basenji puppies, and low levels of both in cocker spaniels, did not reflect a single underlying variable. An F_2 generation showed a variety of associated levels (Scott & Fuller, 1965).

Artificial selection has been used to generate a whole variety of changed, one might call them specialized or even distorted, behavioural phenotypes. Amongst many other, *Drosophila* have been bred for mating speed, general activity and geotaxis, rats for open-field emotionality and maze learning, and mice for aggressiveness. In all cases which have been adequately studied, the change selected has had repercussions on other aspects of behaviour and such 'correlated characters' as animal and plant breeders call them, also present us with causal problems. The same behavioural phenotype can often be achieved in more than one way and the associations do not reveal an obligatory but merely one possible causal relationship. Thus in one selection experiment I found slow mating speed in *Drosophila* was associated with raised general activity (Manning, 1961) but in a second and comparable experiment slow mating and lowered activity were associated (Manning, 1963). In other cases an association seems more stable and suggests that changing the one aspect of behaviour cannot avoid changing the other. Hall & Klein (1942) found that Hall's line of rats selected for high open field emotionality were markedly non-aggressive, whereas the non-emotional line fought actively. Data from an F_2 hybrid population would be interesting here, but instead we have the results of a converse selection experiment performed on mice by Lagerspetz (1964). Mice selected for high aggression show low open field emotionality and vice versa: this, taken together with the evidence from rats, strongly supports the idea of a causal association.

THE UNITS PROBLEM

One of the most persistent difficulties facing students of behaviour is the selection of units which both have some behavioural validity and serve to illuminate the particular problem being studied. There is some justification for claiming, following Darwin's definition of a species, that a unit is what any competent behavioural worker chooses to call one.

Units differ greatly in their scale. Response to gravity or to light is in one sense a unit, but a unit with general and pervasive effects on all behaviour. Other units, such as the specialized movements made by a young bird when hatching from its eggshell, are unique and behaviourally quite isolated. Most units will be somewhat intermediate in their scale and isolation from others. Certainly we must be prepared to lump, split or re-classify units according to their usefulness as work proceeds. It is tempting to look to genetics for help. Presumably many people would agree that a behavioural element which can be shown to have a particular gene or set of genes exercising unique control over its development, has one kind of biological validity as a unit. In his valuable discussion of behavioural development, Bateson (1976) classifies developmental determinants as being inherited or environmental on the one hand, and having specific or non-specific effects on the other. Inevitably behaviour patterns have several determinants of different types and this blurs the picture, but we can sometimes catch glimpses of relatively simple processes which give encouragement.

Rothenbuhler's (1964) famous analysis of nest-cleaning behaviour in honey-bees shows that a classification into two units (1. uncapping cells; 2. removing contents), which makes intuitive sense in behavioural terms, has also a genetical equivalent – i.e. each of these units has a major gene with specific effects on its development. It will require further behavioural analysis to discover the extent of such units in the bees. Do they involve sensory elements and motor elements in addition to central nervous mechanisms, for example?

We can rarely expect to find such a clear-cut example, but certainly we can often identify behavioural elements whose genetic determinants are largely distinct from those of others. Goy & Jackway's (1962) analysis of male sexual behaviour in guinea pigs revealed separate patterns of inheritance for genes affecting general activity and mounting on the one hand and for intromission frequency and ejaculation latency on the other. Von Schilcher & Manning (1975) measured the flight wing-beat frequency and the inter-pulse-interval of the courtship song in males of *Drosophila melanogaster*, *D. simulans* and the hybrid between them. Wing beat frequency showed intermediate inheritance in the hybrid, but courtship song complete dominance of *simulans*, whose X chromosome the hybrid had inherited. Such results suggest a marked degree of genetic 'isolation' between the elements and makes it very improbable that they share any genetic determinants of major effect. In both evolutionary and behavioural terms they may constitute valid units.

I certainly do not wish to imply that units must have a high degree of genetic determinancy. Certainly environmental determinants play a part in the development of most behavioural elements that we measure; they will play an over-riding part in some. My contention is only that when we have some genetic evidence of the type outlined above, we have grounds *a priori* for designating a unit and we can then proceed to more behavioural studies to check its validity.

The evidence from genetics is not always clear-cut because the pattern of inheritance of a behavioural measure – a potential unit – sometimes changes according to the genotypes involved in hybridization. McGill's (1970) studies of the inheritance of sexual behaviour elements in mice is a case in point. He showed that when one inbred strain (DBA/2J) was crossed with two other strains (C57BL/6J and AKR/J) the mode of inheritance of 14 measures agreed in only four cases. This certainly makes the identification of natural units more difficult, and it would require a whole spectrum of crosses to identify those measures which showed independent segregation from others.

In fact one sexual behaviour pattern (rooting) in which the male pushes his snout beneath the female's body and jerks upwards – although not used by McGill in this study – probably would meet the necessary criteria. It is particularly characteristic of two strains, A/J and BALB/C, and Vale & Ray (1972) found it to have a far higher degree of genetic determinancy than any other measure of male sexual behaviour.

In the absence of any evidence of this type, there is little agreement amongst behaviour geneticists on how units may best be defined. Ginsburg (1958 and later papers) has argued that we should accept any mutant phenotype as representing a natural unit of change from the normal. Thus he regards phenyl-ketonuria in man as a unit because its symptoms, although diverse, can be traced back to one enzyme deficiency. It is incontrovertible that a single gene substitution does represent the basic unit of genetic change. As we shall discuss later, a number of groups (notably Benzer's and Brenner's) are studying mutants for just this reason. However, their approach relies on the capacity to pick up very small behavioural changes, whilst many mutants produce unmanageably large effects. If we take Ginsburg's injunction literally then we shall often be faced with gross distortions of behaviour to analyse. It will not be easy to use phenyl-ketonuric patients to study the development of intellectual functioning – the damage is too pervasive. More complex genetic changes, such as those involved in Turner's syndrome (when all the body's cells contain only a single X chromosome – the XO configuration) may nevertheless result in altered intellectual development of a much more subtle kind which provides a better foothold on the problem.

Hirsch (1967*a*, *b*) has argued that we should begin our search for the units of phenotypic variation with the study of individual differences. He attacks classifications which label individuals as trait bearers or non-trait bearers on rather gross characters and suggests that a far finer scale of behaviour genetic analysis will be possible when we examine in detail how individuals vary and how such variations relate across families. Hirsch (1967*a*) claims that whilst some simple traits, e.g. ability or inability to taste phenyl-thio-urea, lend themselves to a straight classification others, such as handedness or colour vision, do not. He wants to construct a set of dimensions along which such traits may vary and then to look for resemblances between relatives. This seems a reasonable suggestion, although I find it rather hard to distinguish from the efforts of others to 'factorize' complex traits. I think, however, that Hirsch's call for a really detailed look at phenotypic variation is one which should appeal to ethologists and may yield fine scale units for study.

Ethologists, for obvious reasons, have tended to concentrate on fixed action patterns as their most readily accessible units. But here the argument proceeds from behavioural observation to genetics, rather than the reverse. Thus the fact that fixed action patterns show very considerable developmental constancy is usually taken as evidence for a high degree of genetic determinancy. In the absence of suitable variations for analysis there cannot be any firm evidence for this conclusion. However, fixed action patterns certainly represent natural units in the phylogenetic sense and much of the evidence from inter-species hybrids is very hard to explain if the development of their highly specific features were not under genetic control. Why else should Dilger's (1962) *Agapornis* hybrids tear strips of leaves normally (a unit present in both parent species) but perform truncated versions of *both* parental carrying movements (in which unit the two differ markedly)?

Most of our evidence about genetic influences on the development of fixed action patterns comes from interspecies hybrids, and only rarely do we have F_2 and backcross generations available. Franck (1974) provides a very good review. As expected most of the evidence suggests polygenic inheritance, even down to the quite simple units of motor function in the songs of cricket (*Teleogryllus*) hybrids studied by Bentley (1971).

There are some cases where determination of a pattern is claimed to be controlled by a single gene (e.g. Hörmann-Heck, 1957), but such cases certainly depend on one locus with major effect on a performance threshold, as in Rothenbuhler's honey-bees. In general, we must suppose that the action of many genes is responsible for organizing the elaborate patterns of muscle action which are involved in even the simplest fixed patterns. How many genes is anybody's guess and not even a very pressing question. It is far more important to ask, as does Brenner (1973), what is the strategy involved in their deployment. The earthworm has a developmental instruction in its genome which says, in effect, 'build a segment'. Are there modules of behaviour in the same way that there are modules of morphological development? In the nematode *Caenorhabditis*, Brenner's group have discovered some mutants which affect a single unit of the central nervous system, but others affect a whole series of homologous units. Brenner suggests such genes must encode an instruction of the type 'build a neurone whose axon runs posteriorly and which has a dorsal process'.

Is it possible that fixed action patterns have some kind of modular basis in their organization? This would mean that some genes can be shared between different patterns with other genes of more detailed effect to determine the specialized features of each pattern. We can only speculate what such modules might be – perhaps a basic oscillator unit, to be built in to the control mechanism of any rhythmic response. Dawkins argues persuasively in this volume for hierarchical organization as a candidate principle in ethology. Perhaps some kind of basic hierarchical unit – a neurone or neuronal cluster which has the potential to accept a simple input and deliver a more complex output – might also be a genetically controlled module. Brenner argues that modular control of development is a necessity because there is not enough genetic material to code for every process separately, and his argument has equal force for the behavioural repertoire.

We lack suitable hybrid material to explore these questions, yet we have no examples of segregation into such units, nor can we have any confidence that we should recognize them if they occurred. While incomplete morphological elements are easily recognized as such, behavioural ones may not be. For the present we have tantalizing hints which suggest that there may be strong linkage between the genes controlling some fixed action patterns, and that the control of sequences of such patterns is more labile genetically than that of the patterns themselves (see Ewing & Manning, 1967; Manning, 1975). We also know a good deal about the action of genes which affect the performance frequency and intensity of homologous patterns; their effects are similar in closely related species, in mutant strains and in selected lines. Brenner would call these 'optimizing genes' and their role in the microevolution of behaviour can quite clearly be recognized.

Observations on hybrids, for all their genetical shortcomings, provide us with some interesting behavioural questions. Franck's (1974) hybrid *Xiphophorus* males perform the courtship patterns of one parent species when their sexual excitement is low, but switch to those of the other when it is high. Davies' (1970) hybrid *Streptopelia* males show the form of one parent at the head low phase of their bow–coo display, that of the other when the head rises to the top. Then we have the completely incomprehensible result of Sharpe & Johnsgard's (1966) study on mallard (*Anas platyrhynchos*)/pintail (*A. acuta*) hybrids. Sexual behaviour patterns and plumage characters proved to be highly correlated in the recombining F_2 generation.

We certainly need more evidence from hybrids, but a complete search for behavioural units must extend beyond fixed action patterns. Often we wish to measure behavioural differences which involve motivation, sensory capacities, arousal, emotionality, learning or others which show continuous variation between individuals or populations.

Motivational scales are not likely to be divisible into meaningful units. Sensory units are certainly possible – colour-blindness is a good example of a Ginsburg-type unit – and can often be related directly to physiology, since most of them are detected peripherally. Thus Benzer (1973), Hotta & Benzer (1972) and Pak, Grossfield & Arnold (1970), have described a range of mutants which affect the processing of visual stimuli by the eyes of *Drosophila*. The accumulation of mutants with distinct effects and the collation of such results will eventually enable us to make a map of the developmental units of the eye's construction. It will be more difficult to move away from the periphery. Genetics may help, but the identification of central units of processing may have to await better methods of behavioural and physiological measurement.

The more complex continuously varying traits, such as learning and intelligence, absolutely demand subdivision if we are to understand their nature.

We shall certainly need to retain compound entities because they may well be the kind of compromise unit upon which natural selection operates. However, it was one of the earliest of all behaviour genetics experiments – Tryon's selection for maze-bright and maze-dull rats – which showed the striking advantages of 'parsing the phenotype' of learning. The major units which Tryon extracted, from the many that constitute maze learning, proved to relate to cue attentiveness, and much subsequent work on learning has shown them to be valuable ones.

Factor analysis has often been applied to such complex traits and there have been some studies which combine this approach with genetics. Much the most complete is Scott & Fuller's (1965) work on dogs. In their Chapter 14 they examine the cross correlations between a whole battery of tests given to cocker spaniels, basenjis and the hybrids between them. The factorial method relies on correlations between diverse measurements but can make no distinction between correlations which result from common environmental determinants and those that result from genes. Scott & Fuller look for correlations that appear only in the segregating populations – the F_2 and backcrosses. An emotional factor, probably best labelled 'timidity', proved to be the only one with clear genetic determinance which affected measurements on the majority of tests. This effect is not surprising since many tests involved responding to novel situations in a variety of social and learning contexts. There were other correlations between measurements in different tests in the segregating populations – many more than would have been expected by chance – but all of them were of low value. Quite often factors had major effects on one test but relatively minor effects on the others. This is a difficult situation to analyse further because we may have trouble identifying factors if they can be reliably defined only in terms of a single test.

Clearly Scott & Fuller's battery of tests are aimed to split up a series of compound behavioural entities – physiological/emotional responsiveness, emotionality, sociability, trainability and problem solving. It is attractive to speculate that factors that we extract from a complex of tests may represent relatively 'pure' units with more distinct genetic determinance.

In an admirable discussion of the problem of units Thompson (1967) describes a number of attempts to parse complex traits and relate the results to genetics. For example, Royce (1957) suggests that whilst complex traits will be under the control of numerous genes, their component factors will each be affected by fewer loci of more specific effect. Royce's model (Fig. 1) includes a simple hierarchy of levels. An ethologist might be tempted to apply similar arguments when speculating on the genetic control of the different levels of a hierarchy of instincts as in Tinbergen's (1950) model. Cattell (1960) has argued along somewhat parallel lines that whereas com-

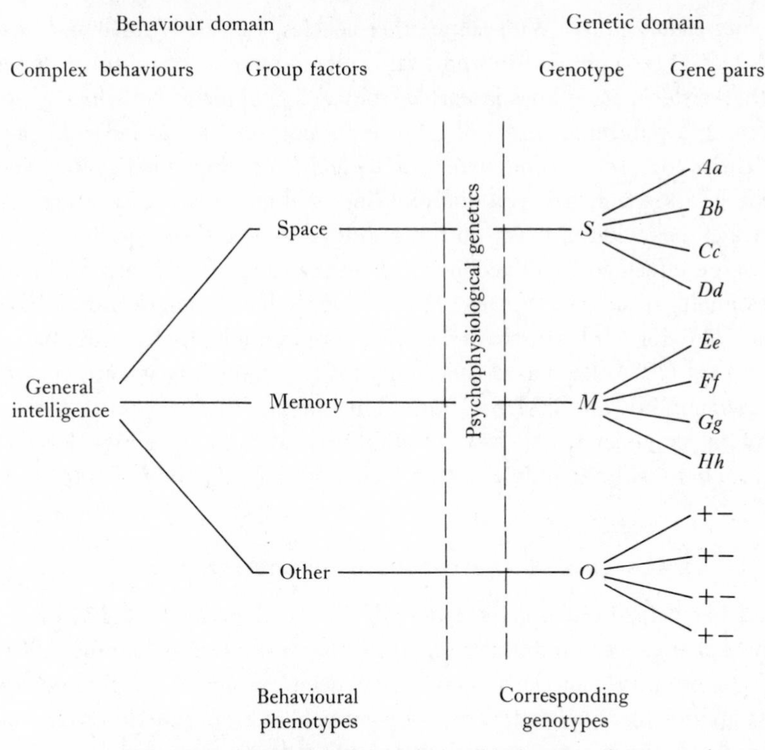

Fig. 1. Royce's model of the relationship between factors and genes using intelligence as an example (from Royce, 1957).

pound traits will always have a complex of both genetic and environmental determinants of variance, their factors are more likely to show a predominance of one determinant or the other.

Thompson (1967) points out that there is no real evidence to support Royce's or Cattell's hypotheses, and some evidence against. He cites heritability estimates for composite tests, primary and second-order factors of various intelligence scales and there is no clear trend to be discerned, certainly not one towards higher genetic determinancy for factors.

An honest summary must admit that there are few valid generalizations or even hopeful speculations to be made about the genetic basis of behavioural units. However, even the few examples we have of the genetic 'independence' of some behavioural elements gives us encouragement to explore further. Almost always we shall find polygenic inheritance and this has sometimes been regarded as a barrier to further analysis of how genes operate. It need not be. Thoday (1961) has described methods for isolating loci with relatively large effects from a mass of polygenes. The method is laborious and relies on having a good source of marker genes and this virtually means using

Drosophila melanogaster. With some other species, especially mice and rats, Bailey's (1971) recombinant-inbred strains technique may be useful. A series of brother–sister mating lines is started from an F_2 population in which genes determining a polygenic trait will be recombining (at least as far as linkage allows them to). Inbreeding will lead to loci becoming homozygous and differences in the trait between different lines will not only reveal something of the traits' behavioural make-up but also give some estimate of how many loci of large effect are involved in its determination. It will always be an underestimate, of course, because one can never have enough inbred lines to be sure of fixing a full sample of the genes. For example, Eleftheriou, Bailey & Denenberg (1974) have used this technique to examine the genetic control of aggression in C57 and BALB/C mice. They suggest there are two dominant genes of major effect which can act additively to produce a high level of fighting. There will certainly be other loci which modify their effect.

GENES AND BEHAVIOURAL DEVELOPMENT

Genetics has helped our understanding of the development of behaviour in the evolutionary sense, and in particular its microevolution (Manning, 1965, 1975). Quantitative genetic analyses have been employed to disclose the 'genetic architecture' behind a trait – how much additive genetic variance is concerned, whether control is predominantly by dominant genes etc. – from which estimates can be made of the nature of past selection on the trait (see Broadhurst & Jinks, 1974, for a useful review).

It is a more challenging task to try to analyse the effects of genes during the development of behaviour within an individual. Such a study will bring us close to that fundamental aim of behaviour genetics, to understand the expression of behavioural potential in genetic terms.

Certainly we have a number of examples of genes acting to affect particular stages of development. Benzer (1973) describes a number of loci which affect developmental processes in the eye of *Drosophila*. Gwadz (1970) has discovered a single gene in *Aëdes* mosquitoes which determines the stage, following eclosion, at which the corpus allatum becomes active and the females sexually receptive. Both examples of gene action reveal clear cut behavioural consequences, but they concern action which is too peripheral or too late in development to illuminate the main issue.

As Bateson implies in his chapter to this volume, there must be a *strategy* to behavioural development, and different sets of genes may be operating at different phases. In Waddington's (1957) model of the epigenetic landscape it was the genes which carved out the general pattern and form of the valleys down which the developing organism moves. Behavioural development, particularly in vertebrates, can continue long into an organism's adult life and genes will still set some limits to, and determine some of the directions of, its response to its environment.

There is clear evidence of the genetic control of the precise course of behavioural development. The classical ethological studies on bird song development are of importance here, revealing striking differences between species which must be the result of natural selection operating on inherited differences. The development of sexual attachments has also been modified during evolution – sexual imprinting is common in birds especially in those species with cryptically patterned females, but must be avoided in brood parasites such as the European cuckoo (*Cuculus canorus*) and cowbirds (*Molothrus ater*).

The detailed timing of developmental stages also varies between species and populations exposed to the same environment. King (1967) has described work on two subspecies of the deer mouse, *Peromyscus maniculatus*, *bairdii* and *gracilis*. The latter is more arboreal and showed an enhanced tendency to cling at all stages during various repeated tests of the agility and locomotor development of young mice. *Bairdii* showed less clinging at the corresponding stages but developed escape swimming and locomotion more rapidly than *gracilis*.

Some of the most detailed developmental studies have, of course, used human subjects. Freedman & Keller (1963) showed that concordance for the timing of social smiling and the transitory fear of strangers was far higher in monozygotic than in dizygotic twins. Wilson's (1972) studies of the mental development of children up to 24 months reveals a fairly complex pattern of slow periods and spurts. He found that the synchronization of these growth rate changes was almost perfect in monozygotic twins, and considerably higher than in dizygotic pairs. There is a clear presumption of different gene-controlled processes being called into action as development proceeds.

To take analysis further will require suitable populations whose behavioural development varies, in some features of timing, sensitivity to environmental factors etc., in order that their genetics can be investigated. Perhaps the most general question concerns whether genetical differences in a fully developed behaviour trait can already be detected at earlier stages of its development. The measures of intelligence quotient (IQ) in infants show that they cannot, or rather that they become more difficult to discern. The correlation between child and parental IQ becomes 0.5 at about 24 months of age and stays close to this value thereafter. However, the correlation is much lower at earlier ages – indeed it is zero at around 12 months when tests first become possible (Jones, cited by Thompson, 1967). There are obvious measurement problems involved, but such a result could imply a good deal of developmental 'noise' in young infants with the gene controlled processes only beginning to assert themselves after about 2 years.

A similar result was obtained by Vicari (1929), whose data on maze running in mice were re-examined by Broadhurst & Jinks (1963). The heritability of the differences between Vicari's strains increased over successive trials in the maze. They suggest that 'this represents a progressive release of a genetically determined response from the effect of environmental stimuli irrelevant to it, but which tend to obscure its action in the early stages of training'. This is an interesting but not a universal result in learning situations. Fuller & Thompson (1960) discuss tests of learning in different dog breeds which show no change in heritability estimates over successive trials. Kerbusch (1974) found a decrease in heritability over trials for maze learning in a situation similar to Vicari's, though with some different mouse strains. We badly need more examples.

But apart from heritability estimates changing progressively over time, we must also look for genetic discontinuities in the course of development. Such discontinuities could indicate the existence of distinct sets of genes becoming activated that would in turn have bearing on the units problem. Although this shows promise of being one clear way ahead for behaviour genetics, and was discussed by Broadhurst & Jinks (1963) and most fully by Thompson (1967), we still have very few examples to call on.

Wilcock & Fulker (1973) examined fairly detailed aspects of the genetic architecture of shuttle-box avoidance conditioning across eight strains of rats. At trial 20 they found a discontinuity in the directional dominance associated with successful escape. Up to this point dominance was associated with low levels of responding, beyond it with high. This suggests that selection has favoured these distinct types of response in analogous situations and the shift after a certain amount of experience. Wilcock & Fulker link their findings with Mowrer's (1960) two factor escape conditioning theory. This proposes that during early trials the animal is becoming classically conditioned to the warning stimulus whilst later it responds instrumentally to avoid the shock. It is certainly plausible to link the genetics and the behavioural changes in this way because selection will presumably require the animal's attention to be focussed on the relevant stimuli during the early stage of avoidance, before efficient escape can be developed.

It will probably be more difficult, but even more valuable, to extend this kind of analysis to behavioural development other than learning. Much will depend on locating suitable material, and mice are the most hopeful. We have evidence from studies by Ginsburg (1967), Henderson (1970) and Manning & Thompson (1976) that, for example, genotype/environment interactions are strong, and may be related to different stages of the development of emotionality, learning ability, aggressivity and sexual behaviour.

I find Bateson's (1976) view of the genotypic and environmental axes of development helpful here. He points out that whilst we may be able to classify the specific determinants of a stage in the development of a pattern in this way, the values we ascribe along such axes represent only a still picture of a continuous process. Other determinants will emerge as development proceeds, and not only may different genes become activated, but new environmental determinants may play a part in such activation. Bateson cites as an example how the level of crowding of locust nymphs may determine which of a whole set of genetically programmed responses are activated and thus lead to *solitaria* or *gregaria* phases.

THE DIRECT APPROACH TO GENE ACTION

Observing the effects of single gene mutations directly has always been an attractive approach to the genetics of behaviour. I shall not discuss it at length here although I consider it of great importance. It is, after all, straightforward in its aims although, as Wilcock (1969) has pointed out, much of the earlier work was ill-controlled. It also dealt with mutants whose effects were often rather trivial behaviourally and this was generally because they were identified on other effects and the behaviour came second.

Most recent attention has rightly focussed on the groups working with Brenner, Benzer and Kaplan. Their approach has usually involved screening populations which have been exposed to mutagens. Benzer's group, in particular, has developed a number of elegant and ingenious systems for identifying mutants affecting the phototaxis, circadian rhythms, locomotion and learning of *Drosophila*. A large number of mutant stocks have now been isolated and their behavioural interest often goes far beyond the apparent simplicity of their phenotypes. A whole series of 'sluggish' mutants has been isolated, in which for example, some show a universal sluggishness whilst others, though showing greatly reduced general activity, have undiminished sexual behaviour (S. Benzer, personal communication). Such results resemble those from selection experiments (Manning, 1961) and it is of great interest that such highly specific effects on activity can be pinned down to a single gene.

Information in this field is now accumulating rapidly. Whilst analysis sometimes has to remain at a behavioural level, we can sometimes get physiological and even neurophysiological information on the mutants. Benzer's group (see Hotta & Benzer, 1972) has devised techniques using mosaic individuals that enable the primary site of action of mutants to be closely localized. With new neurochemical markers (Benzer, 1973), it is possible to determine the genotype of individual neurones in the brain. It is also possible to tackle the problem of primary action from the other end by producing phenocopies of behavioural mutants by environmental means. Burnet, Connolly & Harrison (1973) have begun this with *Drosophila* and Burnet & Connolly (1974) provide a valuable review of their group's work on the behaviour, physiology and neurochemistry of *Drosophila* activity and sexual behaviour.

Such advances provide powerful tools, but to rely on mutants to build up a picture of the genetic control of behaviour one has to be something of an optimist. It is rather like trying to complete a jigsaw puzzle (for which you have no picture as a guide) with pieces being thrown to you at random. One's hope is that key pieces will arrive and that a pattern will begin to emerge long before the picture is complete.

CONCLUSION

It must have become obvious that I do not wish to advocate a single line of approach in this field. Brenner (1973), for example, has argued that only point mutations can lead us to the core of the problem. I see the force of his argument, but as someone interested in behaviour for its own sake, I consider that there are other ways in which the links between genetics and behaviour must be studied, and other questions I want to ask of genetics. Some of them I have outlined above, and I look forward to progress on all fronts. There is more study of behaviour genetics now than ever before and though there remain formidable barriers, we perhaps see more clearly to which parts we should direct our attack.

SUMMARY

The literature and research strategies of behaviour genetics are reviewed with special reference to three issues upon which it may throw light. In all cases, we badly need more information, and conclusions must remain tentative for the present.

(1) *Causal associations.* Genetic analysis can, by use of segregation in F_2 or backcross generations, determine how far behavioural elements are free to recombine and how far they are causally associated.

(2) *Behavioural units.* It is argued that if a particular element or pattern can be shown to have genetic determinants uniquely associated with it, these are grounds *a priori* for identifying a behavioural unit. Examples of different types of unit are given, and the relation between genetics and the factorization of complex traits such as intelligence is discussed.

(3) *Development.* There are good examples of precise genetic control over behavioural development. The possibilities for elucidating developmental strategies by genetics are considered. One approach is via the study of single

gene mutations, another uses quantitative methods to estimate if and when new sets of genetic determinants are activated during development.

REFERENCES

Bailey, D. W. (1971). Recombinant-inbred strains. An aid to finding identity, linkage and function of histocompatability and other genes. *Transplantation*, **11**, 325–327.

Bateson, P. P. G. (1976). Specificity and the origins of behavior. *Advances in the Study of Behavior*, **6**, 1–20.

Bentley, D. R. (1971). Genetic control of an insect neuronal network. *Science, Washington*, **174**, 1139–1141.

Benzer, S. (1973). Genetic dissection of behavior. *Scientific American*, **229**(6), 24–37.

Brenner, S. (1973). The genetics of behaviour. *British medical Bulletin*, **29**, 269–271.

Broadhurst, P. L. & Jinks, J. L. (1963). The inheritance of mammalian behavior re-examined. *Heredity*, **54**, 170–176.

Broadhurst, P. L. & Jinks, J. L. (1974). What genetical architecture can tell us about the natural selection of behavioural traits. In *The Genetics of Behaviour*, ed. J. H. F. Van Abeelen, pp. 43–64. North-Holland: Amsterdam.

Burnet, B. & Connolly, K. (1974). Activity and sexual behaviour in *Drosophila melanogaster*. In *The Genetics of Behaviour*, ed. J. H. F. Van Abeelen, pp. 201–258. North-Holland: Amsterdam.

Burnet, B., Connolly, K. & Harrison, B. (1973). Phenocopies of pigmentary and behavioural effects of the yellow mutant in *Drosophila* induced by α-dimethyl-tyrosine. *Science, Washington*, **181**, 159–160.

Cattell, R. B. (1960). The multiple abstract variance analysis equations and solutions: for nature nurture research on continuous variables. *Psychological Review*, **67**, 353–372.

Davies, S. J. J. F. (1970). Patterns of inheritance in the bowing display and associated behaviour of some hybrid *Streptopelia* doves. *Behaviour*, **36**, 187–214.

Dilger, W. C. (1962). The behavior of lovebirds. *Scientific American*, **206**, 88–98.

Eleftheriou, B. E., Bailey, D. W. & Denenberg, V. H. (1974). Genetic analysis of fighting behavior in mice. *Physiology and Behavior*, **13**, 773–777.

Ewing, A. W. & Manning, A. (1967). The evolution and genetics of insect behavior. *Annual Review of Entomology*, **12**, 471–494.

Franck, D. (1974). The genetic basis of evolutionary changes in behaviour patterns. In *The Genetics of Behaviour*, ed. J. H. F. Van Abeelen, pp. 119–140. North-Holland: Amsterdam.

Freedman, D. G. & Keller, B. (1963). Inheritance of behavior in infants. *Science, Washington*, **140**, 196–198.

Fuller, J. L. & Thompson, W. R. (1960). *Behavior Genetics*. Wiley: New York.

Ginsburg, B. E. (1958). Genetics as a tool in the study of behavior. *Perspectives in Biology and Medicine*, **1**, 397–424.

Ginsburg, B. E. (1967). Genetic parameters in behavioral research. In *Behavior-genetic Analysis*, ed. J. Hirsch, pp. 135–153. McGraw-Hill: New York.

Goy, R. W. & Jakway, J. S. (1962). Role of inheritance in determination of sexual behavior patterns. In *Roots of Behavior*, ed. E. L. Bliss, pp. 96–112. Harper: New York.

Gwadz, R. W. (1970). Monofactorial inheritance of early sexual receptivity in the mosquito. *Aëdes atropalpus. Animal Behaviour*, **18**, 358–361.

Hall, C. S. & Klein, S. J. (1942). Individual differences in aggressiveness in rats. *Journal of comparative Psychology*, **33**, 371–383.

Henderson, N. D. (1970). Genetic influences on the behavior of mice can be obscured by laboratory rearing. *Journal of comparative and physiological Psychology*, **72**, 505–511.

Hirsch, J. (1967*a*). Intellectual functioning and the dimensions of human variation. In *Genetic Diversity and Human Behavior*, ed. J. Spuhler, pp. 19–31. Aldine: Chicago.

Hirsch, J. (1967*b*). Behavior-genetic, or 'experimental' analysis; the challenge of science versus the lure of technology. *American Psychologist*, **22**, 118–130.

Hörmann-Heck, S. V. (1957). Untersuchungen über den Erbgang einiger Verhaltensweisen bei Grillenbastarden (*Gryllus compestris–Gryllus bimaculatus*). *Zeitschrift für Tierpsychologie*, **14**, 137–183.

Hotta, Y. & Benzer, S. (1972). Mapping of behaviour in *Drosophila* mosaics. *Nature, London*, **240**, 527–535.

Kerbusch, J. M. L. (1974). A diallel study of exploratory behaviour and learning performances in mice. In *The Genetics of Behaviour*, ed. J. H. F. Van Abeelen, pp. 65–89. North-Holland: Amsterdam.

King, J. A. (1967). Behavioural modification of the gene pool. In *Behavior-genetic Analysis*, ed. J. Hirsch, pp. 22–43. McGraw-Hill: New York.

Lagerspetz, K. M. J. (1964). Studies on the aggressive behaviour of mice. *Annals of the Academy of Sciences, Fennica, Series B*, **131**, 1–131.

Manning, A. (1961). The effects of artificial selection for mating speed in *Drosophila melanogaster. Animal Behaviour*, **9**, 82–92.

Manning, A. (1963). Selection for mating speed in *Drosophila melanogaster* based on the behaviour of one sex. *Animal Behaviour*, **11**, 116–120.

Manning, A. (1965). *Drosophila* and the evolution of behaviour. *Viewpoints in Biology*, **4**, 125–169.

Manning, A. (1975). Behaviour genetics and the study of behavioural evolution. In *Function and Evolution in Behaviour*, ed. G. Baerends, C. Beer & A. Manning, pp. 71–91. The Clarendon Press: Oxford.

Manning, A. & Thompson, M. L. (1976). Postcastration retention of sexual behaviour in the male BDF_1 mouse: the role of experience. *Animal Behaviour*, **24**, 519–531.

McGill, T. E. (1970). Genetic analysis of male sexual behavior. In *Contributions to Behavior-genetic Analysis: the Mouse as a Prototype*, ed. G. Lindzey & D. D. Thiessen, pp. 57–88. Appleton-Century-Crofts: New York.

Mowrer, O. H. (1960). *Learning Theory and Behavior*. Wiley: New York.

Pak, W. L., Grossfield, J. & Arnold, K. S. (1970). Mutants of the visual pathway of *Drosophila melanogaster. Nature, London*, **227**, 518–520.

Roderick, T. H. (1960). Selection for cholinesterase activity in the cerebral cortex of the rat. *Genetics*, **45**, 1123–1140.

Rothenbuhler, W. C. (1964). Behavior genetics of nest cleaning in honey-bees. IV. Responses of F_1 and backcross generations to disease-killed brood. *American Zoologist*, **4**, 111–123.

Royce, J. B. (1957). Factor theory and genetics. *Education and Psychological Measurement*, **17**, 361–376.

Schilcher, F. von & Manning, A. (1975). Some aspects of sexual behavior in hybrids between *Drosophila melanogaster* and *Drosophila simulans*. *Behaviour Genetics*, **5**, 395–404.

Scott, J. P. & Fuller, J. L. (1965). *Genetics and the Social Behavior of the Dog.* University of Chicago Press: Chicago.

Sharpe, R. S. & Johnsgard, P. A. (1966). Inheritance of behavioral characters in F_2 mallard×pintail (*Anas platyrhynchos*×*Anas acuta*) hybrids. *Behaviour*, **27**, 259–272.

Spuhler, J. N. (ed.) (1967). *Genetic Diversity and Human Behavior.* Aldine: Chicago.

Thoday, J. M. (1961). Location of polygenes. *Nature, London*, **191**, 368–370.

Thompson, W. R. (1967). Some problems in the genetic study of personality and intelligence. In *Behavior-genetic Analysis*, ed. J. Hirsch, pp. 344–365. McGraw-Hill: New York.

Tinbergen, N. (1950). The hierarchical organization of nervous mechanisms underlying instinctive behaviour. *Symposia of the Society for experimental Biology*, **4**, 305–312.

Vale, J. R. & Ray, D. (1972). A diallel analysis of male mouse sex behavior. *Behaviour Genetics*, **2**, 199–209.

Vicari, E. M. (1929). Mode of inheritance of reaction time and degree of learning in mice. *Journal of experimental Zoology*, **54**, 139–144.

Waddington, C. H. (1957). *The Strategy of the Genes.* Allen & Unwin: London.

Wilcock, J. (1969). Gene action and behavior: an evaluation of major gene pleiotropism. *Psychological Bulletin*, **72**, 1–29.

Wilcock, J. & Fulker, D. W. (1973). Avoidance learning in rats: genetic evidence for two distinct behavioral processes in the shuttle box. *Journal of comparative and physiological Psychology*, **82**, 247–253.

Wilson, R. S. (1972). Twins: early mental development. *Science, Washington*, **175**, 914–917.

11

Stages in the early behavioural development of altricial young of selected species of non-primate mammals

JAY S. ROSENBLATT

Altricial young provide a good opportunity to study early stages of behavioural development among mammals that might be more difficult to analyse in the seemingly more complex precocial young that possess advanced sensory and motor capacities from birth. Among altricial young, there are relatively distinct periods of thermotactile, olfactory and visual control of behavioural responses and underlying motivational processes, and the relatively simple forms of motor activity can be studied as they become organized into more complex action patterns. While it is true that limited motor abilities may hide advanced sensory and integrative processes in altricial young, it is equally true that advanced sensory and motor abilities may mask simple behavioural capacities in precocial young. The problem is, therefore, to determine, through analytical experimental procedures and theoretical synthesis of the findings, the nature of behavioural organization at each stage of development.

We are not yet prepared to develop detailed models of the behavioural organization of young at the various stages but we can sketch the major lines along which development proceeds. Among altricial young these lines are clear from many good descriptions of the development of behaviour toward the mother, siblings and nest or home site in a variety of mammalian species (Rheingold, 1963*a*; Hafez, 1969; Rosenblatt, 1971).

Suckling and non-suckling contact with the mother, and huddling with siblings in the nest, take an early lead in behavioural development among altricial young. At first the mother takes the initiative in suckling, but gradually the young contribute increasingly to the initiation and termination of feeding sessions, until the relative contribution of each shifts and the young play the dominant role in the feeding interaction until weaning intervenes. Similarly, huddling among siblings is a relatively simple behaviour pattern while the young are still all confined to the nest. As young become able to move about more freely, behavioural interactions among siblings become more complex: young become capable of many more responses to one another and respond increasingly to individuals and their characteristics rather than to group characteristics.

Initially the young are either passively confined to the home area or are actively oriented to it but as development proceeds they become able to move more freely in relation to it at a distance. This early form of orientation to the home area wanes after a period and the young's relationship to its socially conditioned environment is based upon its relationship to its companions and to various specific functions that emerge as development proceeds.

Schneirla (1959, 1965), Schneirla & Rosenblatt (1961) and Rosenblatt, Turkewitz & Schneirla (1969) have described the genesis of early approach responses to the low intensity stimulation provided by mother, littermates, and socially conditioned home area. In the present article I shall attempt to trace further developments which arise from these early general and local approach responses. The changing relationship between central control processes and locally stimulated responses will be discussed along with the gradual specialization of central arousal processes, as sensory discrimination and the organization of motor activity into functionally adaptive action patterns proceeds. This view of behavioural development, as a continuous process, only arbitrarily divided into stages in which changes occur during each stage which contribute to succeeding stages (Schneirla & Rosenblatt, 1963), is completely at odds with the 'critical period' theory of Scott (Scott, 1962; Scott, Stewart & DeGhett, 1974) which emphasizes only one or several periods as critical.

The first three sections of this article will review studies on the development of suckling, huddling, and home orientation, respectively. The fourth section will analyse the stages of early behavioural development that can be distinguished in the studies reviewed, and the fifth will briefly point out the functional aspects of each of these stages.

DEVELOPMENT OF SUCKLING

Among all species of altricial mammals suckling is initiated shortly after birth. The newborn actually begin to nuzzle the mother's nipple region during parturition, when given the opportunity in the intervals between births. Kittens, however, are only rarely successful in attaching to nipples and this is very likely the case in other altricial species (Schneirla, Rosenblatt & Tobach, 1963). Ewer's (1961) suggestion that the delay of an hour or so *post partum* in the onset of suckling in kittens is based upon the absence of a special releasing stimulus for nipple grasping seems unnecessarily elaborate. Kittens have little opportunity to attach to nipples during parturition even when the mother is not actively giving birth. The mother is engaged in a variety of activities such as licking the kittens, eating the placentae, licking birth fluids from her fur and the delivery site, etc. which make it difficult for kittens to attach to nipples (Schneirla *et al.*, 1963).

Not until parturition is completed does the female make herself available for feeding. The position assumed after parturition by the exhausted mother among cats, dogs and other small mammals facilitates nipple searching and grasping by the newborn: she lies on her side, her nipple region facing toward the newborn, her limbs outstretched forming an oval-shaped corral in which the young are confined. Comparable positions are assumed by rat, hamster, and rabbit mothers. Most successful artificial mothers have been modelled after the immediately post-partum nursing female.

In this position the mother presents a variety of attractive thermal, tactile and olfactory stimuli. The effects of these stimuli on early suckling have been investigated through the presentation of one or several of the component stimuli under controlled conditions in artificial brooders.

Among the important features of the mother's body that stimulate approach and nuzzling appear to be the fur-textured surface, and the rounded shape with few entrapping crevices or projections that might distract newborn from locating the nipples and surrounding areola. Body surface temperature in the range of 32 to 39 °C, as measured in various species, and as provided in artificial mothers (Toropova, 1961; Thoman & Arnold, 1968; Jeddi, 1970 and unpublished data; Alberts, 1974), has proved attractive to newborn rats, rabbits, puppies, and kittens. The moist surface of a pulsating plastic tube was most attractive to brooder-reared rat pups in the study by Thoman & Arnold (1968).

In addition to being attracted to the mother's nipple region by these stimuli, newborn kittens are led along certain paths in their nuzzling of her body by patterns of thermal and tactile stimuli. N. Freeman (unpublished) found a broad but well-defined thermal gradient extending over the entire body surface of the lactating cat that could provide a basis for locating the nipples in newborn kittens. When on a floor surface at 30.2 °C a thermal gradient extended from the mother's limbs, which had a surface temperature of 33.3 °C, to her nipples which were at 37 °C.

The pattern of fur growth offers additional directional stimulation: when the mother is lying on her side, crawling against the grain leads kittens upward from the midline to the lateral nipples. Moreover, the areola surrounding the nipples, with its bare, warm moist surface, attracts kittens. It has been suggested that the pattern of hair growth on the belly of the pig provides paths which piglets follow in finding nipples. Static features of the mother and more active ones, such as her licking that steers kittens to her ventrum, offer attractive stimuli which initially bring newborn into contact with her. More detailed and patterned features lead them to the nipples.

Suckling and thermotactile stimuli

Thermal and tactile stimuli provide the basis for the earliest suckling approaches to the mother in kittens, rat pups, and newly born rabbits and puppies. In the kitten, the snout and lips are supplied with low threshold thermotactile receptors (as indicated in studies of adult cats, Kenshalo, Duncan & Weymark, 1967; Hensel & Kenshalo, 1969; Kenshalo, 1970; Kenshalo & Brearley, 1970; Kenshalo, Hensel, Graziadei & Fruhstorfer, 1971; Rowe & Sessle, 1972) and greater anterior-end sensitivity to such stimuli appears to be the general rule among newly born altricial young (Carmichael, 1946; Welker, 1959; Jeddi, 1970 and unpublished data). The snout functions as a probe as it is moved by side-to-side head movements during crawling, a characteristic of kittens and other altricial young that move in this manner. These movements enable the newborn effectively to scan a fan-shaped region in front of it (Tilney & Casamajor, 1924; Langworthy, 1929; Windle, 1930; James, 1952*a*; Prechtl, 1952; Rosenblatt *et al.*, 1969; Fox, 1970). Contact with a low-intensity tactile or thermal stimulus slows the return head swing (Gard, Hard, Larsson & Petersson, 1967) and the newborn veers in the direction of the stimulus by repeated movements of the contralateral forelimb and cessation of ipsilateral forelimb movement.

On the other hand, contact with a strong tactile stimulus (e.g. wire textured floor surface, Stanley, 1970; Stanley, Bacon & Fehr, 1970; Bacon, 1973*a*) or excessively warm or cool stimulus (e.g. overheated or cooled floor surface or overly warm or cool littermate, Tilney & Casamajor, 1924; Welker, 1959; Bacon, 1973) initiates head withdrawal followed by cessation of forward movement and veering away. If the stimulus persists, the newborn pivots a half circle and crawls away from it. This withdrawal response, which can also be elicited by a puff of air directed at the newborn's face, has been used to establish early escape and avoidance learning in newborn of several altricial species (Bacon & Stanley, 1970; Misanin, Nagy & Weiss, 1970; Nagy, Misanin & Newman, 1970; Bacon, 1973*a*; Stanley, Barrett & Bacon, 1974).

At this early stage, thermal and tactile stimuli in the litter situation also stimulate newborns to activity which eventuates in their crawling to the mother. Okon (1971, 1972) has shown the distress-inducing effects of strong tactile stimulation and extreme thermal conditions upon hamster, rat, and mouse pups, as measured by a high rate of audible and ultrasonic calling. This builds up during the first weeks after birth then declines at the end of the second and beginning of the third week. Although rough handling or tail pinching are most effective in eliciting ultrasonic calling, it is particularly significant that loss of contact with tactile stimulation (i.e. even nesting material) stimulates calling in three day old rat pups (Oswalt & Meier, 1975) and young hamsters (Goldman & Swanson, 1975). Similarly, cooling caused by exposing three to nine day old pups to a 22 °C ambient temperature stimulates a significant rise in the proportion of animals that vocalize and in the rate of ultrasonic vocalization (Allin & Banks, 1971). Newborn rabbits are stimulated to activity at 20–25 °C ambient temperatures which cause a decline in body temperature of 9 °C (E. Jeddi, unpublished).

Among puppies and kittens loss of contact with tactile stimulation from the mother and littermates stimulates an increase in activity that usually consists of pivoting and calling (Tilney & Casamajor, 1924; Windle, 1930; James, 1952*a*, *b*; N. Freeman, unpublished). In puppies, quiescence is restored when head and body contact is made with a soft tactile stimulus (Scott *et al.*, 1974), and a most effective stimulus for quieting kittens that have lost contact with the mother and litter is the weak tactile stimulation of the face and forehead provided by a canopy of soft bunting (G. Turkewitz, unpublished).

Cooling is as effective as tactile stimulation in arousing puppies and kittens to activity and in stimulating calling. In an ambient temperature of 30 °C puppies remain quiescent, breathing lightly in a relaxed sleeping posture, but if the ambient temperature falls to 27 °C or rises to 32.5 °C, they become alert and restless, begin to call, and exhibit rapid breathing (Welker, 1959; Crighton & Pownall, 1974). Kittens placed on a cool surface (18 °C) become active and call during the first week. Many crawl to a warmer area if this is available, but others become so active they simply walk off the table surface (N. Freeman, unpublished).

Newborns that lose contact with the mother therefore become active as a result of both the loss of tactile stimulation and exposure to a lowered ambient temperature. They begin to pivot and circle, emitting calls to which the mother often responds by bringing her body closer and into contact with the young or by licking them. The newborn is likely to make contact with the floor surface around the mother which is warmed by her body and with her extended fore- and hindlimbs as well as the warm furry surface of her nipple region. These ensure that the newborn will regain contact and warmth and if it has not fed for a time, that it will initiate nipple-locating movements and finally suckle from the mother.

Nipple grasping and sucking: specialized responses

Nipple grasping and sucking are specialized responses of the newborn to features of the nipples and the surrounding areola. The specific nipple characteristics which enable newborns to grasp them have not yet been established for any altricial species and so newborns do not initially attach spontaneously to the rubber nipples of artificial brooders (Rosenblatt *et al.*, 1961; Toropova, 1961; Kovach & Kling, 1967; Thoman & Arnold, 1968).

The projection of the nipple from the surrounding mound of bare skin (areola) is an important feature enabling kittens to locate and grasp the nipples. While nuzzling the mother's fur they crawl upward on her body but upon reaching the areola region their nuzzling abruptly changes to gentle nose tapping and forward crawling stops. As they contact the projecting nipple with the nose and lips, the head is withdrawn and raised and the mouth is opened; the nipple is grasped by a forward head lunge with mouth open, and often several such lunges are made before the nipple is centred in the kitten's mouth.

This is a variable response even in newborn kittens. Brooder-reared kittens develop a different pattern in adapting to the longer and more flexible rubber nipple used (Rosenblatt *et al.*, 1961). They approach in a similar way but upon making contact they position the nipple at one side of the mouth, move the head sideways, thus bending the nipple, then allow it to spring back into the mouth. If brooder nipples are placed in rounded depressions on the brooder surface, kittens find it difficult to attach to them. The reason appears to be that when probing the depression, the kittens are stimulated more strongly around the face by the edges of the depression than on the nose and mouth by the nipple. Face stimulation stimulates crawling and strong pushing into the depression but appears to inhibit a response to the nipple. The location of the nipple on a mound, as in the mother, therefore has the opposite effects: crawling is inhibited by the nose stimulation and nipple grasping is elicited.

Each kitten rapidly develops a preference for either a single nipple or a pair of nipples (Wodinsky, Rosenblatt, Turkewitz & Schneirla, 1955; Ewer, 1959, 1961; Rosenblatt, 1971). Textural differences on the surface surrounding the nipples may play a role in this discrimination, since R. J. Woll (unpublished) has shown, in brooder-reared kittens, that a discrimination between differently textured nipple flanges, one providing milk and the other without milk, can be learned by the second or third day.

In rat pups tactile stimulation of the upper lip appears to be the stimulus for nipple grasping (W. G. Hall, unpublished). Pups, whose upper lips had been desensitized, failed to grasp the nipples although they nuzzled them when placed on the nipple region of anaesthetized mothers. When they were placed on nipples, however, they suckled normally.

Once the nipple is grasped and sucking begins, young are extremely sensitive to tactile and perhaps thermal stimuli within the oral cavity. A series of postural changes during suckling has been described in rat pups, in which changing nipple stimulation and finally milk ejection stimulate first treading then stretching responses. Treading anticipates actual milk ejection and may be in response to subtle oral tactile or thermal stimuli resulting from milk let-down within the mammary gland (Vorherr, Kleeman & Lehman, 1967; Lincoln, Hill & Wakerley, 1973; Drewett, Statham & Wakerley, 1974). In human mothers, oxytocin released in response to infant crying raises breast temperature by 1 °C as a result of milk let-down (Vuorenkoski, Wasz-Hökert, Koivisto & Lind, 1969).

The likelihood of thermal stimulation playing a role in nipple grasping arises from N. Freeman's (unpublished) finding in the lactating cat that nipple temperature is higher (37 °C) than the nearby surrounding skin (34.7 °C) and G. H. Rose's (unpublished) observation that warming the nipple and flange of an artificial mother aids the newly born kitten to find and grasp the nipple. Stanley and his associates (Stanley, 1970) have shown that the newborn puppy is able to modify its sucking in accordance with the rate and pattern of milk flow, probably on the basis of oral tactile stimulation.

Suckling and intraorganic stimulation: early stage of development

The initiation and termination of suckling cannot be explained entirely by reference to exteroceptive thermal and tactile stimuli from the mother even in the early stage of development. Young of all altricial species exhibit a periodicity of feeding that appears to depend on internal sources of stimulation generally labelled, in aggregate, as hunger motivation. The most obvious during suckling is the milk which fills the newborn's stomach. It has been assumed that filling of the stomach terminates suckling as a result of stimuli arising from muscular distention or sensory receptors in the stomach, and, conversely, that an empty stomach stimulates the young to initiate suckling and to be responsive to nursing stimuli.

James (1957) was the first to show that in young puppies (i.e. less than 26 days) suckling was not inhibited by preloading their stomachs with adequate amounts of milk. Latencies to initiate suckling and the duration of sucking were not different in preloaded and non-preloaded puppies which had been separated from their mothers for up to 3 hours, although preloaded puppies consumed less milk. Further study (Satinoff & Stanley, 1963; James & Rollins, 1965) showed that preloading to the extent of overloading (i.e. milk regurgitated through the mouth) could inhibit sucking and reduce milk intake, but strong efforts had to be made with these fully preloaded puppies to keep them awake. Even though they attached to nipples, once they were released after being hand held and stimulated they released the nipples and fell asleep. Thus the stomach probably does not directly regulate the responsiveness of young puppies to the exteroceptive nursing stimuli of the mother.

Hall (1975) has recently shown this even more clearly with rat pups during the first 10 days, using the anaesthetized mother to test for suckling. Hall deprived one group of pups of suckling for 22 hours and removed a second group from the mother just before testing and anaesthetization. Both deprived and non-deprived pups were placed on the mother's exposed nipple region and their readiness to attach and suckle was measured. Their level of general activity was also measured. During the first 10 days, deprived and non-deprived pups attached with the same latencies. Latencies for attaching declined in both groups from just below 200 seconds to between 50 and 75 seconds; activity levels did not differ substantially between the two groups for most of the 10 day period. This study shows that readiness to suckle was not dependent upon any differences in stomach loading that resulted from differences in the recency of suckling.

Anaesthetized rat mothers do not release milk in response to sucking because of the inhibition of oxytocin release necessary for milk let-down. In Hall's study, therefore, pups sucked from dry nipples yet they continued to suck for the remainder of the five minute test period. In this early period, therefore, ingestion of milk is not an essential stimulus for sucking; Hall (1975) reports that pups eight days of age and younger suck for at least six hours, continuously, despite the fact that they obtain no milk throughout this period.

Similar findings have been described for kittens during the first three weeks after birth (Koepke & Pribram, 1971). Two groups of kittens were studied: one was presented with a normal lactating mother and the other with an anaesthetized non-lactating female in whom sucking did not, of course, cause milk release. Both groups were fed adequate amounts of milk by tube directly into the stomach at four hour intervals and were exposed to the 'mothers' for six hours each day, the exposure starting two hours before a scheduled tube-feeding.

During the first three weeks the dry-sucking group sucked as frequently and for as long as the group that obtained milk during sucking. Both groups sucked (i.e. were attached to nipples) for between 31 and 42 minutes of the first hour, indicating that latencies to attach were rather short and did not differ between groups. At the end of two hours, both groups were tube fed according to schedule and the effect was to reduce suckling during the third hour, equally in both groups, but not to eliminate it.

This study suggests that milk intake is not essential either for the initiation of suckling during the first three weeks or for maintaining sucking. We shall see that in puppies, rat pups, and kittens, the situation changes as the young become older and milk intake plays an important role in sucking. Yet there is a periodicity in feeding even during the early period that is not based entirely upon the mother's behaviour since it appears in brooder-reared kittens as well. Among kittens and rat pups the young sleep during sucking though they may remain attached to the nipples, and when the mother rises to leave, some are so strongly attached that they are dragged from the nest.

The ingestion of milk plays a role in sucking but its role is indirect. Following stomach filling with milk the young may be lulled to sleep, as a result of a sharp reduction in general arousal perhaps through inhibitory effects of stimuli from the distended stomach. Sucking may be terminated, therefore, or fail to be initiated or only weakly initiated, in young preloaded with excessive amounts of milk (Satinoff & Stanley, 1963; James & Rollins, 1965).

During the first 10 to 15 minutes after preloading with milk, kittens exhibited an overall lethargy that made them unresponsive to all forms of stimulation and precluded any response to a nursing mother. There was then gradual recovery and the kittens became more responsive to stimuli and often began to initiate suckling approaches with nipple grasping and sucking (J. S. Rosenblatt, unpublished).

Suckling has many intraorganic effects some of which derive from milk ingestion and others from exteroceptive stimuli received during contact with the mother. Hofer and his associates (Hofer & Reiser, 1969; Hofer, 1970, 1971; Hofer & Grabie, 1971; Hofer & Weiner, 1971) have uncovered an effect of milk intake on cardiac function. Two week old rat pups, separated from their mother overnight, show a decline in heart rate that cannot be attributed to a decline in body temperature. These investigators have shown that only tube feeding at regular intervals (i.e. one or four hour intervals) prevents this decline: Koch & Arnold (1976) have extended these findings to pups four to 14 days of age. In addition they have found in four to 10 day old pups that even non-suckling contact with the mother by tube-fed pups accelerates cardiac functioning more than tube feeding alone. Since rat pups in this age range are unable to thermoregulate, the effect may be based upon the thermal insulation provided by contact with the non-suckling female as well as contact with a source of warmth well above the pups' body temperature. Rabbit young suffer a loss in body temperature when isolated in 20–25 °C ambient temperature and contact with the mother produces an immediate rise in the young's body temperature. This suggests that suckling exerts an important influence on the body temperature of young partly through body contact and partly through the warming effects of milk; full body contact with the mother has similar or even greater effects (E. Jeddi, unpublished).

It is perhaps significant that others (e.g. Hopson, 1973) from a different perspective have proposed that maternal behaviour among mammals has evolved primarily in relation to the thermoregulatory needs of altricial young and that nursing is a secondary development which evolved later (Long, 1972).

Nevertheless a number of investigators have shown that altricial newborn of a number of species can learn discriminations of various sorts on the basis of milk reinforcement of sucking (reviewed Rosenblatt, 1971). Tactile, taste, and olfactory discriminations have been established and sucking has been 'shaped' by the presentation of milk (Thoman, Wetzel & Levine, 1968; Stanley, 1970; Stanley *et al.*, 1970; R. J. Woll, unpublished). If the effects of reinforcement are related to the underlying motivational conditions, then in these young, reinforcement must act in a different way than in adult animals with well-defined motivational systems. The nature of this difference, however, is not yet known.

Suckling and olfaction

Olfaction enters early into the suckling pattern of altricial newborns but it may be several days before it plays a significant role in the approach to the mother and in nipple searching. As early as three days of age, rat pups show evidence of responding to the odour of a lactating female or of nest deposits, and on the fifth day pups show behavioural arrest and EEG synchronization in response to maternal odour (Tobach, Rouger & Schneirla, 1967; Salas, Schapiro & Guzman-Flores, 1970; Math & Desor, 1974). By the ninth day the evidence of response to maternal odour is clear (Nyakas & Endroczi, 1970; Schapiro & Salas, 1970; Gregory & Pfaff, 1971).

Bilateral bulbectomy on the second day has a severe and irreversible effect on suckling in rat pups, resulting in a gradual weight loss beginning one or two days after surgery and 80% mortality by the sixth day (Singh & Tobach, 1974). Pup approaches to the mother when she enters the nest and interactions with littermates are only mildly affected by bulbectomy (see below). At later ages also, bulbectomy produces a severe reduction in suckling and consequent weight loss, with high mortality but mild effects on social interactions with the mother and the littermates. The effect of bulbectomy appears to be specific to nipple grasping since approaches to the mother and nuzzling in her fur are not different in bulbectomized and normal or sham-operated control pups. This suggests an early role for olfaction in nipple grasping. Kovach & Kling (1967) similarly found that bulbectomized kittens could not suckle with the mother but were able to suckle when placed on an artificial nipple.

The rabbit pup, kitten and puppy show evidence of use of olfaction in their response to the mother during the first week. Artificial odours have been applied to the mother's nipple region and later tested alone to determine whether, in the course of suckling, young rabbit pups and puppies are influenced in their behaviour by the presence of the odour. Rabbit young thus exposed to odours during the first five days of suckling began to respond positively to the odour alone on the second or third day (Ivanitskii, n.d.) and puppies exposed similarly responded immediately when tested for the first time on the sixth day (Fox, 1970).

Among kittens any of the three functional pairs of nipples may be suckled by each of the kittens of a litter during the first day *post partum*. By the second and third day, however, each kitten confines its suckling to either a single nipple or a pair of nipples (Wodinsky *et al.*, 1955; Ewer, 1959, 1961; Rosenblatt, 1971). Once nipple position preferences are established nose contact with non-preferred nipples does not elicit nipple grasping as it did on the first day. To determine whether olfactory cues could be the basis for these preferences, brooder-reared kittens were presented with two nipples and flanges, one of which provided milk while the other was blind. Two-day old kittens were tested for their ability to establish an olfactory discrimination. Within a day or two the kittens were able to discriminate between two artificial odours: they approached and suckled almost exclusively the nipple–flange combination with the positive odour.

The ability of rat pups to approach the mother on the basis of an olfactory stimulus has been clearly established by Moltz & Leon (1973) and Leon (1975). From 14 days pups respond to the odour of lactating mothers in preference to that of non-lactating females. The odour emanates from a volatile component of the mother's faeces which the young ingest, and if mothers are fed different diets the offspring can distinguish the odour of their own mother. Under these conditions, the odour of a strange mother has no special attraction for them, indicating that maternal odour acquires its attractiveness through experience.

Olfactory stimuli appear gradually to become a component of the young's approach and suckling response to the mother in these altricial species. From the beginning, olfactory stimuli are able to influence the newborn's behaviour. Ammonia fumes (which may in fact have irritating tactile effects) stimulate pivoting and vigorous crawling in newborn rats (Gard *et al.*, 1967). There are, undoubtedly, odours to which newborn respond positively from birth onward, particularly those related to birth fluids etc., but, in the main, odours appear to acquire their significance through association with thermotactile stimuli that accompany them.

With the development of olfactory responses to the mother's body and particularly to those regions involved in nursing, certain changes occur in the suckling pattern:

(1) Suckling becomes more specifically a response to the newborn's own mother and to specific aspects of the mother's nipple region (e.g. development of nipple position preferences).

(2) Since olfactory stimuli from the mother may reach the young before

actual contact is made with her, the young may initiate the suckling approach while the mother is still at a short distance and thus they may anticipate her approach.

(3) Since olfactory stimuli may be spread outside the nest or home site, familiarity with maternal odour may enable young to widen the range of their activity to these areas.

(4) Since olfactory stimuli may also be deposited on littermates, young may also begin to respond to their siblings on this basis.

Suckling and vision

At the time vision becomes functional in altricial young, the suckling pattern is well established on the basis of olfactory and thermotactile stimuli. These stimuli, however, can only be received by young when the mother is close by or in actual contact with them. Use of vision enables young to perceive the mother at a distance and approach her for suckling (Walters & Parke, 1965).

The onset of visually stimulated suckling behaviour is usually taken as the beginning of distant approaches to the mother. In the rat this begins between the fourteenth and sixteenth days, in the hamster around the second week, in the puppy around the fifteenth day, in the rabbit around the tenth to fourteenth days, and in the kitten around the seventeenth day. It must be kept in mind, however, that vision may begin to function earlier in suckling but at short distances where it would be difficult without specific examination to distinguish between a thermotactile, olfactory, or visual basis for the young's approach.

While there is considerable descriptive evidence of visually guided approaches to the mother beginning sometime after eye-opening in the species we have been discussing, experimental evidence is often lacking. In the kitten, eye-opening occurs around the seventh to ninth day and there is evidence of the onset of visual functioning between the fourteenth and seventeenth days (Karmel, Miller, Dettweiler & Anderson, 1970; Rosenblatt *et al.*, 1971). Visually guided approaches to the mother from a distance start around the seventeenth day and are fully established by the twenty-first (Rosenblatt, 1971). In the 10 day old rabbit pup (Ivanitskii, n.d.) an odour associated with the mother initiates approach to the mother which is then guided by visual stimulation. The implication of the studies on maternal pheromonal stimulation in the rat by Moltz & Leon (1973) and Leon (1975) is that, from around the fourteenth day, pups identify a lactating female by odour but their approach to her is guided by visual stimulation. Altman and his colleagues (Altman, Brunner, Bulut & Sudarshan, 1974; Bulut & Altman,

1974) have shown that rats orient to the mother and littermates in the home cage, presumably on the basis of visual supplemented by olfactory stimuli, during the third week. The beginning of visual functioning in the rat pup has been found on the fourteenth day by Turner (1935) who studied pups' ability to find an exit door, marked by a vertical black stripe, in a circular field. Pups improved rapidly from the fourteenth day onward in their ability to go from the centre of the field directly toward the door, a distance of 13 inches (33 cm).

Visually guided approaches to the mother in puppies described by Rheingold (1963*b*) as starting around the fifteenth day and increasing thereafter, are correlated with visually evoked responses in the cortex which assume the adult form at around the same age (Fox, 1968). By four weeks of age, puppies will approach a two-dimensional drawing of a female, paying particular attention to the head and flank, and at a slightly older age to the mammary region, when they are hungry (Fox & Weisman, 1970).

The onset of visually stimulated approaches to the mother marks an important stage in the young's development with implications for the organization of suckling behaviour. It is no doubt significant that the ages at which young increasingly initiate suckling approaches to the mother from a distance in the rat pup, puppy, and kitten correspond roughly with the ages at which intraorganic stimuli become increasingly important in the regulation of feeding.

If the young approach the mother when she is unresponsive or not easily available it is likely to result in either failure to suckle or in rejection by the mother. It is during this stage that maternal rejection or evasion of the young become prominent. Sucking approaches made when the mother is receptive result, on the other hand, in her cooperating with them by assuming a nursing position. Young gradually become sensitive to subtle features of the mother's behaviour indicating her readiness or unwillingness to nurse and adjust their suckling approaches accordingly. Moreover, each young approaches the mother to suckle individually, whereas earlier the mother's approach had stimulated the entire litter simultaneously and all young suckled together. Once nursing is initiated with one offspring other young often join it, but each young acts on the basis of its own perception of the situation rather than to the stimuli that initiated suckling in its sibling.

Suckling and intraorganic stimulation: late stage

Between the tenth and fourteenth day an important change takes place in the intraorganic stimulation of suckling among rat pups (Hall, 1975). Around this age the latency to initiate suckling (with an anaesthetized mother) begins to differ between pups that have been deprived of suckling for 22 hours and those that have only recently been removed from their mothers. Latencies are very short for the deprived pups but range above 200 seconds for recently fed pups. Moreover, failure to receive milk from the anaesthetized mother, which had little effect on maintenance of suckling earlier, now begins to affect suckling: pups shift from one nipple to another, within the five minute suckling test, presumably in response to the dry nipple.

Similarly, around the third week of age, kittens suckling from a dry mother initiate nuzzling of the mother's nipples with the same frequency as kittens suckling from a lactating mother, but the frequency of attaching and sucking declines rapidly from the earlier period and the duration of sucking also declines to 20% of the previous period. Meanwhile kittens suckling from a lactating mother continue to suckle for extended periods (Koepke & Pribram, 1971).

After 30 days of age, preloading the stomach of puppies with a milk–chow mixture has the same effect of reducing or eliminating further eating from a dish as the same amount of food consumed voluntarily by control puppies. This contrasts with the earlier finding that preloading the stomach with either milk alone or a similar mixture has no effect on suckling (James & Gilbert, 1957; James, 1957, 1969; James & Rollins, 1965).

Indications are, therefore, that arousal produced by intraorganic stimulation has become more specific than earlier so that exteroceptive stimulation received before or during suckling is not sufficient to initiate or maintain suckling. Milk intake and its intraorganic effects have become an essential part of the suckling pattern. At this stage of development, stomach filling or its absence has become a more focal source of arousal than earlier and it determines to a larger extent the young's response to the exteroceptive stimuli associated with suckling.

The kinds of exteroceptive stimuli that arouse suckling approaches undergo a change during this period. Among rat pups this is the period during which suckling approaches are increasingly initiated by the pups from a distance, in response to olfactory and visual stimuli from the mother. Among kittens there is the beginning of kitten-initiated suckling approaches to the mother, starting around three weeks of age. At this stage suckling approaches are increasingly determined by the past experience which the young has had with the mother and less by immediate stimuli for suckling.

This was shown in a study by Kovach & Kling (1967) in which they found that kittens reared in isolation from an early age until about three weeks and fed by stomach tube to prevent suckling, were unable to initiate suckling when returned to their mothers. While they attributed this to the waning of the sucking reflex, our analysis indicates that at three weeks of age the sucking reflex is only a small part of the suckling pattern. Moreover, as part of an organized suckling pattern it is initiated mainly on the basis of intraorganic stimuli rather than the previously effective proximal exteroceptive stimuli.

In a study comparable to that of Kovach & Kling (1967) kittens were reared in isolation from one to three weeks of age with feeding by suckling from an artificial rubber nipple mounted on a brooder (Rosenblatt *et al.*, 1961). When these kittens, in whom suckling had been maintained as an active component of the feeding pattern, were returned to their mothers there was a delay of about 20 hours in initiating suckling. The previously isolated kittens had no difficulty in establishing sustained contact with the mother, which in all cases took less than three hours and in most kittens was accomplished within the first hour. Their difficulty lay in perceiving the mother as an object to be suckled. They crawled over her body, slept in close contact with her nipple region and often nuzzled, but not until they had been without milk for many hours did they finally exhibit nipple grasping and sucking. In these brooder-reared kittens the suckling pattern had been organized in relation to the brooder and artificial nipple, which differed in many ways from the mother. Under comparable conditions in the brooder situation, these kittens would initiate suckling within seconds after being returned to it.

As young develop, therefore, the relative influence of exteroceptive and intraorganic factors in the regulation of suckling shifts in favour of intraorganic factors. Moreover arousal undergoes changes: it can no longer be described simply in terms of general arousal. Arousal has become more specific as a result of experience in the litter situation and of neural maturation which gives rise to new physiological relationships between visceral and somatic sensorimotor processes. When aroused, young are channelled into particular patterns of behaviour not only by the prevailing exteroceptive stimulation, as earlier, but by intraorganic factors among which previous suckling experience plays a crucial role.

DEVELOPMENT OF HUDDLING AND RELATED BEHAVIOUR

Newborn huddle from birth onward: after nursing has been completed and the mother leaves the nest or home site the young crawl into contact with one another and form a closely knit huddle. Since heat loss is dependent upon body surface exposure to the environment in young with poor thermoregulation, the huddle maintains body temperature and metabolic activity (Cosnier, 1965; Alberts, 1974). Huddling is an early form of social behaviour among littermates that may be closely related in development to later forms of group activity (e.g. play).

Thermotactile basis of early huddling

Cosnier (1965) and Alberts (1974) have shown that early huddling among rat pups is based upon thermal and tactile stimuli provided by littermates. Pups up to 10 days of age are more attracted to a warm live pup (i.e. 37 °C) than a cool dead one (25 °C) but the warm pup can be replaced by warm tubing or a warm fur-lined nylon sleeve (Cosnier, 1965; Alberts, 1974). Pups are particularly stimulated to huddle when the surrounding floor is cool (Cosnier, 1965).

James (1952*a*, *b*), Welker (1959) and Crighton & Pownall (1974) analysed the behaviour of individual puppies, when separated from their littermates in warm and cool environments, in an attempt to understand the thermotactile basis of huddling. Separation stimulates pivoting and vigorous crawling until the puppy makes head contact with its littermates. The essential feature of the contact is thermal, however, since contact with cooled puppies or contact with littermates in an overly warm environment elicits withdrawal rather than huddling. Within the huddle temperature is maintained at 30 °C and puppies remain quiescent, breathing quietly and in a relaxed sleeping posture. Warming the huddle to 32.5 °C causes dispersal as each puppy responds to contact with its neighbour by withdrawing; cooling the huddle intensifies huddling as each puppy responds to the warmth of its neighbour by approaching it. The greater thermotactile sensitivity of the face and snout over the remainder of the body assures that puppies face into the huddle. The few observations of kittens indicate that similar factors operate.

The role of tactile stimuli in huddling receives some confirmation from a study by Scott *et al.* (1974) in which puppies that were isolated in a warm environment but nevertheless vocalized were induced to become quiet by confining them in a section of stove pipe the floor of which was lined with soft material. The tactile stimulation, and perhaps the insulation provided, are similar to that provided by neighbouring puppies during huddling.

Huddling and olfaction

Singh & Tobach (1974) found that bulbectomized rat pups, from the second day onward, tended to be apart from the litter for longer periods than their normal littermates, although in general their social behaviour was not particularly deviant. The suggestion that olfaction plays a role in huddling was more clearly established by Alberts (1974). Using intranasal infusion of zinc sulphate to make pups anosmic he found that huddling was completely eliminated in pups 10 days of age and older, but was retained by five day old pups.

Fox (1970) has shown that olfaction may begin to play a role in the puppy's response to its littermates as early as the first week after birth. An odour applied to the mother's body, and therefore spread to the littermates, elicited strong approach responses in puppies exposed to the odour during the first five days of life, but strong avoidance responses from puppies presented with odour for the first time on the sixth day. Similarly, among rabbits, maternal odour, which is shared by the littermates, elicits a strong searching response when held close to the young's nose without tactile contact as early as the second to fifth day after birth. Evidence suggests also that hamster pups may begin to respond to litter odours at the end of the first week (Devor & Schneider, 1974).

Orientation to littermates and vision

Huddles become somewhat less compact as young develop thermoregulation and are less in need of insulation from heat loss. Nevertheless they continue to aggregate in groups which may consist of several young instead of the entire litter as earlier. Of greater interest is the fact that they orient to littermates from a distance suggesting that vision is beginning to play a role.

As late as 17 to 20 days of age, about a week after eye-opening, puppies still do not orient to littermates in the home cage from a distance. This does not mean that they do not react to them but they continue to show circular movements similar to those seen earlier. Starting around 25 days of age, however, they begin to approach littermates from a distance directly (James, 1952*b*). Rheingold & Eckerman (1971) showed showed that kittens respond to littermates at a distance even when they do not orient directly to them. They studied the earliest age at which kittens would be comforted by the presence of a littermate in a strange environment. At two weeks of age, the earliest age at which testing was done, kittens emitted 50% fewer vocalizations when a littermate was present compared with when it was absent.

Despite the fact that only sketchy evidence is available to trace the development of huddling and other relationships with littermates among these altricial species, the main outlines of this development are clear. Stages in the development of responses to littermates are similar to those found in the development of suckling: there is an initial stage of dependence upon thermotactile stimulation, a later stage of olfaction based responses and a final stage of visually guided responses to littermates. Moreover, early responses to littermates are dominated by exteroceptive stimuli while later stages show evidence of an increase in self-initiated approaches to siblings based upon the developing social relationships. Huddling is a relatively simple behaviour pattern compared to suckling, but it represents only the earliest form of interaction among littermates that ultimately develops into the complex patterns of play.

DEVELOPMENT OF HOME ORIENTATION

The nest or home site is the socially conditioned environment in which the young undergo the early stages of their development. The young, from an early age, develop a pattern of home orientation that plays an important role in their relationship to the mother, and the organization of their behaviour in relation to their surroundings and littermates. Since the original discovery of home orientation in the kitten (Rosenblatt & Schneirla, 1962; Rosenblatt *et al.*, 1969) orientation to the nest site has been found in the hamster, rat, and puppy, in one form or another, suggesting that further research will reveal that this is a widespread and basic phenomenon among altricial newborn.

Thermotactile basis of home orientation

In all altricial species the nest or home site is usually provided with a soft insulating material and is warmer than the surrounding regions, largely as a result of warmth from the mother's body, the huddling young, and the insulation. Tobach *et al.* (1967) found that three day old rat pups were more active in shavings, the usual nest material used in laboratories, than on the surrounding bare floor, on which they remained immobilized. The soft material used to line the nest site in many species is likely to be more attractive to newborns than coarse material because of its tactile properties, as the studies on newborn puppies by Stanley *et al.* (1970) have shown. While these tactile properties are attractive they are of little use in orienting to the nest from a distance; they function mainly to keep the newborn in the nest or to bring to rest a newborn that has been wandering.

The situation with respect to the thermal stimulation provided by the nest or home site is quite different. N. Freeman (unpublished) measured the floor temperature at various regions of the home cage of a mother cat and litter just after they were removed in preparation for testing kittens in home orientation. The floor temperature of the home site was 35.3 °C at its centre, 33.6 °C at its border, about 25 cm away from the centre, and it decreased from 25.2 to 23 °C at a distance of 15 cm. This thermal gradient persisted for some time after removal of the litter, but, of course, if the mother and litter remained in the home region it would be maintained constantly.

On the basis of the thermal gradient kittens could distinguish the home from the adjacent corner when placed there during a test. They vocalized loudly and moved about actively in the adjacent corner while they soon became quiet and fell asleep when placed in the home site. Odours are not involved in this response initially, since the thermal gradient produced by a heat lamp on a floor surface that had not been occupied by the mother and litter or any other cats had the same effect.

The existence of the thermal gradient between the home site and the adjacent corner also enabled kittens of a slightly older age to crawl to the home from this outlying region. Also, when tested on a thermal gradient ranging from 15.6 °C to 48.3 °C over a distance of 56 cm, kittens up to the age of one week or so, crawled in the direction of the higher temperature if they were placed at a region below their preferred temperature.

Similarly, starting at one day of age and continuing through the eighth or ninth day, hamster pups orient to the warm end of a thermal gradient, crawling distances as great as 45 cm in one minute (Leonard, 1974). Okon (1971) has shown that hamster pups are distressed in an ambient temperature of 28 °C and emit audible and ultrasonic vocalizations. They also become responsive to a thermal gradient and crawl rapidly to the region of their preferred temperature of 33 °C.

The age at which thermal orientation wanes in hamster pups is variable and depends upon ambient temperature. As pups become older (i.e. 11 days and older) cooler temperatures are required to stimulate thermal orientation: at warm ambient temperatures, still below their preferred temperature, thermal orientation may wane as early as the sixth day. It is clear that thermal orientation is based in part upon a contrast between the pup's body temperature and the thermal gradient and also between the floor surface under the pup and the floor temperature of a neighbouring region (see also Cosnier, 1965).

In rat pups the ability to orient to a thermal source develops over the first two weeks with an important change occurring around the end of the first week (Fowler & Kellogg, 1975). Until that age pups react to contact with a thermal source by remaining in contact with it but they do not orient to it from a distance. This is in agreement with Cosnier's (1965) findings, based upon tests in which pups were placed directly in contact with the heat source. Between birth and the fifth day, pups become increasingly active when the ambient temperature is reduced. If they make contact with a warm area they come to rest but there is no indication that they orient to the warm area from a distance. Alberts (1974) reported that wandering in response to cool ambient temperatures was the basis of huddling among five day old pups and Fowler & Kellogg (1975) have confirmed this. After the fifth day, pups will orient toward a heat source provided they can detect it from a distance through a thermal gradient.

Fowler & Kellogg (1975) found that during the first five days pups exposed to an ambient temperature of 21–23 °C for one hour declined in body temperature to between 28 to 29 °C, indicating failure of thermoregulation with respect to the usual body temperature of pups at an older age (i.e. 36–37 °C). Although body temperature declined during this early period, thermal orientation was not evident until after the fifth day.

To test thermal orientation Fowler & Kellogg (1975) placed rat pups between two chambers only one of which was warmed. Between the starting point and the warmed chamber, located about 15 cm away, floor temperature increased from 23 to 36–37 °C with a 1 degC increase as close as 5 cm from the starting point. The unheated chamber was at 22 °C, the ambient temperature during testing. Thus, in effect, the pups followed a rather steep thermal gradient (i.e. 14 °C over 14 cm) in crawling from the starting point to the warmed chamber. In a less steep thermal gradient pups are able to turn in the direction of the warmed floor surface but are unable to follow it for any appreciable distance (M. Grabon & J. S. Rosenblatt, unpublished).

Thus orientation to the nest or home site arises early in development among altricial newborn: they respond to lowered temperature (within limits) by calling and increased locomotion. At the earliest age they are poorly equipped to respond to thermal gradients but they do respond to entering or being placed directly in a warmed region: they become calmed, vocalization ceases, they come to rest and usually fall asleep. The activation caused by cooling is reduced by warmth. At a slightly older age they respond to cooling by adopting a path along a thermal gradient, leading to the warmer region. This ability improves over the next period but, as we shall see, it often gives way to orientation based upon olfaction. However, the calming effect of warm temperatures remains an important feature of the nest or home site beyond the period when orientation to these regions is based upon thermal gradients.

Home orientation and olfaction

The nest or home site differs from the surrounding regions in the odours deposited there by the mother and litter. These too provide a basis for orientation to the home site. Olfactory orientation to the home site develops after thermotactile orientation has been established and it may be based upon this earlier form of orientation.

Towards the end of the first week, a warmed floor surface is not sufficient to quiet kittens and a thermal gradient between two cage regions, lacking home cage odours, does not result in kittens adopting a path from the cooler to the warmer region. Only the presence of home cage odours quiets kittens and only if the home site has the usual odours and, presumably, an olfactory gradient, can kittens find their way back to it (Rosenblatt *et al.*, 1969; Rosenblatt, 1971; N. Freeman, unpublished). From the seventh day onward, home orientation in kittens is based mainly upon olfaction, although at the home site, the combination of olfactory and thermal stimuli is more effective than either alone.

In the hamster thermal orientation declines as orientation based upon odours from the nest becomes established (Devor & Schneider, 1974). Starting around the seventh and eighth day, hamster pups turn toward nest shavings and away from fresh shavings when placed on a screen above the border between the shavings. They spend 80% to 100% of their time crawling in the nest shavings. Shavings from nests inhabited by a mother and litter are preferred to shavings from nests inhabited by either a male or a non-pregnant female, and nest shavings from four day old litters are as effective as shavings from older litters.

Earlier, hamster pups can distinguish between odours, preferring, for example, the odour of cedar to pine shavings; this appears on the third day. Nest odours must require an additional period of four or five days to become associated with thermotactile stimuli from the mother and littermates to provide a basis for orientation to the nest (Devor & Schneider, 1974).

As indicated earlier, rabbit pups and puppies begin to respond to odours from the mother before the end of the first week and it is likely that the same odours deposited in the nest would provide a basis for home orientation. It is clear that the odour acquires its attractiveness through association with other stimuli from the mother during the first five days (Ivanitskii, n.d.; Fox, 1970).

Home orientation in puppies that appears to be based upon olfaction has recently been reported by Scott *et al.* (1974). Puppies were placed either in their own nest boxes, emptied of the mother and littermates, or in a similarly constructed nest box that was lined with fresh towelling; vocalizations were recorded in each. One precaution was taken, which, however, may have affected the results: to exclude the possibility that puppies might respond to the thermal difference between the litter nest box and the unused one, both boxes were warmed to about 29 °C before testing. Despite this, vocalizations were more frequent in the unused nest box than in the litter nest box from around the sixth to eighth day in several litters and on the eleventh–twelfth day in all litters. Vocalizations were infrequent in both conditions at first but they rose sharply after the twelfth day in the unused nest box.

The distress caused in puppies by placing them in a strange nest box is similar to the distress caused in kittens by placing them in a strange cage: the rate and intensity of vocalizations rises during tests, and with age they become more frequent and more intense. The locomotory behaviour of the puppies in the above study was not described but, in an earlier paper (Elliot & Scott, 1961) with older puppies, vocalization was accompanied by an increase in activity in a strange pen and presumably the same occurred here.

Rat pups exhibit the first signs of orientation to the home cage from a distance of less than 30 cm on the third day and even more clearly on the fifth and sixth days (Altman *et al.*, 1974). Placed on a 11.4 cm diameter circular platform, located midway between the home cage and a strange cage, pups point themselves toward the home during most of a three minute test but they are not yet able to crawl to it. Between the ninth and twelfth days pups become better able to locomote over long distances by crawling (Gard *et al.*, 1967; Altman *et al.*, 1971). On the ninth day, pups placed in a neighbouring cage reach the home cage in 50% of the tests by passing through a narrow alley, with latencies ranging from two to three minutes (Altman *et al.*, 1974). Latencies to traverse the 15 to 30 cm distance between the starting chamber and the home cage become shorter on the twelfth day (i.e. one to two minutes) and 85% of the pups are successful and on the sixteenth day, around the time of eye-opening, nearly all pups orient to the home with latencies ranging from 30 to 40 seconds. At 19 and 21 days latencies are 10 seconds or shorter.

Neither the paths taken to the home cage nor the sensory cues utilized by pups were analysed in the above studies. Turkewitz (1966) reported that successful orientation from a neighbouring cage to the nest site in the home cage by 9 to 12 day old rat pups was accomplished indirectly through wall-hugging. The long latencies to reach the home cage in the above studies suggests that a similar mode of home orientation occurred at 12 days of age. Old pups, however, appear to adopt direct paths to the home cage and as a consequence latencies are considerably shorter.

On the basis of Turkewitz's (1966) study, it would appear that olfactory and perhaps thermal stimuli are involved in home orientation by rat pups before their eyes open on the fifteenth to seventeenth day. The long latencies suggest pups adopt an indirect path to the home and this is more characteristic of young that orient on a nonvisual basis: the shorter latencies after eye-opening certainly contrast with the longer latencies earlier. Findings reported earlier, that rat pups show evidence of olfactory discrimination of nest shavings as against fresh shavings around the ninth day and respond to maternal odours by the fourteenth day, would also support this interpretation of the above home orientation (Nyakas & Endroczi, 1970; Schapiro & Salas, 1970; Gregory & Pfaff, 1971; Moltz & Leon, 1973; Leon, 1975).

Home orientation and vision

The onset of vision marks the beginning of the decline of home orientation among altricial young. With the ability to view the home from a distance it is no longer necessary to return to it; instead young tend to remain in the vicinity of the home or nest site for a period before they disperse. Moreover, orientation to the home declines also because the young begin to follow the mother and littermates and these become centres of orientation rather than the nest or home site. There is, however, a short period after eye-opening and the beginning of vision when home orientation continues and vision is used to enable the young to return to the home from a distance.

In the rat, as we have seen, the latency for reaching the home from a nearby region is reduced to a few seconds shortly after eye-opening (Altman *et al.*, 1971; Bulut & Altman, 1974). This indicates that the young begin to take direct routes to the home, and this probably depends on visual stimulation (Turner, 1935). The hamster pup shows a decline in olfaction-based orientation beginning around the thirteenth day, shortly after eye-opening, and it is completely absent by the nineteenth day. This may be based upon the gradual increase in visually based orientation.

Among kittens the use of vision in home orientation begins around the fourteenth day (Rosenblatt *et al.*, 1969; Rosenblatt, 1971). Testing kittens in the dark results in some decrement in performance with indications that the kittens are seeking the visual stimuli to which they are accustomed. However they are still capable of using olfaction and most of them reach home. If olfactory cues are removed, some kittens are still able to reach the home and this increases after the seventeenth day.

The use of vision in home orientation among kittens eventually leads to the decline of this behaviour at around the eighteenth to twenty-first day. Instead of returning to the home region, kittens look toward it, or they may actually enter the home region briefly then leave it. The ability to see the home often enables kittens to wander more freely around the entire cage since the home region is always in view. However, when the home was made especially prominent visually by placing a fur piece there, home orientation persisted until 45 days of age. With home odours absent, kittens began to show this visually based home orientation around 25 days of age and gradually perfected it, traversing the cage, a distance of about 76 cm, in less than 20 seconds as they moved directly toward the home and came to rest after entering it (J. S. Rosenblatt, unpublished).

STAGES IN THE EARLY BEHAVIOURAL DEVELOPMENT OF ALTRICIAL YOUNG

The early behavioural development of altricial young can be divided into three stages: in the following section we shall analyse these stages and the transitions from one stage to the next.

First stage: thermotactile stimulation

The neonates' earliest behaviour is organized predominantly in relation to low intensity thermotactile stimulation provided by the mother, littermates and nest or home site. These elicit from the newborn either general or specialized approach responses (Schneirla, 1959; 1965). The general response of forward crawling is elicited by centrally applied head and face stimulation over a broad area and turning is elicited by laterally applied stimulation, through the close relationship that exists between anterior end stimulation and movement of the forelimbs in an alternating paddling motion and the hindlimbs in a joint pushing action. Specialized responses are elicited by localized stimulation of the snout, mouth, and lips; the specialized responses take their character from the peripherally organized pattern of movement characteristic of the local region and the restricted locus of stimulation.

The thermal and tactile characteristics of the mother, littermates and nest or home site provide patterned thermotactile stimuli which elicit from newborn the general and special approach responses necessary for the initiation of suckling, and huddling, and for remaining within the confines of the home. These responses are only loosely patterned sequences at first, depending upon kinds of stimulation the newborn encounters in succession as it crawls forward. Stimulated initially to approach the mother by contact with the attractive thermo-tactile stimulation of her body, the newborn subsequently encounters the more detailed stimuli of the mother's fur around the nipple, the areola, and finally the nipple itself.

At this early of development the central control of behaviour is not yet strongly developed beyond the mediation of the admittedly complex sensorimotor integrations underlying general and specific approach and withdrawal responses to stimulation. It is doubtful whether, for at least a short time after birth, the functionally distinct behavioural responses to the mother, littermates and nest or home site are represented by equally distinct central regulatory processes. The underlying similarities between the behavioural adaptations to the mother, littermates, and nest or home site, arise from the fact that they share in common responses to similar thermotactile stimuli. The observable differences arise from the fact that in each of these responses the neonate also responds to specific stimulus properties of the mother, littermates, and the nest or home site.

While the relative contributions of central regulatory processes and peripherally elicited responses favour the latter in the neonate (as compared with older animals in which central regulation plays the dominant role) central processes are not without effect on the neonate's behaviour. Exteroceptive stimuli (i.e. thermotactile) appear to have two concurrent effects upon the neonate. They arouse the neonate to activity, as when the mother's licking activates the sleeping neonate to raise its head and begin to crawl. Once the newborn is aroused, stimuli also elicit approach responses of a general and specialized nature. Thermotactile stimuli are therefore both arousal inducing and response-eliciting. Neonates that have been aroused are more sensitive to exteroceptive stimuli that follow and exhibit more vigorous responses to these stimuli. The central component of exteroceptive stimulation therefore is an arousal that in turn potentiates the responsiveness of peripheral processes.

Early in development central arousal is relatively non-specific in its afferent and efferent relationships and this contrasts with the specificity of the peripheral responses that are elicited by thermotactile stimuli. It is this contrast between the greater specificity of peripheral responses – their direct relationship to thermotactile stimuli – and the non-specific character of arousal and its contribution to early behaviour, that has led many to characterize the neonate's behaviour with some justification as simply reflexive in nature (Scott, 1958; Kovach & Kling, 1967).

Early suckling among newborn rat pups and puppies exemplifies the relationship between central and peripheral processes in the regulation of feeding. The compelling effect of thermotactile nipple stimuli in eliciting suckling is shown by the fact that prior feeding or stomach loading cannot prevent the young from grasping the nipple and sucking. In the rat pup, if milk is not forthcoming sucking may continue indefinitely (Hall, 1975); suckling does not continue indefinitely in the kitten but its duration is not shorter than that of kittens that obtain milk (Koepke & Pribram, 1971). This difference may reflect the fact that the rat pups were tested with anaesthetized mothers while the kittens were tested with awake, non-lactating females.

Yet normally rat pups and kittens do not suckle interminably once they attach to nipples. After a period they began to loosen their grip on the nipples and slide off, or they are detached from them when the mother rises to leave. The intake of milk has an effect on the peripherally organized response of sucking, but the effect appears to be mediated by a lowered arousal, to the point of sleep. All overt activity ceases, not only suckling, and the newborn's behaviour alternates between sleep and feeding for a considerable period after birth.

Preloading of newborn with milk has an effect on suckling when the amount preloaded is sufficient to induce sleep (Satinoff & Stanley, 1963; Stanley, 1970). With effort, even under these conditions, sucking can be elicited by bringing the newborn's mouth to the nipple but the central effect of intraorganic stimulation arising from the distended stomach or sensory receptors in the stomach, usually prevails and the newborn releases the nipple soon after grasping it. Preloading also reduces milk intake but the basis for this is not clear.

Varying levels of central arousal appear to influence the responsiveness of newborns to exteroceptive stimulation: low-level arousal, that characteristic of newborn shortly after feeding, dampens the effect of exteroceptive stimulation, while very high levels of arousal, following long periods without food or upon exposure to low ambient temperature, produces an excessive response to exteroceptive stimulation. There is therefore an optimal level of central arousal at which the newborn exhibits its typical responses to exteroceptive stimuli.

The sources of central arousal in the newborn are interoceptive as well as exteroceptive, as studies on stomach preloading indicate. These sources have not been studied to any great extent, but interoceptive stimuli have a history in prenatal ontogeny that predates exteroceptive stimuli (Schneirla, 1965). Exteroceptive stimuli may in fact exert their influence, in part, through their effect upon interoceptive processes, as for example in responses to ambient thermal stimulation.

During the earliest postnatal stage there is a close relationship between those exteroceptive stimuli (i.e. thermotactile) to which newborn are most responsive in their behaviour toward the mother, littermates, and nest or home site, and the conditions which stimulate it to activity. Thus, newborns are stimulated by loss of contact with the mother or littermates and by displacement from the nest or home site; in the latter cases it is probably exposure to lower ambient temperatures than are normally present in these situations which stimulates the newborn as studies on kittens have shown (N. Freeman, unpublished). At this early age many different objects having attractive thermotactile properties may be equivalent in their arousal and calming effect upon newborn because the special properties of the mother, littermates and nest or home site do not yet play a role. This fact has allowed investigators to ignore the social nature of the newborn's responses to species mates and to the socially conditioned home and to introduce special criteria for when the young's responses are to be considered social (Rheingold & Eckerman, 1971; Scott *et al.*, 1974). Since the usual criteria are based upon visual responses to species mates, responses that appear during the third stage of early development are more likely to be called social responses and social behaviour is therefore viewed as a later development. Such a view ignores the ontogenetic basis of social responses and their relationship at each stage to the behavioural capacities and behavioural organization of the young (Schneirla & Rosenblatt, 1963).

The earliest changes in the newborn's behaviour are the formation of extended action patterns, incorporating the earlier hesitating and variable crawling approach response and the specialized feeding and huddling responses into more smoothly coordinated patterns. In brooder-reared kittens, after a day or two of feeding, the kitten crawls to the correctly textured flange (i.e. that associated with a nipple that gives milk) in a smoother more patterned movement, with short pauses at crucial points such as the edge between the brooder and the floor and between the brooder and cover and the textured flange. At these points the kitten sniffs the brooder cover and rubs its snout against its surface as it does when it reaches the flange. A short period of flange contact is followed by nipple localization and nipple grasping and sucking (R. J. Woll, unpublished).

Among puppies approaches to the nipple can be associated with either soft- or hard-textured paths with gradual improvement in turning toward one or the other and heading rapidly for the nipple, grasping it and sucking (Stanley, 1970). Repeated experience with either warm or cool thermal stimuli along alleys leading to a nipple results in a gradually more rapid crawling approach to either (Bacon, 1973*b*).

The formation of these specialized action patterns based upon thermotactile stimuli indicate progress along several lines: approach responses become specialized in relation to the young's discrimination among thermotactile stimuli. In this respect approach responses are formed more rapidly when a low intensity, approach-eliciting stimulus is the positive one than when a stimulus that is initially either weakly approach-eliciting, or is actually withdrawal-eliciting, is used as the positive stimulus (Stanley, 1970; Bacon, 1973*b*). In addition, components of the approach response are integrated into a pattern that appears to be less dependent upon continuous guidance by exteroceptive stimulation and more dependent upon central regulation. In suckling approaches, striving toward the nipple indicates an early appearance of anticipatory responses in advance of actual contact with the nipple region.

These changes channel the earlier general arousal along specific lines. Thermotactile stimuli are no longer equally effective in eliciting approach responses: the newborn turns away from certain thermal stimuli and becomes highly excited when it makes contact with others. There are indications in the study of Bacon (1974) that negative stimuli do not acquire inhibitory effects at this stage but rather that positive stimuli acquire heightened arousal effects. These may have their origin in the heightened arousal that occurs when the young are stimulated by milk during sucking (Stanley, 1970).

At this stage, new sources of disturbance may arise when the usual incentive is absent or is altered; this is in fact evidence that action patterns of a broader nature already exist in the young and that they are highly specific to the stimulus conditions in which they were formed (Papoušek & Papoušek, 1975).

Second stage: olfactory stimulation

Although in altricial young the second stage of early behavioural development is characterized by the growing influence of olfactory stimulation, its main feature is the increasing specificity of behavioural responses in relation to the familiar objects in the environment. The young use olfactory stimuli to differentiate between familiar and unfamiliar objects.

The initial exposure to olfactory stimuli occurs during responses to thermotactile stimuli, but in the beginning olfactory stimuli can provide little guidance to the newborn. Except when it encounters strong aversive olfactory stimuli, which induce withdrawal responses (Fox, 1970), olfactory stimuli cannot elicit either general approach or specialized responses in the newborn. Not until responses have begun to become organized into specific action patterns based upon thermotactile discrimination, and the central arousal processes are channelled, can olfaction begin to play a role. The suggestion is that olfaction advances further the process of discrimination among objects, along lines that are necessarily specific to the mother, littermates and nest or home site. More important, however, is the ability of olfactory stimuli to arouse the initiation of action patterns that are then guided by combined olfactory and thermotactile stimuli. Olfactory stimuli may therefore determine whether or not a response will occur.

There is associated with the onset of olfaction the appearance of olfactory orientation behaviour by means of which young explore their olfactory environment. Welker (1964) has described the development of sniffing as an olfactory exploration pattern in the young rat, and Komisaruk (1970) has added important details to the analysis of olfactory exploratory behaviour in this species. When placed in a new environment young initiate non-specific olfactory exploratory behaviour which soon gives way either to specific behaviour patterns upon identification of the odour or to further tactile exploration with the vibrissa. These exploratory patterns range from air sniffing to sniffing of the floor surface and approach and sniffing of objects in great detail. Thus, for example, by sniffing the floor surface kittens find their way to the home region and rat pups by sniffing the air are able to locate the mother at a distance of more than 30 cm (Nyakas & Endroczi, 1970; Moltz & Leon, 1973; Leon, 1975). Depending upon the distribution of odours, therefore, olfaction enables young to develop distance perception of objects and to initiate movements toward objects before actual contact is made with their thermotactile properties. In this sense, therefore, olfaction plays a large role in developing central control of action patterns begun with respect to thermotactile stimuli. Anticipation of forthcoming stimulation and the associated action pattern, exhibited during the latter phase of the first stage of development, is gradually transformed into self-initiated approaches to objects with anticipatory action patterns almost entirely centrally aroused, ready to be performed when the object is reached. Rat pups deprived of olfaction during this phase appear much less oriented to significant social stimuli and to the nest site (Singh & Tobach, 1974). The stage is set during this second stage of early behavioural development for the incorporation of vision into the central control of action patterns during the third stage of early development.

As olfaction begins to contribute to central arousal processes and the action patterns to which they give rise, it begins to play an increasing role as a motivating condition and as a possible source of distress. Rat and hamster pups, kittens and puppies begin to show distress when removed from their familiar olfactory environment. In kittens and puppies this is evident at the end of the first week and it appears at a slightly older age in rat and hamster pups; it provides the basis for these young to initiate movements toward littermates, resulting in huddling, and toward the home or nest site during the development of home orientation. Moreover, olfaction serves as a goal for these behaviours in the sense that upon reaching the familiar olfactory situation in the huddle or in the home region, young are calmed and soon come to rest. It is apparent, therefore, that not only perceptual and motor processes become more complex as development proceeds but motivational processes share in this growing complexity.

Third stage: visual stimulation

Vision greatly enlarges the young's capacity to differentiate perceptually between the significant objects in its environment. Neither thermotactile nor olfactory stimuli are able to convey to young the specific actions of their mother or littermates and the specific location of their nest or home site. While olfaction may indicate that the mother is nearby it does not indicate what she is doing, while vision conveys both. This stage is marked by an acceleration in the young's development of social interactions based upon vision. Initially social interactions consist of visual approaches to species mates at a distance, at which point action patterns based upon olfactory and thermotactile stimuli take over: vision therefore contributes little to the character of the interactions but does contribute to their occurrence. Thus, for example, kittens at three weeks of age, approach the mother at a distance for suckling but when they reach her, they adopt their earlier mode of nipple searching with their eyes closed. Gradually, however, vision comes to trigger not only the approach to the mother, but also those action patterns that were formerly triggered by non-visual sensory systems. Kittens at this stage, approach the mother and reach up from beneath her as she remains standing, locating and grasping a nipple without any significant preliminary nuzzling. Brooder-reared kittens walk directly to the nipple and grasp it instead of following a path of nuzzling on the brooder surface, and cage-reared kittens walk directly across the cage from the diagonal corner to the home region instead of crawling along a path that passes through the adjacent corner (Rosenblatt *et al.*, 1969).

Vision accelerates the process of increasing central regulation of behaviour and with this furthers self-initiated behaviour with well-defined goals. Perceptual differentiation and the multiplication of action patterns, and their growing complexity, imply highly specific central states of arousal and complex interactions, both facilitating and inhibitory, among these different states.

During the period of transition from olfactory to visual control of behaviour in the young, interoceptive stimulation appears also to become more influential and specific in its effect upon the central state of arousal. At this age, Hall (1975) found that food-deprived and non-food-deprived rat pups began to differ in their suckling behaviour, the former being faster to initiate suckling with an anaesthetized mother and abandoning suckling if no milk was forthcoming. Similar findings have been reported in three to four week old puppies and kittens whose suckling approach to the mother was influenced by stomach preloading with relatively small amounts of milk (James & Gilbert, 1957; J. S. Rosenblatt, unpublished).

The goal-directed character of behaviour at this stage is exemplified by the persistence kittens show in following the mother around the cage, while they look for an opportunity to suckle. At the slightest pause, they immediately reach up to her nipples and if they are detached by her movement they resume following her. At other times, when tired, they walk across the cage to join another kitten that has settled in a corner to sleep and huddle against it, falling asleep. While they remain highly responsive to environmental stimulation, their behaviour is not directly elicited by this stimulation: rather, they are capable of making perceptual discriminations among the various stimuli and which stimuli they respond to depends upon the central state operative at the time.

During this third stage of development kittens and puppies that are placed in novel environments exhibit signs of distress (i.e. vocalization and agitated movement) that are relieved when littermates or the mother are placed in the same environment (Rheingold & Eckerman, 1971; Scott *et al.*, 1974). Rat and hamster pups orient to the mother and littermates when displaced to a neighbouring, strange cage (Altman *et al.*, 1974; Devor & Schneider, 1974). Thus the absence of familiar visual stimuli becomes a source of disturbance that motivates young to regain visual contact with social companions. Often the experimenter serves a similar role if he has become familiar to the young. Since visual stimuli are likely to be a principal source of social stimuli from this age on into adulthood, and social responses of a clearly recognizable nature are particularly evident during this third stage, this stage has been singled out by Scott (1958) as the beginning of true socialization. This analysis has shown, however, that social responsiveness arises almost immediately after birth and continues to grow in depth and complexity during each stage of ontogeny. Visually based social responses have their ontogenetic origin in earlier non-visual stages and they retain their close relationship to these earlier stages throughout life.

FUNCTIONAL ASPECTS OF EACH STAGE OF BEHAVIOURAL DEVELOPMENT

The neonate's dependence upon thermal and tactile stimulation in its behavioural responses to its surroundings is based upon its need to maintain an optimal thermal environment for adequate physiological functioning in the face of its inability to regulate fully its own body temperature (Jeddi, 1970, and unpublished data; Crighton & Pownall, 1974; Leonard, 1974). Contact with the mother's body causes a rise in body temperature and huddling with the littermates, shown by Albert (1974) to result in minimal exposure of the group to ambient thermal conditions, maintains body temperature, and reduces metabolic activity (Cosnier, 1965). Leonard (1974) has suggested that overall rate of heat loss is the stimulus to which newborn hamsters respond in the thermotaxic orientation, and the effect of prolonged exposure to low ambient temperature on ultrasonic and audible sound emission by rat and hamster pups (Okon, 1971) supports this view. However, the short latency with which neonates respond to altered thermal conditions, particularly during nipple searching, indicates that thermal sensory receptors located in the snout, lips and face also play an important role (Welker, 1959; Leonard, 1974).

Neonates whose behaviour is organized in relation to thermotactile stimulation are necessarily confined to the close proximity of heat sources and their tactile representatives, and they must possess means of reaching these heat sources from short distances if they are displaced. The behaviour of neonates is therefore limited in spatial scope and in the range of thermal conditions under which they can function adequately.

As these thermal limitations are relaxed with the development of thermoregulatory mechanisms, neonates become capable of extending the scope of their functioning, both spatially and with respect to environmental conditions. Olfactory stimuli play an important role in this process. They arise from substances deposited by the mother in the vicinity of the home or nest and have special significance for the newborn, since they have been experienced in conjunction with thermotactile stimulation from the mother, and therefore they can provide means for the newborn to extend its scope of activity. The mother creates an olfactory zone around the nest or home (as well as in the home itself when she is absent) in which the newborn can function because of its growing responsiveness to olfactory stimulation. Moreover, odours are species- and individual-specific (Leon, 1975): they provide a basis for the specialization of responses to particular kinds of animals and particularly to the individual mother and littermates. Evidence of olfaction-based individual attachments by neonates have been reported for rats and kittens and will, very likely, be found in many different altricial species.

Home or nest orientation develops in relation to the differential distribution of odours in the vicinity of the nest or home site. It requires a responsiveness to gradients of olfactory stimulation or to orientated deposits which are the products of the mother's own position in relation to the home or nest as the centre of her maternal activity.

The maturation of vision increases even further the scope of the young animal's activity and the period of combined olfactory and visual functioning ensures that early visual functioning will be in relation to the most significant features of the young's social environment and within the zone of its socially conditioned physical environment. The specialization of the young's response to qualitative features of stimuli (rather than simply to quantitative features of the earlier phases) advances further with the introduction of vision. A wider range and greater variety of social signals are displayed visually than through odours or thermotactile stimuli. Play arises during the period of visual functioning and is based largely upon the young's response to the variety of visually perceived actions on the part of siblings and other familiar contemporaries.

Earlier sensory systems do not, of course, fall out of use when newer ones mature and become functional. They do, however, play a different role in behaviour than during the phase of their predominance.

SUMMARY

Three stages can be distinguished in the early behavioural development of altricial young of selected species of mammals. During the first stage, behavioural adaptations to the principal features of their environment, the mother, littermates and nest or home site, are based upon low intensity thermotactile stimuli which elicit general and local approach responses. During this stage peripheral control of behaviour is more important than central: the latter is exerted through levels of non-specific arousal that have widespread effects on behavioural responsiveness. During the latter part of this stage there is increasing specialization of perceptual and motor processes and central arousal begins to become specialized. During the second phase, olfactory stimuli begin to play an important role in the young's behavioural adjustments. They introduce increasing specialization of action patterns in relation to familiar stimuli through perceptual discrimination and increasing specialization of arousal processes. Central control processes begin to initiate specialized action patterns in relation to the prevailing peripheral stimulation. During the third stage, visual stimulation begins to play an important role in the young's response to its environment. This leads to further specialization of central control processes through the acquisition of distance perception and through the greater variety of socially significant stimuli that become available to the young. Functional aspects of each stage of behavioural development are discussed.

The preparation of this article was supported by USPHS Grant MH 08604 and a grant from the Alfred P. Sloan Foundation. I wish to thank Natalie Freeman and Robert W. Woll for allowing me to cite their research and Dr Warren G. Hall for the opportunity to discuss many of the issues raised in this article and for making available to me his unpublished research. Publication Number 216 of the Institute of Animal Behavior, Rutgers, The State University.

REFERENCES

Alberts, J. R. (1974). Sensory controls and physiological regulation of huddling in the developing rat. Ph.D. thesis, Princeton University.

Allin, J. T. & Banks, E. M. (1971). Effects of temperature on ultrasound production by infant albino rats. *Developmental Psychobiology*, **4**, 149–156.

Altman, J., Brunner, R. L., Bulut, F. G. & Sudarshan, K. (1974). The development of behavior in normal and brain-damaged infant rats, studied with homing (nest-seeking) as motivation. In *Drugs and the Developing Brain*, ed. A. Vernadakis & N. Weiner, pp. 321–346. Plenum Press: New York.

Altman, J., Sudarshan, K., Das, G. D., McCormick, N. & Barnes, D. (1971). The influence of nutrition on neural and behavioral development. III. Development of some motor, particular locomotor patterns during infancy. *Developmental Psychobiology*, **4**, 97–114.

Bacon, W. E. (1973*a*). Aversive conditioning in neonatal kittens. *Journal of comparative and physiological Psychology*, **83**, 306–313.

Bacon, W. E. (1973*b*). Acquisition of temperature discrimination in neonatal dogs. *Bulletin of the Psychonomic Society*, **1**, 65–67.

Bacon, W. E. (1974). Stimulus control of discriminated behavior in neonatal dogs. *Journal of comparative and physiological Psychology*, **76**, 424–433.

Bacon, W. E. & Stanley, W. C. (1970). Avoidance learning in neonatal dogs. *Journal of comparative and physiological Psychology*, **71**, 448–452.

Bulut, F. G. & Altman, J. (1974). Spatial and tactile discrimination learning in infant rats motivated by homing. *Developmental Psychobiology*, **7**, 465–473.

Carmichael, L. (1946). The onset and early development of behavior. In *Manual of Child Psychology*, ed. L. Carmichael, pp. 43–166. Wiley: New York.

Cosnier, J. (1965). Le comportement grégaire du rat d'élevage (Etude éthologique). Doctoral thesis, University of Lyon, France.

Crighton, G. W. & Pownall, R. (1974). The homeothermic status of the neonatal dog. *Nature, London*, **251**, 142–144.

Devor, M. & Schneider, G. E. (1974). Attraction to home-cage odor in hamster pups: Specificity and changes with age. *Behavioral Biology*, **10**, 211–221.

Drewett, R. F., Statham, C. & Wakerley, J. B. (1974). A quantitative analysis of the feeding behaviour of suckling rats. *Animal Behaviour*, **22**, 907–913.

Elliot, O. & Scott, J. P. (1961). The development of emotional distress reactions to separation in puppies. *Journal of Genetic Psychology*, **99**, 3–22.

Ewer, R. F. (1959). Sucking behavior in kittens. *Behaviour*, **15**, 146–162.

Ewer, R. F. (1961). Further observations on suckling behaviour in kittens together with some general consideration of the interrelations of innate and acquired responses. *Behaviour*, **17**, 247–260.

Fowler, S. J. & Kellogg, C. (1975). Ontogeny of thermoregulatory mechanisms in the rat. *Journal of comparative and physiological Psychology*, **89**, 738–746.

Fox, M. W. (1968). Neuronal development and ontogeny of evoked potentials in auditory and visual cortex of the dog. *Electroencephalography and clinical Neurophysiology*, **24**, 213–226.

Fox, M. W. (1970). Reflex development and behavioral organization. In *Developmental Neurobiology*, ed. W. A. Himwich, pp. 553–580. Thomas: Springfield.

Fox, M. W. & Weisman, R. (1970). Development of responsiveness to a social releaser in the dog: effects of age and hunger. *Developmental Psychobiology*, **2**, 277–280.

Gard, C., Hard, E., Larsson, K. & Petersson, V.-A. (1967). The relationship between sensory stimulation and gross motor behavior during the postnatal development in the rat. *Animal Behaviour*, **15**, 563–567.

Gregory, E. H. & Pfaff, D. W. (1971). Development of olfactory-guided behavior in infant rats. *Physiology and Behavior*, **6**, 573–576.

Goldman, L. & Swanson, H. H. (1975). Developmental changes in pre-adult behavior in confined colonies of golden hamsters. *Developmental Psychobiology*, **8**, 137–150.

Hafez, E. S. E. (1969). *The Behaviour of Domestic Animals*, 2nd edn. Williams & Wilkins: Baltimore.

Hall, W. G. (1975). The ontogeny of ingestive behavior in the rat. Doctoral thesis, The Johns Hopkins University.

Hensel, H. & Kenshalo, D. R. (1969). Warm receptors in the nasal region of cats. *Journal of Physiology, London*, **204**, 99–112.

Hofer, M. A. (1970). Physiological responses of infant rats to separation from their mothers. *Science, Washington*, **168**, 871–873.

Hofer, M. A. (1971). Cardiac rate regulated by nutritional factor in young rats. *Science, Washington*, **172**, 1039–1041.

Hofer, M. A. & Grabie, M. (1971). Cardiorespiratory regulation and activity patterns of rat pups studied with their mothers during the nursing cycle. *Developmental Psychobiology*, **4**, 169–180.

Hofer, M. A. & Reiser, M. F. (1969). The development of cardiac rate regulation in preweanling rats. *Psychosomatic Medicine*, **31**, 372–388.

Hofer, M. A. & Weiner, H. (1971). Development and mechanisms of cardiorespiratory responses to maternal deprivation in rat pups. *Psychosomatic Medicine*, **33**, 353–362.

Hopson, J. A. (1973). Endothermy, small size, and the origin of mammalian reproduction. *American Naturalist*, **107**, 446–452.

Ivanitskii, A. M. (n.d.). The morphophysiological investigation of development of conditioned alimentary reaction in rabbits during ontogenesis. *Works Higher Nervous Activity, Physiology Series*, **4**, 126–141.

James, W. T. (1952*a*). Observations on the behavior of new born puppies. I. Method of measurement and types of behavior involved. *Journal of genetic Psychology*, **80**, 65–73.

James, W. T. (1952*b*). Observations on the behavior of new-born puppies. II. Summary of movements involved in group orientation. *Journal of comparative and physiological Psychology*, **45**, 329–335.

James, W. T. (1957). The effect of satiation on the sucking response in puppies. *Journal of comparative and physiological Psychology*, **50**, 375–378.

James, W. J. (1959). A further analysis of the effect of satiation on the sucking response in puppies. *Psychological Record*, **9**, 1–6.

James, W. T. & Gilbert, T. F. (1957). Elimination of eating behavior by food injected in weaned puppies. *Psychological Reports*, **3**, 167–168.

James, W. T. & Rollins, J. (1965). Effect of various degrees of stomach loading on the sucking response in puppies. *Psychological Reports*, **17**, 844–846.

Jeddi, E. (1970). Confort de contact et thermoregulation comportementale. *Physiology and Behavior*, **5**, 1487–1493.

Karmel, B. Z., Miller, P. N., Dettweiler, L. & Anderson, G. (1970). Texture density and normal development of visual depth avoidance. *Developmental Psychobiology*, **3**, 73–90.

Kenshalo, D. R. (1970). Psychophysiological studies of temperature sensitivity. In *Contributions to Sensory Physiology*, vol. **4**, ed. W. D. Neff, pp. 19–74. Academic Press: New York.

Kenshalo, D. R. & Brearley, E. A. (1970). Electrophysiological measurements of the sensitivity of the cat's upper lip to warm and cool stimuli. *Journal of comparative and physiological Psychology*, **70**, 5–14.

Kenshalo, D. R., Duncan, D. G. & Weymark, C. (1967). Thresholds for thermal stimulation of the inner thigh, footpad, and face of cats. *Journal of comparative and physiological Psychology*, **63**, 133–138.

Kenshalo, D. R., Hensel, H., Graziadei, P. & Fruhstorfer, H. (1971). On the anatomy and physiology and psychophysics of the cat's temperature-sensing system. In *Oral–Facial Sensory and Motor Mechanisms*, ed. R. Dubner & Y. Kawamura, pp. 23–54. Appleton-Century-Crofts: New York.

Koch, M. D. & Arnold, W. J. (1976). Maternal and nutritional factors in maintenance of infant rat cardiac rate following maternal separation. *Physiology and Behavior*, in press.

Koepke, J. E. & Pribram, K. H. (1971). Effect of milk on the maintenance of sucking behavior in kittens from birth to six months. *Journal of comparative and physiological Psychology*, **75**, 363–377.

Komisaruk, B. R. (1970). Synchrony between limbic system theta activity and rhythmical behavior in rats. *Journal of comparative and physiological Psychology*, **70**, 482–492.

Kovach, J. A. & Kling, A. (1967). Mechanisms of neonate sucking behavior in the kitten. *Animal Behaviour*, **15**, 91–101.

Langworthy, O. R. (1929). A correlated study of the development of reflex activity in foetal and young kittens and the myelinization of tracts of the nervous system. *Contribution to Embryology, Carnegie Institute, Washington*, **20**, 127–171.

Leon, M. (1975). Dietary control of maternal pheromone in the lactating rat. *Physiology and Behavior*, **14**, 311–319.

Leonard, C. M. (1974). Thermotaxis in golden hamster pups. *Journal of comparative and physiological Psychology*, **86**, 458–469.

Lincoln, D. W., Hill, A. & Wakerley, J. B. (1973). The milk-ejection reflex of the rat: An intermittent function not abolished by surgical levels of anaesthesia. *Journal of Endocrinology*, **57**, 459–476.

Long, C. A. (1972). Two hypotheses on the origin of lactation. *American Naturalist*, **106**, 141–144.

Math, F. & Desor, D. (1974). Evolution de la reaction d'arrêt de l'electroencephalogramme consecutive aux stimulations olfactives naturelles chez le Rat blanc élève en semi-liberté. *Comptes rendus hebdomadaire des scéances de l'Académie des sciences, Paris*, **279**, 931–934.

Misanin, J. R., Nagy, Z. M. & Weiss, E. M. (1970). Escape behavior in neonatal rats. *Psychonomic Science*, **18**, 191–192.

Moltz, H. & Leon, M. (1973). Stimulus control of the maternal pheromone in the lactating rat. *Physiology and Behavior*, **10**, 69–71.

Nagy, Z. M., Misanin, J. R. & Newman, J. (1970). The anatomy of escape behavior in neonatal mice. *Journal of comparative and physiological Psychology*, **72**, 116–124.

Nyakas, C. & Endroczi, E. (1970). Olfaction guided approaching behavior of infantile rats to the mother in maze box. *Acta physiologica Academiae scientiarum hungaricae*, **38**, 59–65.

Okon, E. E. (1971). The temperature relations of vocalization in infant golden hamsters and wistar rats. *Journal of Zoology*, **164**, 227–237.

Okon, E. E. (1972). Factors affecting ultrasound production in infant rodents. *Journal of Zoology*, **168**, 139–148.

Oswalt, G. L. & Meier, G. W. (1975). Olfactory, thermal, and tactual influences on infantile ultrastonic vocalization in rats. *Developmental Psychobiology*, **8**, 129–135.

Papoušek, H. & Papoušek, M. (1975). Cognitive aspects of preverbal social interaction between human infants and adults. In *Parent–Infant Interaction*, CIBA Foundation Symposium, no. **33**, pp. 241–269. Elsevier: Amsterdam.

Prechtl, H. F. R. (1952). Angeborenen Bewegungsweisen junger Katzen. *Experientia*, **8**, 220–221.

Rheingold, H. L. (1963*a*). *Maternal Behavior in Mammals*. Wiley: New York.

Rheingold, H. L. (1963*b*). Maternal behavior in the dog. In *Maternal Behavior in Mammals*, ed. H. L. Rheingold, pp. 169–202. Academic Press: New York.

Rheingold, H. L. & Eckerman, C. O. (1971). Familiar social and nonsocial stimuli and the kitten's response to a strange environment. *Developmental Psychobiology*, **4**, 71–89.

Rosenblatt, J. S. (1971). Suckling and home orientation in the kitten: A comparative developmental study. In *The Biopsychology of Development*, ed. E. Tobach, L. R. Aronson & E. Shaw, pp. 345–410. Academic Press: New York.

Rosenblatt, J. S. & Schneirla, T. C. (1962). The behaviour of cats. In *The Behaviour of Domestic Animals*, ed. E. S. E. Hafez, pp. 453–488. Bailliere, Tindall & Cox: London.

Rosenblatt, J. S., Turkewitz, G. & Schneirla, T. C. (1961). Early socialization in the domestic cat as based on feeding and other relationships between female and young. In *Determinants of Infant Behaviour*, ed. B. F. Foss, pp. 51–74. Methuen: London.

Rosenblatt, J. S., Turkewitz, G. & Schneirla, T. C. (1969). Development of home orientation in newly born kitten. *Transactions of the New York Academy of Sciences*, **31**, 231–250.

Rowe, M. J. & Sessle, B. J. (1972). Responses of trigeminal ganglion and brain stem neurones in the cat to mechanical and thermal stimulation of the face. *Brain Research*, **42**, 367–384.

Salas, M., Schapiro, S. & Guzman-Flores, C. (1970). Development of olfactory bulb discrimination between maternal and food odors. *Physiological Behavior*, **5**, 1261–1264.

Satinoff, E. & Stanley, W. C. (1963). Effect of stomach loading on sucking behavior in neonatal puppies. *Journal of comparative and physiological Psychology*, **56**, 66–68.

Schapiro, S. & Salas, M. (1970). Behavioral response of infant rats to maternal odor. *Physiological Behavior*, **5**, 815–817.

Schneirla, T. C. (1959). An evolutionary and developmental theory of biphasic processes underlying approach and withdrawal. In *Nebraska Symposium on Motivation*, ed. M. R. Jones, pp. 1–42. University of Nebraska Press: Lincoln, Nebraska.

Schneirla, T. C. (1965). Aspects of stimulation and organization in approach/withdrawal processes underlying vertebrate behavioral development. *Advances in the Study of Behavior*, **1**, 1–74.

Schneirla, T. C. & Rosenblatt, J. S. (1961). Behavioral organization and genesis of the social bond in insects and mammals. *American Journal of Orthopsychiatry*, **31**, 223–253.

Schneirla, T. C. & Rosenblatt, J. S. (1963). 'Critical periods' in the development of behavior. *Science, Washington*, **139**, 1110–1115.

Schneirla, T. C., Rosenblatt, J. S. & Tobach, E. (1963). Maternal behavior in the cat. In *Maternal Behavior in Mammals*, ed. H. L. Rheingold, pp. 122–168. Wiley: New York.

Scott, J. P. (1958). Critical periods in the development of social behavior in puppies. *Psychosomatic Medicine*, **20**, 42–54.

Scott, J. P. (1962). Critical periods in behavioral development. *Science, Washington*, **138**, 949–958.

Scott, J. P., Stewart, J. M. & DeGhett, V. J. (1974). Critical periods in the organization of systems. *Developmental Psychobiology*, **7**, 489–513.

Singh, P. J. & Tobach, E. (1974). Olfactory bulbectomy and nursing behavior in rat pups (Wistar DAB). *Developmental Psychobiology*, **8**, 151–164.

Stanley, W. C. (1970). Feeding behavior and learning in neonatal dogs. In *The Second Symposium on Oral Sensation and Perception*, ed. J. F. Bosma, pp. 242–290. Thomas: Springfield.

Stanley, W. C., Bacon, W. E. & Fehr, C. (1970). Discriminated instrumental learning in neonatal dogs. *Journal of comparative and physiological Psychology*, **70**, 335–343.

Stanley, W. C., Barrett, J. E. & Bacon, W. E. (1974). Conditioning and extinction of avoidance and escape behavior in neonatal dogs. *Journal of comparative and physiological Psychology*, **87**, 163–172.

Thoman, E. B. & Arnold, W. J. (1968). Incubator rearing of infant rats without the mother: effects on adult emotionality and learning. *Developmental Psychobiology*, **1**, 219–222.

Thoman, E., Wetzel, A. & Levine, S. (1968). Learning in the neonatal rat. *Animal Behaviour*, **16**, 54–57.

Tilney, F. & Casamajor, L. (1924). Myolenogeny as applied to the study of behaviour. *Archives of neurological Psychiatry*, **12**, 1–66.

Tobach, E., Rouger, Y. & Schneirla, T. C. (1967). Development of olfactory function in the rat pup. *American Zoologist*, **7**, 792–793.

Toropova, N. V. (1961). Technique of artificial feeding of puppies in the early postnatal period. *Pavlov Journal of higher nervous Activity*, **11**, 137–138.

Turkewitz, G. (1966). The development of spatial orientation in relation to the effective perceptual environment in neonate rats. Doctoral dissertation, New York University.

Turner, W. D. (1935). The development of perception: I. Visual direction; the first eidoscopic orientations of the albino rat. *Journal of genetic Psychology*, **47**, 121–140.

Vorherr, H., Kleeman, C. R. & Lehman, E. (1967). Oxytocin induced stretch reaction in suckling mice and rats: a semi-quantitative bio-assay for oxytocin. *Endocrinology*, **81**, 711–715.

Vuorenkoski, V., Wasz-Höckert, O., Koivisto, E. & Lind, J. (1969). The effect of cry stimulus on the temperature of the lactating breast of primipara. A thermographic study. *Experientia*, **25**, 1286–1287.

Walters, R. H. & Parke, R. D. (1965). The role of the distance receptors in the development of social responsiveness. *Advances in Child Development and Behavior*, **2**, 59–96.

Welker, W. I. (1959). Factors influencing aggregation of neonatal puppies. *Journal of comparative and physiological Psychology*, **52**, 376–380.

Welker, W. I. (1964). Analysis of sniffing in the albino rat. *Behaviour*, **22**, 223–244.

Windle, W. F. (1930). Normal behavioral reactions of kittens correlated with postnatal development of nerve-fibre density in the spinal grey matter. *Journal of comparative Neurology*, **50**, 479–503.

Wodinsky, J., Rosenblatt, J. S., Turkewitz, G. & Schneirla, T. C. (1955). The development of individual nursing position habits in new born kittens. Paper presented to the Eastern Psychological Association.

Note added in proof

Two recent studies have reported substances, on the nipples of lactating rats, to which pups respond by grasping the nipples for suckling (Hofer, Shair & Singh, 1976; Teicher & Blass, 1976). These substances are present already on the fourth day and appear to be deposits of pup saliva.

REFERENCES

Hofer, M. A., Shair, H. & Singh, P. (1976). Evidence that maternal ventral skin substances promote suckling in infant rats. *Developmental Psychobiology*, in press.

Teicher, M. H. & Blass, E. M. (1976). Suckling in neonatal rats: eliminated by nipple lavage; reinstated by pup saliva. *Science, Washington*, in press.

12

The study of animal play

M. J. A. SIMPSON

This essay presents some anecdotal observations which have stimulated me to think in new ways. I describe infant rhesus monkeys repeating relatively complex patterns of activity, which I call 'projects', and discuss the implications of such patterns for studies of the development of animals' skills in and knowledge of their particular surroundings. I believe that detailed longitudinal studies of playing individuals can tell us about how some of their skills develop and I look forward to animal studies of the kind that Bruner (1971) is pioneering with human infants.

This discussion of repeated playful actions poses the question of how we, as observers, discern attentional and learning processes in free-living individuals. An infant directs his behaviour and his attention to his surroundings so as to frame certain groups of events involving himself and the surroundings. Thereby he limits the situation so that what his surroundings do back to him is limited. Control is thus distributed between an animal and his surroundings, where the distribution of control is analogous to that in one of our experiments: we control conditions, but within those conditions the subjects control their situation and thereby tell us something new.

Medawar (1967) has reminded us of the difference between completed projects and those still fragmentary and growing. The former can be presented as a logical progression of ideas, but this progression seldom reflects the ontogeny of those ideas. This essay presents only some fragments, and such progression as occurs is irregular and not always logical. But the task of pointing to research problems which cannot yet be formulated in fully operational ways is inevitably difficult: the words of this essay, like Graves' (1957) cabbage white butterflies, 'lurch here and here by guess' towards ideas which finally may be expressed in visual and topological rather than verbal terms.

AN EXAMPLE OF AN INFANT'S PROJECT

Captive rhesus monkey infants of six to 12 weeks repeat sequences of activity involving particular leaps, particular climbs and particular routes. Thus the leap may be to one object, and from one starting point. And such recognizable actions may be repeated in series sufficiently often for observers to talk of an infant 'having a project'. While the term 'project' is convenient shorthand for conveying an impression of practice toward some perfect version of the particular leap, we should beware of its connotations of goal-directedness. What we describe, once we have suitable methods, may be successive instances (leaps in this case) becoming more similar to each other, and also more smoothly executed. Our interpretation of such a progression need not include any idea, in the animal's head, of the kind of leap it finally does. Such a series could occur if repetition with slight variation was interesting (e.g. discussion by Humphrey, 1973) and if clumsy performances, including heavy landings, were punishing. The series would end when all the slight variations, consistent with comfortable performance, had been tried and had become boring. One common early project is leaping vertically up to a low branch; later the monkey may leap horizontally from a post to the mesh side of the cage, and still later horizontally from the mesh side to some limited target like a pole. The 'project' may also become complicated: climb up a sloping pole, jump across to the wire mesh, climb down, fiddle with some fixture like a bolt, climb up the sloping pole again. By 12 weeks, such sequences may involve other animals. The leaps may now be at peers, who must jump out of the way or be hit. The mother is usually incorporated in such sequences, in the rather passive sense that the infant regularly returns to her side. Such regularity can often be seen in 'runs' of excursions away from mother, which have surprisingly constant durations.

Such patterns suggest that rhesus monkey infants can repeat what they did before. This ability becomes surprising when we compare actual sequences for an infant with all the possible ones he might perform. On leaving its mother, there are many possible outgoing courses the infant could take, and many starting points for leaps. But particular courses and particular leaps seem to be repeated.

Obviously we need to develop methods for confirming the robustness and and reality of such repetitions. Are they robust, in the sense of being relatively unaffected by moment-to-moment changes in the infant's surroundings? For example, if the infant finds a peer sitting at the point where he starts his current leap, does he push that peer out of the way?

Can we devise measures of 'similarity' that allow us to judge whether successive cases of a presumed pattern constitute repetitions? And can we discover, perhaps by computer simulation techniques, or by calculating *a priori*, the probability that a pair or series of instances have a particular degree of similarity? These are two separate statistical operations. The first is part of any taxonomic technique, such as cluster analysis, which is used for sorting 'objects' (e.g. excursions or leaps) into groups according to their similarity with each other (Morgan, Simpson, Hanby & Hall-Craggs, 1976) – though of course our definition of similarity must suit the case being considered. The second operation, discovering the probability of the occurrence of a particular grouping, is more difficult (see e.g. Morgan *et al.*, 1976), and our tactics for doing this must vary according to the particular case. Appropriate methods are at present being devised: here we must return to the biological implications of repetition.

We started by defining our 'unit' before beginning the study. Obviously we have to decide on *some* unit before we can begin at all. But an infant could be making its excursions in *pairs*, with its mother at the centre of the same figure-of-eight. Then comparing successive single excursions (the two halves of the figure) would result in dissimilarity, while comparing excursions in pairs would result in similarity. Alternatively, we might find that smaller units were being repeated within a single excursion. For each time it left mother, the infant might repeat three successive leaps from the same place.

When we take smaller units, like leaps, as our starting point, we face new problems of description; we should perhaps describe not only the movement of the infant relative to its surroundings but also the movements of all those parts of his body that are free to move relative to each other: eyes relative to head, head to trunk, limbs relative to trunk, and so on. There is a form of movement notation (the Eshkol–Wachmann movement notation used by Golani, 1973 and 1976) which can be used to describe such movements, which may prove useful.

FRAMING

Elaborating our methods for describing patterns of movement compounds the problem of which units to take as starting points in our study. This problem must also be faced by the infant contributing and responding to the stream of events in which he is involved. The question of how to impose some kind of frame on the stream of events should perhaps be seen from the infant's point of view.

If we use the Eshkol–Wachmann movement notation to record the continuous torrent of events in which the infant is involved, our record comes to look rather like a musical score: different groups of horizontal lines representing individuals, different lines within those groups representing the movements of parts of those individuals (e.g. Golani 1973 and 1976). We can confine our attention to particular parts of this record by applying a frame to it. For example, we can frame our record of excursions from the mother's side by fitting the frame's left- and right-hand margins over the beginnings and ends of excursions. Or, reading vertically, we can apply a frame whenever the infant's mother puts her hand on an object, and we could look in our record for the line representing the object's properties, including its colour.

Our use of the frame is a possible model for attentional processes in the infant. This model emphasizes how attention can stretch back in time to link events into groups (e.g. Lashley, 1951), and it can also group aspects of a situation simultaneously present.

In learning experiments, we are made especially aware of attentional processes whenever our animal subjects subsequently show us that they had learned patterns and stimuli which were present but not necessary for their performance in that particular experiment. Perceptual learning (e.g. Gibson, 1969) and latent learning (Mackintosh, 1974), and transfer of training experiments (Mackintosh, 1974) all show us how much incidental detail animals notice. Bateson & Chantrey (1972) describe experiments which suggest that monkeys and chicks do not merely notice and learn stimuli left on their home cages, they may classify them together. This idea of classifying together explains why prior exposure to two stimuli makes discrimination between those stimuli, in subsequent learning tasks, more difficult than it is for those individuals previously exposed to only one of them. Bateson & Chantrey (1972) (see also Bateson, 1973) also describe more natural situations for chicks, where we would expect them to notice, learn and classify together such things as the different parts of their mother hen, presented to their view at the same time (vertical framing in the terms of the present discussion), and different views of the mother, presented at successive moments (horizontal framing).

This line of thought suggests a host of observations and experiments by which we could analyse how animals look at their surroundings, such as placing two stimuli on the same or different walls of their square home pens (e.g. Chantrey, 1973), or placing the two objects so that the subjects could not see both from the same place. But this essay is concerned with the problems of observing the attentive behaviour of animals moving freely in relatively unrestricted and complex situations.

The idea of a person using a frame or ruler to look at parts of a movement notation record or an orchestral score reminds us of two ways in which attention can be seen as an active process. First, hold in mind what you have already seen or heard: remember enough notes to be able to recognize a melody. Second, take enough control of the situation so as to limit the extent and complexity of what happens: ask the pianist to repeat the first eight bars, or walk out if it is all too complicated.

If we imagine a monkey infant as analogous to a participant in a jazz session, we can see how these two kinds of active process are combined. Shared conventions about rhythm and beat (represented on paper by bar lines, time signature etc.) help him to frame the on-going music; and as the session proceeds he will at different times be listening to and 'recognizing' melodies, joining in, or adding something new to the emerging composition. In adding something new, he will be framing in an actively controlling way.

It is all too easy to stun verbal thought with paradox. How can an individual be both in control of what occurs and responding to what occurs? The solution lies in how we analyse 'what occurs', and such analysis must begin with a description of 'what occurs' sufficiently detailed to allow us to look for both the fine details and the larger patterns (see also discussion in Miller, Galanter & Pribram, 1960, and Dawkins, this volume). Within the category 'what occurs', are occasions when the infant is in control of its surroundings, and occasions when the infant is allowing its surroundings to affect it. Sometimes the occasions will be separated in time, and such occasions will be the easiest for us as observers to distinguish.

In the analysis of behaviour, we should look for some distribution of control between the infant and its surroundings, analogous to the distribution we arrange in our own scientific experiments. As experimenters, we restrict the responses available to our animal subjects, and we further confine ourselves by recording only some of those responses. But our animal subject is not totally under our control; we arrange the situation so that his response can tell us something new.

In this view, learning experiments do much framing and controlling for the animal, and such experiments must be designed with some idea (implicit or explicit) about how the animal does these things. For example, to discover whether the colour blue is a cue that an animal can use, we restrict the animal's situation so that only a few things can happen: a few different colours appear on a panel, a lever moves when the animal presses it, a pellet of food appears when the lever is pressed in the presence of a blue panel. We also make the animal hungry.

In a free-living situation such restriction and simplification must come from elsewhere. For example, a free-living rhesus monkey may learn something about the properties of objects eaten by his particular troop by observing what his mother puts in her mouth (e.g. Hall & Goswell, 1964). As his mother picks blue berries, she may also intermittently groom her sister, using hands and lips as she does so; grunt in pleasure as the sun comes out; lip-smack at an approaching male, and her infant may be sitting in her lap, fiddling with a pebble, and probably not hungry. Somehow that infant must attend to berries not pebbles, and to the relationship between mother's hand and her berry, not her sister. Then he must perhaps follow the hand with the berry in it to her mouth, and associate the berry with her chewing movements after (not before) its arrival, and avoid associating it with all the other things her mouth has just done and will do again in a few seconds (grunting, licking, lip-smacking).

How does this infant take over the experimenter's role? Perhaps, whenever his mother's hand moves to a non-monkey object, he follows her hand with his gaze for the next five seconds. In this 'experiment', the infant's control consists of holding mother's hand on his fovea, and mother's control is where her hand goes within the next five seconds. Other events within this time span are ignored by the infant.

We could discover much of this from the infant's behaviour in the successive seconds after his mother puts her hand on objects: for instance, does he stop other activities, or watch her hand?

Other regularities in the infant's behaviour could also give us hints. For example, if we discovered that an infant away from its mother's side looked at her every 15 seconds, we might expect him to frame groups of events falling between pairs of successive glances at mother, while members of series of events separated by a glance at mother might be framed in two groups by the infant. He might, for example, associate his approach to a peer and being hit by that peer only if both events occurred within the same 15 second interval between glances at mother.

CATEGORIZATION AND CALIBRATION

Table 1 refers to pairs of events, framed by a rhesus monkey infant. Event 1 is a controlling action by the infant, while Event 2 is the response from another part of the infant himself, or from his surroundings. Case A has already been described. Case B is most like an experiment: what does a peer do after being hit? The table shows only the results of *single* trials.

A *series* of trials can produce a particular result with more or less predictability. Our current understanding of exploration and novelty (e.g. Berlyne, 1960; Hinde, 1970; Bateson, 1973; Humphrey, 1973) suggests that the infant will repeat actions of the category 'Event 1' only so long as the relationship between Event 1 and Event 2 is unpredictable for the infant.

I assume that a predictable outcome for us is also an outcome that the infant has come to predict, or has learned. When he has seen mother quickly put blue berries into her mouth on five successive occasions he will no longer follow her hand with his gaze when she is near that particular bush. If, however,

Table 1. *Pairs of events: Event 1 under the individual's control; Event 2 to some degree under the control of the surroundings*

The examples describe *single* trials. In other trials, Event 1 will remain the same, but Event 2 may be different (see text).

Pair	Event 1	Event 2
A	Infant looks at non-monkey object in mother's hand for 5 sec after her hand touches object	Mother's hand reaches her mouth within 5 sec
B	Infant approaches and hits another infant, waits 5 sec	Other infant hits back within 5 sec
C	Infant, on all fours, suddenly extends hind legs (lurch)	Extra weight shifts to infant's arms, he retains balance and does not fall
D	Infant, on all fours, leans back, and suddenly extends hind legs (hop)	Infant leaves ground, all four limbs braced for landing – does not topple
E	Infant glances at a branch 30 cm up, leaps towards it	Infant reaches branch

she begins to pick green berries, he will watch closely again, and if she sometimes hesitates, and sometimes drops them, he may watch her through more trials.

There is an additional and more dangerous assumption underlying this model: that the infant's recording of the situation is effective. An infant watching his mother eating berries may be so distractable that although mother put a continuous series of seven into her mouth, the infant only saw three go in: what *we* saw as seven out of seven was only three out of seven for the infant.

One could make a further assumption: that natural selection ensures that an infant embarks only on those experiments whose results he is competent to register. The appropriate skills could emerge in due course as the infant matures, according to this assumption. However, it is also possible that infants could be building up their skills. Students of infant play (e.g. Bruner, 1971) distinguished between the practising of an action or movement regardless of context, and the application of that movement in a variety of particular and recognizable contexts (e.g. Loizos, 1967). Table 1 shows three pairs of Events, which I call lurch (C), hop (D) and leap (E). The leap often clearly involves a very particular context: it is repeatedly directed from a particular take-off point to a particular spot on a branch. Lurches and hops are seen in younger infants, and often seem to occur almost at random.

In lurching and hopping, Event 2 involves the infant's own body as intimately as Event 1: in these cases the 'experiments' involve discovering the consequences of movements in one part of the infant for other parts of the same infant. If he suddenly extends his hind legs, what happens to his front limbs?

Pairs B and A fall easily into learning models: one peer can be hit with impunity, another not; blue berries are eaten by mother, purple not. Objects and situations are discriminated amongst, or categorized.

Pairs C, D and E might be classified as skills: the successful lurches, hops and leaps are ones where the particular configuration of movements in the limbs comes to bear a predictable relationship to what those limbs do at the end. At the end of a hop, the limbs are suitably placed for the landing. In performing a leap to some particular target, there is a relationship between how far that target looks, the propulsive force exerted by the limbs, and the result – the target is reached and not overshot. An infant which can judge the leap required for *any* target within his range could be said to have achieved some state of *calibration* between the apparent distance of that target, and the force exerted to reach it. Note that any skill we recognize can involve both calibration and categorization: there may be a group of targets which the infant would class as outside his range; and of those within range only those requiring a leap through free space may involve calibration.

CALIBRATION

My discussion of calibration was inspired by studies showing the importance of stimulation consequent upon an animal's own movements and activities as it developed its sensorimotor skills. For example, kittens allowed to move themselves through a patterned environment learned to avoid visual cliffs more quickly than kittens which were passively moved through the same environment (Held & Hein, 1963). Human adults with spectacles that reverse visual feedback are at first confused and unable to walk easily with their eyes open. If they are allowed to move of their own accord, rather than being passively moved in a wheel-chair, they adjust more quickly. In general, an ability to relate feedback to motor performances seems to be affected by practice (discussion in Hinde, 1970).

I suggest that, in addition tc imposing some condition that requires the subject to calibrate, we can sometimes observe calibration as it occurs in developing individuals. Of course, we should not expect all animals to become calibrated through practice. For example, Wells (1958) showed that newly hatched cuttlefish are more selective about the size of crustacean prey that they attempt to take, than they become later. On hatching, they can categorize and avoid over-large prey animals, and they are sufficiently well calibrated to be able to seize those within the correct size-range. The praying mantis' prey-seizing behaviour, which has been analysed in detail (Mittelstaedt, 1962), seems rather fixed in its calibration.

Miller *et al.* (1960) have reminded us that as we learn skills, we can deliberately use special tricks to make calibration more easy. They describe the problems of learning to control a single-engined aircraft as it begins its take-off run. When the throttle lever is pushed forwards into the dashboard to provide the full power needed, the torque from the airscrew tends to cause the aircraft to veer left as it begins to move down the runaway. This turning tendency must be anticipated and corrected by putting pressure with the foot on the right rudder bar. The coordination between hand and foot can be made easier if the hand movement is standardized. For that reason (and others to do with the working of aircraft engines), the throttle should be pushed steadily into the dashboard, and the operation should take five seconds. But counting five seconds as you are trying to control a craft that may be travelling at 30 or 50 km per hour by the end of those five seconds is not easy for beginners. One solution is to place four fingers at regular intervals down the throttle shaft, and push at such a speed as to feel the dashboard with the successive fingers at intervals of one second. The finger-spacing strategy economizes on attention 'used-up' in operating the throttle lever. Connolly (1973) and Bruner (1971) give analogous accounts of the problems faced by children as they master skills.

If he is lucky, a human pilot has instructors who can articulate the tricks by which they build up their skills. A growing rhesus monkey infant does not, and his very growth, and the fact that most of his playmates will be changing in size relative to him, adds to his problems. (A monkey's peers are not all born at exactly the same time as him, so that they may be either growing more quickly or more slowly than he is.) I suggest that play may ensure not only the initial calibration of the infant's skills, but also a continuous re-calibration of those skills as he changes in size relative to the fixtures of his surroundings, and relative to his peers, who are also changing in size and strength.

RESTRICTING MOVEMENTS IN ORDER TO FOCUS ON PARTICULAR DIFFICULTIES

That repeating particular kinds of action is important is suggested by Connolly's (1973) and McFarland's (1973, pp. 334–337) discussions of how complex skills may be learned. Skills could be progressively 'built up' so that, at any stage in their practice, the variation in the animal's success can be related to some limited aspect of the skill: the animal is always focussing on some particular area of difficulty. For example, a monkey leaping to a branch uses many of the muscles and joints between its feet, which direct the force, and its hands, which grasp the branch. Failure could be the consequence of uncontrolled variation in any of a score of joints and muscles: the potential number of degrees of freedom in the activity is immense. We must presume that, when the infant begins to try leaping for objects, it has already gained sufficient control over most of those muscles and joints, so that (to oversimplify) he may hold everything, except the force in his gluteal muscles, constant. In varying that force, he may become able to relate variation in success to variation in that limited aspect of his leaping behaviour.

According to this idea, a free-living animal should repeat actions in its environment in such a way as to practise progressively more complex skills. Once a skill of a particular level of difficulty has been mastered it will no longer be practised. The skill, and not necessarily the precise movement, will be incorporated into more difficult and complex activities, which will in their turn be practised.

ACCOMPLISHED PERFORMANCES

Imagine a monkey extending its home range into a group of trees that it has never entered before. The monkey moves fairly smoothly and decisively through three of the five trees, looks at the other two, and returns to the more familiar parts of its range. The adult monkey may not need to move systematically through every branch of every new tree in its extended range; it is presumably quite easy for him to 'fill in' the details of those he does enter, without trying every branch, and he may even be able to learn much of those he never touches.

In short, this is the banal point that he can to some degree build up his knowledge by judging distances by eye, rather than actually pacing or leaping them. Such abilities could be the results of calibration (knowing how much effort to put into a leap that 'looks' so far), and the following comparison, between an experienced monkey and a musician, takes this point further.

An adult monkey is about to move quickly to a food plant 30 metres away, or a musician is about to play a passage by sight. For the musician, the score dictates that he play certain notes in a certain order. The monkey's route to the food-plant requires the use of certain branches, separated by certain leaps, and he may also need to plan his route in order to avoid other members of his social group. At first glance, this comparison may not seem fair, for the musician seems to have fewer possible ways of reaching the end of his passage than the monkey. But there are many instruments (e.g. clarinet, bassoon), where the same note can be produced in more than one way, and for a particular sequence of notes, some fingerings are more economical. Where there is a choice of finger positions for a note, one should choose that pattern which makes the easiest transition to the finger positions dictated by the next note.

There are two general solutions for monkey and musician: the first is to try every step, and adjust it until it is 'right'. The monkey tries to run along particular branches, and if he fails, tries another; he tries a certain leap until he succeeds, and so on. The musician listens to each note he blows, and adjusts his lips and diaphragm until it is 'right' in terms of the score. Neither would complete his passage quickly, if at all.

We imagine that the monkey is able to judge particular leaps, and perhaps even to judge whole sequences of moves in advance, including the possible consequences of moving close to a particular irritable male sitting on some intervening branch. In playing a rapid passage, a musician has insufficient time to listen to and correct each note (see also Bateson, this volume), he must judge in advance the necessary fingering, lip position, diaphragm pressure, etc.

The adult monkey and the accomplished clarinettist can progress smoothly through new 'passages' on sight. The smooth execution of their behaviour now betrays nothing: neither what now goes on in their heads, nor what went on between them and their surroundings during their practice sessions.

When we turn to the infant monkey, and watch him practise, we may discover something about his control of his experience in the ways in which he limits what he does when he practises.

OBSERVABLE CALIBRATION PROCESSES

Leaping to a branch

We shall regard a series of leaps by a seven week rhesus infant from directly beneath a branch as a series of relatively standard moves, applied in this particular situation. The infant pauses under the branch, looks up, then leaps. Such a leap can have the following outcomes: (*a*) branch not reached; (*b*) branch touched but not grasped; (*c*) branch gained with a scramble, and (*d*) branch easily reached.

The leap is repeated three times. Three failures (*a*), or three easy successes (*d*), might constitute predictability for observers and monkey infant alike. The result for such a series might be that the infant did something else. After a run of failures he might try a lower leap. One failure (*a*), and two near misses (*b*), might lead the infant to persist for another four attempts, but the last three might, through fatigue, be near missses. Two near misses and one partial success (*c*) might also lead to more trials; resulting in two more partial successes, and then a run of three successes (*d*). In this case, the result might be the infant's ability to perform that leap correctly on every subsequent occasion, given that particular starting point from directly under the branch. That particular leap now belongs to the class of leaps that the infant can do.

But the force of the term *calibration* lies in its emphasis on the ability to relate the 'look' of any new leap to the action required to achieve it. After a successful run of trials at a particular kind of leap, we would expect the infant to begin to vary his leaps: apply a vertical leap in new places, or vary the leap to the original target by varying the take-off point.

For example, we might now see a series of oblique leaps from *one* place proceed until their outcomes were predictably successful. This series could involve alternating between an oblique take-off point, and one vertically underneath. As observers, we would find that, as the infant became more skilled, it was more difficult for us to recognize and classify the elements in the series. To return to a musical analogy: a beginner with a flute or clarinet can often play only one note, and it will be easy at that stage to discern 'series of trials'. By the time he can play the eight notes of one scale, he will not only play series of eight-note scales from top to bottom, or bottom to top, but he will also play his scales of notes in more complicated patterns: e.g. in the fours C D E F, D E F G, E F G A, . . ., G A B C.

In the monkey example, an 'element' or 'trial' has become 'any leap from somewhere beneath the branch'. In general, once a range of leaps has been successfully calibrated, we should no longer expect series of leaps within that range. Such runs or series that we do find may now involve another kind of leap, perhaps a parabolic one between two points one metre above the ground.

Mounting behaviour

The idea of calibration can also be applied to a monkey's interaction with its companions. For example, the process whereby a male rhesus monkey might become adept at making the correct orientations for mounting could work as follows. First, he may calibrate his own body. The first surface providing tactile frictional stimuli consequent upon his own movements is provided by his mother's body. When she is stretched out on a warm day, he will be crawling all over her, and sliding his body, including his relatively large scrotum, over her back. For him, she will be a large and relatively uniform surface. As he crawls and slides over her, different parts of his body will feel different, and, in so far as her surface is large and uniform relative to his, such differences in feeling can only reflect differences in his own body, and the different kinds of movement he makes.

So far, the 'calibration' is also a differentiation: genital area feels different; sliding movements enhance this feeling, but bring it to an end because they take him to the edge of his mother, where he falls off; to and fro thrusting movements of the pelvis both enhance and maintain the feeling.

At about eight weeks (see Harlow, 1969, 1974), male infants begin to play actively with others, in the course of which the males chase and the females run and stand (Harlow, 1974). Unlike his mother, female peers are as small as the infant, and different parts of them will feel very different to the infant: small changes in his position may lead to large consequent changes in sensory feedback from his genital area. If, however, the male infant is by now able to locate the feeling in a limited area on his own body, he will be able to focus that area onto the female peer, and use it to discover how different parts of her feel. It is possible that an infant without the prior experience of calibrating himself on the relatively uniform area provided by his mother could be confused by the sensations coming from a small peer, and fail to learn to mount correctly.

CONCLUSION

This essay presented approaches which could throw light on developmental processes, including those called 'learning'. However, to discover what kinds of 'learning experiments' an animal of a particular age is performing is only to take a step which could guide the more closely controlled studies that are then necessary.

Our understanding of how a developing individual attends to his surroundings and acts on them could also help us to see how particular social conditions affect the course and outcome of that development. For example, if we know that a rhesus monkey infant learns nothing about his social companions unless he can play with them in bouts of at least 15 seconds long, then we could predict the effects of styles of mothering that interrupted the infant's activities more frequently than once every 15 seconds.

In this essay, the rhesus monkey examples have been described loosely: given current methods for describing playful behaviour we could hardly do otherwise. And the separate sketchy observations are not even very interesting as isolated facts: they will become interesting only if they can be combined into some larger whole.

SUMMARY

In so far as learning and attention are bound up with each other, and in so far as attention involves active processes which are observable in behaviour, a study of developing animals in relatively complex and natural surroundings may reveal what they are attending to as they build up their skills and knowledge of their surroundings. Such a study, in contrast to those involving tight experimental control, sees the infant performing his own experiments on his surroundings. The animal, rather than an experimental situation contrived by us, takes care of what he shall attend to, and for how long, and we attempt to observe the processes involved.

As a focus for discussion, I used the example of a rhesus monkey infant practising a leap of a particular level of difficulty.

The word 'framing' was used to refer to attention which is active in two senses:

(i) actively reviewing events already passed, for example by grouping and regrouping them in order to discern patterns in them;

(ii) staging situations so as to limit the possible patterns of events that can emerge (e.g. asking a particular question, arranging a particular controlled experiment).

Young animals may so constrain their actions as to simplify the responses that the world, sometimes including parts of themselves, gives back to them. Often that response is not consistent from trial to trial, and so long as it is not consistent, they may repeat their actions until the response becomes predictable. This consistency may lie in the quality of the response obtained: one peer hits on being approached, another avoids, and those peers may be correspondingly categorized by us and the young animal. There may also be a quantitative relationship: a gap that looks so wide requires such a leap. Then the relationship between the individual's behaviour and the situation could be said to be calibrated.

An action which a young animal persistently repeats may tell us about his current stage of development, and it may become possible to describe the emergence and disappearance of particular patterns of activity in such a way that we can follow the development of that animal's skills and learning abilities.

I was supported by the Medical Research Council when I wrote this essay. I thank Pat Bateson, John Cowley, Robert Hinde, Anne Simpson and Ann Weisler for their generous comments and advice at all stages during its preparation.

REFERENCES

Bateson, P. P. G. (1973). Internal influences on early learning in birds. In *Constraints on Learning: Limitations and Predispositions*, ed. R. A. Hinde and J. G. Stevenson-Hinde, pp. 101–116. Academic Press: London & New York.

Bateson, P. P. G. & Chantrey, D. F. (1972). Retardation of discrimination learning in monkeys and chicks previously exposed to both stimuli. *Nature, London*, **237**, 173–174.

Berlyne, D. E. (1960). *Conflict, Arousal and Curiosity*. McGraw-Hill: Toronto & London.

Bruner, J. S. (1971). The growth and structure of skill. In *Motor Skills in Infancy*.

ed. K. J. Connolly. Academic Press: London & New York. Reprinted (1974) in *Beyond the Information Given: Studies in the Psychology of Knowing*, ed. J. M. Anglin. Allen & Unwin: London.

Chantrey, D. F. (1973). Imprinting and perceptual learning in chicks. Ph.D. thesis, Cambridge University.

Connolly, K. (1973). Factors influencing the learning of manual skills by young children. In *Constraints on Learning: Limitations and Predispositions*, ed. R. A. Hinde & J. G. Stevenson-Hinde, pp. 337–365. Academic Press: London & New York.

Gibson, E. J. (1969). *Principles of Perceptual Learning and Development.* Appleton-Century-Crofts: New York.

Golani, I. (1973). Non-metric analysis of behavioural interaction sequences in captive jackals. *Behaviour*, **44**, 98–112.

Golani, I. (1976). Homeostatic motor processes in mammalian interactions – a choreography of display. In *Perspectives in Ethology*, vol. **2**, ed. P. P. G. Bateson & P. H. Klopfer. Plenum Press: New York & London, in press.

Graves, R. (1957). *Poems Selected by Himself.* Penguin: London.

Hall, K. R. L. & Goswell, M. J. (1964). Aspects of social learning in captive patas monkeys. *Primates*, **5**, 59–70.

Harlow, H. F. (1969). Age-mate or peer affectional system. In *Advances in the Study of Behavior*, **2**, 333–383. Academic Press: New York & London.

Harlow, H. F. (1974). Sexual behaviour in the rhesus monkey. In *Sex and Behavior*, ed. F. A. Beach. R. E. Frieger Publishing Co.: New York.

Held, R. & Hein, A. V. (1963). Movement-produced stimulation in the development of visually guided behaviour. *Journal of comparative and physiological Psychology*, **56**, 872–876.

Hinde, R. A. (1970). *Animal Behaviour: A Synthesis of Ethology and Comparative Psychology.* McGraw-Hill: New York.

Humprey, N. K. (1973). The illusion of beauty. *Perception*, **2**, 429–439.

Lashley, K. S. (1951). The problem of serial order in behaviour. In *Cerebral Mechanisms in Behaviour: The Hixon Symposium*, ed. L. A. Jeffress, pp. 112–138. Wiley: New York.

Loizos, C. (1967). Play behaviour in higher primates: a review. In *Primate Ethology*, ed. D. Morris, pp. 176–218. Weidenfeld & Nicholson: London.

Mackintosh, N. J. (1974). *The Psychology of Animal Learning.* Academic Press: London, New York & San Francisco.

McFarland, D. J. (1973). Discussion of Connolly's contribution (q.v.). In *Constraints on Learning: Limitations and Predispositions*, ed. R. A. Hinde & J. G. Stevenson-Hinde, pp. 334–337. Academic Press: London & New York.

Medawar, P. B. (1967). *The Art of the Soluble.* Methuen: London.

Miller, G. A., Galanter, E. & Pribram, K. H. (1960). *Plans and the Structure of Behaviour.* Holt, Rinehart & Winston: New York.

Mittelstaedt, H. (1962). Control systems of orientation in insects. *Annual Review of Entomology*, **7**, 177–198.

Morgan, B. J. T., Simpson, M. J. A., Hanby, J. P. & Hall-Craggs, J. (1976). Visualising interaction and sequential data in animal behaviour: theory and application of cluster analysis. *Behaviour*, **56**, 1–43.

Wells, M. J. (1958). Factors affecting reactions to *Mysis* by newly hatched *Sepia*. *Behaviour*, **13**, 96–111.

13

Rules and reciprocity in behavioural development

P. P. G. BATESON

A major preoccupation in the study of behavioural development has been with what makes individuals different. Attempts to uncover sources of differences in behaviour have required an experimental approach and gave rise to techniques of great ingenuity and elegance. Ideally, any coordinated research programme ought to operate within a broad and coherent theoretical framework if the experimenting is not to become piecemeal or pedestrian. However, theories of behavioural development have not been strikingly coherent. It is possibly for this reason that much of the extensive evidence gathered on the subject appears to be fragmented into esoteric little pockets of knowledge.

As far as most experimenters are concerned, theory seems to be subservient to methodology at the moment. The characteristics of the workable theories are determined not so much by what is plausible as by what is most easily analysed. The theories offer useful if limited tactics but little in the way of an overall research strategy. It is not hard to think of reasons why this might have happened. The success of an experimental manipulation is conventionally judged in terms of whether it exerts an influence on the end state. If it does not, the experimenter is faced with the unprofitable task of proving a negative, and often the results are quickly forgotten. Inevitably this breeds a strong commitment to studying sources of positive differences in behaviour. Since the sources which are most easily manipulated are those external to the animal, it is hardly surprising that the principal current theories of behavioural development make a virtue of experimental convenience and place as much as possible of the control outside the animal in its environment.

It would be quite wrong to present existing ontogenetic theories as simple-minded or unsubtle. The thinking of Schnierla (1965), for example, was complex and rich. Nevertheless, in his developmental theory the underlying assumption was that, to begin with at least, the developing animal is essentially reactive, even though he postulated considerable intrinsic constraints on responsiveness. Initially, environmental conditions are held to change the young animal's behaviour, which in turn changes the animal's relation to its environment. The new stimuli from the environment produce fresh changes in behaviour and so on in chain-like fashion. This account oversimplifies a complex argument but does not misrepresent its spirit, which is to reduce to a minimum the number of postulated internal processes on which development depends. The emphasis is placed squarely on the interactions taking place between the animal and its environment. My purpose is not to discredit views which, I shall argue, have an important part to play in understanding the complex mediating role of the animal's environment in its behavioural development. Nevertheless, I believe that the relentless hunt for yet further external sources of individual variation, and the shyness of accepting substantial internal control over developmental processes, have led to a certain sterility of thinking and an incapacity to comprehend more than a small fraction of the rapidly accumulating data.

The complexity of the evidence and the lack of useful explanatory principles may, of course, be something we simply have to live with. Instead of modelling ourselves on classical physicists in search of unifying principles, we may have to treat the subject like a classical chemist and hunt for the equivalent of a Periodic table. However, the apparent conceptual difficulties presented by masses of seemingly unrelated evidence may also arise from the tight focus which experimental analysis so often requires. There may well be a case for softening the focus or, better still, changing to a lower magnification. To use a different metaphor, maybe vision outside the wood will no longer be cluttered by the tangled undergrowth of endless interactions.

Other views of behavioural development do, of course, exist. But they often consist of nothing more than vague appeals to fashionable technologies. The patterning of development is attributed, for example, to 'internal programming' without any indication being given of what the programme might be like and what precisely is being controlled. Nevertheless, the intuition lying behind such points of view may be right. Furthermore, apposite analogies may be drawn not only from machine intelligence but also from other areas of biology. Certainly, the embryologists have accepted the need to postulate considerable internal regulation of development, and the relevance of some of their concepts to the work on behavioural development has, over the years, become increasingly apparent (see for example Gesell, 1954, and Oppenheim, 1974). The first part of this chapter is devoted, therefore, to examining in some detail an idea from embryology used to explain self-stabilising developmental pathways. The second half of the chapter attempts to relate the idea about guided and buffered systems governed by rules back to the evidence on the reciprocity between the developing animal and its environment.

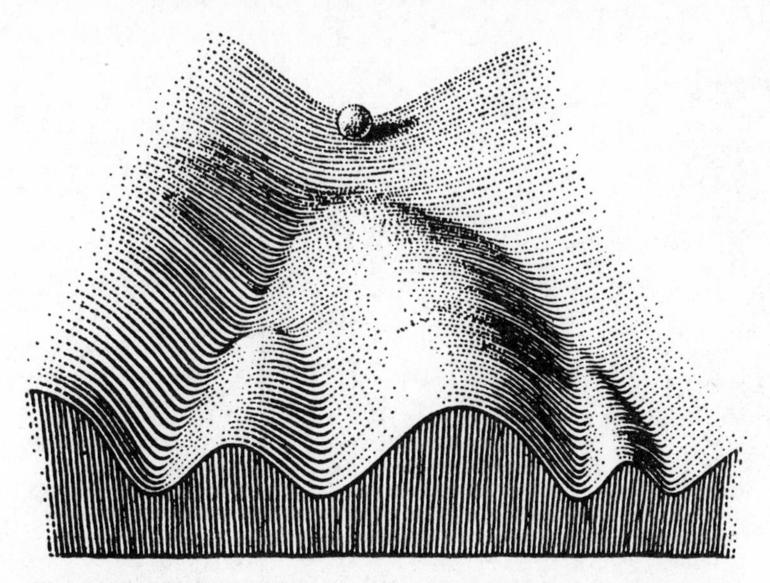

Fig. 1. Waddington's (1957) epigenetic landscape. The ball represents some tissue at an early stage in ontogeny. Development is represented by the ball descending through the landscape. The mechanisms regulating development are represented by the position and shape of the valleys.

MODELS FOR THE CONTROL OF DEVELOPMENT

In a delightful visual aid to the biologist who has difficulty in grasping the abstractions of a purely mathematical model, Waddington (1957) represented the development of a particular part of a fertilised egg as a ball rolling down a tilted plane which is increasingly furrowed by valleys (Fig. 1). He called the surface down which the ball rolls the 'epigenetic landscape'. For present purposes it is not necessary to be concerned with the embryological analogue of gravity which draws the ball downwards in the model; nor is it necessary to bother with the precise conditions that make the ball go down one valley rather than another. The essential point is that the mounting constraints on the way tissue can develop are pictured by the increasing restriction on the sideways movement of the ball as it rolls towards the front lower edge of the landscape. The landscape represents, therefore, the mechanisms that regulate development. Waddington's model is attractive to the visually minded because it provides a way of thinking about developmental pathways and the astonishing capacity of the developing system to right itself after a perturbation and return to its former track. To take a specific example from post-embryonic development, if a juvenile rat is starved during its development, its weight curve falls off while it is being deprived (Fig. 2). When it is put back onto

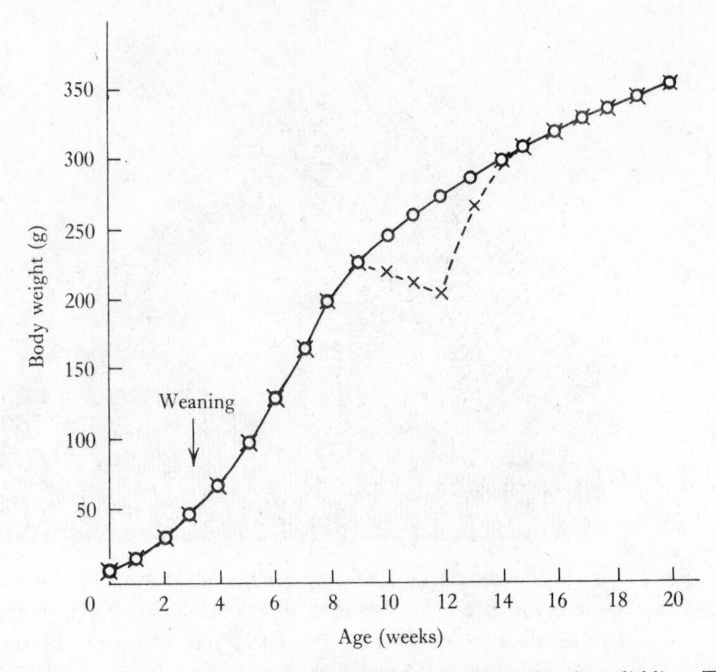

Fig. 2. The gain in weight of normal rats is shown by the solid line. The dotted line shows the weights of rats which were undernourished from 9 to 12 weeks (McCance, 1962).

a normal diet, its weight curve rapidly picks up and rejoins the curve of the rat that has not been deprived (McCance, 1962). Similar examples of growth spurts after illness are well known in humans (e.g. Prader, Tanner & Von Harnack, 1963). For the moment the possibility that the individuals showing the catching-up phenomenon may differ in undetected ways from normal individuals can be ignored. The prime question is how weight-gain is controlled and two individuals with quite different dietary histories end up weighing the same.

The systems theorists in general, and von Bertalanffy (1968) in particular, have laid considerable emphasis on the self-correcting features of development, and have called the convergence of different routes on the same steady state 'equifinality'. While it is easy to become a little mystical about this (and von Bertalanffy did not always resist the temptation), Waddington's model immediately suggests a way of handling 'equifinality'. If the ball rolling down the epigenetic landscape encounters an obstacle in one of the valleys and is not stopped dead, it would ride up round the obstacle and fall back into the valley down which it had been rolling.

Waddington's model is, of course, informal and he would have been the

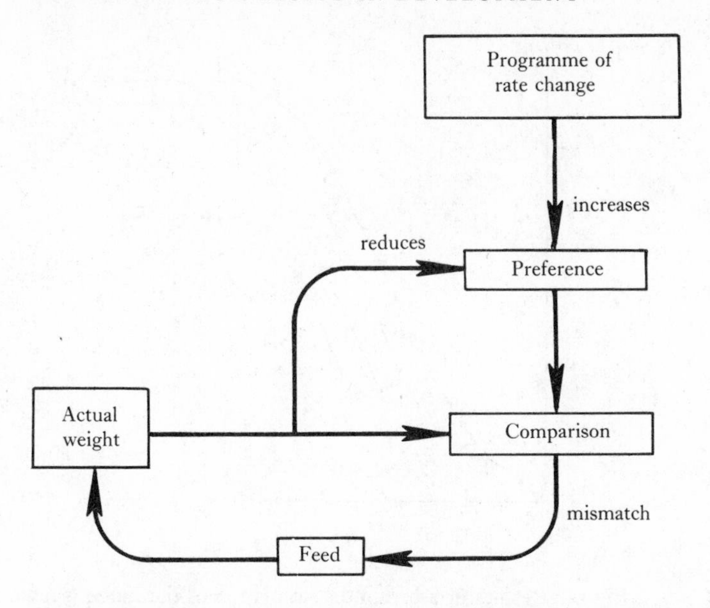

Fig. 3. A model for weight control in the developing rat. The actual weight has no influence on the preference unless the discrepancy between actual and preferred weights is more than a predetermined amount.

first to point to its limitations. Nevertheless it is not difficult to simulate with greater rigour a system which compensates for short periods of food deprivation during development. If the amount of food which the animal attempts to eat is determined by a comparison between a predetermined setting and the actual weight of the animal and if the value of the preferred weight is increased as the animal ages (see Fig. 3), something very similar to the data shown in Fig. 2 can be obtained. To make things more realistic the predetermined increments in the preferred value first increase and then decrease as the hypothetical animal gets older. As a consequence the weight curve of the undeprived animal is sigmoid. Furthermore, an upper limit is set on how much the model rat can eat at any one time and small periodic decrements in weight are imposed to simulate the costs of metabolism. When all this is done, the performance of the model can be readily simulated on a computer (see top line in Fig. 4 and also the results of a rather similar negative feedback model proposed by Weiss & Kavanau in 1957).

It is important not to gloss over a significant feature of this model. By arranging for the preferred value of the closed feedback loop to be changed according to some predetermined plan, the system has been opened up. In the simplest case the developmental process is essentially ballistic – its pathway is determined in advance and does not depend on a dynamic interaction

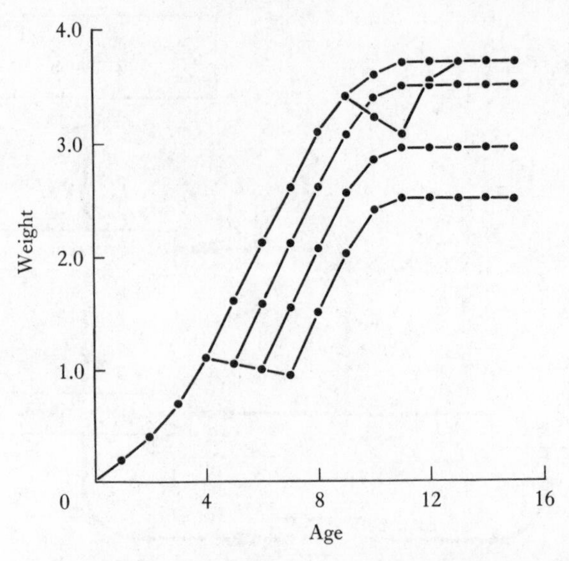

Fig. 4. Results of a computer simulation of the model shown in Fig. 3 showing the interaction between environmental conditions and the predetermined programme for the rate of weight gain. At Age 9 some hypothetical rats were deprived of food for 2 units of age; when allowed to feed again their weight returned to the normal value. Three other groups of rats were deprived of food at Age 4 for 1, 2 or 3 units of age. The longer the period of deprivation the greater the reduction of adult weight.

between the system and other factors which might change during the course of development. The model can, of course, be elaborated so that the preferred weight is modifiable by environmental conditions, and I shall consider a few of the possible elaborations later. This model is not presented in the service of analysing the detailed mechanisms by which rats regulate their weight during development. In all probability such mechanisms are diffuse; possibly the change in 'preferred value' may be mediated indirectly by an increase in the size of the body; and many other factors may be involved besides those included in my model. None of this matters. At this stage the major point I wish to establish is that even rather simple models can account for different developmental routes leading to precisely the same steady state. The phenomena, which were so entrancing to an old-fashioned vitalist, do not pose inordinate conceptual problems.

I now wish to examine how the models based on the development of body tissue may be extended to the ontogeny of behaviour. I cannot claim to be the first to do this. Piaget, among others, has been concerned with the general issue of autoregulation for many years. In particular he has attempted to relate the ideas of thinkers such as Waddington and von Bertalanffy to behavioural development (Piaget, 1967).

EQUIFINALITY IN BEHAVIOURAL DEVELOPMENT

Children differ astonishingly in the age at which language development begins. Some may begin before the end of their first year and others may not utter a recognisable word until they are three or more. Furthermore, during acquisition, styles of forming word patterns may be markedly different. Despite these enormous differences it is remarkably difficult to pick out the early developers when the children are older (Nelson, 1973). Put very cautiously, behaviour at one stage of development is an exceedingly poor predictor of behaviour at another. More boldly one could argue that a child which has been initially slow to develop can demonstrate the catch-up effect seen in tissue growth and reach roughly the same steady state in one aspect of language ability as a much more precocious child.

The example from language development can be matched by many others from child development (see Dunn, this volume), but it is sufficient to demonstrate not only the advantages but also the difficulties of employing the concept of equifinality in developmental studies of behaviour. For any definitions of steady state presuppose an adequate measure of change in that state and an ability to calculate its rate; a steady state is achieved when the rate of change approaches zero. Changes in behaviour patterns pose formidable methodological problems for this kind of analysis, and so do the taxonomic and philosophical problems of specifying when two steady states are the same (which is required if the concept of equifinality is to be of any use). Even with adequate measures, how can one be sure that the absences of differences are not caused by ceiling effects on the scale of measurement? Above all, the level of description of the steady state is critically important (see Hinde & Stevenson-Hinde, this volume). In the example of language development, it is obviously untrue that the words uttered by an adult French person are the same as those made by an adult English person. If one is to talk about equifinality in language development one must be thinking about some quality of language that is more universal and more abstract than the detailed sounds people make. Finally the concept of equifinality can mislead in two distinct ways. First, it can falsely imply that a steady state can only be achieved by a goal-directed system – a steady population size could for example be achieved when births and immigration were balanced by deaths and emigration. Secondly, a steady state may not be the 'final state'; appreciation that a steady state may be followed by one or more periods of rapid change is especially important in the study of development. In any event the final state is presumably death.

Despite these very considerable practical and philosophical difficulties, achieving equifinality in behavioural development does not pose insuperable problems of principle. The parts of the model used to account for the control of weight can be readily adapted to behavioural examples. The preferred value against which the actual state is compared can be for, say, the proprioceptive feedback from a certain action or, at another level, the feedback provided by the behaviour of a parent (see for example Bowlby, 1969; Bischoff, 1975). The justification for thinking in these terms is that it provides a different perspective from the more conventional interactional approach and suggests new ways of looking at the data. It makes sense of experimental studies in which a manipulation may produce obvious short-term effects on behaviour but have no apparent lasting influences (see Dunn, this volume).

Even the simple model which I have proposed so far could, with a few embellishments, lead to marked differences in the pattern of development even though the final outcome was the same. Just as two rats with different food preferences can put on weight at the same rate, so different types of action can lead to the same behavioural end-point. A feature of a system dependent for its control on feedback is that it need not be fussy about how a match between the actual value and the preferred value is achieved. It is the consequences of an action that count not the precise form and patterning of that action. Admittedly possible courses of action may be so constrained that the system is likely to do only one thing when a mismatch between the actual state and the preferred value is detected – such as an electric fire controlled by a thermostat. However, the constraints need not be so great. A remarkable illustration is given by Hoyle (1964, 1970), who found that quite different combinations of muscles in a locust's leg could contract to produce the same overall movement of the leg – the offered explanation being that movement is controlled by means of sensory feedback.

Although a considerable number of possible courses of action may be available to the animal, it may come to repeat the one it performed first, either by a process akin to 'functional validation' (Jacobson, 1969), or by conventional reinforcement when a match between the preferred and actual values is achieved. For reasons which were initially determined by chance or by local peculiarities of its environment, an animal may adopt an individually distinctive style in order to reach the same goal as other members of its species. A possible illustration of this comes from experimental work originally done by Vicari (1929) and re-analysed by Broadhurst & Jinks (1963). Vicari showed that in the early stages of learning to run through a maze, heritability of performance in the maze from one generation of mice to the next was low. However, heritability rose steadily as the mice were given more trials in the maze (see also Manning, this volume). One interpretation of these data is that, because the number of options initially open to the mice was relatively large, environmental conditions were important during the development of a solution to the task but were much less influential on the final performance, which was probably limited by such factors as speed of movement.

Different routes to the same goal may be achieved even more dramatically than in the cases already considered if the young animal is equipped with two or more alternative systems controlling development of the same pattern of behaviour. Redundancy of this kind is common enough in man-made machines when lives are at stake, as in a space capsule. Clearly redundant developmental systems could be highly adaptive for an animal, particularly if the alternative control systems were suited to different environmental conditions to which they were appropriate – the provision of special horses for particular courses. The provision of other systems protects against failure, and from time to time animals must be faced with the situation where no amount of tactical manoeuvring will enable one of their developing systems to proceed along a particular route. Such an animal is a bit like a traveller who arrives at a station only to find that the trains have been cancelled. He can still reach his destination but only by choosing a quite different method of getting there.

If contingency arrangements of this kind have been selected during evolution, Waddington's epigenetic landscape would have to be redrawn so that some valleys ran together again. It could, of course, be argued that a ball that had descended by one valley had had a different history from one that had descended by another so that even though the balls ended up in the same place, the concept of equifinality was valueless. The importance of this objection would depend on whether the different histories did indeed leave distinctive traces on the metaphorical ball. Even if they did, the objection might still not be serious since the resulting differences might be biologically trivial by comparison with the ultimate similarities.

A particularly interesting example of descriptively different developmental pathways to the same pattern of adult behaviour has been given me by Peter Marler and Mark Konishi (personal communication). Isolated white-crowned sparrows (*Zonotrichia leucophrys*) were treated in one of two ways. One group were trained during the sensitive period for song development with the typical song produced by an adult white-crowned sparrow reared in isolation (Marler, 1970). The other group were not trained. The trained group progressed rapidly from sub-song to a 'plastic' song with one theme which crystallised into the typical isolate song. The untrained birds showed a slow and variable progression from sub-song to plastic song which had several themes. Finally, however, the plastic song of the untrained birds crystallised into the typical isolate song. This may provide an example of independent developmental control mechanisms coming into operation under different environmental conditions. Much as I would like to argue for this interpretation, I am inclined to think that the experiments reveal something about the ontogeny of a single underlying mechanism regulating the development of song.

In other contexts, inputs which may be relatively non-specific are frequently required to facilitate the development of particular systems (Bateson, 1976). The inputs may be provided by external environmental conditions or by feedback from the animal's activities such as its own vocalisation (e.g. Gottlieb, 1971). Now, as Marler (1970) argues, it seems highly plausible that the white-crowned sparrow learns to produce a song by matching feedback from its own vocal output against a preferred 'template'. The development of that template may be facilitated by external input, as was the case in the trained birds; and subsequent song development would then occur smoothly. On this view, the untrained birds would have found comparison of their vocal output with their template difficult because the template was comparatively ill-formed. As a result their output would, indeed, have been variable. Later, when feedback from their own sub-song and plastic song had had the same effects on the developing template as those of external training, the untrained song sparrows would catch up the trained birds and ultimately produce a single song.

It may not yet be possible to give a clear instance where different developmental control mechanisms generate the same behavioural end-product. Nevertheless, this could well be an area where good cases will emerge once we start to look out for them.

Rules for changing rules

Up to this point I have implicitly set in opposition the two views of development presented by the interaction theorists on the one hand and the control theorists on the other. However, I believe that if the subject is to progress satisfactorily some synthesis must be achieved between them. I do not merely mean that a trivial compromise should be worked out whereby it is admitted that some patterns of behaviour develop totally as a consequence of interactions between the animal and its environment and others are the products of predetermined, self-correcting developmental processes.

It is already clear that, even with the highly predetermined models which I have been discussing, distinctive styles of individual development could be selected or even induced by particular environmental conditions. This implies a certain interacting relationship between the developing animal and its environment. I now want to consider the possibility that many developmental control mechanisms are themselves modifiable by the environmental conditions in which the animal grows up.

Before giving the first example it may be useful to distinguish between the *development of behaviour* and *behaviour used in development* (cf. Hinde, 1959, on evolution). The biological function of some of the behavioural mechanisms found in many developing animals, particularly higher vertebrates, seems to be the gathering of information. Their predispositions to learn the characteristics of certain things can be highly specific (see Hinde & Stevenson-Hinde, 1973). Such proclivity can be extremely important in directing the course of development. A good example is provided by the behaviour which young precocial birds normally direct towards their parents. This example also illustrates the more general point about the modifiability of control mechanisms.

When a day-old domestic chick is removed from a dark incubator where it hatched and is placed in a plain environment, it soon starts to walk about and emit piercing calls. If a visually conspicuous object, particularly something resembling the size, shape and colour of an adult hen, is presented to the chick, its behaviour changes dramatically. The disorientated movements stop and the piercing peeps are replaced by soft twittering calls. The bird orientates towards and approaches the object. It is difficult (and probably unnecessary) to avoid the anthropomorphic impression that the chick is happier. Indeed, if the chick is placed in a situation in which the presentation of the visually conspicuous object depends on the bird performing a particular act such as pressing a pedal, the bird learns with astonishing rapidity to press the pedal (Bateson & Reese, 1968, 1969). The object can evidently be used as a reward for the animal and as a means of strengthening arbitrarily selected activities. The control of the bird's behaviour can be interpreted in terms of the model shown in Fig. 5. The preferred setting consists of some broad representation of visually conspicuous objects. This is matched against incoming visual input. A mismatch leads to behaviour which, in a natural situation at any rate, increases the chances of the bird encountering something more appropriate. By scanning and moving around, it is more likely to encounter its mother than if it stood stock still and, of course, the piercing calls cause the mother to emit calls herself and move towards the chick. In terms of the model, a match terminates all this and allows the chick to change its behaviour to that of approaching the stimulus achieving the match. This brings the chick underneath, or at least close to, the mother or her substitute. A match also has another effect which is crucial to the development of the chick's behaviour.

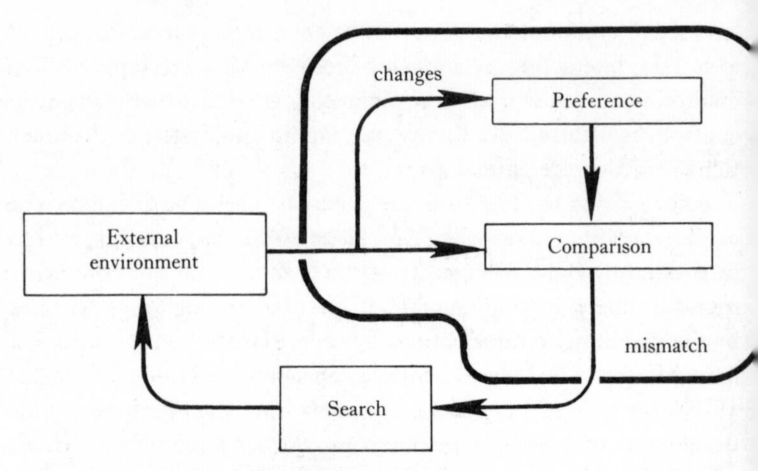

Fig. 5. A model for imprinting in which the preferred value in the negative feedback loop is changed by the input from the external environment.

Before a chick has seen a conspicuous object, any one of a large number of different things will be effective. However, after being exposed to one of those objects, a bird will increasingly tend to scan for that object in preference to all others. The process which restricts the bird's preferences to an object which is familiar is, of course, known as 'imprinting'. Its relevance to the argument is that, in as much as the closed-loop model for behaviour in development is justified, the model has to incorporate a self-organising principle. When a match is achieved the preferred setting is more tightly specified so as to correspond in greater detail to a representation of the particular object which achieved that match. After modification the control loop will maintain 'searching' until something resembling the familiar object is encountered.

A single rule for changing the characteristics of the control system is probably not enough to account for the developing chick's behaviour in relation to its mother (or the flashing light, moving box or whatever else is substituting for her). Since the mother in back view presents physically quite different stimuli to the young than she does in front view, it is probably necessary for the young to acquaint themselves with those different views. Considerable evidence now suggests that the young chicks do indeed prefer and work for slight novelty having learned some of the characteristics of one object (Bateson, 1973*a*; Jackson & Bateson, 1974; Bateson & Jaeckel, 1976). I have suggested elsewhere ways in which the chicks might narrow down their preferences to familiar objects while at the same time showing a preference for slight novelty (Bateson, 1973*a*, *b*). This mechanism coupled with a means for classifying together stimuli which have been encountered in the same temporal context (see Bateson & Chantrey, 1972; Chantrey, 1972, 1974) provides the animal with a powerful yet flexible way of developing constancies within a perceptual category such as 'mother'. It must be emphasised that such mechanisms add to but do not substantially alter the model for a self-modifying system whose eventual steady state is determined by environmental conditions.

Another kind of modification dependent on environmental conditions is suggested by the stunting obtained if animals or humans are starved for long enough during development. The simple model which I have already used for the control of weight can be readily adapted to cope with such evidence by making the extent of the increments in preferred weight dependent on the difference between the preferred weight and the actual weight (see Fig. 3). If the discrepancy is large the increment in preferred weight is less than if the discrepancy is small (see Fig. 4). This one rule, which could be specified in advance, would greatly enhance the dynamic interaction with the environment. It would have one interesting consequence which would be particularly striking if the normal growth curve were sigmoid with the period of maximum growth occurring mid-way through development. The stunting effects of starvation would be particularly marked at times of rapid growth. As can be seen from Fig. 6 this would give rise to periods in development when the animal was especially vulnerable to environmental disturbance. The biological advantage of a rule which allows for a change in the preferred value is that the animal does not endlessly attempt to reach a state which may never be achievable in the particular conditions in which it is developing. Deprivation of optimal conditions for one system does not necessarily imply that conditions are bad all round and normal development of the animal's other systems may still be possible. Although the animal may be handicapped, its chances of surviving and leaving offspring may not be reduced to nothing.

One behavioural example of settling for less than the best may be provided by the nest-site selection of the blue tit (*Parus caeruleus*). Hinde (1952) found that in the spring the tits would visit a large variety of crannies many of which were obviously unsuitable. One way of interpreting their behaviour would be that, if the actual site did not match up to the characteristics of an optimal nest-site, they kept searching – to begin with at least. However, if optimal sites were unavailable or already occupied, the birds would ultimately nest in places which they had previously rejected. It would make good sense if they were equipped with a rule for gradually relaxing the conditions under which searching for a nest-site was brought to an end and nest-building began. It may seem like a glimpse of the obvious that, metaphorically speaking, a starving man is not fussy about what he eats. Nevertheless, the relevance of the blue tit example to a discussion of development is that once the bird has

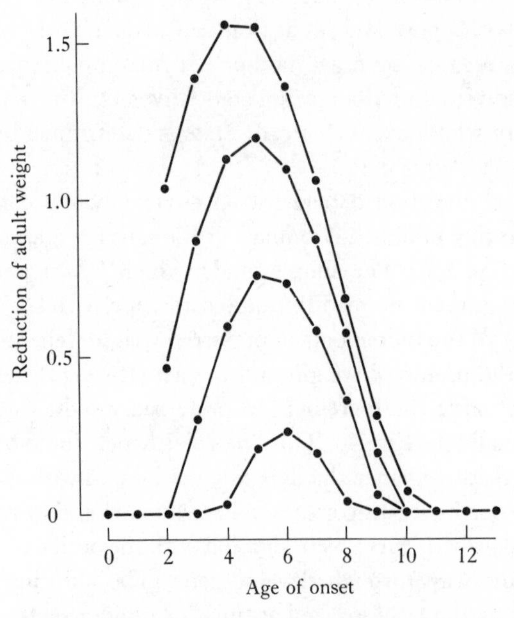

Fig. 6. The reduction of adult weight of hypothetical rats deprived of food for varying amounts of time (1–4 units of age) at different ages. All units are arbitrary. The data are derived from a computer simulation of the model shown in Fig. 3.

selected a sub-optimal site it will, for that breeding season at least, presumably prefer it even if an optimal site should subsequently become available.

It may prove particularly profitable to examine the modification of preferred values in the context of emerging social relationships. Suppose that it is important for the maintenance of a relationship between two individuals that they both have the same general pattern of behaviour – the same activity rhythm, for example. Now in the early stages of a relationship, differences in pattern might well exist but these might reflect nothing more than the relatively unimportant peculiarities of personal history. Without cost it might be possible for one or both of them to change their preferred patterns. One simple way of doing this is shown in Fig. 7. The idea is that if a pattern of behaviour is achieved by comparison with a preferred standard, that same standard could also be used for judging a companion's behaviour. What is central to this argument, A's standard could be changed by B's and vice versa. As the model is shown in Fig. 7 any mismatch would immediately lead to the individuals breaking-off contact with each other. It would, therefore, be necessary to provide for a mechanism which would, in the early stages of a developing relationship, over-ride or inhibit the consequence of a mismatch.

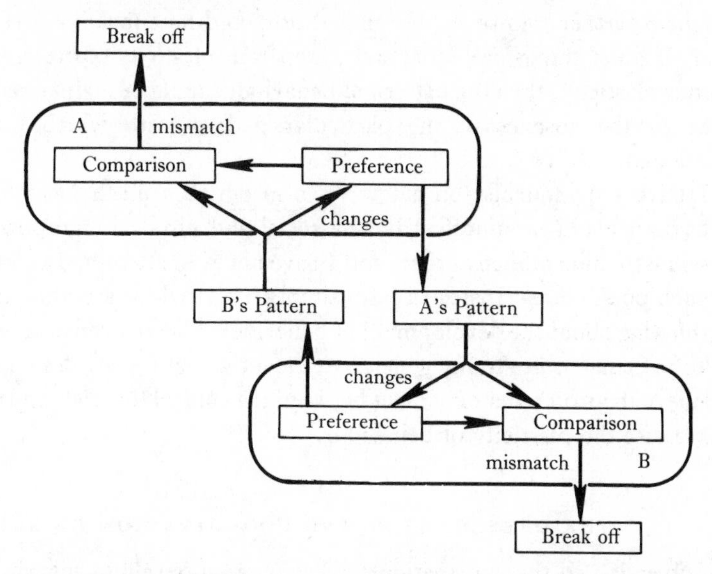

Fig. 7. A model for the meshing of behaviour of two interacting animals. An animal's preferred value determines not only its own pattern of behaviour but also provides the standard by which it judges the behaviour of its partner. As the model is shown, a mismatch will lead to a break-off in the relationship. But if that is prevented, each animal's preferred value may be modified so that it resembles more closely that of its companion.

For example, two individuals might be drawn together by the physical appearance of the other. During the 'honeymoon' period the relatively subtle differences of behaviour would be ignored. It would only be later, when the effects of physical appearances had started to wane that a mismatch of behaviour would become important and lead to a disruption of the relationship. But in the intervening period one or both of the individuals could have changed its pattern of behaviour so as to correspond to that of the other. Of course, the flexibility of an animal might be constrained by some social roles and facilitated by others.

An example of the behavioural meshing I have in mind could be the relationship that develops between mother and infant rhesus monkeys (*Macaca mulatta*) described by Hinde & Simpson (1975). Simpson had been able to measure independently when the mother left her infant and when the infant left its mother. In some pairs the probability that the mother would leave the infant at any particular moment after they had come together was closely related to the probability that the infant would leave the mother. Such meshing could, of course, be obtained in a variety of ways. For example, the two individuals might become highly sensitive to the immediate cues provided by their partner. However, the model proposed here makes a useful prediction. If other things are equal and if apparent plasticity of preference is not merely elasticity, then the pattern of behaviour should be maintained for some time in the absence of the particular partner with which the pattern developed.

I have only touched on a few ways in which a mechanism controlling behaviour could be modified by the social and physical environment. It is possible to think of many others and I have not even attempted a classification of such possibilities. Instead I have simply tried to draw attention to one way of thinking about the development of behaviour. The relevance of postulating rules to change rules to my general argument is that it provides a major and, I believe, fruitful point of contact between the control theorists and those who emphasise the plasticity of behaviour.

THE DEVELOPMENT AND COORDINATION OF RULES

An objection to the explanation for behavioural development which I have presented here is that it seems to be heading for an infinite regress with rules for changing the rules that change the rules and so on. Ultimately one must ask how the starting rules developed in the first place. While the issue ought not to be evaded, it need not necessarily be the immediate concern of somebody whose major research interests and competence lie at the behavioural level. If the models postulated here have some validity, the nexus of events necessary for the development of the starting rules required by the models may be largely cellular and beyond analysis at the level of the whole animal. Models of development embodying certain rules can perform a useful service in integrating evidence and offering certain predictions without ever explaining how those rules came about. But having made that disclaimer, I should not want to shovel all responsibilities for analysis of the early stages of ontogeny onto the developmental neurobiologists. Even this work may have a 'whole animal' component to it in as much as a rule governing behaviour has been selected during evolution for its adaptive outcome rather than its cellular origins. So long as the rule reliably develops in the individual, one presumes that natural selection is indifferent to how that came about. The selection may have taken place when certain environmental conditions were invariant and successful ontogeny may depend on the maintenance of those conditions. If they are changed for whatever reason (experimental, social), development may be disrupted. Knowledge of such disruption (however non-specific) is relevant to an understanding of the mechanisms involved in development (Bateson, 1976). My guess is, though, that major advances will not be made here without detailed knowledge of the cellular machinery and I prefer to focus attention on the later stages of development where behavioural evidence is likely to make most impact.

If internal mechanisms have developed, by some means or another, to control later stages of behavioural ontogeny, one would expect a considerable degree of coordination to exist between different mechanisms. For example, the rates of development of two patterns of behaviour may be independently influenced by interactions with the environment and it may be important that the development of one does not outrun the development of the other. Alternatively the order in which behaviour patterns develop may be important; for example, exploration of the environment may be disastrous if it occurs before the young animal has established some standards of what is familiar. In such cases acquisition of information must precede performance. This is particularly obvious in the development of complex motor patterns such as those used in bird song (see Marler, 1970; Marler & Mundinger, 1971). If the song is to be finely tuned to match some predetermined pattern (which may or may not be learned), feedback from the performance must be compared with the preferred value. Once a motor pattern producing the appropriate feedback has been established dependence on feedback can be reduced or even eliminated and the animal can accelerate the output rate. This is a bit like a musician learning a new part. While he has to monitor the individual sounds he is making to ensure their accuracy he must allow at least 100 msec between each note. However, in the final performance when such control is no longer needed, the gaps between notes can be greatly reduced to 50 msec or less (see Lashley, 1951).

The implication of examples such as these is that when certain conditions have been satisfied, new mechanisms of control can be brought into operation. In self-modifying systems, for instance, the conditions necessary for progressing to the next stage of development could be the levelling-off of modification – in other words the achievement of a steady state. This type of explanation would side-step an unprofitable debate about the precise chronology of developmental stages (see for example Hutt's (1973) discussion of sensitive periods). It would focus attention on the environmental conditions and on the state of the animal associated with a transition from one stage of development to the next rather than on age as such.

A great deal more can be said about the interactions between mechanisms controlling different aspects of behaviour (cf. Fentress, this volume). For the moment, I just want to add one point. Such interlocking systems, far from providing yet another way to complicate the lives of those investigating development, could be usefully harnessed for research purposes. Control mechanisms which are not directly influenced by environmental conditions may nevertheless be linked to control mechanisms that are. It may therefore be possible to gain access to the first by manipulating the second.

CONCLUSION

At the beginning of this chapter I argued that recent experimental work on the development of behaviour had been dominated by a view that admitted little in the way of internal regulation. The developing animal has been treated as though it were a billiard ball whose path is the resultant of the various external forces that acted on it. Consequently such organisation and patterning in development as has been recognised has usually been attributed to invariant features of the environment rather than to any kind of internal control.

The problem with this position is not so much that it tends to play down the development of behaviour patterns which are buffered against environmental influence but that it offers no clear strategy as to how research on the relatively labile systems should proceed. What is needed is not a comfortable compromise between two extreme theoretical view-points whereby it is admitted that some behaviour is stable and some is not. The need is for a theoretical framework that encompasses the different ways in which behaviour develops. I cannot claim to have done that but I have attempted to marry two points of view which have sometimes seemed incompatible. I have tried to show that, far from being irreconcilable, the approaches of theorists interested in interactions and those interested in control mechanisms usefully complement each other. A failure to show that some experimental manipulation of the environment has long-term effects on behaviour is not a reason for despair. It may be a useful pointer to a self-correcting feature of the underlying processes. Conversely, a developmental control mechanism whose preferred value drifts should not be dismissed as an irritating example of variability in biological material. Instead, the environmental conditions in which those drifts take place should be closely scrutinised since the modification in preferred value may have some obvious functional significance. In brief, I wish to argue that the development of behaviour very often requires internal rules for its guidance but that reciprocity between the animal and its environment is also needed in needed in order to give those rules greater flexibility and definition.

SUMMARY

Two seemingly incompatible views of behavioural development are outlined. The dominant view in recent years has been that an adult's behaviour is the product of continuous interaction between the individual and its environment throughout ontogeny; the analytical task is seen, therefore, as uncovering the sources of individual differences by close examination of those interactions. An alternative view, much less common in studies of behavioural development, emphasises the predetermined, self-correcting features of developmental processes; this places emphasis on control systems which allow the same steady state in the adult to be reached by apparently different developmental routes. I argue that this second approach provides some valuable perspective in the study of behavioural ontogeny because it offers coherent principles by which development might be regulated. This perspective can, however, be easily reconciled with the interactionists' point of view and I go on to consider some simple ways in which it is possible to bring together the seemingly opposed positions. For example, plasticity in response to different environmental conditions may often usefully reside in those mechanisms that determine action by matching actual input values with preferred values. The main point is that if animals have rules by which their behaviour is controlled, functional reciprocity between the developing animal and its environment can be usefully achieved by equipping the animal with rules for changing the rules.

In addition to the members of the conference, whose comments helped me a great deal, I should like to thank Ariane Etienne, G. Horn and R. Oppenheim for their guidance.

REFERENCES

Bateson, P. P. G. (1973*a*). Internal influences on early learning in birds. In *Constraints on Learning: Limitations and Predispositions*, ed. R. A. Hinde & J. Stevenson-Hinde, pp. 101–116. Academic Press: London & New York.

Bateson, P. P. G. (1973*b*). Preference for familiarity and novelty: a model for the simultaneous development of both. *Journal of theoretical Biology*, **41**, 249–259.

Bateson, P. P. G. (1976). Specificity and the origins of behavior. *Advances in the Study of Behavior*, **6**, 1–20.

Bateson, P. P. G. & Chantrey, D. F. (1972). Retardation of discrimination learning in monkeys and chicks previously exposed to both stimuli. *Nature, London*, **237**, 173–174.

Bateson, P. P. G. & Jaeckel, J. (1976). Chicks' preferences for familiar and novel conspicuous objects after different periods of exposure. *Animal Behaviour*, **24**, 386–390.

Bateson, P. P. G. & Reese, E. P. (1968). Reinforcing properties of conspicuous objects before imprinting has occurred. *Psychonomic Science*, **10**, 379–380.

Bateson, P. P. G. & Reese, E. P. (1969). The reinforcing properties of conspicuous stimuli in the imprinting situation. *Animal Behaviour*, **17**, 692–699.

Bertalanffy, L. von (1968). *General System Theory*. Braziller: New York.

Bischof, N. (1975). A systems approach towards the functional connections of attachment and fear. *Child Development*, **46**, 801–817.

Bowlby, J. (1969). *Attachment*. Hogarth Press: London.

Broadhurst, P. L. & Jinks, J. L. (1963). The inheritance of mammalian behavior re-examined. *Journal of Heredity*, **54**, 170–176.

Chantrey, D. F. (1972). Enhancement and retardation of discrimination learning in chicks after exposure to the discriminanda. *Journal of comparative and physiological Psychology*, **81**, 256–261.

Chantrey, D. F. (1974). Stimulus pre-exposure and discrimination learning by domestic chicks: effect of varying inter-stimulus time. *Journal of comparative and physiological Psychology*, **87**, 517–525.

Gesell, A. L. (1954). The ontogenesis of infant behavior. In *Manual of Child Psychology*, ed. L. Carmichael, pp. 335–373. Wiley: New York.

Gottlieb, G. (1971). *Development of Species Identification in Birds*. University of Chicago Press: Chicago.

Hinde, R. A, (1952). The behaviour of the Great Tit (*Parus major*) and some other related species. *Behaviour Supplement*, **2**, 1–201.

Hinde, R. A. (1959). Behaviour and speciation in birds and lower vertebrates. *Biological Reviews*, **34**, 85–128.

Hinde, R. A. & Simpson, M. J. A. (1975). Qualities of mother–infant relationships in monkeys. In *Parent–Infant Interaction*, Ciba Foundation Symposium, vol. **33**, pp. 39–67. Elsevier: Amsterdam.

Hinde, R. A. & Stevenson-Hinde, J. (ed.) (1973). *Constraints on Learning: Limitations and Predisposition*. Academic Press: London & New York.

Hoyle, G. (1964). Exploration of neuronal mechanisms underlying behaviour in insects. In *Neural Theory and Modelling*, ed. R. F. Reiss, pp. 346–376. Stanford University Press: Stanford.

Hoyle, G. (1970). Cellular mechanisms underlying behavior – neuroethology. *Advances in Insect Physiology*, **7**, 349–444.

Hutt, S. J. (1973). Constraints upon learning: some developmental considerations. In *Constraints on Learning: Limitations and Predispositions*, ed. R. A. Hinde & J. Stevenson-Hinde, pp. 457–467. Academic Press: London & New York.

Jackson, P. S. & Bateson, P. P. G. (1974). Imprinting and exploration of slight novelty in chicks. *Nature, London*, **251**, 609–610.

Jacobson, M. (1969). Development of specific neuronal connections. *Science, Washington*, **163**, 543–547.

Lashley, K. S. (1951). The problem of serial order in behavior. In *Cerebral Mechanisms in Behavior*, ed. L. A. Jeffries, pp. 112–136. Wiley: New York.

Marler, P. (1970). A comparative approach to vocal learning: Song development in white-crowned sparrows. *Journal of comparative physiological Psychology Monographs*, **71**, 1–25.

Marler, P. & Mundinger, P. (1971). Vocal learning in birds. In *The Ontogeny of Vertebrate Behavior*, ed. H. Moltz, pp. 389–450. Academic Press: New York.

McCance, R. A. (1962). Food, growth and time. *Lancet*, **2**, 671–676.

Nelson, K. (1973). Structure and strategy in learning to talk. *Monographs Society for Research in Child Development*, **38**.

Oppenheim, R. W. (1974). The ontogeny of behavior in the chick embryo. *Advances in the Study of Behavior*, **5**, 133–172.

Piaget, J. (1967). *Biologie et Connaissance*, Gallimard: Paris.

Prader, A., Tanner, J. M. & Von Harnack, G. A. (1963). Catch-up growth following illness or starvation. *Journal of Pediatrics*, **62**, 646–659.

Schneirla, T. C. (1965). Aspects of stimulation and organisation in approach/withdrawal processes underlying vertebrate behavioural development. *Advances in the Study of Behavior*, **1**, 1–74.

Vicari, E. M. (1929). Mode of inheritance of reaction time and degrees of learning in mice. *Journal of experimental Zoology*, **54**, 31–88.

Waddington, C. H. (1957). *The Strategy of the Genes*. Allen & Unwin: London.

Weiss, P. & Kavanau, J. L. (1957). A model of growth and growth control in mathematical terms. *Journal of General Physiology*, **41**, 1–47.

EDITORIAL: 6

Three problems recur throughout this section on development: How is the behaviour of a developing animal to be described? How are the developmental origins of behaviour to be uncovered? How are the developmental processes to be analysed? Though inter-related, it is important to consider these issues separately.

Any study of behaviour raises the problem of finding stable units for description and measurement. As Fentress pointed out in Part A, the boundaries of behavioural systems may change as the animal's state alters. So the possibility, discussed by Manning, that the genotype eventually expresses itself in the form of stable and easily recognisable phenotypic characters, gives only limited scope for optimism. One might go further and argue that it will only rarely be possible to unscramble behaviour into a series of absolute units, and that the appropriateness of a measure is much more dependent on the question that is being asked than it is on some invariant feature of the animal under study.

Given that definition of the end-products of behavioural ontogeny is controversial, agreement about the *changing* characteristics of behaviour patterns during ontogeny is even more elusive; the difficulties of obtaining universal units are greatly compounded in developmental research. Over and above this problem of description looms another question: Does physical similarity of behaviour patterns observed at different times in an animal's life justify the use of the same label for each pattern? When the answer is uncertain, is it enough merely to be alert to the possibility that a common description may not entail a common mechanism? Yet other questions arise when description involves the identification of different developmental stages. Do those stages merely represent the drawing of arbitrary lines across continuous change for descriptive convenience? Or does the classification represent recognisable discontinuities in rates of change, or different clusters of behaviour? A related issue is whether the study of learning in terms of numbers of trials falsifies nature, imposing an episodic character on processes more properly represented as continuous (Hinde, 1973). These issues are central to both Rosenblatt's and Simpson's chapters.

The second general problem, that of uncovering the origins of behaviour, has been responsible for some of the most persistently fog-bound controversies in the past. However, much of this evaporated when it was appreciated that a concern with the origins of the adaptiveness of behaviour is not the same as a concern with the processes of development. A behaviour pattern may have been selected during the evolution of a species and yet be acquired in the life of each individual – the social preferences of some birds are well-known examples. Since this valuable clarification, the focus of most developmental research has been on the determinants of particular behavioural characters. The studies reviewed in Rosenblatt's chapter provide one of the most complete and elegant examples of the systematic use of differences to analyse the sources of behavioural change.

Inevitably the subtlety and complexity of the developmental influences acting on behaviour raise the third general issue – the problem of discovering how those influences produce their effects. What processes are involved in the interactions between the developing animal and its environment? General principles are not easy to abstract at the moment. But the possibility emerging from this section is that the development of behaviour is not always ballistic, and may be in some measure self-regulatory. From rather different standpoints, Manning and Bateson raise this issue. The implication is that behind the complexity of continuous interactions some over-riding strategy to behavioural development can be detected. The value of considering the constraints on learning and the predispositions to learn in this broader perspective is also discussed by Simpson. Both he and Bateson emphasise the behavioural mechanisms involved in the overall process of the development of behaviour.

The emerging concern with the mechanisms regulating the course of development throws into a new light the long-standing preoccupation with sources of differences. A factor which can exert a short-term influence on behaviour may have no lasting effect if the underlying processes are indeed self-stabilising. This issue is raised also by Dunn in the next section, where it has important practical implications in the context of child development.

REFERENCE

Hinde, R. A. (1973). *Constraints on learning: An introduction to the problems.* In *Constraints on Learning: Limitations and Predispositions*, ed. R. A. Hinde & J. Stevenson-Hinde. Academic Press: London & New York.

PART D

Human social relationships

EDITORIAL: 7

Some of the most important early advances in ethology came from studies of insects, fish and birds. However, many ethologists have had their eyes on the application of their work to man. Perhaps this had to wait until the way was paved by studies of non-human primates; although these were pioneered in the thirties by Carpenter (e.g. 1964) and Zuckerman (1932), really substantial progress did not come until the sixties. Be that as it may, it is now clear that ethologists are in a position to make a significant contribution to the understanding of human behaviour.

In the early stages of an ethological approach to human behaviour the basis has not unexpectedly been descriptive. It is no easy task, though, to find the right perspective and emphasis. A search for human universals needs to be tempered with a deep awareness of cultural differences, and ethologists have had much to learn from those approaching the subject from a different background (compare e.g. Eibl-Eibesfeldt, 1972, with Leach, 1972; and see Ekman & Friesen, 1969). A balance must also be struck between fine-grained accounts of human expressive movements (e.g. Brannigan & Humphries, 1972) and a broader focus on relationships between behavioural elements (e.g. Blurton Jones, 1972). Finally, ethological descriptions of human behaviour are not likely to be of much value if they disregard context and meaning, or fail to relate to affective and cognitive processes (e.g. Poole, 1975).

Of the chapters in this section, the one by Blurton Jones provides a strong link with Part B. While concerned primarily with the nature of the contributions that ethologists can make to the social sciences, he comes down firmly in favour of a functional approach and makes a plea for strong links with ecology. By contrast, Hinde & Stevenson-Hinde's contribution is concerned primarily with problems of development and causation – with understanding the complex dynamic processes that underlie the formation and maintenance of inter-individual relationships. Dunn's chapter is also concerned with development, and focusses on an issue raised in Part C by Bateson: to what extent is behavioural development such that a given end-point is reached despite vicissitudes along the way? Her discussion is firmly linked to the practical problems of providing an adequate social environment for young children.

The section ends with two over-views of human ethology from different perspectives. Medawar's comes from outside the subject, although he has long been a friend to ethologists. His provocative essay stimulated some of the most lively discussion at the conference. Rather than alter the freshness of its impact for the reader, his manuscript was not subsequently revised; instead Medawar provided an addendum in which he replied to some of the points raised in the discussion. The second over-view is by Tinbergen, who has inspired all of us. Unfortunately he was prevented by illness from attending the conference. Had he done so he would certainly have been in sympathy with Dunn's goals, for he argues forcefully that ethologists should consider taking up problems of immediate practical importance to mankind.

REFERENCES

Brannigan, C. R. & Humphries, D. A. (1972). Human non-verbal behaviour, a means of communication. In *Ethological Studies of Child Behaviour*, ed. N. G. Blurton Jones. Cambridge University Press: London.

Blurton Jones, N. G. (1972). Categories of child–child interaction. In *Ethological Studies of Child Behaviour*, ed. N. G. Blurton Jones. Cambridge University Press: London.

Carpenter, C. R. (1964). *Naturalistic Behaviour of Nonhuman Primates.* The Pennsylvania State University Press: University Park.

Eibl-Eibesfeldt, I. (1972). Similarities and differences between cultures in expressive movements. In *Non-verbal Communication*, ed. R. A. Hinde. Cambridge University Press: London.

Ekman, P. & Friesen, W. V. (1969). The repertoire of nonverbal behavior: categories, origins, usage and coding. *Semiotica*, **1**, 49–98.

Leach, E. R. (1972). The influence of cultural context on non-verbal communication in man. In *Non-verbal Communication*, ed. R. A. Hinde. Cambridge University Press: London.

Poole, R. (1975). Objective sign and subjective meaning. In *The Body as a Medium of Expression*, ed. J. Benthall & T. Polhemus. Allen Lane: London.

Zuckerman, S. (1932). *The Social Life of Monkeys and Apes.* Routledge: London.

14

Growing points in human ethology: another link between ethology and the social sciences?

N. G. BLURTON JONES

I have understood my invitation to participate in this symposium as an invitation to speculate and predict about the next few years of ethological studies of man. Consequently I shall be concerned primarily with some issues in the study of human behaviour in which, I predict, biologists could usefully involve themselves. Hopefully, biologists making excursions into the study of human behaviour will retain the rigour and clear operational thinking that has characterised the Madingley subdepartment's first 25 years. This, and an understanding of evolution by natural selection, are the most valuable things that biologists can bring with them into the study of human behaviour.

The last ten years have seen several classes of interaction between ethology and the traditional studies of human behaviour. The popular books on comparisons of man and animals led to a lively clash. We can all criticise these but I should like to compliment the authors on their courage. Caution and modesty have held back too many good biologists from getting involved in the study of human behaviour. But courage needs to be allied to a great deal of reading and discussion aimed at finding out what the issues are in established human behaviour research, and even more important, at finding out what the data are and what they are worth. Too many biologists remain either totally unaware of enormous areas of social science literature, or else treat it much less critically than do their social science colleagues.

Comparative studies can probably be carried out in a more systematic manner than the popular books might seem to suggest, and can have some use, though many of the claims made cannot be supported. (For instance, the presence and survival value of a kind of behaviour in animals as well as in man does not imply that it is 'innate' nor that it has a particular kind of motivation (e.g. that it is 'spontaneous').) There are several ways in which comparative studies can be useful. I have attempted in two other papers (Blurton Jones, 1972*a*, 1975) to illustrate some of the possibilities and limitations of systematic comparative studies, taking mother–infant contact as the central example. My argument was that the comparative method was basically one of correlation of behaviour with, say, anatomy or ecology. Statistical regressions can be used to extrapolate to 'missing' data, such as some features of the behaviour of man. To give but one example, very short intervals between feeds are predicted for human newborns from the low fat and protein content of human milk, the slow rate of sucking, and the high growth rate of babies compared to the non-human primates. But the inductive approach illustrated in those papers is much less powerful and elegant than the applications of the deductive approach of sociobiology to questions of parent–offspring relationships, life histories and development (Trivers, 1974; Gadgil & Bossert, 1970).

Ethology has fruitfully interacted with developmental psychology in the convergence towards studies of children by direct observation. In these studies, definable, observable items of behaviour are recorded in situations approximating to real life (or with aspects of real life such as presence or absence of mother, sex and age of social partner, systematically varied). Several of these studies have been concerned with testing the reality of the categorisations of behaviour given to us by our everyday view of life. The promotion of these kinds of direct study was the main aim of the book of such studies which I edited (Blurton-Jones, 1972*a*). Important recent examples are the papers by Hinde & Simpson (1975) and Dunn (1975), and the chapter by Hinde & Stevenson-Hinde in this volume. In many respects the meeting between ethology and developmental psychology is happily so complete that it no longer makes any sense to compare and contrast approaches nor to ask about the academic origins of any individual researcher.

But the task of challenging the untested large categorisations of behaviour remains nowhere near complete. If our aim is a scientific study of behaviour, we must be able to test and if necessary reject our everyday culture's categorisations of behaviour. Though these are valuable sources of ideas they are likely to be too simple, too imprecise and insufficiently explicit. (They may even be designed to mislead, see discussions below.) It is too easy to pay only lip service to science and to continue on the assumption that we know how people really work. (Perhaps we do, but if so why bother with trying the scientific approach at all?)

The result is that many studies of human behaviour work with a dangerous mixture of the 'emic' and 'etic' research options. The distinction between these options has become an important issue in anthropology (Harris, 1968), and ethology can continue to have a useful influence by provoking continued consideration of this distinction in studies of the behaviour of, for example, parent and child. By Harris's definition (slightly different from the original usage of Pike, 1954) emic statements concern categories, 'Contrasts and discriminations significant, meaningful, real, accurate, or in some other fashion regarded as appropriate by the actors themselves' (Harris, 1968, p. 571). Etic statements 'depend upon phenomenal distinctions judged appropriate by the community of scientific observers' (*ibid.*, p. 576). They cannot be falsified by the actors' beliefs, but by the observations of independent observers.

Until recently, psychology has wavered between etic and emic approaches, possibly because the distinction is not understood and certainly it is not maintained. This means that there is little good emic research, and in some areas little etic research. The work on parental attitudes has perhaps been the most complete muddle between the two. This is unfortunate because the study of the pure emic aspects of parental behaviour (what parents believe they do or should do and their expressed reasons for what they do) would be interesting and rewarding, just as is the study of the etic aspects of parental behaviour (what parents do do and under what circumstances).

The grip that our cultural categorisations have on our minds is strong, and seems to me to be a major obstacle to progress in the human behavioural sciences. Thus it is very difficult, if one has already thought of maternal behaviour in terms such as adequate or inadequate mothering, warm or cold mothering etc, to attach full significance to the important distinctions that Dunn, and Hinde and Simpson have described for aspects of maternal behaviour. Very general categories of good and bad mothers will keep coming back whatever the data say. Even something so evidently clearly defined as maternal responsiveness (to behaviour *x* or *y*) gets used to mean 'nice', or criticised as not 'really' being responsiveness. It is a ghastly task trying to communicate to someone who is not a Hinde monkey student the difference between mothers who (*a*) initiate lots of interactions, (*b*) have lots of interactions, (*c*) respond often to child initiations. And these are certainly not features of mothers readily distinguished and picked up by most people's intuitions. Yet for predicting child behaviour they are absolutely crucial. It is of course equally important to test whether distinctions about maternal behaviour that help us predict child behaviour are the same or different from distinctions about maternal behaviour that help us predict about maternal behaviour.

The continuing attempt to 'operationalise' in the strictest sense becomes even more important when we attempt to study more complex information processing. The study of 'cognition' has plenty of mentalistic corners, so much so that one wonders if the opportunity to preserve mentalism is what attracts some researchers to that field.

Another important aspect of recent child development research, which has to do with the wide ranging approach of the biologists and is seen in the survey studies of medically trained psychiatrists (Rutter, Tizard & Whitmore, 1970), and in the more microscopic studies of ethologists, concerns the realisation that many different kinds of important variables influence the child's development. Dunn and Richards (Richards & Bernal, 1972; Bernal & Richards, 1973) have shown clearly the importance of a multi-level approach, the necessity of data at levels from the biochemical to the sociological for explaining even simple aspects of child development. The application of this approach is spreading but it is still far from generally accepted.

Important gains are to be made in the study of child development and mother–infant interaction by coming to terms with the size and complexity of the system we are dealing with. It is to be hoped that the problems of dealing with large systems will not distract us into a new phase of mysticism in child development but will force us to use the rigorous techniques already available for the study of large systems. The apparent ability of the child to correct its developmental trajectory as described by Bateson (this volume) and Dunn (this volume) is one major challenge.

The development of evolution-based interpretations and direct studies of human behaviour has been another growing link between biology and the social sciences. The discussion of mother–infant attachment by Bowlby (1969, 1973) and theory and research by Ainsworth (1969) is one important example. The research on population differences in newborns by Freedman is another (Freedman & Keller, 1971). There have been a number of moves towards ethology from within anthropology, mainly in the USA. One part of this has been a move towards studies of primate social organisation (DeVore, 1965; Barkow, 1973) and towards behavioural ecology (Rappaport, 1968; Vayda, 1969).

In this paper, rather than discuss the entire range of growing links between biology and the study of human behaviour, I wish to concentrate on the relationship between two important developments of the last few years, one in biology (socio-biology) and the other in anthropology (ecological anthropology or cultural materialism). I then attempt to explore the prospects for a link between these two fields.

A LINK FROM BIOLOGY: DEDUCTIONS FROM NATURAL SELECTION

The study of the implications of natural selection for social behaviour, described in the papers by Bertram and by Clutton-Brock & Harvey in this volume, is one of the most exciting areas of contemporaneous biology, although its value for direct studies of causation and development has yet to be proved. I. DeVore & R. L. Trivers (unpublished) stressed two aspects of this deductive approach: (1) that given a very few basic facts about an animal one can predict many of its features by working out the full implications of natural selection; (2) that the behavioural and social sciences have suffered from the lack of any explicit theory which allows such prediction, and that contrary to all one has come to expect about human social behaviour, natural selection may be capable of providing such a theory. (There has of course been much theory in the social sciences, but much of it has been untestable, yielding at best only general predictions, or being really a substitute for missing data rather than a theory.) I suspect that DeVore may be right on both these issues, but at first sight the arguments against the applicability of natural selection theory to human behaviour appear overwhelming. In this section I want to discuss one or two of these arguments before moving on to a discussion of ecological anthropology, finally returning to natural selection to discuss ways in which sociobiology and ecological–economic anthropology might come together.

What could the theory of natural selection tell us about such a variable, fast-learning creature such as man? The claims range from an understanding of the way natural selection produced our species (e.g. understanding human evolution up to the point when agriculture became the predominant way of life), through to explaining every single facet of all human behaviour. But what does natural selection really imply; what are we saying about a piece of human behaviour when we say that it is adaptive? What does it mean when, for example, Hamilton (1975) describes how natural selection will make altruism less frequent and aggression more frequent between small segregated populations in proportion to the degree of genetic isolation between them? Let me say straight away that the implication for motivation and development may be nothing more than that all people tend to be nicer to their kin or familiars and nastier to unrelated and unknown people. It does *not* imply that the people in those three cultures must innately differ in their levels of aggression or altruism or in their selection of targets for these behaviours. When Trivers (1972) discusses the conditions (recognition of individuals, long life, limited mobility, lack of dominance hierachy, etc.) under which reciprocal altruism could evolve (conditions most of which apply to evolving man) and the elaborations to protect the system against cheating (which provides such a good description of Bushman morality and aggression), it is important to remember that he is saying these are *the conditions under which* it *could* evolve by natural selection, and these are the elaborations for which selection will press. Both are important to those whose concern is the direct study of human behaviour and in two ways: (1) as an ultimate explanation (the third and fourth of Tinbergen's (1951, 1963) four 'whys'), when we have observed a system like the system of Bushman food sharing and dispute. (2) As a prediction:

that reciprocal altruism is more likely to occur where the facilitating conditions are present, less likely where they are absent; that some proximal causal and developmental mechanisms are required to mediate this variability, and proximal mechanisms such as aggression are required to perform the anti-cheating functions. It is up to us, the direct observers of human behaviour to find and describe these mechanisms.

It is as well for biologists to be forewarned about the misunderstandings that can arise about deductions from natural selection. With respect to altruism, parental investment, parent–offspring conflict, and 'spiteful' behaviour (Hamilton, 1970), it is essential to look closely at the definitions and to remember that they are radically different from the definitions of both layman and psychologist. Thus altruistic behaviour is behaviour which in the short-term reduces the acting individual's chance of reproductive success while increasing that of another individual. This is a definition by effects, and more than that, by effects on reproductive success. The layman and the psychologist's definitions are much more like definitions in terms of motive. This implies that they are not definitions but labels, so a definition in terms of causation could be radically different from a definition by effect. This again implies that Trivers' (1972) theory of altruism is not telling us directly about causation, or about ontogeny. (Of course further elaborations of it might: for instance, the early development of reciprocal altruism has to be compatible with the different set of factors influencing the distribution of altruistic behaviour between siblings.) It is not telling us that this is 'innate' behaviour, nor does the argument depend on the 'innateness' of altruistic behaviour. The same conclusion arises from the fact that natural selection is interested in the finished product, not how the product arises. So Trivers' theory is telling us, sometimes in detail, about the tasks which causation and ontogeny have to fulfil. Thus in the case of reciprocal altruism the theory gives us good reason to believe that altruistic behaviour would from quite early on have played the major part in human adaptation that it appears to play in surviving hunter–gatherers, and thus be an ancient and notable characteristic of man. But it does not tell us whether we behave this way by calculating the pay-offs in our head, or by growing up to be someone who performs altruistic behaviour in response to certain simple stimuli such as a cry for help. Linked to kin selection it tells us that there is pressure to be selective about altruism (to degrees depending on the relatedness etc. of the population). Sometimes, as in the basic parts of parent–offspring conflict (Trivers, 1974) this role of theory is unimportant – we know that siblings sometimes compete and parents try to stop them, but we didn't realise that this was to the adaptive advantage of both (i.e. that much of 'socialisation' is an inevitable conflict not necessarily a teaching process) or that the strategy for maximising reproductive success is bound to be different for parent and for offspring; nor did we realise that the conflict should differ between species depending on whether siblings had the same father or different fathers. What is left for us to discover is the short-term mechanism, and why the mechanisms work.

The next point is of peculiar but daily concern to those who work on human behaviour. It is strictly nothing to do with science, but it has a lot to do with life as a student of human behaviour. Deductions from natural selection theory do not give us views that we need to accept as moral judgements. This seems obvious but many people, modern equivalents of the much maligned 'Social Darwinists', could see natural selection as justifying their actions. Thus it provides a splendid explanation of the 'dual standard' of sexual behaviour (see Clutton-Brock & Harvey in this volume). People could also make use of the theory of reciprocal altruism to justify their 'what's in it for me' approach to social interaction, or they could make use of the fact that natural selection acts on individuals to justify (unadaptive) extremes of competitiveness or for that matter justify their own altruism by the theory and the likely role of altruism in human evolution. It would be very unfortunate for the progress of our understanding of social behaviour if the justifiable opposition to some of these *non sequiturs* was to manifest itself (as it would by analogy with other areas of research) as opposition to deductive studies concerned with natural selection.

Thus when we consider the relevance of the deductive studies of natural selection to human behaviour we shall not suppose that their predictions concern only 'innate' behaviour. Such studies give us very general laws governing the behaviour of organisms and its variation between species and individuals. So long as we remember that it neither implies immutability of behaviour, nor provides moral judgements on behaviour, it is well worth following up the implications of natural selection for the understanding of human behaviour. The natural selection studies so far perhaps imply disappointingly little about specific mechanisms of motivation and development. However, many of us do believe that they play a useful role in giving us a framework for studying motivation and development as well as, of course, providing ultimate explanations. It is probably much easier to study a mechanism if you know that it is there. And it is probably even easier if you know what it is designed to achieve. (McFarland puts this argument more completely elsewhere in this volume with respect to his studies of optimisation strategies of behaviour, which ultimately are aimed at showing what evaluations the physiological mechanisms of behaviour have to make.)

The application of the deductive natural selection approach to communication would seem to be one area that holds great promise for the study of human behaviour, in particular for the solution to the problem of the function or adaptive value of many of the less tangible aspects of social behaviour. Communications are selected for the advantage of the communicator. Thus they may not always be for the dissemination of truth, nor for the well-being of society at large. Communications may range from the purely altruistic (e.g. Trivers' (1972) view of alarm calls) through those that are of some advantage to other individuals but of greater advantage to the message-giver by virtue of its effect on the other individuals (Charnov & Krebs' (1974) view of alarm calls), to those that are of advantage to the message-giver and of no advantage or even of disadvantage to the receiver (e.g. protective mimicry as a communication from insects to birds). I. DeVore & R. L. Trivers (unpublished) believe that the principles of natural selection will apply from the simplest of animal signals right through to human belief systems. It will be interesting to see what their detailed argument will do to the way we think about the mechanisms and organisation of human cultures.

I have already described the distinction between emic and etic approaches to human behaviour. Biologists should not entirely neglect the emic, for beliefs and category systems must have evolved for something, and it might be fruitful to consider emic phenomena in the framework of the communicative function of belief systems. Harris (1968) implies that we can follow up the aspects of ritual ceremony and belief systems that expend measurable energy using 'the etic research option': for instance, we can examine the sources and sinks of the energy, the effects of the observable actions, and the events that are associated with them. In other words we can apply all the usual measurement and analytical techniques of non-experimental scientific research. But I would suggest in addition that if DeVore and Trivers were to follow up their suggestions about communication and belief systems (that communications are to increase the inclusive fitness of the message-giver) they would come out with some highly testable predictions, and some clear and important questions about short-term causes and development. And I suspect these would probably be rather similar to, but more testable than, some of Marx's and Harris's views on the role of ideals in human societies. Indeed this would be no coincidence, I am told (by Lionel Tiger, and others independently) that there is a letter from Marx asking Darwin to allow him to dedicate *Das Kapital* to Darwin. Darwin did not give his permission. Thus the ecological, 'adaptive radiation', model lost out to the phylogenetic 'ladder of life' model of Spencer and the Social Darwinists. Something of this history is still with us. In the recent history of comparisons between man and animal by biologists, the phylogenetic was at first emphasised at the expense of the adaptive, though this is least true of Morris (1967) who is still outstanding among this particular tradition. Evolutionary work by American anthropologists (e.g. Lee & DeVore, 1968) has been much more concerned with ecology and adaptation. In the next section I argue for still more interest in ecology and adaptation.

A LINK FROM ANTHROPOLOGY AND ARCHAEOLOGY: ECOLOGICAL ANTHROPOLOGY

In 1968 Marvin Harris of Columbia University published a contentious, probably unpopular but very powerful book called *The Rise of Anthropological Theory*. I think this is the most important book for the behavioural sciences since *The Origin of Species* and it must eventually come to stand as one of the major landmarks in the development of the social sciences. Harris argues that the history of anthropology has been bedevilled by (1) a flight from the search for generalisations about societies; (2) a flight from the attempt to find causal mechanisms for temporal change, long (evolutionary) and short-term, in human social behaviour and organisation; (3) the search for causes among the social data that the causes are intended to explain, rather than in the material world; (4) too much emphasis on the study of the emic (subjective) aspects of human behaviour. He clears the scene for the growth of a science of human society and he implies that he is putting his money on something close to American ecological anthropology. In later books (Harris, 1971, 1975) he gives a more positive indication of the direction in which he feels that anthropology should develop.

Although there are other powerful new developments in anthropology I think that it is to this branch of anthropology that ethologists should look for their natural allies, for the possibility of really close links between behavioural ecology and studies of human ecology and social organisation.

Not only behavioural ecologists but any of us who wish to get involved in studying human behaviour will do well to digest thoroughly Harris' work. His critical review of the historical succession of theories in anthropology and sociology (and to a lesser extent history) is of great value to the newcomer, presenting the range of views in those subjects and the criticisms which many of us from biology would make but which come so much better from an insider.

The Rise of Anthropological Theory is about the history of the combat between (in White's (1949) words not Harris') the 'spiritualistic, vitalistic, or idealistic' strand in sociology and anthropology and the more broken 'mechanistic, materialistic' strands. No single author has stood completely on one or other side but there have been some stalwart extremists. Harris characterises the conflict as between particularist (every culture is a law unto itself, unique in every aspect), cultural idealist (ideas determine culture and behaviour), non-deterministic (there are no general laws of culture or culture change) non-theories of cultural differences, culture change and history, and the nomothetic (seeking generalisations), diachronic (seeking long-term causal factors), cultural materialist (techno-economic and ecological factors are the most causal) theories. The latter were early on represented in much of the writing of Marx, and in some crucial passages from Tylor and Morgan, but were swamped by the particularist, non-causal outlook of Boas and Benedict, and in Europe by Durkheim and the later Lévi-Strauss, and the English tradition that arose from Malinowski, Radcliffe-Brown and Evans-Pritchard. Thus the history of anthropology has been dominated by those whose basic belief ultimately boils down to the view that cultural diversity is such as almost to forbid explanation, and that culture and idealogy have their own life independent of economic, physical or biological factors.

Some of the major contemporary figures in British social anthropology may not fit completely into this simple categorisation. For instance, Leach (1961) and Douglas (1971) do both seem to be seeking generalisations but currently to give minor or non-existent roles to ecological or biological factors. Their structuralist generalisations seem to concern the immediate contemporary motivational or cognitive state of people: human beings have brains that classify and make patterns, organising their world into systems of binary opposites; features of this patterning explain the meaning of human behaviour, whether food preferences, kinship, non-verbal communication or whatever; observable behaviour has no meaning, except as an illustration of the classificatory habits of the *observer's* culture. These authors may be making no statement about long-term influences, except that these can act only by changing the entire cognitive system or structure. Thus they may be in conflict with some psychological theories of motivation but perhaps do not have to be in conflict with cultural materialism (techno-economic, ecological factors predominate and ultimately determine the culture's ideals and cognitive world as well as its observable behaviour).

British biologists attempting to relate to anthropology and sociology are not going to meet many Harris fans. They will meet people involved in some form of structuralism, or in ethnosemantics or ethnomethodology. What these people are saying possibly need not totally conflict with what Harris is recommending, although they are likely to say that it does. Harris is talking about what may be primarily long-term causes of cultural similarities and differences. Structuralists and ethnomethodologists seem to be talking mainly about the here and now situation of the minds of their subjects. Some of them will claim to be concerned with culture change, and while some of these will approve of a view of a drifting history, a few may be interested in changes outside their culture's system of thought patterns affecting a sensitive point in the system and thus changing the rest of the system. They all assume that ideas cause behaviour.

The view that ideas are the prime causes was in the field before the view that economic–ecological factors are causes of behaviour. Whenever the latter view has been proposed, opposition has arisen in reaction to (*a*) the spectre of 'rational economic man' (someone who consciously calculates his economic gains from moment to moment), an unrealistic automaton that ecological anthropologists also will not support; and (*b*) the obvious inadequacies of early attempts to relate culture to hopelessly simplified ecological categories – the analogy with Crook & Gartlan (1966) versus Clutton-Brock (1974) is close (but unfair to Crook & Gartlan). Clutton-Brock suggests we refine our ecology, but anthropologists decided to drop ecology altogether – contemplate the implications of that for the study of primate social organisation! Now Harris, and others such as Rappaport (1968) and Vayda (1969), are coming back with refined, detailed, ecology with great success.

In the area of economics and ecology the following extended quotation from Harris is a good indication of what he is criticising:

If there is any single theme pervading The Argonauts, it is precisely that the motives and feelings which arise from noneconomic needs dominate the whole Kula enterprise. The fact of the matter is that Malinowski deliberately refrained from carrying out a genuine economic study of the productive and distributive system of the Trobriand Islands because his ethnographic orientation was overwhelmingly opposed to any such option. Malinowski himself is my authority for this statement, for he was perfectly aware of what he was doing and of how his monographs would have looked had he really been concerned to examine the 'material foundation' of Trobriand culture...

The Kula, that splendid example of commerce inscrutable in its risk of life on long sea voyages all for the sake of a few spondylus and *Conus millepunctatus* shells, is to the functionalists what potlatch is to the Boasians. But economic analysis is properly a matter of a system of production and distribution: of energy, of time and labor in-put, of the transformation, transportation, mechanical and chemical interplay between a human population and their habitat and of the distribution of the products of this interaction in terms of energy, especially food energy, and the mechanical and biological apparata upon which all these processes depend. Malinowski treats us instead to a vastly elaborate account of the ritual aspects of the preparations for open-sea expeditions in which the subjective motivations of the actors in terms of prestige and magical aspirations dominate every aspect of the ethnography. We learn only incidentally, never in detail, that the whimsical voyagers circulate not only armbands and necklaces but coconuts, sago, fish, vegetables, baskets, mats, sword clubs, green stone (formerly essential for tools), mussel shells (for knives), and creepers (essential for lashings).

Harris, 1968, pp. 563–564

In *The Rise of Anthropological Theory* Harris explodes other classic cases where economic explanations have hitherto seemed inadequate, such as the potlatch, the question of who milks their cattle and who doesn't, who eats their cattle and who doesn't (*The Myth of the Silly Bantu*, p. 368). 'Morton

Fried (personal communication) offers a simple explanation for why the Chinese in the village he studied in 1949 would not milk their cows: there weren't any worth milking' (Harris, 1968, pp. 369–370). 'Emic, short-run and impressionistic data have produced many cases of sensational mismanagement of pig-raising. Few ethnographers have hitherto concerned themselves to assess the nutritional importance of relatively small amounts of animal fat and protein in diets which consist of root crops' (*ibid.*, p. 567). Rappaport's (1968) *Pigs for the Ancestors* is a detailed and successful remedy of this situation. 'It would have been a real coup for the camp of the antideterminists and culture-by-whimsy school if Malinowski had discovered the Trobriand Islanders practising a custom whereby, among the commoners, the man who produced the least number of yams had the biggest house and the most wives.' An extended account of the ecological explanations of some of these famous 'mysteries' of culture is given by Harris (1975).

But Harris does not give all the answers, he clears the ground, describes the seeds that have been planted and ends with an exhortation, quoting first from Adams (1966, pp. 174–175):

'What seems overwhelmingly most important about these differences is how small they bulk even in aggregate, when considered against the mass of similarities in form and process. In short, the parallels in the Mesopotamian and Mexican "careers to statehood", in the forms that institutions ultimately assumed as well as in the processes leading to them, suggest that both instances are most significantly characterized by a common core of regularly occurring features. We discover anew that social behavior conforms not merely to laws but to a limited number of laws, which perhaps has always been taken for granted in the case of cultural subsystems (e.g. kinship) and among "primitives" (e.g. hunting bands). Not merely as an abstract article of faith but as a valid starting point for detailed, empirical analysis, it applies equally well to some of the most complex and creative of human societies.'

We may note that Adams, who deals with time scales on the order of millenia, appears to have misunderstood recent anthropological history in suggesting that elements of the nomothetic revival were always 'taken for granted'. It has been a long hard struggle all the way. In concluding our review of this centuries-old struggle to achieve a science of history, it cannot be overemphasized that the vindication of the strategy of cultural materialism does not depend on the verification of the hydraulic hypothesis [*irrigation*] or of any other particular techno-environmental, techno-economic theory. Rather, it lies in the capacity of the approach to generate major explanatory hypotheses which can be subjected to the tests of ethnographic and archaeological research, modified if necessary, and made part of a corpus of theory equally capable of explaining the most generalized features of universal history and the most exotic specialities of particular cultures.

But some exciting questions are tackled much less successfully both by Harris and by subsequent writers (Harris, 1971; Ruyle, 1973; Cloak, 1975).

Thus 'We have every good logico-empirical reason to suppose that most human populations reach an adaptive equilibrium within their ecosystem' (Harris, 1971, p. 370). Given that we do, and we do now have enough studies to suggest that something *like* this is likely, as well as good reason to believe that this is a heuristically valuable theory for a while, how does it come about? What are the feedback routes? How is culture modified in such a way that this occurs? Surely we are not here dealing with the directing effects of natural selection? Or are we? So are we to postulate competition between subcultural groups with successful replacing less successful, or imitation of successful traits by other groups, in which case how do they recognise the successfulness of the group they imitate? Or, more likely, something like this but between families rather than groups. How do *adaptive* cultural changes follow from the ecological pressures?

CAN THERE BE AN INTEGRATION OF BEHAVIOURAL ECOLOGY WITH ECOLOGICAL ANTHROPOLOGY?

Exciting and elegant as modern ecological anthropology may be, it still leaves many gaps and uncertainties. The ecology is primarily in terms of partly worked-out energy budgets. The functional interpretations are very like group selection interpretations even though the benefits to individuals are often perfectly clear. Often a group selection explanation is given for behaviour which indeed seems to conflict with the individual's interests: to the biologist this suggests that the ecological–economic analyses had stopped short of completion. In many accounts society remains the unit to be explained, not individuals: individuals do things that function for the good of society, the culture, the population or the control of its level. There is little or no discussion of how people come to behave in such adaptive ways, how ecological requirements produce adaptive behaviour. Are we seeing the results of selective processes? If so, what are the units selected: society, cultures, populations, families or individuals? How does behaviour change? Are we concerned with replacement of one set of people by another, or with changes in the behaviour of any one set of people, or with what kind of mixture of both?

These questions are beginning to be answered but seldom with any cross-reference from one kind of answer to another. For instance, Hamilton (1975) has discussed the application of natural selection to aggression and altruism in human groups with greater or lesser isolation from each other. His analysis makes good sense of the differences in number of fatalities in inter-group contact between the !Kung Kalahari Bushmen, and the Tsembaga New

Guinea swidden agriculturalists and the Yanamamo of South America. This series of increasing fatalities goes with decreasing migration and food exchange between groups. Feldman & Cavalli-Sforza (1975) have applied the mathematical techniques of population genetics to the phenotypic transmission of traits from one class of individuals to another ('cultural inheritance'), one of the few serious works on what biologists call 'cultural evolution' (this term is used in a quite different sense by ecological anthropologists).

Cloak (1975) proposed an ingenious theory of cultural evolution in which a cultural 'instruction' is treated like a gene, and thus presumably selected for its transmittability, one component of which will be that it is not a handicap to survival. Another level of mechanism is discussed by Lee (1972), who described some of the ecological pressures behind the shortening of Bushman birth-spacing as people settle and begin to get involved with agriculture, and proposed immediate physiological mechanisms by which this change can occur in the life of any one woman.

Cultural inheritance is not a complete replacement for the theory of natural selection. It is merely a means of transmission of behaviour. Some essential features of evolution by natural selection are not paralleled. The source of specificity, a substitute for differential mortality, must be specified. If cultural inheritance is to be adaptive, then only certain individuals must get their traits transmitted to subsequent generations and they must be traits that are in some way adapative: something must happen to individuals who do not acquire/possess traits that does not happen to individuals that do possess/acquire these traits. Cultural inheritance as formulated at present is merely a means of transmission and is raw material only for a revival of the 'culture-by-whimsy' (Harris, 1968) school.

Ecological anthropology need not answer questions about the short-term causal links but only a few of its proponents (e.g. Freeman (1971) discussing reasons for infanticide in Pelly Bay Eskimos) show an awareness of the distinction between functional and causal explanations. There seems to be little awareness in anthropology as a whole of the potential frame-setting value of the functional approach for those whose central concern is short-term causation.

At present much ecological anthropology leans rather heavily on what can now be seen to have been a half-baked animal ecology, the ecology of a period when the relationship of the animal to its environment was only just beginning to be seen to have a relationship to evolutionary theory, a period when group selection lurked unnoticed under every explanation and when the really remarkable predictive power of the theory of natural selection acting on individuals was not widely known. The chapters in this volume by Bertram, Clutton-Brock & Harvey and McFarland show what startling developments have been underway. If ecological anthropology is not to wither into a quantified version of early British functionalism surely it must continue to see whether it can make any use of developments in animal ecology.

Ecological anthropology has come a very long way, struggling out of a very unpromising environment. The arguments against the particularist, idealist approaches are overwhelming and there can be no turning back. But at the same time we must recognise that ecological anthropology has further to go. Perhaps biologists can help it along, if only by playing devil's advocate from time to time. It seems to me that the time is ripe for a cooperation between animal behavioural ecology and ecological anthropology comparable to the meeting in the last decade of ethology and developmental psychology.

There are other arguments in favour of a cautious but determined attempt to integrate socio-biology and ecological anthropology. Natural selection is an all pervading process from which no organism is immune. No general theory of human social behaviour can be *incompatible* with the implications of natural selection (Hamilton, 1975). At the very least we have to look at the implications for our social behaviour of the selection pressures of the hunting and gathering niche which created our species. Probably we have to look further, at the range of strategies which a species has. Perhaps we have to consider the possibility that natural selection has exerted a significant effect even in man's short agricultural past.

Natural selection seeks the best compromise between a multitude of ecological factors. The way in which these factors are related to each other, and weighed off one against the other, is the most important set of facts about any species. As McFarland (this volume) argues, natural selection is in the optimisation business, and biologists are now trying to exploit optimisation techniques for furthering their understanding of the massive system that is an animal.

If for no other reason than that these are good, disciplined, explicit and thoughtful techniques for organising a wide range of data and understanding the relationships between these data, they should be of interest to ecological anthropologists. Indeed they offer us some chance to try to decide what, if anything, an individual or a culture is optimising. A biologist's bet, when he has seen the power of natural selection theory in explaining social behaviour and organisation of other animals, is that inclusive fitness is the outcome whose optimisation would repay examination *even* in human social behaviour. Thus it seems to me that the McFarland approach holds the key to a useful integration of socio-biology and ecological anthropology, even if the result is that we discover sharp limits to the extent to which human beings are optimising their inclusive fitness. At any rate it seems that ecological anthropology has reached a point where it can no longer evade the questions: whence comes the adaptedness of behaviour, and *what* is it that people's behaviour maximises?

One's intuitive objections to these propositions are innumerable, but in many areas the outcome of such an approach might not be so outrageously far from the basic thoughts of ecological anthropologists as one would think at first sight. Communication, even belief systems, function to increase the inclusive fitness of the message sender. Natural selection operates on individuals (on their inclusive fitness), not on societies. These may be preposterous contradictions of countless writings in sociology and anthropology. But they are surely not very far from some of the views (for instance conflict between classes and groups within society, and that ideals and beliefs serve the interests of one class more than another) of Marx, whose influence in these subjects is welcomed by many in sociology and anthropology. And no distance from the interpretations Harris (1966, 1975) gives for the functions of beliefs such as the Sacred Cow in India or the need for burning witches in medieval Europe.

The idea of the Sacred Cow is both directly useful to the peasant in protecting himself from the temptation to eat his tractor (bull) or tractor factory (cow), and indirectly by allowing his cows to forage on the property of other people, including wealthier people. Somehow the peasants have made this belief stick with those who gain least from it. The importance of burning 'henbane-trippers' and torturing them into confessing the names of other harmless people was, under cover of religious doctrines which claimed to be doing it all to protect the people from evil, to attack and destroy the messianic (heretical) popular revolts whose occurrence correlates so closely with periods and locations of intensive witch persecution, and which threatened to overthrow the secular power of the church. One could suggest other examples that might be worth examining in this way. A witch doctor 'heals' his patients, gaining prestige and acquiring material 'credit' from the belief system he helps promote, so long as he can keep his clients going along with the system: surely there is a close parallel here with protective mimicry in animals, for the correct balance of occasional successful cures and high risks has to be maintained for the clients to continue to go along with the beliefs. Just as in the study of monkey groups, where Williams (1966) and Wilson (1975) suggest that the traditional group selection explanations of dominance hierarchies can be discarded and replaced with the much more fruitful analyses of the kind described by Clutton-Brock & Harvey (p. 195), so can this be attempted for human groups. Maybe the witch doctor really does do rather well out of his clients or even better than they do. But they probably do as well as they can manage. The feudal lord and his people had the relationship that they had, not because it preserved society (a myth that is usually of greatest benefit to the ruling class) but because each was getting the best out of their particular circumstance that way. The feudal lord had to balance short-term gains against long-term stability, maintaining a balance of costs and benefits in his subjects which they would endure, keeping them in a situation where their best course was to go along with him. This could be done by benevolence, or, probably less enduringly, by terror.

Almost every theory of societies, both animal and human, has failed to consider 'society' as an outcome of the interaction between the behaviour strategies of individuals. Given that 'society' or a 'culture' is nothing more than 'the statistical consequence of a compromise made by each individual in its competition for food, mates and other resources. Each compromise is adaptive but not the statistical summation' (Williams, 1966, as quoted by Clutton-Brock & Harvey, this volume) then there is little reason to expect societies or cultures to remain stable or constant. It is usual to think of primitive cultures as being very stable (until they contact industrial cultures), and thus of this or that trait as functioning to preserve stability, or to aid the survival of 'the culture'. But both the behaviour of the people in the culture, and the representation of different families in subsequent generations, may change rapidly and continually.

One consequence of this probable state of affairs (that cultures may never have been stable) is that we must re-examine our suppositions about selection pressures exerted by society or culture. It is altogether too crude to talk of this culture selecting for peaceful individuals, that for aggressive, and so on, there will always be a fluctuating compromise and it is upon this compromise that natural selection probably acts. This implies a good deal of adaptive variability in each individual. So perhaps we should *not* be thinking in terms of the sudden change in selection pressures from hunting and gathering to farming to industry, but rather in terms of the much smaller but more complex changes in the strategies by which individuals maximise their inclusive fitness in competition and cooperation (where it aids inclusive fitness) with other more and less related individuals. This brings us very close to Humphrey's (p. 303) suggestions about the selection pressure for intelligence, but we should not forget the role of the 'irrational' in human behaviour: ecological or economic man is not the same as rational economic man. We can separate the question of rational or irrational from the question of economic or uneconomic or ecological. Human brain power must have had some adaptive value for its effects on the success with which individuals maximised their inclusive fitness by computing the costs and benefits of different behaviour. But sheer thought is unlikely to be the only proximal mechanism involved; and the word 'economic' also raises the separate question of what are they maximising? 'Economic' suggests they maximise their short- or long-term personal wealth either in money, capital, goods or food. There is of course no necessary survival value in maximising personal wealth. The 'Zen road to affluence' (Sahlins (1974) 'that human material wants are finite and few') is just as likely to lead to high reproductive success as any other. For this reason I would not regard 'economic' as interchangeable with 'ecological'. And in the argument between formalist and substantivist economic anthropology I would not want to support the classic type of either school. ('Formalist' economic anthropology attempts to apply the formal analyses of western economics to other cultures, 'Substantivists' tend towards treating economics as just another category system of each culture.) Again the really crucial question is: what, if anything, is human behaviour designed to maximise?

Those of us who have been interested in the study of human evolution for the light it may or may not shed on contemporary industrial man have been too much concerned with momentarily observable behaviour and not sufficiently concerned with strategies of survival and the mechanisms which subserve them. Instead of debating whether a tendency to set up hierarchies was a characteristic of early man or not we should have paid more attention to the circumstances in which more or less marked hierarchies arose and to the actual costs and benefits of the various forms of individual relationships involved and to the mechanisms by which the animal evaluated the costs and benefits. The lesson available in the variability of social organisation within a non-human primate species (Crook, 1970) was not really absorbed.

Future research both on animals and man might usefully concentrate on: (*a*) the range of different strategies that any one animal species or individual has for maximising its inclusive fitness, e.g. the ability to vary sex ratio at birth (Trivers, 1973), the ability to vary clutch size from season to season (Perrins & Moss, 1974) and other variations reviewed by Sadleir (1969); and (*b*) the exact ecological reasons for the differences in foraging and social organisation of different hunter–gatherer groups, e.g. why do Bushmen (Lee, 1968) exceed Hadza (Woodburn, 1968) in such things as length of gathering excursions, amount of food exchange between men and women, and Hadza exceed Bushmen in amount of food eaten out of camp rather than in camp, and amount of foraging by children. The first place to look is in the quantitative relationships between localisation of high yield plant foods, abundance of animal food, aggressiveness and timidity of lions, distances between water sources, and their interaction with the theory of parental investment and parent–offspring conflict.

Some questions that seem to this biologist to be important for ecological anthropology are

(1) Can we show whether cultural inheritance somehow makes group selection more likely to occur than it is with natural selection?

(2) If not, what happens if we try to replace the group selection explanations of cultural adaptation with individual selection explanations?
(3) What, if anything, do people maximise (by the consequences of their behaviour not by the intentions in their hearts!)?
(4) How will we measure or demonstrate 'adaptation'? Can we find anything as convincing as Tinbergen's work on egg-shells (Tinbergen, 1965; Tinbergen, Impekoven & Franck, 1967), Tinbergen and Kettlewell on differential predation on colour forms of moths, Patterson (1965) and Kruuk's (1964) work on differential reproductive success of gulls showing more or less of various kinds of behaviour?
(5) What do we mean by 'adaptation' when we apply it to the behaviour of individual people or of cultures?

Harris (1971) and Ruyle (1973) have made the most serious attempts to examine the term 'adaptation' in ecological anthropology. Ruyle attempts to relate it to current biological thinking and has pointed out its heavy reliance on group selection, and the problems of time scale that are involved. He does a valuable service in stressing the importance of examining the strategies of individuals and not reifying society. He then concludes that one cannot explain the adaptiveness of cultures by group selection but he also seems to grant individual selection a negligible role, understandably in view of the rapidity with which cultures can change. He argues that the individuals change to more satisfying forms of behaviour. 'Individuals find the prospect of a secure future satisfying; individuals find the prospect of watching their children grow to maturity satisfying; individuals find the prospect of more food for less work satisfying' (Ruyle, 1973). Pleading the principle of parsimony he concludes that long-term factors do not need to be called upon until the short-term factors that he proposes are proved incapable of explaining the phenomena. While there are many ways in which his suggestions could be developed it is not clear that the 'struggle for satisfaction' as he calls it can have the predictive power that recent work on natural selection shows can be derived from 'the struggle for survival'. He also phrases the 'satisfactions' in terms much too close to motivational units to allow us to cover the range of short-term causal mechanisms that go to achieving a maximisation of inclusive fitness. He seems to be heading rapidly back to cultural idealism ('it is not just the material needs of the individual which must be satisfied but social and ideological needs as well') and to a position incapable of explaining 'riddles of culture' such as the Indian Sacred Cow which ecological anthropologists have so successfully explained (Harris, 1966, 1975). His stress on individuals allows Ruyle to get nearer than most but it seems to me that he, like the rest of us, stops short of a complete description of the proper meaning of adaptation when applied to human behaviour, or of the relationship between ultimate and proximal causes of human behaviour. His position seems near to the logical but hard to use argument that people learn to do what is rewarding, and that from this fact one should be able to explain and predict all of human behaviour.

Harris (1971), as one might expect, seems to be the author who best appreciates the distinction between proximate and ultimate causes when discussing general principles, even if his ultimate causes happen to be phrased in terms of group selection. The position indicated in the following quotation (see Ruyle, 1973, p. 213)

> The most successful innovations are those that tend to increase population size, population density, and per capita energy production. The reason for this is that, in the long run, larger and more powerful sociocultural systems tend to replace or absorb smaller and less powerful sociocultural systems.
>
> The mechanism of innovation does not always require actual testing of one trait against another to determine which contributes most in the long run to sociocultural survival. Given a choice of bow and arrow versus a high-powered rifle, the Eskimo adopts the rifle long before there is any change in the rate of population growth. In the short-run, the rifle spreads among more and more people not because one group expands and engulfs the rest, but because individuals regularly accept innovations that seem to offer them more security, greater reproductive efficiency and higher energy yields from lower energy inputs. Yet it cannot be denied that the ultimate test of any innovation is in the crunch of competing systems and differential survival and reproduction. But that crunch may sometimes be delayed for hundreds of years...

seems but a short distance from the argument that natural selection will have produced short-term mechanisms that mediate extensive adaptive modification of the behaviour of an individual (not simply pliable behaviour but usefully flexible behaviour systems). This may well be achieved, in any animal, by designing a number of separate short-term goals, whose settings may or may not be modified only by natural selection. What seems of prime importance is to reserve judgement on many of these issues and think up, while doing research on both animals and people, the widest possible range of mechanisms that natural selection could have designed to maximise inclusive fitness in the face of rapid short-term changes of the environment. That natural selection has not perfected this design is shown by the fact that extinctions occur, and indeed threaten our own culture if not our species. But this is a long way from saying, as many have done, that the drives evolved in the Pleistocene now threaten our survival. There is an enormous difference between 'drives' and survival strategies, both in animals and man. I contend that the functional evolutionary approach, a combination of deductions from natural selection theory, Oxford-style optimisation studies, and ecological anthropology could lead us in the next few years to a position where we can make precise statements and predictions about human social behaviour, that are far more specific than anything we could derive from non-teleonomic approaches based on concepts of drive, instinct, reinforcement, satisfactions, or cultural drift.

SUMMARY

Developments in ecological–economic anthropology suggest that it will soon be useful to examine the use of the term adaptation when applied to cultures or the social behaviour of people, and relate it to current developments in evolutionary theory and behavioural ecology. I argue that ecological anthropology is a school of anthropology to which biologists could look for data and theories which might be compatible with biological ways of thought, because of its emphasis on observable behavioural and economic data as well as its emphasis on the causal influence of the material world on social behaviour. A central problem will be the speed with which human behaviour changes. 'Cultural inheritance' is an incomplete theory of culture change because it is only a means of transmission. The source of selectivity, the counterpart of differential mortality, has to be specified.

I wish to thank all those social scientists who have had the patience and toleration to discuss their subjects and mine, thus enabling me to just begin to understand their point of view. Particular thanks should go to Michael Mair of University College London for explaining British structuralism to me more successfully than the many who have tried before; to M. J. Konner of Harvard University for telling me to read Harris's *The Rise of Anthropological Theory* – an influence on my thinking as profound and pleasant as that of my visit to Konner's fieldwork on !Kung Bushman infancy; and to R. L. Trivers also of Harvard University for his prompt and patient answers to my letters containing wild speculations about some implications of his work.

REFERENCES

Adams, R. McC. (1966). *The Evolution of Urban Society*. Aldine: Chicago.

Ainsworth, M. D. S. (1969). Object relations, dependency and attachment: a theoretical review of the infant–mother relationship. *Child Development*, **40**, 969–1027.

Barkow, J. (1973). Darwinian psychological anthropology: a biosocial approach. *Current Anthropology*, **14**, 373–388.

Bernal, J. F. & Richards, M. P. M. (1973). What can zoologists tell us about human development. In *Ethology and Development*, ed. S. A. Barnett, Clinics in Developmental Medicine, no. **47**. Heinemann Medical Books: London.

Blurton Jones, N. G. (ed.) (1972*a*). *Ethological Studies of Child Behaviour*. Cambridge University Press: London.

Blurton Jones, N. G. (1972*b*). Comparative aspects of mother–child contact. In *Ethological Studies of Child Behaviour*, ed. N. G. Blurton Jones. Cambridge University Press: London.

Blurton Jones, N. G. (1975). Ethology, anthropology and childhood. In *ASA Studies – Biosocial Anthropology*, ed. R. Fox. Dent: London.

Bowlby, J. (1969). *Attachment and Loss*, vol. **1** *Attachment*. Hogarth: London.

Bowlby, J. (1973). *Separation: Anxiety and Anger*. Hogarth: London.

Charnov, E. L. & Krebs, J. R. (1974). The evolution of alarm calls: altruism or manipulation? *American Naturalist*, **109**, 107–112.

Cloak, F. T. Jr. (1975). Is a cultural ethology possible? *Human Ecology*, **3**, 161–182.

Clutton-Brock, T. H. (1974). Primate social organisation and ecology. *Nature, London*, **250**, 539–542.

Crook, J. H. (1970). The socio-ecology of primates. In *Social Behaviour in Birds and Mammals*, ed. J. H. Crook. Academic Press: London & New York.

Crook, J. H. & Gartlan, J. S. (1966). Evolution of primate societies. *Nature, London*, **210**, 1200–1203.

DeVore, I. (ed.) (1965). *Primate Behavior*. Holt, Rinehart & Winston: New York.

Douglas, M. (1971). Do dogs laugh? A cross-cultural approach to body symbolism. *Journal of psychosomatic Research*, **15**, 387–390.

Dunn, J. (1975). Consistency and change in styles of mothering. In *Parent–Infant Interaction*, ed. M. O'Connor. Elsevier: Amsterdam.

Feldman, M. W. & Cavalli-Sforza, L. L. (1975). Models for cultural inheritance: a general linear model. *Annals of Human Biology*, **2**, 215–226.

Freedman, D. G. & Keller, B. (1971). Genetic influences on the development of behaviour. In *Normal and Abnormal Development of Brain and Behaviour*, ed. G. B. A. Stoelinga & J. J. Van der Werff ten Bosch. Boerhaave Series. Leiden University Press: Leiden.

Freeman, M. M. R. (1971). A social and ecological analysis of systematic female infanticide among the Netsilik Eskimo. *American Anthropologist*, **73**, 1011–1018.

Gadgil, M. & Bossert, W. H. (1970). Life historical consequences of natural selection. *American Naturalist*, **104**, 1–24.

Hamilton, W. D. (1970). Selfish and spiteful behaviour in an evolutionary model. *Nature, London*, **228**, 1218–1220.

Hamilton, W. D. (1975). Innate social aptitudes of man: an approach from evolutionary genetics. In *ASA Studies – Biosocial Anthropology*, ed. R. Fox. Dent: London.

Harris, M. (1966). The cultural ecology of India's sacred cattle. *Current Anthropology*, **7**, 51–66.

Harris, M. (1968). *The Rise of Anthropological Theory. A history of theories of culture*. Crowell: New York.

Harris, M. (1971). *Culture, People, Nature: an Introduction to General Anthropology*. Crowell: New York.

Harris, M. (1975). *Cows, Pigs, Wars and Witches. The Riddles of Culture*. Hutchinson: London.

Hinde, R. A. & Simpson, M. J. A. (1975). Qualities of mother–infant relationships in monkeys. *The Parent–Infant Interaction*, Ciba Foundation Symposium, no. **33**. Associated Scientific Publishers: Amsterdam.

Kruuk, H. (1964). Predators and anti-predator behaviour of the Black-headed Gull (*Larus ridibundus* L.). *Behaviour Supplement*, **11**.

Leach, E. (1961). Rethinking anthropology. *L.S.E. Monographs on Social Anthropology*, no. **22**. Athlone: London.

Lee, R. B. (1968). What hunters do for a living, or, how to make out on scarce resources. In *Man the Hunter*, ed. R. B. Lee & I. DeVore. Aldine: Chicago.

Lee, R. B. (1972). Population growth and the beginnings of sedentary life among the !Kung Bushmen. In *Population Growth: Anthropological Implications*, ed. B. Spooner. MIT Press: Cambridge, Mass.

Lee, R. B. & DeVore, I. (ed.) (1968). *Man the Hunter*. Aldine: Chicago.

Morris, D. (1967). *The Naked Ape*. Cape: London.

Patterson, I. J. (1965). Timing and spacing of broods in the Black-headed Gull (*Larus ridibundus* L.). *Ibis*, **107**, 433–460.

Perrins, C. M. (1965). Population fluctuations and clutch-size in the Great Tit *Parus major* L. *Journal of Animal Ecology*, **34**, 601–647.

Perrins, C. M. & Moss, D. (1974). Survival of young Great Tits in relation to age of female parent. *Ibis*, **116**, 220–224.

Pike, K. (1954). *Language in Relation to a Unified Theory of the Structure of Human Behaviour*, vol. **1**. Summer Institute of Linguistics: Glendale.

Rappaport, R. A. (1968). *Pigs for the Ancestors*. Yale University Press: London.

Richards, M. P. M. & Bernal, J. F. (1972). An observational study of mother–infant interaction. In *Ethological Studies of Child Behaviour*, ed. N. G. Blurton Jones. Cambridge University Press: London.

Rutter, M., Tizard, J. & Whitmore, K. (1970). *Education, Health and Behaviour*. Longmans: London.

Ruyle, E. E. (1973). Genetic and cultural pools: some suggestions for a unified theory of biocultural evolution. *Human Ecology*, **1**, 201–215.

Sadleir, R. M. F. S. (1969). *The Ecology of Reproduction in Wild and Domestic Mammals*. Methuen: London.

Sahlins, M. (1974). *Stone Age Economics*. Tavistock: London.

Tinbergen, N. (1951). *The Study of Instinct*. Oxford University Press: London.

Tinbergen, N. (1963). On aims and methods of ethology. *Zeitschrift für Tierpsychologie*, **20**, 410–433.

Tinbergen, N. (1965). Behaviour and natural selection. In *Ideas in Modern Biology*, Proceedings of the XVI International Zoological Congress, Washington, 1963, vol. **6**, ed. J. A. Moore. Doubleday: New York.

Tinbergen, N., Impekoven, M. & Franck, D. (1967). An experiment on spacing-out as a defence against predation. *Behaviour*, **28**, 307–321.

Trivers, R. L. (1972). The evolution of reciprocal altruism. *Quarterly Review of Biology*, **46**, 35–57.

Trivers, R. L. (1973). Natural selection of parental ability to vary the sex ratio of offspring. *Science, Washington*, **179**, 90–92.

Trivers, R. L. (1974). Parent–offspring conflict. *American Zoologist*, **14**, 249–264.

Vayda, A. P. (1969). *Environment and Cultural Behaviour*. The Natural History Press: New York.

White, L. (1949). Ethnological theory. In *Philosophy for the Future: the Quest of Modern Materialism*, ed. R. W. Sellars, V. J. McGill & M. Farber. Macmillan: New York.

Williams, G. C. (1966). *Adaptation and Natural Selection: a Critique of Some Current Evolutionary Thought.* Princeton University Press: Princeton.

Wilson, E. O. (1975). *Sociobiology: The New Synthesis.* Belknap: Cambridge, Mass.

Woodburn, J. (1968). An introduction to Hadza ecology. In *Man the Hunter*, ed. R. B. Lee & I. DeVore. Aldine: Chicago.

15

Towards understanding relationships: dynamic stability

R. A. HINDE AND JOAN STEVENSON-HINDE

INTERACTIONS, RELATIONSHIPS AND SOCIAL STRUCTURE

Throughout its development, ethology has been closely concerned with social behaviour. Armed with the concepts of social releaser and fixed action pattern, ethologists have gained considerable understanding of the interactions between individuals in a variety of invertebrate and vertebrate phyla. In particular, ethologists have been especially successful in analysing the complexities of vertebrate social displays (e.g. Tinbergen, 1959).

Perhaps just because ethologists were so successful in understanding such seemingly complex behaviour, the view that all social behaviour can be understood in terms of interactions has often been accepted unthinkingly. Although some of the earliest ethological papers were concerned with relationships (e.g. Lorenz, 1931, on bonding in jackdaws (*Corvus monedula*); and 1935, on parent–offspring relationships), surprisingly little work since then has focussed on long-term relationships between individuals. Only now is it becoming recognized that interactions between individuals known to each other cannot be regarded as isolated events, but must be seen as parts of a series of mutual interactions extending over time – that is, as parts of a relationship (e.g. Simpson, 1973*a*).

In an attempt to increase conceptual clarity in this area, the framework shown in Fig. 1 was elaborated (after Hinde, 1976*a*). The study of social behaviour is seen as involving three levels – interactions, relationships, and social structure. To describe an interaction, it is necessary to specify what the participants are doing together (its content), and how they are doing it (its quality). Description of a relationship involves description of the content and quality of the component interactions, and also of how they are patterned over time. Description of social structure requires description of the content, quality and patterning of the constituent relationships.

The distinction between three levels in the conceptual framework of course

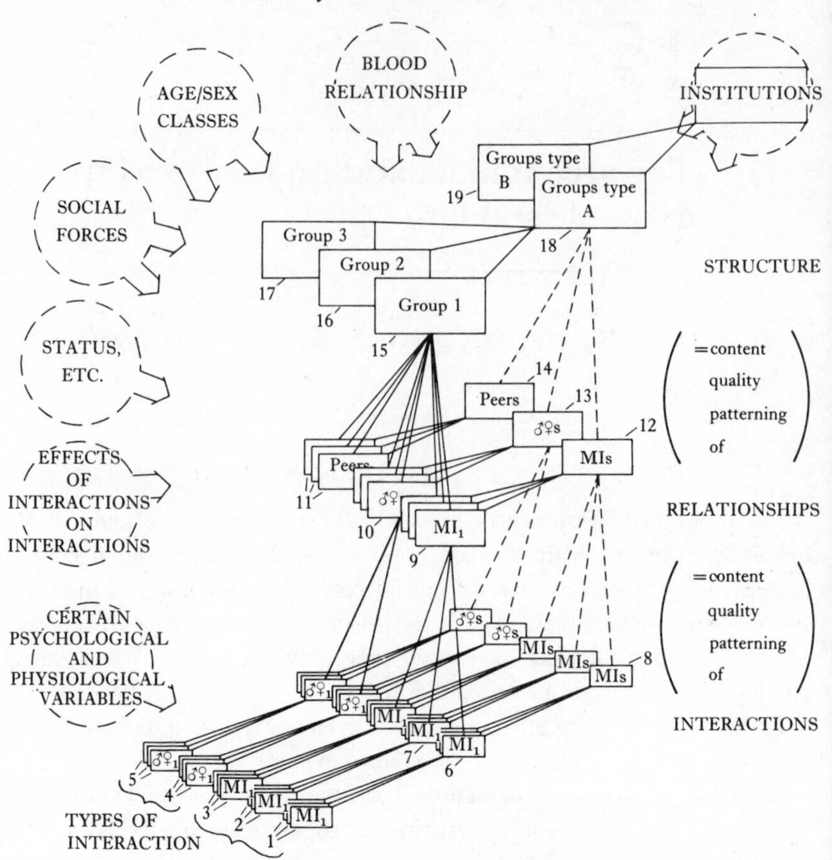

Fig. 1. Diagrammatic representation of the relations between interactions, relationships and social structure. Interactions, relationships and social structure are shown as rectangles on three levels, with successive stages of abstraction from left to right. The discontinuous circles represent independent or intervening variables operating at each level. Institutions, having a dual role, are shown in both a rectangle and a circle. In the specific instance of a non-human primate, the rectangles might represent

(1) Instances of grooming interactions between a mother A and her infant B.

(2) Instances of nursing interactions between A and B.

(3) Instances of play between A and B.

(4) Instances of grooming between female A and male C.

(5) Instances of copulation between A and C.

(6) First stage abstraction – schematic grooming interactions between A and B. Abstractions of grooming interactions between other mother–infant pairs are shown behind, but the specific instances from which they were abstracted are not shown.

(7) First stage abstraction – schematic nursing interactions between

carries no implications about the direction of causal relationships: interactions influence, and are influenced by, relationships and social structure.

At each of these levels it is necessary to proceed in two ways. First, from data concerning particular interactions, relationships or structures, we must make generalizations about the same characteristics over a wider range of occasions, individuals, circumstances, cultures or species. This is represented in Fig. 1 by successive blocks of rectangles from left to right. In general, the wider the generalization, the more superficial it will be in the sense that more information about particulars must be added to relate it to an empirical instance.

Second, at each level, we must seek principles involving variables which will 'explain', in a causal or functional sense, the empirical data. Some examples of such variables are represented by the discontinuous circles in Fig. 1. At each level it is necessary to employ concepts additional to those required at the level below it. At each level more than one principle operates, so that way in which the principles interact must also be specified.

A and B. Abstractions of nursing interactions of other mother–infant pairs are shown behind.

(8) Second stage abstraction – schematic grooming interactions between all mother–infant pairs in troop.

(9) Mother–infant relationship between A and B. Mother–infant relationships of other mother–infant pairs are shown behind (but connections to grooming, nursing etc. interactions are not shown).

(10) Consort relationship between A and C. Other consort relationships are shown behind.

(11) Specific relationship of another type (e.g. peer–peer).

(12) (13) (14) Abstraction of mother–infant, consort and peer–peer relationships. These may depend on abstractions of the contributing interactions.

(15) Surface structure of troop containing A, B, C etc.

(16) (17) Surface structures of other troops (contributing relationships not shown).

(18) Abstraction of structure of troops including that containing A, B, C etc. This may depend on abstractions of mother–infant etc. relationships.

(19) Abstraction of structure of a different set of troops (from another environment, species etc.).

Rectangles labelled MI_1 refer to behaviour of dyad female A and her infant B. Rectangles labelled ♂♀$_1$ refer to consort pair female A and male C. Rectangles labelled MIs, ♂♀s refer to generalizations about behaviour of mother–infant dyads and consort pairs respectively (after Hinde, 1976*a*).

Although these two methods of proceeding – seeking for descriptions with increasing degrees of generality, and seeking for principles with which to understand causation – appear to be orthogonal, they may meet in the study of human institutions. This is therefore shown by both symbols in Fig. 1.

This conceptual framework has been discussed in more detail elsewhere (Hinde, 1976*a*), and particular aspects have been examined more closely in other papers (Hinde & Simpson, 1975; Hinde, 1976*b*, *d*). The present discussion is concerned with the dynamics of relationships between individuals.

THE STUDY OF RELATIONSHIPS

We believe that the time is now especially propitious for progress in the study of relationships, for a number of reasons:

(1) Relationships consist of a series of interactions extended in time. Understanding of the constituent interactions was an essential first step, and much progress has been made.

(2) Whilst ethologists have studied social communication primarily in the context of social releasers producing either immediate effects on behaviour or longer-term motivational (e.g. endocrinological) consequences, there is a growing awareness that subtle and often idiosyncratic signals may be important and may mould the future course of a relationship (Menzel, 1971, 1972; Simpson, 1973*a*, *b*).

(3) The several interrelated ways in which learning can operate in the development of a relationship have been analysed in some detail in the relatively simple (and in some ways special, see below) case of parent–offspring relationships in birds (Bateson, 1966, 1973).

(4) There is increasing appreciation of the species- and even individual-specific constraints and predispositions that may guide the course of learning (Seligman & Hager, 1972; Hinde & Stevenson-Hinde, 1973).

(5) Finally, and perhaps most important, ethologists can profit greatly from the considerable literature on human relationships. Whilst ethologists have always been ready – some would say over-ready – to apply their findings to man, they have perhaps been too reticent about engaging in traffic in the opposite direction. The literature is scattered in diverse disciplines and, just because of the prepotent influence of culture, much of it seems to be of limited generality. But it seems to us likely not only that judicious use of this literature could aid the ethologists' studies of relationships between individuals of non-human species but also that such studies could provide a testing ground for concepts which could then be useful for study of the more complex human case.

Now it may immediately be said that human relationships involve affective and cognitive phenomena as well as behavioural interactions. Though this may also be true of animals, it is true to a much lesser extent. Does this render the study of animal relationships irrelevant to the human case? There are a number of issues to be considered here:

(1) It may be argued that a relationship involves more than interactions because it can continue when the partners are separated from each other, and thus are not interacting. However, such a statement implies that A, in the absence of B, feels or behaves differently by virtue of the fact that he has interacted with B. Thus the relationship is ultimately based on interactions.

(2) To understand a human relationship, we must understand not only each participant's view of himself, of his partner and of their relationship but also each participant's view of the other's views, and even each participant's views of the other's views of his own views of himself, and of his partner and of their relationship (Laing, 1962, 1969). Returning to behavioural terms for a moment, A may show behaviour X because he knows that B expects him to show behaviour Y. He may even show behaviour Y because he knows that B knows that he initially intended to show Y and will expect him to change to X, and he wishes to thwart B's expectations. Furthermore, the behaviour of A to B may be influenced by moral standards at levels of complexity far beyond the reach of non-human species. We do not wish for one moment to belittle the gap between the complexity of the relationships shown by human and non-human species. For many purposes it may be as well to regard that gap as absolute. And yet there are hints that it is bridgeable. Thus the relationships formed by young human infants lack some of the sophistication that we ascribe to relationships between adults, and may not be so very different from those of monkeys and apes at some stages of their development. In any case too liberal a use of Lloyd Morgan's canon can cause us to underestimate the complexity of the relationships in non-human species. The evidence for this view is in part anecdotal but here are two examples:

(i) Anecdotal data indicate considerable cognitive complexity in the social behaviour of non-human primates. For instance Maxim & Buettner-Janish (1963) record adult baboons (*Papio* sp.) sitting near the entrance of a trap and threatening away juveniles who were about to enter.

(ii) In a series of experiments Menzel (e.g. 1971, 1972) has shown that an acquired pattern of behaviour can be transmitted across successive groupings of chimpanzees (*Pan troglodytes*), and that chimpanzees have a remarkable ability to communicate to each other the location of hidden food or a dangerous object. Of particular interest is Menzel's evidence for the deception, apparently deliberate, of one animal by another.

(3) In any case we are not arguing that relationships consist of nothing but interactions. For one thing they may possess qualities emergent from the patterning of the constituent interactions. And while it is likely that they possess subtle qualities that are beyond the reach of the behavioural scientist, it remains at least possible that those qualities are correlated with behavioural properties that are accessible. For example, it has been argued elsewhere that 'intersubjectivity' (Asch, 1952) and 'penetration' (Altman & Taylor, 1973) are correlated with 'behavioural meshing' (Hinde, 1976*b*). For many practical purposes, it is the extent of such correlations that are important. For example, if we can specify behavioural indices of an affectionate relationship, and can then discover the conditions necessary to promote relationships having high values of such indices, we may thereby be specifying the conditions necessary to promote the more intangible properties of such relationships.

(4) Human inter-personal relationships are often institutionalized – that is, each participant attempts to behave in such a way that the relationship will conform to a culturally ordained pattern. Indeed relationships as defined here, involving individuals known personally to each other, are in one sense a special case. We also have relationships with categories of people – we behave as citizens to policemen, drivers to traffic wardens, patients to doctors: in each case our behaviour is determined in part by previous experience of other individuals in the same category, and in part by knowledge acquired in other ways of the characteristic behaviour of people occupying those positions in society. At a superficial level this need not involve a difference in kind from non-human species: the behaviour of a subadult monkey to a dominant male may be affected by his experience of other members of the category 'dominant male' in the past. But, in the human case, verbal language permits traditional and cultural influences to be more potent and more pervasive: the policeman behaves as he does in part because he is attempting to exhibit behaviour that he has learnt is appropriate to his role in society not only from previous interactions with policemen and non-policemen, but from oral instruction and police manuals. Cultural determinants of the behaviour of non-human species lack the complexity that verbal language permits. Similar differences are to be found in the factors affecting relationships between individuals: the cultural factors affecting the behaviour of husband and wife in a human society must be incomparably more potent that those affecting the behaviour of the members of a consort pair of chimpanzees.

In brief, then, we do not believe that a strictly behavioural approach to relationships is sufficient for solving every question that will come up, but we do believe it provides a starting point. When the task is important we must use the tools that are to hand – though of course this need not prevent our also searching for new ones. In any case, the aim of the comparative psychologist should be not to explain the human case in terms of data derived from animals but to abstract principles whose applicability to the human case can subsequently be assessed (Hinde, 1976*c*). It is just because non-human primates are different from man, because they are simpler and less influenced by culture, that principles with a degree of generality may be easier to find.

THE NATURE OF RELATIONSHIPS

In this paper our main concern is with principles related to the development and stability of relationships. First, however, it is necessary to clarify what is meant by a relationship. The ethologist must approach inter-individual relationships by way of behaviour. From this point of view, a relationship involves a series of interactions between two participants known to each other. We will not attempt to define the concept of 'interaction' precisely here, but we are referring to such incidents as 'A does X (to B)' or 'A does X (to B), and B responds with Y'. (Strictly it might be argued that an *inter*action must involve actions by both partners. However, in the case 'A does X (to B)', A must be responding to stimuli at least broadcast by B, if not specifically directed towards A. The distinction between broadcast signals and those specifically directed is in practice not easy to make. In any case, it will be noted that 'interaction', like 'relationship', refers to events that could not be fully understood by observation of one partner alone.)

To describe an interaction it is necessary to describe first what A did to B (and B to A). They may, for instance, be talking together, or fighting, or one may be chasing the other. In addition we must specify how they are doing it – are they talking in an animated fashion, or are they fighting savagely? We shall refer to these latter properties as qualities, without of course intending to imply that they could not be subjected to quantitative treatment.

A relationship* involves a series of such interactions between participants who are known to each other. To describe a relationship, it is necessary to specify not only what the individuals do together and how they do it (i.e. the content and quality of their interactions) over a specified time interval, but also how those interactions are patterned (i.e. their absolute and relative frequencies and how interactions of one type are related to interactions of other types).

It is important to emphasize that, whereas each interaction takes place in

* Whereas 'interaction' applies unambiguously to behavioural data, 'relationship' is often used also as an intervening variable. Thus we may say that 'attachment behaviour results in the formation of a relationship', or that 'punishment changes the course of the parent–child relationship': in each case 'relationship' refers not to an actual pattern of interactions, but to a potential for patterns which may be of a certain general type but whose precise form depends on other events as yet in the future.

a strictly limited span of time, relationships can be discussed only in relation to a much longer time period (see also Simpson, 1973*a*). (That there may be instances where it is difficult to specify when it is profitable to stop talking in terms of an interaction, and start talking about a relationship, is not our present concern; the distinction is in most instances clear enough.) In part just because relationships involve a long period of time, the study of relationships involves complications additional to those involved in the study of interactions. Thus

(*a*) As already mentioned, each interaction may be influenced by those that preceded it, and perhaps by expectations of those in the future. Because present interactions may be directed towards producing effects on interactions in the future, the reinforcers that operate currently may derive their effectiveness from expectations of future effects.

(*b*) Because a relationship involves particular individuals, interactions of one type may affect interactions of another type within the same relationship.

(*c*) Indeed, interactions within a relationship may be affected also by interactions of a similar type entered into by one of the participants in a quite different relationship involving a third party.

(*d*) Furthermore, a relationship may possess emergent qualities not present in its constituent interactions but derived from the relative frequency and/or sequential relations between those interactions (see Hinde & Simpson, 1975; Hinde, 1976*b*).

THE DYNAMIC STABILITY OF RELATIONSHIPS

By the definition we have used, relationships exist in time, and each interaction within a relationship may be influenced by past interactions as well as by expectations of future ones. In addition, the participants in a relationship are likely to change with time, through the natural processes of growth and development or as a consequence of events within or outside the relationship. All this implies that a relationship is seldom static, and that its stability is essentially dynamic in nature (see e.g. Altman & Taylor, 1973, p. 25). Thus whilst the properties of a relationship between two individuals may remain fairly constant over time, they are more likely to vary whilst remaining within certain limits, or change progressively with time, or change from one temporarily stable state to another. Clearly when a relationship continues but changes in nature, the changes may be anything from trivial to so great that the new relationship has little in common with the old. An example of the latter would be a parent–child relationship which becomes, with time, more like a relationship between peers. The extent of change, which we allow without describing the relationship as a new one, is of course an arbitrary matter.

In analysing the stability of relationships it becomes necessary to distinguish between different types of stability, and control systems analysis provides some guidelines. In the case of *global stability* the state tends always to aim towards a particular goal or ideal, no matter what the current parameters may be. With *asymptotic stability*, by contrast, the state heads for a goal only when it is within a specifiable region near the goal. Thus in the case of a relationship, marriage partners might always strive towards an ideal relationship, no matter what the circumstances; or they might do so only as long as the nature of their relationship remains within certain limits (D. J. McFarland, personal communication).

With both global and asymptotic stability, a goal must be postulated. In the human case, especially, culture provides goals which may constrain the behaviour of the participants in a way which promotes development in a particular direction: they may seek to build a marriage or a parent–child relationship which their society will view with approval. Such cultural determinants appear to be all-important in cross-cultural comparisons, where individual idiosyncracies are neglected. But within a culture, where we are interested in the multitudinous influences on the course of relationships of a particular type, such cultural constancies become part of the background and the multiplicity of influences and possible end points becomes the important issue. Thus whether we lay emphasis on equifinality (Bertalanffy, 1952) or on multifinality (Buckley, 1967) depends in part on the level of analysis with which we are concerned (cf. Bateson, this volume).

With relationships in which a goal state seems unlikely, one may postulate a *stability boundary* which is avoided, as when marriage partners retreat from states which make divorce a real possibility. Such a mechanism could permit considerable flexibility in the content of the relationship and could operate, for instance, in the relationship between parent and adolescent child cited above: the important issue to one or both partners is the avoidance of interactions which would terminate the possibility of any future interactions, not stability of content. Indeed it may be profitable to distinguish mechanisms promoting stability (i.e. constancy of content) from mechanisms promoting continuity (i.e. continued association).

It must be emphasized that neither approach towards a goal state nor avoidance of an undesirable state need be based on culturally determined or conscious recognition of those states. For instance, the individuals may be so constituted as automatically to avoid states in which aggression between them is probable. In practice, of course, many relationships involve acknowledged goals or boundaries. One aspect of the term 'commitment' is the acceptance of such goals and/or boundaries.

Finally, this reference to possible goals and/or boundaries must not be taken to imply that such a model will necessarily fit every relationship. Even where it does fit, it may not tell the whole story. We have already suggested that each interaction may be shaped by past and expected interactions, and in this way shape the course of a relationship. Furthermore, each relationship is set in a nexus of other relationships, and is inevitably affected by them. All this could contribute to 'stability' in the descriptive sense of continuation over time, with progressive rather than dramatic change in content.

To understand the stability of relationships we must come to terms both with how relationships are preserved in the face of internal changes and the buffetings of the external world; and with how they may change progressively, again as a consequence of factors internal and external to them, and yet preserve their integrity. These two aspects of dynamic stability are, of course, interdependent. In the remainder of this paper we attempt to define the areas in which it will be profitable to seek for principles to aid our understanding of the dynamics of relationships.

THREE LEARNING PARADIGMS

The patterning of interactions within a relationship must involve individual learning. How far can our general knowledge of learning help us here? For the sake of simplicity, and as a starting point only, we shall consider briefly three paradigms of learning (for a more thorough discussion see Kling, 1971). The distinctions between them are purely operational. That one can fit diverse examples into a particular paradigm does not imply that only one underlying learning process is involved, nor does it ignore the constraints and predispositions acting in any learning situation (see e.g. Seligman & Hager, 1972; Hinde & Stevenson-Hinde, 1973). Learning paradigms are useful in that they provide a convenient framework for diverse examples of learning: of course additional concepts can be added as necessary where complexity or diversity requires them, though this may dilute the value of the initial paradigm.

Exposure learning refers to changes in behaviour that result from an individual being exposed to a situation, with no consistent response (except perhaps investigatory and exploratory responses) elicited by that situation, and no obvious reinforcement. For example, Gibson, Walk & Tighe (1959) found that if stimulus objects were present in rats' home cages, then those rats learned a discrimination based on those objects more quickly than rats that had not been exposed to them in their home cages (see also Bateson & Chantrey, 1972). In addition to affecting discrimination learning, exposure to stimuli can also promote a preference. For example, Bateson (1964) found that chicks reared in pens whose walls were painted with vertical stripes subsequently followed models painted with vertical stripes more readily than models painted with horizontal stripes, and vice versa. Similar ideas have been applied to inter-individual attraction in man. Homans (1950) postulated that how much two people liked each other was likely to increase with the frequency of their interactions. More recently, Zajonc (1971) has cited both correlational and experimental evidence to support the view that repeated exposure of an individual to a stimulus object, including a member of his own species, enhances his attraction to it.

Another example that fits the exposure learning paradigm is the observation of the behaviour shown by another individual, its retention in a coded form, and its subsequent reproduction. Whilst the precise nature of such 'modelling' is a matter of some discussion (e.g. Berger & Lambert, 1968), its importance in social contexts is beyond dispute (Bandura, 1971).

Classical conditioning involves the pairing of a so-called 'unconditioned stimulus', which already elicits a response, with a (to be) 'conditioned stimulus', which does not initially elicit that response. After one or more pairings the conditioned stimulus, even if presented on its own, comes to elicit a response closely similar to that initially elicited only by the unconditioned stimulus. For example, if tactile stimulation elicits relaxation in an infant, it is possible that visual and/or auditory stimuli associated with this event (e.g. sight and voice of parent) will come to do this on their own. Indeed, such a paradigm can readily be applied to emotional responses. Thus Mowrer (1960) postulated that, through a classical conditioning paradigm, conditioned stimuli can elicit emotional states (e.g. hope, disappointment, fear, and relief). He then used an operant conditioning paradigm (see below) to describe behaviour directed towards (or away from) these conditioned stimuli. Such a model has been used by a number of investigators concerned with human inter-personal relations (e.g. Byrne & Clore, 1970; Lott & Lott, 1972, 1974). The former authors suppose

(*a*) Various inter-personal events can be classed as reinforcing or punishing.

(*b*) These generate respectively positive and negative affect.

(*c*) Stimuli associated with positive or negative affect develop the capacity to evoke that affect.

(*d*) Stimuli that evoke positive affect are liked and vice versa.

With *operant conditioning*, the emphasis is on the manner in which responses are shaped and maintained by their environmental consequences. If a stimulus event which is contingent upon a response increases the future probability of that response, the stimulus is called a reinforcer. If reinforcement has occurred in a particular situation, then that situation will set the occasion for the repetition of that response.

Gewirtz (1961) used reinforcement to account for the development of an infant's attachment to its mother. Assuming that relationships are made up of behaviour systems directed at and maintained by reinforcers, Gewirtz argues that an infant will form an attachment with the person 'who mediates most of the important environmental consequences of his behaviours' (*ibid.*, p. 237).

Thibaut & Kelly (1959) and Homans (1961) made extensive use of operant terminology to erect a theoretical scheme accounting for many aspects of human social relationships. In brief, they suggested that individuals continue to perform activities in so far as they produce positive reinforcement or values from others, and in so far as they do not produce punishment or involve excessive cost (which may include other potential reinforcers foregone in the performance of the activity). Homans argues that no interaction will continue unless both parties make a 'psychic profit', this being defined as reward less cost. To suggest a non-human example that exemplifies the view, the theory would suggest that two chimpanzees would continue grooming until the reinforcement provided by the interaction less the expenditure it involved ceased to outweigh the potential reinforcement that would be provided by other possible activities less the expenditure they would involve (see also Blau, 1964; Bott, 1966; Bailey, 1969).

PARADIGMS AND PRACTICE

As Kling (1971) points out, although thousands of learning situations have been studied, only a small number of paradigms are needed to describe the experimental procedures involved. In studying the role of learning in social interactions, application of one or more of the above paradigms often shows up similarities between different situations. As with most tools of investigation, the dangers lie not in the tools themselves, but in how they are used. If one uses a paradigm at more than a descriptive level, or with too heavy a hand, then its advantages will be lost.

Paradigms not processes. We cannot stress too strongly that we are considering paradigms of learning, not learning processes. There is no doubt that even though learning in different situations, and by different species, can be described in terms of these paradigms, learning processes will differ. These differences need not be ignored if one is prepared to build additional concepts around the paradigms. For example, in considering human emotional behaviour, Schachter (1964) goes a step beyond a classical conditioning paradigm, to propose that what a person actually experiences will depend not only on physiological arousal but also on how he labels the situation. Such a formulation generates experiments in which subjects' expectancies, as well as physiological responses, are manipulated. Thus, a classical conditioning paradigm could be used to describe that part of Schachter & Singer's (1962) experiment in which epinephrine (unconditioned stimulus) was administered to elicit physiological arousal (unconditioned response) in different situations (conditioned stimuli). However, they found that the subjects' emotional responses depended not only on the above manipulations, but also on what the experimenter had told the subject would happen as a result of the injection. Thus a flexible paradigm which leaves room for cognitive variables can go a long way towards understanding the development of interpersonal emotional responses (see e.g. Berscheid & Walster, 1974, for a sensitive application to the development of 'passionate love').

Again, Homans has been criticized (perhaps unfairly) because his theory can be regarded as reductionist, implying that the complex phenomena of human interactions require for their understanding nothing but the principles necessary for understanding limited aspects of the operant behaviour of pigeons, and neglecting the qualitative differences that emerge with increasing complexity (cf. Schneirla, 1949). Stebbins (1969) therefore advocates the need for propositions concerned with how the actor 'defines the situation' to bridge the gap (see also Davis, 1973).

Interdigitation of learning paradigms. The paradigms of exposure learning and classical conditioning might seem to apply most particularly to the relative attractiveness of potential partners and the early development of a relationship, and that of operant conditioning to its subsequent course. In practice all three are closely interdigitated (cf. Thorpe, 1956). For example, while the exposure learning paradigm is pleasing in its simplicity, it is usually possible to argue that a classical or operant conditioning paradigm might also apply in any particular case. On this view it is not mere exposure that is the crucial issue, but the conditions under which that exposure takes place. Although, as we have said, modelling fits an exposure learning paradigm, reinforcement may also be involved. For instance, reinforcement received from another individual may affect the probability that that individual is taken as a model (e.g. Grusec & Mischel, 1966). Of course this does not suggest that reinforcement is necessary for modelling to occur (Bandura, 1971), although the neutral conditions of exposure necessary to resolve this issue are difficult to achieve.

The interdigitation of these three paradigms has been demonstrated with particular elegance in studies of the imprinting of the following response of nidifugous birds (see also Bateson, 1971):

(*a*) As mentioned above, the paradigm for exposure learning is very similar to the procedure used to study imprinting – a period of exposure to stimuli followed by a test to see what has been learned (e.g. Gibson *et al.*, 1959; Bateson, 1964; Bateson & Chantrey, 1972).

(*b*) A classical conditioning paradigm has been applied to imprinting by James (1959). He proposes that retinal flicker acts as an unconditioned stimulus for approach behaviour, and that stimuli associated with the unconditioned stimulus can become conditioned stimuli for approach.

(*c*) An operant conditioning paradigm is also appropriate, as shown by Bateson & Reese (1969). Dark-reared chicks will perform an operant response (stepping on a pedal) if that response produces a conspicuous object, which they in turn approach. In addition, an object on which imprinting has already occurred can act as a reinforcer (Peterson, 1960; Hoffman, Searle, Toffey & Kozma, 1966).

Thus each of the three paradigms that we have considered can be applied to the imprinting situation. This suggests that, when a chick is establishing its relationship with its natural mother, all three paradigms are applicable, and interdigitation of paradigms is ubiquitous.

REINFORCEMENT

We have already discussed reinforcement in the context of paradigms of learning, and have mentioned two ways in which the concept has been applied to social interactions (Gewirtz, 1961; Homans, 1961). It is now unarguable that its application can bring order to data about social interactions and suggest further ways of studying them. Numerous examples of this are given in Homans' book. We may extend these by considering an example of the application of his concepts to non-human species.

As Homans emphasized, an important element in nearly all our relationships is the social approval that we obtain from others – either from the interactant, or from third parties. We must add quickly that the concept of social approval is being used loosely here, and covers a spectrum from the condescension of a superior through the approval of peers to adulation or obeisance. Now the obvious ways of expressing social approval in man involve verbal interchange, and it might seem improbable that social approval could be of much importance in non-verbal species. However, in practice, non-verbal communication is much used for the expression of social approval in man – certain kinds of smile, soft voice, and so on. Is it possible that social approval, expressed non-verbally, is important also in monkeys and apes?

Such a view would bring order to certain anomalies in the literature on non-human primate communication. Consider, for instance, the still unsolved problem of sociosexual presenting. This is usually described as a sort of submissive posture, serving to switch off the aggression of a dominant individual. But there are some cases that hardly seem to fit this. Rhesus mothers may present to infants which are temporarily inaccessible to them as a means of making the infants return to them (Hansen, 1966). Dominant females will sometimes present as they approach a subordinate mother with her infant, and males will sometimes present to strange females (e.g. van Lawick-Goodall, 1968; Okano *et al.*, 1973). One way of describing all these cases is to say that presenting is an expression of social approval: whether this means that this is a merely superficial description, or that it is fundamental and basic, is an open issue.

Another case is grooming. While grooming no doubt has some function in care of the body surface, many observers have the impression that monkeys and apes groom each other more than that requires. Grooming of another individual is more frequent than self-grooming, and is accompanied by signs of excitation and/or relaxation. It does not occur at random between individuals in a group, but rather each individual tends to groom some partners more than others. For instance, it occurs more between animals that have kinship ties (Sade, 1965, 1972; Oki & Maeda, 1973), a consort relationship (e.g. Rowell, 1972), or are close to each other in a dominance hierarchy (Lindberg, 1973), than between others. Where there is patterning in the grooming relationships, the higher status individuals or families are usually groomed more, and groom less, than lower status ones (e.g. Oki & Maeda, 1973; Simpson, 1973*b* and references cited therein). Perhaps grooming also is partly to be understood as an expression of social approval.

That other types of behaviour could also be included here will be apparent to any primatologist. Greeting gestures, astonishingly diverse in some species (e.g. chimpanzee, van Lawick-Goodall, 1968; Nishida, 1970), are an obvious example. Even the desire for, or permitting of, proximity could be regarded in this light (N. K. Humphrey, personal communication).

That there may be some validity in a category broad enough to include both expressive movements made by a subordinate in the presence of a superior and 'reassurance' gestures made by a superior to an inferior is lent support by the fact that the same signal movement may be used in both contexts in some species: van Hooff (1972) reports that the 'silent bared-teeth display' is used by *Mandrillus* as both a submissive and a reassurance signal. In man, also, related forms of smiling appear in the two contexts, and over the primates as a whole there is evidence that some reassurance signals have evolved from submissive ones.

That the broad category has some validity does not of course imply that it should not be subdivided. Indeed this discussion of social approval in non-human primates is intended only as a first step, and is as yet far from satisfactory. 'Social approval' itself has so far not been adequately defined, and to some extent the use of the concept is circular. But the anomalies of sociosexual presenting and grooming have been about in the non-human primate literature for too long; this may be in part because concepts derived from studies of lower species are inadequate to explain them, so that it is worthwhile to try out some concepts derived from human experience. The enterprise may or may not be successful.

A related problem concerns variations in the amount of social approval that is reinforcing between individuals. Homans (1961) points out that, in any human interaction, each participant will expect his rewards to increase with his costs. Thus for each individual the more valuable to the other (and/or costly to himself) is his activity, the more valuable to him (and/or costly to the other) must be the activity that the other gives him. And if each is being reinforced by a third party, each will expect to receive rewards commensurate with his costs (Law of Distributive Justice). As Homans (1961, p. 76) puts it 'For with men the heart of these situations is a comparison.' Besides exchanging rewards with each other, people appraise the rewards and the costs they incur in relation to those received and incurred by those with whom they interact, and by others in comparable situations.

To what extent individuals of non-human species compare themselves with others in this way is quite uncertain. But there is a related issue in which there is at least an empirical similarity between the organization of human and non-human relationships, and for that it is necessary to introduce the related concepts of 'investment' and 'status'. Homans suggested that distributive justice is realized when the profit (i.e. reward less cost) of each man is directly proportional to his 'investments' (i.e. that with which he has been invested or acquired). What counts as an investment is to a large degree culture-specific, but such things as age, maleness, seniority, wealth, wisdom and acquired skill are recognized (rightly or wrongly) in many societies. The concept of 'investment' is clearly related to that of 'status', which is determined by similar independent variables, may be symbolized in the form of address (e.g. *tu* and *vous*; Brown & Gilman, 1970), and pervades all forms of interaction possible in the relationship. Now since investments may be reckoned on criteria which are independent of any particular interaction, this has the result that if one man ranks higher than another in one respect, he may rank higher in others too. For example, in an old-fashioned army, in which commissions depend in large part on accidents of birth, an officer gets more pay, more leave, and better food than a private not simply because he puts more into the army, but because the difference between his rewards and what he puts in 'should be' greater on account of his privileged birth (i.e. his investment).

Now if we suppose that the 'status' or 'investment' of a chimpanzee is related to his 'dominance' as assessed in agonistic interactions, we may then expect those with the higher investment to obtain the greater profit in terms of expressions of social approval. That this is in fact the case in grooming interactions has been shown by Simpson (1973*b*). Amongst the adult male chimpanzees in one community in the Gombe Stream National Park the more dominant individuals tended to be involved in more grooming bouts, and were groomed for longer periods, than their inferiors. (An interesting possible exception concerns the highest ranking male, who may be exceptionally 'generous' in his grooming: this may be related to the fact that in so far as he is also a leader (i.e. determines the behaviour of others) he inevitably earns unpopularity (Homans, 1961) which must be compensated by additional gestures of approval.) Similarly Sade (1972) found a relationship between grooming 'status' and dominance 'status' in female rhesus monkeys (*Macaca mulatta*).

However, at this point the introduction of the concepts of 'investment' and 'status' have added little – we could just as well speak directly of a relation between grooming and dominance rank. The concept of 'status' is useful only if 'status' is determined by more than one sort of factor. There are hints that dominance rank, as indicated primarily by agonistic interactions, may not be the sole factor in determining the total 'investment' of an individual within a non-human primate group. In the male chimpanzees at the Gombe Stream National Park, Bygott (1974, 1976) found that age was correlated more significantly than any agonistic measure with the frequency of grooming and of being groomed, and with the duration of being groomed. Furthermore, the frequencies with which males displayed, though highly correlated with agonistic measures, was also more strongly correlated with a measure of how often individuals were chosen as grooming partners than with age (see also Kummer, 1968). It thus seems that age, as well as dominance rank in agonistic encounters, may contribute to status in some non-human primate groups. Such a fact seems to demand also explanation in functional terms (Clutton-Brock & Harvey, this volume). Why should it increase a young chimpanzee's fitness (in terms of natural selection) to show social approval to an old one? Possibly age is correlated with a knowledge of the habitat which can be shared with younger individuals.

REINFORCEMENT'S DILEMMAS

The reinforcement concept has proved powerful (though not ubiquitously so) for understanding why learning occurs in some situations but not others. Its power depends on our ability to predict which events will and will not have reinforcing properties. In non-social situations predictions are possible for many situations, even though 'Laws of Learning' have now been shown to be less general than was earlier believed (Seligman & Hager, 1972; Hinde & Stevenson-Hinde, 1973). In social situations, the position is much more difficult (Levinger, 1972; Lickona, 1974). While such loose usage as that involved in referring to 'social approval' as a reinforcer can bring considerable insight, too loose a definition of 'reinforcement' renders the concept valueless because incontrovertible (e.g. Firth, 1965; Heath, 1972). The difficulty is exacerbated by the varying degrees of precision with which reinforcement is used by different authors, and even by the same author on different occasions. Thus reinforcement not only is applied precisely to clearly defined operant situations, but also is used to explain why people with similar status or attitudes are attracted to each other (Homans, 1961; Clore & Byrne, 1974), why people with complementary needs are attracted to each other (Winch, 1958), how the stability of a relationship is affected by experiences of the participants in other relationships (Thibaut & Kelly, 1959), and so on.

One way of proving that an event is a reinforcer in a real life situation is to assess its potency in a simplified, laboratory situation and then generalize back to the broader situation. Such a procedure is based on the assumption which made the reinforcement concept useful in the first instance – namely that a stimulus that acts as a reinforcer in one context will do so also in another. But this assumption is not always valid (Seligman & Hager, 1972; Hinde & Stevenson-Hinde, 1973). Hence generalizations about reinforcement must be accompanied by statements concerning the limits of their applicability.

Another approach to this problem is to attempt to classify reinforcers. Foa (1971) has made some progress by distinguishing six categories of interpersonal resources: money, goods, information, services, status, and love. He argues that these vary along two dimensions: concreteness (with money most and love least) and personal specificity (with money least and love most), and is able to go some way towards showing in which situations the various types are likely to be effective reinforcers. But the factors that influence the effectiveness of reinforcers are multitudinous – for instance, motivation, arousal level, developmental level, moral level, social content, cultural content, and stage of the relationship (e.g. Secord & Backman, 1974). Furthermore, in making prognoses about relationships it is necessary to consider both the absolute levels of rewards and of costs, and the relation (ratio?) between them (Altman & Taylor, 1973, p. 92). Thus we are still a long way from being able to predict whether and when a particular event will be reinforcing.

Furthermore, it seems clear that if the reinforcement concept is to be valuable, it must sometimes be related to rather intangible variables. But when the reinforcement theorist, confronted with social interactions in which the reinforcing events are far from obvious, resorts to 'internalized reinforcement', he thereby weakens his case. Yet in so far as he attempts to account for relationships solely in terms of tangible reinforcers, he will be forced to neglect some more subtle, and perhaps important, kinds of reinforcement. The dilemma seems insoluble.

This dilemma means that the learning paradigms must be used with circumspection. On the one hand, they can be used to understand much of the dynamics of relationships. On the other hand, it is easy to be satisfied by their simplicity, and to neglect the complexities of the total situation.

In any case, learning of one sort or another affects relationships in a number of ways, and principles concerned with learning are not in themselves sufficient for a full understanding of their dynamics. In the next section we shall consider a broader framework involving a number of areas where principles must be sought if inter-individual relationships are to be understood.

PRINCIPLES CONCERNING DYNAMIC STABILITY

In this section we shall consider a possible way of classifying principles pertinent to the dynamics of relationships. No way of classifying is wholly satisfactory, and the categories overlap. Furthermore learning, as discussed above, enters at many points. However, the categories may serve as a means to put our thoughts in order.

Principles concerned with the initial nature of the participants

Within any particular species, members of the various age/sex classes form relationships of different sorts. For instance in macaque monkeys young males tend to associate together; females, even when adult, continue to associate with their mothers, but males do not; and adult females may form short or even long-term relationships with particular males. Furthermore, there are consistent differences between the relationships of related individuals and those of unrelated individuals. Comparable generalizations can be made about particular human cultures, though common membership of a kinship group may be more important than blood relationships. Such principles concern the properties of individuals when they come into the relationship, and have been considered elsewhere (Hinde, 1976*a*).

Principles concerned with degrees of compatibility between interactions

Some types of interaction are especially likely to be associated within the same relationship, whilst others are incompatible. Thus in so far as all macaque consort pairs copulate and show mutual grooming but not nursing/suckling, the appropriate relative probabilities of the several types of interaction are inherent in the nature of the relationship. In such a case, this category overlaps with the previous one. However, another important type of case concerns 'status', which in most animal studies is assessed in terms of 'dominance/subordinance'. If 'dominance/subordinance' is applied to only one type of interaction, such as 'A hits B', it is merely descriptively redundant and has no explanatory value. It becomes a concept useful for understanding inter-individual relationships only when a number of different interactions tend to occur together in different relationships – for example if 'A hits B' is associated with 'A threatens B', 'B avoids A', 'B grooms A', and 'B presents to A', but not with 'B mounts A'. In many human societies individuals differing markedly in status are likely to interact in some ways, for example, by gestures of condescension/obeisance, and unlikely to do so in others, such as eating at the same table. In some cases, 'status' can be regarded as an initial property of the individual, so that principles concerned with it would properly belong in the previous category; in others it emerges in the course of the relationship, in which case it should perhaps properly be considered in the next.

Principles concerned with the effects of one type of interaction on the probability of others

Diverse phenomena are included here, and we may consider two ways of subdividing them. The first concerns the temporal nature of the effect:

(*a*) Short-term. Interactions may affect the probability of the interactions that immediately follow them. For example, greeting gestures are an essential preliminary to further interactions in many human and non-human societies, and the nature of the greeting may influence what follows.

(*b*) Medium-term. Sometimes the effects of an interaction are rather long-lasting. In many species courtship produces cumulative effects on the probability of copulation: these may or may not be hormonally mediated. As another example, Kummer (1975) has shown that newly acquainted baboons show first fighting, then presenting, then mounting and then grooming as they establish a relationship. Although a stage may be omitted, the order appears to be invariant, and it seems that the sequence is usually necessary for the establishment of a relationship.

(*c*) Long-term. Learning may of course produce both immediate and long-term effects, but the great majority of the latter must depend upon it. Such effects are especially important in multiplex relationships, that is those that involve more than one type of interaction, for learning in the course of one type of interaction may then affect the course of many others.

Another way of classifying the effects of interactions on subsequent interactions concerns the nature of the effect on the future course of the relationship. We are, of course, not concerned here with relationships in which continued association of the participants depends on external constraints, as when two female monkeys continue to interact intermittently with each other because both seek proximity to the same male. But where continuation of the relationship depends on the effects of interactions on interactions, those effects can be classified according to whether they tend to promote or diminish its stability. If external stresses tend to diminish the association, and this results in a more active striving for association, negative feedback can be said to operate. In other cases positive feedback might occur: if one partner was unwilling to interact in a particular way, the other might show frustration-induced aggression which might enhance the uncooperativeness of the former. Such feedback effects may or may not involve reinforcement, and may or may not depend on influences of one type of interaction on the probability of others.

Negative feedback may permit compensation for vicissitudes in the environment and minor changes in the participants. If an infant becomes more dependent, greater understanding on the part of the parent may lead ultimately to a reduction in the infant's dependence. The importance of this issue became particularly apparent in experiments on the effects of short periods of separation between mother and infant rhesus monkeys. On reunion after a separation lasting one to two weeks, the infant behaved in a more filial and demanding fashion; and our experiments involved procedures in which both the extent of the infant's demands and the readiness of the mother to respond were varied. If the mother responded appropriately to the infant's signals, the relationship was likely soon to be re-stabilized. This could be described in terms of negative feedback. If however she was distracted from responding fully to the infant by the need to re-establish relationships with other social companions, recovery was delayed. The most acute effects were found when the severity of the infant's separation-induced depression reduced his *overt* demands, and the mother's responsiveness was low. Such a finding could be described in terms of positive feedback, the lack of responsiveness of the mother exacerbating the depression of the infant, and vice versa (Hinde & Davies, 1972; McGinnis, 1975; Hinde & McGinnis, unpublished). In most cases both positive and negative feedback presumably operate, the outcome depending on the balance.

The existence of negative feedback mechanisms is of importance for studies of child development. As we learn more and more about the complexity of, say, human mother–infant interactions we cannot fail to be amazed by their subtleties. But this does not mean that their every aspect necessarily *matters*. Absence or excessive presence of a so-called normal aspect of behaviour may be adequately compensated by homeostatic mechanisms (Bell, 1974; Dunn, this volume), or may lead to divergence and/or disruption. We cannot tell, without empirical data, what aspects of a relationship are essential to its nature – and indeed the answer to that question is liable to depend on circumstances. And in any case we must expect a step function – small divergencies may be adequately compensated, whilst slightly larger ones are disruptive. This issue is discussed elsewhere in this volume by Dunn.

We may note here that some relationships, even though apparently well buffered in the short and medium term by negative feedback mechanisms, nevertheless contain the seeds of their own ultimate destruction or transmutation. We refer here to parent–child and teacher–pupil relationships, where the behaviour of one party is directed towards changing that of the other in a manner which in the long run is incompatible with the initial essence of the relationship.

When negative feedback operates, the nature of the changes necessary for a relationship to accommodate external stresses or changes in the participants will depend on the nature of that relationship. It is necessary here first to distinguish between two types of interaction, according to the relative parts played by the two participants. In a reciprocal interaction, the participants behave similarly, either simultaneously or in succession. Thus in rough-and-tumble play the partners take it in turn to chase and be chased, bite and be bitten. In a complementary interaction, by contrast, the two partners play different parts. If one mounts, the other is mounted; if one attacks, the other flees. Now in non-human species it is usually the case that, if some of the interactions within a relationship are complementary, many are and have a similar direction of imbalance. As we have seen, the concepts of 'dominance/subordinance' or 'status' as applied to relationships have some explanatory usefulness in such situations. Of course it does sometimes happen that one partner is dominant in some contexts and the other in others, but it is perhaps only in the most sophisticated of human relationships that the various interactions may exist in all degrees of reciprocity and complementarity, with idiosyncratic patterns of imbalance.

Returning now to the question of stability, small changes in the behavioural propensities of one partner require different changes in the other according to whether the relationship is predominantly reciprocal or complementary. Where reciprocal interactions are involved, if one partner changes, the other should change in a similar fashion to preserve stability. Thus with two peers, if one changes in such a way that he tends to play more, the other should change in a *similar* fashion. By contrast, as we have just seen, the mother–infant relationship is predominantly complementary and, if one partner changes in one direction, stability may be best preserved if the other changes in a complementary fashion. We may note here that special problems are perhaps liable to arise with relationships that are in some contexts reciprocal and other complementary; changes in either partner must be met by appropriate changes in the other.

In either reciprocal or complementary interactions there is a third way of coping with change that must be mentioned: if change occurs in one partner, it can be altered or reversed by appropriate behaviour in the other. Thus if an infant becomes more demanding, the mother might ignore its demands. This would be the equivalent of an extinction procedure, and may be a powerful way to decrease the frequency of an operant response, even in a social situation (e.g. Etzel & Gewirtz, 1967). However, the extinction situation is also one which can produce aggressive behaviour (Azrin, Hutchinson & Hake, 1966). Thus while an extinction procedure may lead to the disappearance of the infant's demands, the relationship may also have been changed in other, perhaps undesirable, ways. Similar considerations apply to punishment (e.g. Feshbach, 1970).

It will be apparent that we have introduced two potentially contradictory principles. On the one hand we have suggested that stability may not be preserved if increased demands by one partner are not met by the other. The evidence here is largely observational or correlational; for instance, Ainsworth & Bell (1974) claim that their data indicate that infants whose mothers most frequently ignore their crying in one quarter year are likely to be amongst the most frequent criers in the sample during the next quarter year. On the other hand, there is evidence that extinction (e.g. by ignoring) or punishment of a response may decrease its frequency, and may also have ramifying effects through the relationship. We suggest that this contradiction is more apparent than real, and that it is probable that in the former case we are dealing with compensatory changes in the whole relationship whose effects are being assessed by a single symptom, while in the latter we are dealing with the treatment of a symptom which has ramifying effects on the relationship (see e.g. Porter, 1968).

The institutionalisation of relationships

Whereas at least most of what we know about relationships between individuals of non-human species can probably be understood in terms of the behavioural propensities of the individuals and their mutual interactions, in man the institutionalization of relationships introduces a new dimension of complexity. We have already mentioned this in discussing the differences between animal and human relationships. Clearly institutionalization may affect the dynamic stability of relationships – to take an obvious example, divorce rate will be affected by cultural conventions about the rights and duties of spouses. Here, then, is another area to search for principles relevant to the stability of relationships.

Social forces

Finally, we must remember that relationships nearly always exist within a nexus of other inter-individual relationships. The patterning of interactions between two individuals may thus be affected by outsiders. For example, if one member of a group persecutes another, other group members may intervene on one side or the other.

CONCLUSION

Understanding the dynamics of relationships between individuals is not only challenging intellectually, it also poses some of the most urgent problems confronting science. We have suggested that the time is propitious for progress with these problems. In advocating a behaviourally oriented approach we are not suggesting that all problems about relationships can be tackled with the tools that are now available, but only that this is a good place to start. We are fully conscious that we have done little more than indicate where some of the problems lie. We hope that recognition that relationships have emergent properties, that reinforcement is a powerful but not all-powerful tool, and that dynamic stability requires new conceptual tools for its understanding, will help to pave the way for future progress.

SUMMARY

(1) A conceptual scheme relating interactions, relationships and group structure is outlined.

(2) The time is propitious for progress in the study of relationships.

(3) The nature of relationships is discussed.

(4) Relationships are seldom static. They are likely to vary whilst remaining within certain limits, or to change progressively with time, or to change from one temporarily stable state to another. Control systems analysis provides some concepts which could be useful in understanding relationships.

(5) The patterning of interactions within a relationship must involve

individual learning. Three paradigms are discussed: exposure learning, classical conditioning and operant conditioning.

(6) The distinction between paradigms and processes must be maintained.

(7) The concept of reinforcement can help us to understand the patterning of social interactions. The particular case of 'social approval' in the behaviour of non-human primates is discussed.

(8) Although the reinforcement concept is valuable to the student of relationships, it is often used loosely. In so far as it is necessary to have recourse to 'internalized reinforcements' the concept loses much of its power since it becomes irrefutable. On the other hand relationships cannot be accounted for solely in terms of tangible reinforcers. The dilemma seems insoluble.

(9) Other principles necessary for understanding the dynamic stability of relationships concern the natures of the individuals (e.g. sex, blood relationship), the degree of compatibility between interactions of different types (including those describable in terms of status), the effects of one type of interaction on others, institutionalization, and social forces from outside the relationship.

REFERENCES

Ainsworth, M. D. & Bell, S. M. (1974). Mother–infant interaction and the development of competence. In *The Growth of Competence*, ed. K. Connolly & J. S. Bruner. Academic Press: London & New York.

Altman, I. & Taylor, D. A. (1973). *Social Penetration: The Development of Interpersonal Relationships.* Holt, Reinhart & Winston: New York.

Asch, S. E. (1952). *Social Psychology.* Prentice-Hall: New York.

Azrin, N. H., Hutchison, R. R. & Hake, D. F. (1966). Extinction-induced aggression. *Journal of the experimental Analysis of Behavior*, **9**, 191–204.

Bailey, F. G. (1969). *Strategems and Spoils: A Social Anthropology of Politics.* Blackwell: Oxford.

Bandura, A. (1971). *Social Learning Theory.* McCalib-Seiler: New York.

Bateson, P. P. G. (1964). Effect of similarity between rearing and testing conditions on chicks' following and avoidance responses. *Journal of comparative physiological Psychology*, **57**, 100–103.

Bateson, P. P. G. (1966). The characteristics and context of imprinting. *Biological Reviews*, **41**, 177–220.

Bateson, P. P. G. (1971). Imprinting. In *Ontogeny of Vertebrate Behavior*, ed. H. Moltz. Academic Press: London & New York.

Bateson, P. P. G. (1973). Internal influences on early learning in birds. In *Constraints on Learning: Limitations and Predispositions*, ed. R. A. Hinde & J. Stevenson-Hinde. Academic Press: London & New York.

Bateson, P. P. G. & Chantrey, D. (1972). Retardation of discrimination learning in monkeys and chicks previously exposed to both stimuli. *Nature, London*, **237**, 173–174.

Bateson, P. P. G. & Reese, E. P. (1969). The reinforcing properties of conspicuous stimuli in an imprinting situation. *Animal Behaviour*, **17**, 692–699.

Bell, R. Q. (1974). Contributions of human infants to caregiving and social interaction. In *The Effect of the Infant on its Caregiver*, ed. M. Lewis & L. A. Rosenblum. Wiley: New York.

Berger, S. & Lambert, W. W. (1968). Stimlus–responses theory in contemporary social psychology. In *Handbook of Social Psychology*, vol. **1**, ed. G. Lindzey & E. Aronson. Addison-Wesley: Reading, Mass.

Berscheid, E. & Walster, E. (1974). A little bit about love. In *Foundations of Interpersonal Attraction*, ed. T. L. Huston. Academic Press: New York.

Bertalanffy, L. von (1952). *Problems of Life*. Harper Torch: New York.

Blau, P. M. (1964). *Exchange and Power in Social Life*. Wiley: London.

Blurton Jones, N. G. (1961). *Ethological Studies of Child Behaviour*. Cambridge University Press: London.

Bott, E. (1966). *Family and Social Networks*. Tavistock: London.

Brown, R. & Gilman, A. (1970). The pronouns of power and solidarity. In *Style in Language*, ed. T. A. Sebeok. MIT: Cambridge, Mass.

Buckley, W. (1967). *Sociology and Modern Systems Theory*. Prentice-Hall: Englewood Cliffs, N.J.

Bygott, D. (1974). Agonistic behaviour and dominance among wild chimpanzees. Ph.D. thesis, University of Cambridge.

Bygott, D. (1976). Agonistic behaviour and dominance among wild chimpanzees. In *Perspectives on Human Evolution*, vol. **4**, ed. D. Hamburg & J. Goodall. Holt, Rinehart & Winston: New York, in press.

Byrne, D. & Clore, G. L. (1970). A reinforcement model of evaluative responses. *Personality: An International Journal*, **1**, 103–128.

Clore, G. L. & Byrne, D. (1974). A reinforcement-affect model of attraction. In *Foundations of Interpersonal Attraction*, ed. T. L. Huston. Academic Press: New York.

Davis, J. (1973). Forms and norms: the economy of social relations. *Man*, **8**, 159–176.

Etzel, B. C. & Gewirtz, J. L. (1967). Experimental modification of caretaker-maintained high-rate operant crying in a 6- and a 20-week-old infant (Infans tyrannotears): Extinction of crying with reinforcement of eye contact and smiling. *Journal of experimental Child Psychology*, **5**, 303–317.

Feshbach, S. (1970). Aggression. In *Carmichael's Manual of Child Psychology*, vol. **2**, ed. P. H. Mussen. Wiley: New York.

Firth, R. (1965). *Primitive Polynesian Economy*. Routledge & Kegan Paul: London.

Foa, U. G. (1971). Interpersonal and economic resources. *Science, Washington*, **171**, 345–351.

Gewirtz, J. L. (1961). A learning analysis of the effects of normal stimulation, privation and deprivation on the acquisition of social motivation and attachment. In *Determinants of Infant Behaviour*, ed. B. M. Foss. Methuen: London.

Gibson, E. J., Walk, R. D. & Tighe, T. J. (1959). Enhancement and deprivation of visual stimulation during rearing as factors in visual discrimination learning. *Journal of comparative and physiological Psychology*, **52**, 74–81.

Grussec, J. & Mischel, W. (1966). Model's characteristics as determinants of social learning. *Journal of Personality and social Psychology*, **4**, 211–215.

Hansen, E. W. (1966). The development of maternal and infant behaviour in the rhesus monkey. *Behaviour*, **27**, 107–149.

Heath, A. (1972). Exchange theory. *British Journal of political Science*, **1**, 91–119.

Hinde, R. A. (1976*a*). Interactions, relationships and social structure. *Man*, **11**, 1–17.

Hinde, R. A. (1976*b*). On describing relationships. *Journal of Child Psychology and Psychiatry*, **17**, 1–19.

Hinde, R. A. (1976*c*). The use of differences and similarities in comparative psychopathology. In *Animal Models in Human Psychobiology*, ed. G. Serban & A. Kling. Plenum Press: New York.

Hinde, R. A. (1976*d*). The nature of social structure. In *Perspectives on human Evolution*, vol. **4**, ed. D. Hamburg & J. Goodall. Holt, Rinehart & Winston: New York.

Hinde, R. A. & Davies, L. (1972). Changes in mother–infant relationship after separation in rhesus monkeys. *Nature, London*, **239**, 41–42.

Hinde, R. A. & Simpson, M. J. A. (1975). Qualities of mother–infant relationships in monkeys. *The Parent–Infant Interaction*, Ciba Foundation Symposium, no. **33**. Associated Scientific Publishers: Amsterdam.

Hinde, R. A. & Stevenson-Hinde, J. (ed.) (1973). *Constraints on Learning: Limitations and Predispositions*. Academic Press: London & New York.

Hoffman, H. S., Searle, J., Toffey, S. & Kozma, F. (1966). Behavioural control by an imprinted stimulus. *Journal of the experimental Analysis of Behavior*, **9**, 177–189.

Homans, G. C. (1950). *The Human Group*. Harcourt Brace: New York.

Homans, G. C. (1961). *Social Behaviour, Its Elementary Forms*. Routledge & Kegan Paul: London.

Hooff, J. A. R. A. M. van (1972). A comparative approach to the phylogeny of laughter and smiling. In *Non-Verbal Communication*, ed. R. A. Hinde. Cambridge University Press: London.

James, H. (1959). Flicker: an unconditioned stimulus for imprinting. *Canadian Journal of Psychology*, **13**, 59–67.

Kling, J. W. (1971). Learning: introductory survey. In *Woodworth & Schlosberg's Experimental Psychology*, 3rd ed, ed. J. W. Kling & L. A. Riggs. Holt, Rinehart & Winston: New York.

Kummer, H. (1968). *Social Organization of Hamadryas Baboons*. University of Chicago Press: Chicago.

Kummer, H. (1975). Rules of dyad and group formation among captive gelada baboons (*Theropithecus gelada*). In *Proceedings of the 5th international Congress of the primatological Society*, ed. S. Kondo, M. Kawai, A. Ehara & S. Kawamura. Japan Science Press: Tokyo.

Laing, R. D. (1962). *The Self and Others*. Quadrangle Press: Chicago.

Laing, R. D. (1969). *The Divided Self*. Pantheon Books: New York.

Lawick-Goodall, J. van (1968). Behaviour of free-living chimpanzees of the Gombe Stream area. *Animal Behaviour Monograph*, no. **3**.

Levinger, G. (1972). Little sand box and big spade. *Representative Research in social Psychology*, **3**, 3–19.

Lickona, T. (1974). A cognitive-developmental approach to interpersonal attraction. In *Foundations of Interpersonal Attraction*, ed. T. L. Huston. Academic Press: New York.

Lindberg, D. G. (1973). Grooming behavior as a regulator of social interactions in rhesus monkeys. In *Behavioral Regulators of Behavior in Primates*, ed. C. R. Carpenter. Bucknell University Press: Lewisburg.

Lorenz, K. (1931). Beiträge zur Ethologie sozialer Corviden. *Journal of Ornithology*, **83**, 137–213. Translated in *Studies in Animal and Human Behaviour* (1970), vol. **1**, ed. R. B. Martin. Harvard University Press: Cambridge, Mass.

Lorenz, K. (1935). Der Kumpan in der Umwelt des Vogels. *Journal für Ornithologie*, **83**, 137–213, 289–413.

Lott, A. J. & Lott, B. E. (1972). The power of liking. In *Advances in experimental social Psychology*, vol. **6**, ed. L. Berkowitz. Academic Press: New York.

Lott, A. J. & Lott, B. E. (1974). The role of reward in the formation of positive interpersonal attitudes. In *Foundations of Interpersonal Attraction*, ed. T. L. Huston. Academic Press: New York.

Maxim, P. E. & Buettner-Janisch, J. (1963). A field study of the Kenya baboon. *American Journal of physical Anthropology*, **21**, 165–180.

McGinnis, L. (1975). Analysis of the factors involved in mother–infant separations in rhesus monkeys. Ph.D. thesis, University of Cambridge.

Menzel, E. W. (1971). Communication about the environment in a group of young chimpanzees. *Folia primatologica*, **15**, 220–232.

Menzel, E. W. (1972). Spontaneous invention of ladders in a group of young chimpanzees. In *Behavior of Non-Human Primates*, vol. **5**, ed. A. Schrier & F. Stollnitz. Academic Press: New York.

Mowrer, O. H. (1960). *Learning Theory and Behavior*. Wiley: New York.

Nishida, (1970). Social behaviour and relationships among wild chimpanzees of the Mahali mountains. *Primates*, **11**, 47–87.

Okano, T., Asami, C., Haruli, Y., Sasaki, M., Itoigawa, N., Shinohara, S. & Tsuzuki, T. (1973). Social relations in a chimpanzee colony. In *Behavioral Regulators of Behavior in Primates*, ed. C. R. Carpenter. Bucknell University Press: Lewisburg.

Oki, J. & Maeda, Y. (1973). Grooming as a regulator of behavior in Japanese macaques. In *Behavioral Regulators of Behavior in Primates*, ed. C. R. Carpenter. Bucknell University Press: Lewisburg.

Peterson, N. (1960). Control of behavior by presentation of an imprinted stimulus. *Science, Washington*, **132**, 1395–1396.

Porter, R. (1968). *The Place of Learning in Psychotherapy*. Churchill: London.

Rowell, T. E. (1972). Female reproduction cycles and social behavior in primates. *Advances in the Study of Behavior*, **4**, 69–105.

Sade, D. S. (1965). Some aspects of parent–offspring and sibling relations in a group of rhesus monkeys, with a discussion of grooming. *American Journal of physical Anthropology*, **23**, 1–18.

Sade, D. S. (1972). Sociometrics of *Macaca mulatta*. I. Linkages and cliques in grooming matrices. *Folia primatologica*, **18**, 196–223.

Schachter, S. (1964). The interaction of cognitive and physiological determinants of emotional state. In *Advances in experimental social Psychology*, vol. **1**, ed. L. Berkowitz. Academic Press: New York.

Schachter, S. & Singer, J. F. (1962). Cognitive, social and physiological determinants of emotional state. *Psychological Review*, **69**, 379–399.

Schneirla, T. C. (1949). Levels in the psychological capacities of animals. In *Philosophy for the Future: the Quest of Modern Materialism*, ed. R. W. Sellars, V. J. McGill & M. Farber. Macmillan: New York.

Secord, P. F. & Backman, C. W. (1974). *Social Psychology*, 2nd ed. McGraw-Hill Kogakusha: Tokyo.

Seligman, M. E. P. & Hager, J. L. (1972). *Biological Boundaries of Learning*. Appleton-Century-Crofts: New York.

Simpson, M. J. A. (1973*a*). Social displays and the recognition of individuals. In *Perspectives in Ethology*, ed. P. P. G. Bateson & P. H. Klopfer. Plenum Press: New York.

Simpson, M. J. A. (1973*b*). The social grooming of male chimpanzees. In *Comparative Ecology and Behaviour of Primates*, ed. R. P. Michael & J. H. Crook. Academic Press: London.

Stebbins, R. A. (1969). On linking Barth and Homans: a theoretical note. *Man*, **4**, 432–437.

Thibaut, J. W. & Kelly, H. H. (1959). *The Social Psychology of Groups*. Wiley: New York.

Thorpe, W. H. (1956). *Learning and Instinct in Animals*. Methuen: London.

Tinbergen, N. (1959). Comparative studies of the behaviour of gulls: a progress report. *Behaviour*, **15**, 1–70.

Winch, R. F. (1958). *Mate-selection: a Study of Complementary Needs*. Harper & Row: New York.

Zajonc, R. B. (1971). Attraction, affiliation, and attachment. In *Man and Beast: Comparative Social Behavior*, ed. J. F. Eisenberg & W. S. Dillon. Smithsonian Institution: Washington, D.C.

16

How far do early differences in mother–child relations affect later development?

JUDY DUNN

The importance of the child's early social experience has usually been discussed in terms of his later emotional adjustment; the consequences of difficulties in the early relationship with the mother have been described in terms of distortions in the ability to form affectional bonds later, or of depression and anxiety. But recently, increasing emphasis has been laid on the importance of the relationship with the mother for the development of communication and understanding – on the functions of the attachment relationships in the development of language and the child's understanding of his social world. There have been two rather different orientations within this research. In one approach, the search is for what is *common* to mother–child relationships, the attempt being to describe the role that interaction with a responsive caretaking figure plays in the development of social and intellectual competence, its contribution to the necessary conditions for normal development. Another approach has been to ask how variations in the ways babies are mothered influence their later development. In this paper I look at some recent work on early social experience, discussing how it might sharpen and clarify the questions we ask about mothering, and examining its implications for the models of development discussed in Patrick Bateson's chapter in this volume.

There is growing evidence that, in order to understand development in infancy, it is essential to see and analyse the baby's behaviour in a social context. By working with audio-visual recordings of mothers and infants together, several researchers are attempting to describe precisely the timing, sequencing and coordination of baby's and mother's gestures and interventions (Brazelton, Koslowski & Main, 1974; Newson, 1974; Stern, 1974; Trevarthen, 1974; J. S. Bruner, unpublished). The picture they give is of an exchange of great subtlety and complexity, even with babies of eight weeks. This has led to the view that 'communication activity' is far more complex than other activities in the two-month old. In tracing the development of this social exchange over the first year, Trevarthen (1974) concludes 'that human intelligence develops from the start as an interpersonal process, and that the maturation of consciousness and the ability to act with voluntary control in the physical world is a product rather than an ingredient of this process'.

What this work has shown is that on the one hand the baby is particularly 'tuned in' to people from the start, and on the other that the mother monitors with great sensitivity and skill the changes in the baby's actions and interests, timing her own interventions to the patterns of his changing behaviour in such a way that she supports and extends his interest in the exchange. Trevarthen and Newson each emphasise the importance of the way in which the mother imputes 'meanings' to the baby's gestures, expressions and actions; they suggest that this influences the timing and manner with which she responds, and that this assumption of the infant's meaning lays the foundation from which the communication of shared meanings begins to grow. Trevarthen argues persuasively that it is into this 'communication framework' that the infant will build his developing understanding of objects and the object world. He does not dismiss the importance of private experiment with the object world, but argues that it is the combination of such experiment with social communication which leads to the development of cognitive mastery.

Throughout these video- or film-based descriptions of the early interchange between infant and mother there are striking examples of how beautifully the mother's response is adjusted and timed to support and develop the exchange. She watches, subdues her own responses, follows and supports the baby's animated mood by imitating his expressions, holding back when it is his 'turn', uses rhythmic touching, calling, and moving to keep him interested and sociable. The fullest expression of the infant's capacities seems to depend on the mother taking a supportive and subordinate place in the exchange, while on his side the baby shows a subtle adaptation to people, and is very effective at encouraging friendly attentions.

In this friendly exchange, the baby's interest and attention may follow the mother's closely, so that her attention to an object brings his own focus on to it (Scaife & Bruner, 1976). Collis & Schaffer (1976) have demonstrated the mirror-image of this close meshing of attention, in their analysis of the synchrony of visual attention to objects, where the mother's direction of gaze follows the baby's with great sensitivity. As the child develops new capacities and interests, the early exchange of smiling and mutual gazing and turn-taking dialogues becomes increasingly focussed on objects and on his new activities. Give-and-take games and playing together with objects begin. Through these joint games and through his mother's responses to his actions, the baby begins to understand what his actions and expressions 'mean' for other people, and begins to use them with intention in a social context. The mother's interest and empathy, emphasised until recently as essential for the child's emotional development, are here stressed as important in enabling her to monitor his interest, and to find ways of making social contact which change as he develops.

These elegant demonstrations of the complexity and importance of the framework the mother provides have led to the suggestion that the 'scaffolding' the mother sets up is essential for the foundation of communication, for extending the child's capacities to bring him to 'the margins of his abilities'. But how far does the ability to take part in prolonged social exchanges, to develop communicative capacities, depend on the experience of the sensitivity and empathy of the mother? How important is it for the baby to have built up a familiarity with the mother over the first few months? We are here faced with the two different orientations of enquiry. What are the necessary conditions for the baby to develop communication skills and intellectual competence? At a different level, what are the consequences of individual differences in the sensitivity with which this framework is built up by the mother?

The work on film and video analysis has focussed on the earliest interaction between baby and adult as being of enormous potential importance for the first question. From the detailed picture of how the mother is behaving in these exchanges – the exaggerated and slowed-down version of normal adult inter-personal behaviour described by Stern (1974) for instance, variations in adult behaviour *elicited* by the infant's behaviour – we can speculate on why and how such specifically 'maternal' behaviour may be important. Stern (1974) suggests for instance that the tempo and exaggeration of adult behaviour are particularly well adapted for the infant's stage of neurological development, and that the variations 'enhance his acquisition of the schemata of human behaviour'. But in terms of the second question we do not know, for instance, what the long-term consequences might be for infants whose mothers have a very restricted range of this behaviour and who elicit less play from their infants.

Brazelton *et al.* (1974) consider such individual differences of possibly very great importance. They suggest (*ibid.*, pp. 70–71) that during the exchange with the mother opportunities are provided for the baby:

> to learn how to contain *himself*, how to control motor responses, and how to attend for longer and longer periods. They amount to a kind of learning about organization of behavior in order to attend. With more disruptive mothering or with none at all, one might expect this kind of learning about self-organization to be delayed. Not only then in a disturbed environment would the experiences be sparse which would contribute to learning in the sphere of social interaction but the cross over to learning the organization necessary to cognitive acquisitions may not be provided and hence learning would be delayed in an infant who had to acquire this organization by maturation alone, without appropriate environmental experiences.

Developmental psychologists have been greatly preoccupied with individual differences in the sensitivity of mothering or caretaking, but it is not clear quite what the relationship is between sensitivity as a clinical term, and the 'sensitivity in communicating' that Newson (1974) refers to when he dismisses the idea that any special 'maternal intuition' is necessary for building up the early communication framework:

> sensitivity in communicating with any particular baby is very much a matter of practice in communicating with that particular baby...a learned skill available to anyone motivated to make the necessary effort to establish a working dialogue with one particular human infant.

But how much skill, how much familiarity? What are the consequences, short term or long term, for the baby of the variation of skill among adults?

If the sensitivity of the adjustment of adult to baby that this research describes is of such fundamental importance, it is difficult to believe it can correspond very closely to the measure of sensitivity that Mary Ainsworth refers to in her studies of attachment (Ainsworth, Bell & Stayton, 1971); her research found that roughly one-third of the sample were 'insensitive'. It would not make very good biological sense if a third of mothers were found wanting in a quality of crucial importance for development; it seems more likely that these aspects of 'insensitivity' are not relevant to the fitness of the offspring in the long term.

Stern's work does have very important clinical implications. He shows (Stern, 1974, p. 210) that even with the most ideally sensitive mother

> the array of stimuli the mother provides and the infant's level of arousal and affect repeatedly fall below some optimal level where interest is lost, and repeatedly climb above some optimal level where active aversion is executed. In either case both infant and mother can readjust their behavior to bring the infant's state temporarily back into an optimal range, where it fluctuates until the boundaries are again exceeded.

It could easily be assumed that such exchanges are important for the infant to develop coping and defensive mechanisms and are early steps on a continuous developmental path. Indeed Stern's case study of twins (Stern, 1971) does tell a story of continuous difficulty in inter-personal relations for the twin whose earliest gaze exchanges with the mother were mutually unsatisfying and difficult. Another important clinical aspect of Stern's work is his demonstration of how misleadingly simplified our impressions of 'controlling' and 'intrusive' mothers may be; his analysis of this mother–twin pair, where the mother was clinically characterised as controlling and insensitive, showed that the behaviour of both mother and baby contributed to the difficulties in the interaction.

It may well be useful, in thinking about the importance of these individual differences, to consider the developmental model Bateson (this volume) describes, with several possible developmental routes to a particular 'steady state'. The embryologists have long stressed the 'self-righting' aspect of the developmental process, and it is rather surprising that developmental psychologists have not paid more attention to this point.* Surprising, in that regulative processes are needed to explain the relatively narrow range of developmental outcomes in a number of areas where there is a wide range of environmental variables. Nelson's (1973) and E. Lieven's (unpublished) work on language has shown that with different mother–child pairs there can be considerable differences in the routes along which language competence develops. And studies of children with various handicaps have clearly shown that conceptual discoveries can be made via a number of routes; examples are Gouin-Decarie's (1969) work with thalidomide children, H. Furth with deaf, C. Urwin with blind children. A system which enables an organism to learn from a wide range of possible environmental situations is clearly more biologically useful than one which specifies a restricted range. Minutely detailed analysis of mother–child interaction has already enabled us to trace differences in the routes by which children develop, but as yet we know very little of how these differences in the course of development relate to later behaviour. Even where we can link short-term effects of differing styles of mothering on the way in which children develop, we have little idea of whether these are immutable, or how much and precisely in what ways they matter in the longer term.

The studies of children raised in residential nurseries provide a useful focus for illustrating these points (Tizard & Rees, 1974, 1975). At two years old, children who had lived all their lives in the nurseries were not significantly different in non-verbal intelligence quotient from working-class family-reared children. That is, the sort of interaction and social exchange provided by multiple caretakers in nurseries run with deliberate policies of discouraging attachments between nurses and children was sufficient for this level of cognitive development. (The nurseries had good staff/child ratios, plenty of good equipment, and the children were taken out and given a variety of experiences outside the nurseries, which in no way resembled the sort of institution on which early deprivation studies were carried out.) The children *were* behind on language development, and there were many differences in their social behaviour. By four years, however, there were no differences in intellectual development, verbal or non-verbal IQ, when the children who had remained in the nurseries for four years were compared with family-reared working-class children. (Children who had spent the first two years in the

* An exception here is Bell's (1974) recent discussion of regulative processes in development.

nursery but were then adopted by middle-class parents had by four years significantly better language development and higher IQ.) The four-year-old nursery children showed no higher incidence of 'behaviour disturbance' than the working-class controls, but differed in their social behaviour with adults. The children who were adopted at two years formed very good attachments to their adoptive parents, in spite of having spent their first two years in nurseries with multiple caretakers, and with deliberate policies of discouraging attachments between child and nurse.

These findings present a real challenge to the idea that sensitive mothering in the first two years is essential for 'normal' development. They suggest that the threshold of familiar interested caretaking required for the establishment of Trevarthen's 'basic communicative network' is very low. But several points of caution must be made. First, in spite of the fact that there were no differences in language performance as assessed on the Wechsler Pre-School and Primary Scale of Intelligence (WPPSI) test, there may well have been differences in language *use* that a performance test would not reveal. Cazden (1970) has stressed that social class differences in language are primarily in use rather than in performance. This illustrates a major problem with using the developmental model that Bateson proposes – that of defining the 'steady state' or the point of 'equifinality'. We just don't know what sort of a difference the varying early verbal experience will make to the children's use of language, and what difference, if any, the variations in their use of language will make to their later intellectual development.

Secondly, it is quite possible that there may be 'sleeper' effects operating; differences between the nursery-reared children and those brought up in families may be apparent later. Thirdly, the differences in social behaviour apparent in the nursery-reared children may in themselves make a great difference to their intellectual achievement later.

Nevertheless, these findings should alert us to the possibility that we have not taken enough account of the *adaptability* of the developmental system. Because our focus has so often been on the mother and first child of middle-class families from America or Western Europe, often in rather artificial circumstances, we may well have blinkered ourselves to the range and adaptability of the course of development. The work of Trevarthen, Bruner, and Newson has shown very elegantly that babies *can* learn and develop effectively this way, through the 'scaffolding' of their mothers, but not that it is the only effective way. Because psychologists demonstrate that something happens between a mother and baby it does not necessarily follow that it is important in the long run. Furthermore, the exclusive focus on the mother, as opposed to the father, grandparent or sibling has certainly been misleading. Schaffer & Emerson (1964) found that one-third of their sample of 60 children directed their most intense attachment behaviour at some individual other than the mother by 18 months. The adaptability of the developmental process would allow too for the fact that mothers are not always 'tuned in' to every phase of their children's development. The mother or the father who enjoys talking with a 2½ year old, who joins in with his games, fantasies and rituals, may well have not enjoyed the first seven months of his infancy.

Such evidence as we do have from cross-cultural studies does support this view of the adaptability of the developmental process – for instance in the Kagan study of Guatemalan children (Kagan & Klein, 1975), and a less extreme and less contested case, Rebelsky's study of Dutch and American babies (Rebelsky, 1967). Rebelsky found that during their first four months Dutch babies spent much more time alone, were given fewer toys to look at and people to interact with, and were fed on schedule, and left to cry. Crying was taken not as a signal for help, to be responded to, but just as something babies did. In the early months the Dutch babies were developmentally behind the American babies and were different in their social behaviour with their mothers, but by the end of the first year the differences had disappeared.

The difficulty of defining a 'steady state', which Tizard's study of the language development of the nursery-reared children raises (Tizard & Rees, 1974, 1975), is a general developmental problem. What criteria are we to use for 'normal' 'adequate' development, at the end point of our enquiry? Bowlby's (1951) work was based on tracing the antecedents of extreme deviation and behavioural disturbance. But within the extremes of disturbance we must be aware that what constitutes normal social development is extremely culture-bound. In the studies by Caldwell and colleagues of the development of children brought up in group day-care (Caldwell, Wright, Honig & Tannenbaum, 1970), it was found that by two years of age the children's attachments to their mothers, on careful assessment, were no different from the attachments of children brought up at home. But interesting differences in other aspects of their social behaviour were noted when these day-care children were seen later on entry to nursery school (Schwarz, Krolich & Strickland, 1973, 1974). The small groups of day-care children brought up together from around six months till 2½ years, were found to be very close to each other, to interact more intensively and exclusively than the other children in the nursery school. They were also found to be less amenable to the adult teachers. The picture of these children's close relationship to each other recalls that of the relationships between kibbutz children, who also maintained close attachments to their mothers (Maccoby & Feldman, 1972). From these studies we have learnt something of the consequences of rearing small groups of children under two years together, as well as something of the necessary conditions for maintaining the attachment bond, and its resilience. But criteria of 'adequate' or 'normal' development and an exclusive focus on the attachment relationship would tell us nothing of the interesting differences in socialisation between the children brought up in different circumstances. It is these differences which are valued differently in different cultures, and which make general criteria of social development of limited use. Clearly, different cultural groups will value a strong adhesive peer group differently (Bronfenbrenner, 1974). And a questioning attitude to authority is obviously seen very differently in different cultures.

These results from the kibbutz and day-care studies support the view put forward by Schaffer & Emerson in 1964 that it is the *quality* of the interaction between the child and adult that is important in maintaining the attachment relationship, rather than the availability of the adult throughout the day. It seems from the kibbutz study that the warmth and intensity of interaction between mother and child in the two evening hours they spent together was sufficient to maintain a warm relationship. This question of distinguishing the *quality* of the exchange between child and other from the issue of *how much time* is spent in interaction is obviously an important one; we know from studies of other aspects of development that it may be quite misleading to assume quantity of interaction is the most important variable. For instance Friedlander and colleagues, in a study of the language development and language environment of two one year olds, found that, for one family, although the father's direct interaction with the child was much less than the mother's, he had a great influence on the child's language development (Friedlander, Jacobs & Davis, 1972). This influence was discernible from the mother's influence, in that the father spoke only Spanish. 'His influence on her language development suggests the need for a major re-evaluation of the implicit assumption that mothers are the primary language model for their babies, simply on the basis of sheer quantity of interaction time.' As the authors point out, their quantitative observations raise but do not answer the critical question of what actually constitutes the significant portions of an infant's language-learning environment.

It is interesting that for those cultural groups where we do have some idea of the daily pattern of interaction and contact between child and adults, the percentage of the child's waking time that is spent in intense play with the mother is very low, even for children at home. In the study by Clarke-Stewart of American children aged between 9 and 18 months (Clarke-Stewart, 1973), the estimated time of playing with the mother was 4%, although at a 'lower' level of intensity which included basic care, the estimated time was 36%. Douglas, Lawson, Cooper & Cooper (1968), in a study of working-class families in London, found one year olds spent a mean of 46 minutes per day in 'concentrated' interaction with adults. The implication of these findings for our views on day-care should be noted.

Many of the issues we have touched on – such as the problems involved in defining a 'steady state' – are highlighted by recent work on child language. In particular this research has provided a useful focus for two issues of general developmental interest: first, the way in which the child's level of capability influences his environment; and secondly, the consequences for the child of differences in the environmental input.

Several studies have documented, for middle-class North American or British children, the way in which mother's speech to children differs from their speech to adults. It is simpler, more redundant, and less confusing (Snow, 1972). (A parallel might be drawn between these differences and Stern's description of the exaggeration and slowed tempo of adult facial expression to very young infants.) These characteristics may well help the child to learn language. But we know that children in different cultural groups are exposed to very different sorts of linguistic environments. In studies such as those of S. Harkness (unpublished), looking at Kipsigis children whose primary caretakers were other children, or the study by Ward (1971) of a rural community in Louisiana, the variation in linguistic environments is beginning to be described. These studies have shown that, on the one hand, both adults and children alter their speech when talking to children, their utterances becoming shorter and simpler, and on the other that aspects of the mother's speech, such as the expansions described by Brown and colleagues (Brown, Cazden & Bellugi, 1969) for Harvard mothers, are by no means universal. Other cross-cultural studies have suggested that some aspects of speech to children are common to all groups studied, and are not exclusive to mothers or those concerned with child care but are apparent in the speech of all adults in the group. Even four year olds change their language when talking to young children. This work, and that of E. Lieven (unpublished) and Snow (1972), illustrate the way in which the child's own language level can affect the sort of interaction he has with adults. One interpretation of these findings is that under some circumstances the child may play a greater part in eliciting what he 'needs' from the environment than has been recognised. The observation that mother's speech increases in complexity in parallel with the developments in the child's language (Phillips, 1973) would be compatible with this view.

What are the consequences for children's language development of the *differences* in the language environments they are exposed to? This is very much a matter of debate. A great deal of detailed research is being carried out; three examples will illustrate the different approaches, which are of potential interest for a wide range of developmental issues. First, experimental alterations of adult speech have been found to have considerable effects on children's language. In the study by Keith Nelson (Nelson, Garskadden & Bonvillain, 1973), after 22 sessions of verbal interaction where the experimenter replied to the child's utterances with recast sentences, maintaining the same meaning but providing new syntactic information, the syntax acquisition of the children was significantly increased, as compared with a control group.

Secondly, in cross-cultural studies the different linguistic environments of young children are being compared. In the Kenyan study, Harkness found that while several of the modifications of adult speech to children were similar to those found in American mothers, some of these 'qualitative' features were not shared by the children caretakers talking to younger children. The study found that children who spent more time with adults spent more time talking, and that those who spent more time talking were linguistically more advanced than those who spent less time talking. These correlations were interpreted as showing that the adult style of speech ('elicitation' speech) which required speech in return, was accelerating language-learning, more than the 'commentary' talk the children heard from other children. While this interpretation tells a plausible story it is by no means the only one possible; it could be that the children whose speech was linguistically more advanced got more attention and interaction with adults as a *result* of their greater language skills. This interpretation would parallel the findings from a study of middle- and working-class children in group care, which showed that the middle-class children were talked to, read to, and played with more often that the working-class children. The staff, without being aware of it, seemed to 'function at a lower level with working- class than with middle-class children' (B. Tizard, unpublished). It seems quite likely that this was because the working-class children made fewer, and different, demands on the staff.

Thirdly, naturalistic studies of the growth of language in individual children are being undertaken which attempt to relate individual differences in maternal style to the child's language. Katherine Nelson (1973), in a longitudinal study of 18 American children, discussed the acquisition of their first words in terms of an interaction model. She distinguished the mothers in her sample along criteria of feedback (acceptance or rejection) and control. She proposed that children vary significantly in particular pre-language concepts, reflected, for example, in the extent to which they are interested in activity, affectivity, things or social relations and that their progress in language development will depend on an interaction or 'match' between their conceptual organisation and the characteristics of their mother's responsiveness and directive style. She suggested that the dimension of control and direction of the language acquisition process may have important consequences for the children's cognitive style.

In asking these questions about the consequences of particular patterns of interaction between child and adult we have moved a great distance from the simple models of 'maternal stimulation' of a few years ago. We have also moved quite a distance from the simple recognition that children can influence maternal behaviour. We are now at a point when we can begin to look much more precisely at the particular ways in which a child influences his environment. The study by Friedlander referred to above (Friedlander *et al.*, 1972) showed that some aspects of both the father's and the mother's speech were altered in similar ways when the adults talked to the child, and were related to the level of the language development of the child. In other ways the father's speech to the child differed from the mother's speech. Parallel studies of other aspects of interaction between young children with different members of their families should be revealing.

We have also moved a long way, in recent years, towards understanding the complexity of the problem of predicting the consequences of particular styles of mothering, or particular early experience. It is now quite clear that we have to take account of the characteristics of the child, and to try to understand the dynamics of the mutual influences of child and mother. However, the idea that there are two sets of variables, child's characteristics and parents' caretaking style, to be identified early on and from which outcome might be predicted, is far too simple and misleading.

First, at no point can we say that the two are independent of each other. We found, in a follow-up study of 80 children in Cambridge (Richards & Bernal, 1972), that individual differences in baby characteristics when the babies were eight days old were correlated with some aspects of maternal behaviour during feeding interaction, and with measures of maternal behaviour several weeks later. These relationships showed that not only was it difficult to assess the direction of effects, but that neither maternal nor infant measures could be assumed to be independent of the course of interaction between them (Dunn, 1975).

Secondly, the developmental outcome for any particular child is part of a giant equation, with social conditions, the child's temperament, health, mother's sensitivity and style of caretaking, the child's social experiences with others all of potential importance. The internal relationships between these variables are very complicated, and we are just beginning to tease apart the interrelations between some of these factors, given that others are constant. Take for instance the consequences of perinatal problems. One of the most comprehensive studies of the consequences of perinatal complications was made by Werner, Bierman & French (1971) who followed up the whole population of 670 children of a Hawaiian island. The infants who had suffered severe perinatal stress were found at 20 months to have lower scores on measurements of physical health and psychological status, but the effects were greatly influenced by social and economic factors. For the babies suffering perinatal stress who were born into families with mothers of high intelligence, or into families of high socio-economic status (SES), the IQ difference between children with and without complications was only 5–7 points. For those with mothers of low intelligence or poor social conditions the IQ difference ranged from 19–37 points. By 10 years there was no correlation with perinatal stress score, but there were correlations between the 20 months and 10 year old data if low SES *and* parental education were taken into account, with stability of low intellectual functioning found for those with IQs below 80 with parents of low SES.

A similar story of the predictive power of social and economic variables emerged from the study by Willerman, Broman & Fiebler (1970), who compared Bayley Developmental scores at eight months and four years. For children from high SES families, there was no relation between scores. With the low SES children, those doing badly at eight months were still doing badly at four years. This picture of poverty amplifying the effects of early trauma or poor development is echoed in the findings of large scale surveys (Davie, Butler & Goldstein, 1972). It is indeed a repeated finding that the effects of extreme trauma on intelligence quotients disappear unless combined with *continuing* severe social conditions (reviewed by Sameroff & Chandler, 1975).

The widely found difficulty in relating 'intelligence' scores in the first two years to later IQ measurements presents another barrier to establishing simple links between mothering and intellectual development. The failure to find connections across the years (McCall, Hogarty & Huilbat, 1972) presents us with two alternatives: either we have very little idea how to assess intellectual competence in the first two years, or, in so far as we *can* assess competence, it looks as if we have no basis for assuming a simple continuity from the early days.

CONCLUSION

There is, of course, a great 'demand pressure' on psychologists to find and tell simple useful truths about mothering, about child rearing and development. Many feel a great concern and responsibility that, even if our knowledge is hopelessly incomplete, we *have* to attempt to provide a helpful picture for those concerned with bringing up children, or who are working on the effects of poverty on children. We do not escape responsibility by standing on the sidelines, since those who have to cope with these problems have to act on some sort of theory or rule of thumb. But we should also be aware of the extent to which, as psychologists, we have, and need 'rules of thumb' for our own work, such as ideas of good mothering or notions of responsiveness. We should certainly question the validity and usefulness of these 'working assumptions'. Words like 'sensitivity' provide deceptively simple links between global clinical assessments and detailed analysis of interaction such as the video work discussed here. What sort of connection there might be between the two is an empirical question. What we have to cope with in looking at mothering is an enormously complicated series of interchanges over a long period of time. To begin to make sense of it, we have to take a long-term view, and a broad view in the sense of looking at different cultures. But we are in a position to do a great deal more than lament the complexity of the enterprise. In summary, I would like to underline a number of the points that seem most useful for beginning to make sense of the complexity.

(*a*) The new interest in how babies learn to communicate, in 'intersubjectivity', has opened up many exciting issues, and promises a real way into understanding problems such as the beginning of intentional behaviour. But it seems important that at this stage it should be described phenomenologically, rather than be taken as providing a biological explanation. Without longitudinal studies into adulthood and much conceptual clarification of what we might measure in adulthood, the significance of such explanations cannot possibly be checked.

(*b*) The extent to which the child is a self-directing actor throughout the process of development is being increasingly precisely defined.

(*c*) The linear model of development, with early trauma producing immutable effects is increasingly questioned; a more flexible model has, of course, positive implications for clinical intervention.

(*d*) The major empirical finding from developmental psychology, the penetrating impact of the class structure on the psychological development of children, plainly underlines the urgency of our learning more about why and how this happens. Its obvious political implications do not need to be developed further here.

(*e*) We will only begin to make sense of points (*a*)–(*d*) if we can take full account of the complexity of these questions; multi-level, long-term research, ethological in the sense that it studies the child in his own social world, will be indispensable if we are to have any hope of mastering this complexity.

SUMMARY

In this paper some accepted ideas on the crucial importance of the child's relationship with the mother in the first two years are questioned. First, the recent emphasis on the importance of the earliest exchange between mother and baby for the foundation of communication and language is discussed, and the extrapolation of general and 'universal' theories from the evidence is criticised in the light of other studies. Secondly, the confusion between the notions of maternal 'sensitivity' used in describing these early exchanges and other clinical or conventional uses of the term is underlined; the consequences of individual differences in sensitivity are discussed and it is argued that, in considering the importance of these individual differences, developmental models which stress the adaptability and self-righting aspect of development are useful. There are, however, considerable difficulties with the notion of 'steady state' or 'equifinality' in these models, and indeed with criteria for 'normal' development. These points are discussed in the context of evidence from cross-cultural studies, studies of group-reared children, and recent work on child language. These studies raise general developmental questions on the extent to which the child's level of capability constrains the interaction he has with others, and on the relative importance of qualitative and quantitative aspects of interaction. Third, the complexity of the problem of predicting the consequences of particular styles of mothering is discussed. The points that seem most useful for beginning to make sense of this complexity are emphasised.

REFERENCES

Ainsworth, M. D. S., Bell, S. M. V. & Stayton, D. J. (1971). Individual differences in strange-situation behaviour of one-year-olds. In *The Origins of Human Social Relations*, ed. H. R. Schaffer. Academic Press: London.

Bell, R. Q. (1974). Contributions of human infants to caregiving and social interaction. In *The Effects of the Infant on its Caregiver*, ed. M. Lewis & L. Rosenblum. Wiley: New York.

Bowlby, J. (1951). *Maternal Care and Mental Health*. World Health Organization: Geneva.

Brazelton, T. B., Koslowski, B. & Main, M. (1974). The origins of reciprocity. In *The Effects of the Infant on its Caregiver*, ed. M. Lewis & L. Rosenblum. Wiley: New York.

Bronfenbrenner, U. (1974). *Two Worlds of Childhood.* Penguin Education: Harmondsworth.

Brown, R., Cazden, C. B. & Bellugi, U. (1969). The child's grammar from one to three. In *Minnesota Symposium on Child Psychology*, vol. **2**, ed. J. P. Hill. University of Minnesota Press: Minneapolis.

Caldwell, B. M., Wright, C. M., Honig, A. S. & Tannenbaum, J. (1970). Infant day care and attachment. *American Journal of Orthopsychiatry*, **40**, 397–412.

Cazden, C. M. (1970). The situation: a neglected source of social class differences in language use. *Journal of social Issues*, **26**(2), 35–60.

Clarke-Stewart, A. K. (1973). Interactions between mothers and their young children: characteristics and consequences. *Monographs of the Society for Research in Child Development*, no. **153**.

Collis, G. M. & Schaffer, H. R. (1976). Synchronisation of visual attention in mother–infant pairs. *Journal of Child Psychology and Psychiatry*, in press.

Davie, R., Butler, N. R. & Goldstein, H. (1972). *From Birth to Seven*. Longmans: London.

Douglas, J. W. B., Lawson, A., Cooper, J. E. & Cooper, E. (1968). Family interaction and the activities of young children. *Journal of Child Psychology and Psychiatry*, **9**, 157–171.

Dunn, J. F. (1975). Consistency and change in styles of mothering. In *Parent–Infant Interaction*, ed. M. O'Connor. Elsevier: Amsterdam.

Friedlander, B. Z., Jacobs, A. C. & Davis, B. B. (1972). Time-sampling analysis of infants natural language environments in the home. *Child Development*, **43**, 730–740.

Gouin-Decarie, T. (1969). A study of the mental and emotional development of the thalidomide child. In *Determinants of Infant Behaviour*, vol. **4**, ed. B. M. Foss. Methuen: London.

Kagan, J. & Klein, R. E. (1975). Cross-cultural perspectives in early development. In *Cultural and Social Influences in Infancy and Early Childhood*, ed. H. Leiderman & S. Tulkin. Stanford University Press: Stanford.

Maccoby, E. E. & Feldman, S. S. (1972). Mother-attachment and stranger reactions in the third year of life. *Monographs of the Society for Research in Child Development*, no. **146**.

McCall, R. B., Hogarty, P. S. & Huilbat, N. (1972). Transitions in infant sensori-motor development and the prediction of childhood I.Q. *American Psychologist*, **27**, 728–748.

Nelson, Katherine (1973). Structure and strategy in learning to talk. *Monographs of the Society for Research in Child Development*, no. **149**.

Nelson, K. E., Garskaddon, G. & Bonvillain, J. D. (1973). Syntax acquisition: impact of experimental variation in adult verbal interaction with the child. *Child Development*, **4**, 497–504.

Newson, J. (1974). Towards a theory of human understanding. *Bulletin of the British psychological Society*, **27**, 251–257.

Phillips, J. R. (1973). Syntax and vocabulary of mothers speech to young children: age and sex comparisons. *Child Development*, **44**, 182–185.

Rebelsky, F. (1967). Infancy in two cultures. *Nederlands Tijdschrift voor da Psychologie*, **22**.

Richards, M. P. M. & Bernal, J. F. (1972). An observational study of mother–infant interaction. In *Ethological Studies of Child Behaviour*, ed. N. Blurton Jones. Cambridge University Press: London.

Sameroff, A. J. & Chandler, M. J. (1975). Reproductive risk and the continuum of caretaking casuality. In *Review of Child Development Research*, vol. **4**, ed. F. D. Horowitz, M. Hetherington, S. Scarr-Salapatek & G. Sregel. University of Chicago Press: Chicago.

Scaife, M. & Bruner, J. S. (1976). The capacity for joint visual attention in the infant. *Nature, London*, in press.

Schaffer, H. R. & Emerson, P. E. (1964). The development of social attachments in infancy. *Monographs of the Society for Research in Child Development*, no. **29**.

Schwarz, J. C., Krolick, G. & Strickland, R. G. (1973). Effects of early day care experience on adjustment to new environment. *American Journal of Orthopsychiatry*, **43**(3), 340–346.

Schwarz, J. C., Strickland, R. G. & Krolick, G. (1974). Infant day care: behavioural effects at preschool age. *Developmental Psychology*, **10**, no. 4, 502–506.

Snow, C. E. (1972). Mothers speech to children learning language. *Child Development*, **43**, 549–565.

Stern, D. N. (1971). A microanalysis of mother–infant interaction: behaviour regulating social contact between a mother and her 3½ month old twins. *Journal of the American Academy of Child Psychiatry*, **10**, 501–517.

Stern, D. N. (1974). Mother and infant at play: the dyadic interaction involving facial, vocal, and gaze behaviors. In *The Effects of the Infant on its Caregiver*, ed. M. Lewis & L. A. Rosenblum. Wiley: New York.

Trevarthen, C. (1974). Conversations with a 2 month old. *New Scientist*, **62**, 230–235.

Tizard, B. & Rees, J. (1974). A comparison of the effects of adoption, restoration to the natural mother, and continued institutionalisation on the cognitive development of 4 yr old children. *Child Development*, **45**, 92–99.

Tizard, B. & Rees, J. (1975). The effect of early institutional rearing on the behaviour problems and affectional relationships of 4 yr old children. *Journal of Child Psychology and Psychiatry*, **16**, 61–73.

Ward, M. C. (1971). *Them Children: A Study in Language Learning*. Holt, Rinehart & Winston: New York.

Werner, E. E., Bierman, J. M. & French, F. E. (1971). *The Children of Kuaii*. University of Hawaii: Honolulu.

Willerman, L., Broman, S. H. & Fiebler, M. (1970). Infant development, pre-school I.Q. and social class. *Child Development*, **41**, 69–77.

17

Does ethology throw any light on human behaviour?

PETER B. MEDAWAR

Although biologists take to it very kindly, the idea that the behaviour of animals can throw any light at all upon the behaviour of men is so far from self-evident that it would have been regarded with the utmost derision by the more tough-minded philosophers and philosophic thinkers of the seventeenth and eighteenth centuries – I mean by men of the stature of Thomas Hobbes and Samuel Johnson.

Dr Johnson, in particular, would have been deeply outraged: 'Sir,' he would have said – he would certainly have said that, but what else he would have said is conjectural – though I think it might have run: 'is not the possession and exercise of moral judgment precisely the distinction between mankind and the brute creation? Show me an earthworm or marmoset that can tell the difference between right and wrong.' Thomas Hobbes would have pointed out – and in Chapter 13 of *Leviathan* did, in effect, point out – that it is only by virtue of characteristics that differentiate civilized mankind from lesser beings that the life of man is anything but 'solitary, poor, nasty, brutish and short'.

Scientists of the twentieth century cannot be expected to quake in their shoes at the thought of the opinion that might have been held of them by eighteenth-century philosophers, however skilful and tough-minded they may have been, for they were simply not in possession of all the information that would have made it possible for them to form a definitive opinion. In particular, they were unaware of the evolutionary descent of man; but even if they had been apprised of it their reaction would probably have been 'what of it?' and Dr Johnson wielding, as ever, the butt end of his pistol, would have demanded to know what great and illuminating new truth about mankind followed from our realization of his having evolved.

No-one would answer this question today with anything like the blandly assertative self-confidence that would have been characteristic of social Darwinists such as Francis Galton and Ernst Haeckel. To them it would have seemed obvious that mankind was stratified into superior and inferior human beings, and that all were engaged in a struggle for ascendancy – it would have seemed obvious, indeed, that 'man is a fighting animal'. It is, therefore, particularly satisfactory to put on record that it was, above all, ethologists who discredited the simple-minded and socially destructive beliefs characteristic of social Darwinism (e.g. Crook and others in Montague, 1968; Hinde, 1970). Some human beings are aggressive, to be sure, but it is by studying ourselves and not by studying animals that we recognize this trait in mankind; indeed, it is perhaps not unfair to say that those who know most about aggression in animals are most cautious in imputing any such thing as aggressive 'instinct' to mankind.

This still leaves us, however, with the rather hectoring question of just what great new truth has been learnt about man as a result of the recognition of his evolutionary descent.

I now believe the question is wrongly put and that it embodies a false conception of the nature of scientific progress. All scientists despise the ideology of 'breakthroughs' – I mean the belief that science proceeds from one revelation to another, each one opening up a new world of understanding and advancing still farther a sharp line of demarcation between what is true and what is false. Everyone actually engaged in scientific research knows that this way of looking at things is altogether misleading, and that the frontier between understanding and bewilderment is rather like the plasma membrane of a cell as it creeps over its substratum, a pushing forward here, a retraction there – an exploratory probing that will eventually move forward the whole body of the cell.

It is, indeed, not a grand ethological revelation that the scientist should seek from his awareness of the evolutionary process, but rather an enlargement of the understanding made possible by a new or wider angle of vision, a clue here and an apt analogy there, and a general sense of evolutionary depth in contexts in which it might otherwise be lacking. In such an exercise as this, as Robert Hinde (1976) points out, we often have as much to learn from the differences as from the similarities between human beings and animal models. In short, I think ethology is one of the many areas of thought in which a philosophic understanding of the nature of the scientific process is salutary: in real life, science does not prance from one mountain top to the next.

A great deal obviously depends upon whether or not it is possible to establish genuine homologies between the behaviour of human beings and that of the collateral descendants of their remote ancestors; to my mind, there can be no question that we can do so: consider, for example, the highly complex and teleonomically related behavioural and physiological activities that go to make up sexually appetitive behaviour, mating, conception, gestation, parturition, lactation, suckling and the care of the young. It is hardly conceivable that this entire complex scenario should have sprung into being for the first time with the evolution of *Homo sapiens*; and there is no reason nowadays why we should put ourselves to the exertion of imagining that it could have done so. The same applies to appetitive behaviour as it relates to the conventional everyday context of seeking food.

Nevertheless, even if the existence of these homologies is conceded, we must not expect too much to follow from them; in particular, we should not expect a great variety of psychologically illuminating insights, for ethology often stops short at just the level at which psychology begins, i.e. stops short of explaining the nature and origin of the *differences* between individuals. As I pointed out in an early essay on this very subject (Medawar, 1957) what is *interesting* about our human propensity for loving or hating is why one person loves a second and hates a third, and that what is interesting about our human dietetic habits is not the psychological need for food or the behaviour that goes with it, but why any one human being will eat this and not that, here and not there, and now and not then. Psychology has had to do with the differences between human beings, ethology – at least in its early days – with the characteristics they have in common. I went on to say:

> It is no great new truth that human beings are ambitious; what is interesting about ambition is why in one person it should take the form of wanting to become a great musician and in another of wanting to raise a large family, and in a third (for this, too, is an ambition) of wanting to do nothing at all.

These questions seem to me to belong to the domain of psychology and not to ethology at all.

I do not think there is any general answer to the question of how exact are the homologies between, say, human and primate behaviour. Scientifically, it would be completely sterile to start with the presumption that homologies were illusory and the correspondences invariably inexact. Methodologically, the sensible thing is to start with the hypothesis that the homologies are fairly close; for such an hypothesis can be made the basis of action and can put us on the right wavelength for making observations of the kind which will either falsify it or strengthen our belief that there may be something in it. For example, if it were to be shown that maternal deprivation had a psychologically damaging effect on rhesus monkey babies, then there would be a case *prima facie* for supposing that the same might be true of human beings. If the idea is wrong it will be shown to be so, for it is the great and distinctive strength of science that we need not persist long in error if there is a genuine determination to expose our ideas to tough critical analysis. In general, it is illuminating to recognize that such human – sometimes all too human – activities as play, showing off and sexual rivalry are not psychic innovations of mankind, but have deep evolutionary roots.

I mention it purely as an aside that the ancient origins and deeply programmatic character of human reproductive behaviour makes it extremely unlikely that a prudential system of sexual morality can be reinstated or the population explosion contained merely by exhortation or appeals to reason, however cogently worded. It also makes it extremely unlikely that there is any easy psychological solution of the problem of family limitation. On the subject of behavioural homologies as they apply to man, I cannot do better than to quote the fine closing paragraph of Darwin's *The Descent of Man* 1871:

> We are not here concerned with hopes or fears, only with the truth as far as our reason permits us to discover it, and I have given evidence to the best of my ability. We must, however, acknowledge as it seems to me, that man with all his noble qualities, with sympathy which he feels for the most debased, with benevolence that extends not only to other men but to the humblest living creature, with his godlike intellect which has penetrated into the movements and constitution of the solar system – with all these exalted powers – man still bears in his bodily frame the indelible stamp of his lowly origin.

For our own purposes the quotation must, of course, be modified so as to read 'behavioural repertoire' instead of 'bodily frame'. With this modification, the only word we could reasonably cavil at is the word 'indelible', for we are by no means at the mercy of our biological inheritance. For example: human evolution has taken such a course that haemolytic disease of the newborn is almost inevitable in certain situations unless we take steps to circumvent it, but we *can* and do take steps to do so. The same applies to human behaviour: our social conventions and institutions do inhibit and to some extent prevent the more exuberant forms of the behaviour so often described as 'bestial' though in reality it is more characteristic of man than of beast.

I should like to quote what has always seemed to me to be a very wise passage by Alfred North Whitehead in his *Introduction to Mathematics* (1911):

> It is a profoundly erroneous truism, repeated by all copybooks and by eminent people when they are making speeches that we should cultivate the habit of thinking of what we are doing. The precise opposite is the case. Civilization advances by extending the number of operations which we can perform without thinking about them.

I find this statement compellingly true of much human behaviour that can be described as skilled, for if somebody asks us what seven sixes are (or anything else from our times tables) the last thing on earth we want to do is to go back to Peano and Frege or to Russell and Whitehead to puzzle out what the answer to such a searching question can be: we want to snap it out without hesitation; and likewise the process of learning to drive a car, obeying traffic signals and avoiding obstacles etc. is in effect learning *not* to think about actions which did, at one time, require anxious deliberation ('now let me see, which of these many pedals will arrest the motion of the car?'). We are learning to give learned behaviour the polish and fitness for purpose which we describe colloquially as instinctive, so that we can drive a car safely even if we are tired, preoccupied and in a hurry.

Paradoxically enough, learning is learning *not* to think about operations that once needed to be thought about; we do, in a sense, strive to make learning 'instinctive', i.e. to give learned behaviour the readiness and aptness and accomplishment that are characteristic of instinctive behaviour. [*But that is only half the story.*] The other half of the half truth is that civilization also advances by a process that is the very converse of that which Whitehead described: by learning to think about, adjust, subdue and redirect activities which are thoughtless to begin with because they are instinctive. Civilization also advances by bringing instinctive activities within the domain of rational thought, by making them reasonable, proper and co-operative. Learning, therefore, is a twofold process: we learn to make the processes of deliberate thought 'instinctive' and automatic, and we learn to make automatic and instinctive processes the subject of discriminating thought. Medawar, 1957

Anybody who professes to discern a moralizing flavour in what I have been writing is perfectly right: it is exactly what I intend. I think we shall have to get used to the idea that moral judgements should intrude into the execution and application of science at every level – and in no context more exigently than the interpretation of science for the benefit of laymen.

Anyhow, these reflections on Whitehead lead me naturally to one or two points I should like to make on the special importance of language and the learning process in human beings and their relevance to the conception and measurement of 'intelligence'.

Having dismissed the idea that we must look to ethology for great revelations about the nature of human behaviour and dismissed also the idea that scientific understanding lurches forward from one revelation to another, I should like to put on record my belief that the great contribution of ethology to the understanding of human behaviour is methodological: indeed, there are already signs that some of the more eclectic psychiatrists, anxious as ever to be in the swim and in the forefront of opinion, are starting to speak about ethology as if they had invented it themselves. This will be all to the good if it has the effect of persuading them to direct upon human beings the intent and candid gaze to which ethologists owe their success in the study of animal behaviour.

The very last thing I have in mind is anything like the at one time popular sociological exercise known as 'mass observation' – a kind of inductivism gone mad, as I remember it – which was dedicated to the proposition that if only one could collect enough information about what people actually do and actually say at home, in pubs and in buses etc., then some great new truth about human behaviour would, of necessity, emerge. So far as I am aware, no truth of any kind emerged, great or small, except perhaps that it takes all sorts to make a world. No: I was allowing myself to cherish the hope that ethological methods might one day make it possible to build up a biologically well-founded psychology or even a psychopathology to take the place of the weird farrago of beliefs which forms the basis of modern psychoanalytic psychotherapy, a system of beliefs which persists because it has never been found wanting, and has never been found wanting because it has never been exposed to any evaluation.

I have defended on several occasions (Medawar, 1957, 1960, 1972) the view that what is characteristic of human beings considered as animals is not the making and use of tools but the communication from one person to another, specially in the next generation, the knowledge and know-how to do so. It is by virtue of this faculty that human beings come to enjoy a kind of cultural evolution which has converted us into animals that are simultaneously aerial, terrestrial and submarine, possessing X-ray eyes and sense organs sensitive enough to feel the heat of a candle at a distance of a mile. In briefest summary, because there is no point in going over old ground again, the main characteristics of this distinctively human form of evolution are:

(1) It is Lamarckian in style, i.e. unlike ordinary or 'endosomatic' evolution it embodies a learning process; for what is learned is passed on and becomes part of the evolutionary heritage.

(2) Cultural heredity is mediated through non-genetic channels – hence the use of the terms 'exogenetic' or 'exosomatic' evolution to describe the process and distinguish it from ordinary genetic or Darwinian evolution.

(3) It is reversible. Not just in theory – as it is possible in theory anyway for a kettle to boil when it is put on ice – but in good and earnest. We could return to the Stone Age in one generation.

This system of evolution is the characteristic to which we owe our clear-cut biological supremacy over all other organisms because it has conferred almost unlimited capabilities upon us, including even that of leaving the earth and going to live elsewhere in the solar system.

This characteristically human system of heredity calls for and depends upon the existence of language and other forms of conceptual communication (one is reminded that Dr George Steiner refers to man as a 'language animal' rather than as a 'tool-making animal'). The existence of and our dependence on exogenetic evolution must place a specially high selective premium upon such capabilities as teachability and imitativeness (a word for which, in this context, the vernacular term 'aping' seems uncannily apt) because these form the causal nexus of cultural heredity – points made very well by Barnett (1973) in a recent lecture entitled *Homo docens*.

With this system of heredity in mind the distinction between programmed and learned behaviour loses some of its force, for an episode of behaviour can be 'programmed' in the sense that the physical plant and the functional capabilities necessary for its execution must be readymade even though the behaviour itself has to be learned. Barnett himself attaches considerable importance to the observations of Kuo (1930) on the killing of rats and mice by cats. It is clear that although a cat's aptitude and general capability for killing mice is 'laid on' developmentally, cats are much more likely to kill mice if they have observed their mothers doing so.

The authors surmise that this kind of behaviour may be characteristic of Felidae generally, noting, however, that no experiments on the subject have yet been done on tigers or lions, but if we reflect on the importance attached to teaching and the entire apparatus of pedagogy in human beings it seems difficult to resist the hypothesis that the same principle is true of human beings (Seligman & Hager, 1972; Hinde & Stevenson-Hinde, 1973).

In the light of these considerations I wonder increasingly at the naiveté of those psychologists who believe that prowess in certain so-called 'intelligence tests' provides a measure of a so-called 'innate intelligence' which is virtually unaffected by the subject's age and which cannot be taught or influenced by experience. If this were true of any human performance, my first reaction would be to dismiss it as relatively unimportant and certainly not one that could be made the basis of a measure of intelligence, for I believe that the endowments that have made human beings what they are, are above all imitativeness and teachability.*

For these and other reasons I believe that the I.Q. concept and some of its practitioners should now be relegated to that dusty, cavernous and ill-lit building called the Museum of Social Darwinism, in which other principal exhibits are Francis Galton, Ernst Haeckel and Alfred Rosenberg, and many others who have misunderstood the bearing of biology on human affairs or who have propagated mischievous views in the name of science.

FURTHER COMMENTS AND AFTERTHOUGHTS

When this conference was planned the title I submitted for a contribution that I thought would come well from a general biologist was: 'Does the study

* These views were the subject of an altercation between Professor H. J. Eysenck and myself in a number of issues of the *New Statesman and Nation* (11 January 1974 and two succeeding issues).

of the behaviour of animals throw any light on the behaviour of Man?' and this is the topic on which I wrote my paper. The title was such a mouthful, however, that I changed it to its present form.

Put colloquially, my answer to the original question can be seen from the text to be: 'Not very much, really – it is by studying human beings themselves that we learn about their behaviour.' If, however, we put the question in the form in which it actually appears at the head of my contribution, then I think the proceedings of this very conference – I am thinking particularly of the work of the Hindes and of Judy Dunn – make it possible to say with some confidence: 'Yes, indeed.' This puts me under an obligation to try to describe what differentiates ethology from psychology in a conventional sense.

I think ethology has two distinctions: in trying to make teleonomic sense of behavioural performances that might seem to inexperienced observers to be a stream of inherent and functionless activities, ethologists are not yet importuned by an insistent and urgent need to find a causal explanation for every phenomenon they observe. Closely related to this is the welcome truth that ethology, unlike some psychological systems, is not yet crabbed and confined by the doctrinal tyranny of any pre-existing explanatory System. These two characteristics give ethology the freshness and spontaneity which other biologists find so enviable, and which are sadly so lacking from many of the older and more conventional branches of zoology.

I turn now to the consideration of a number of comments and criticisms that have been made about the first version of this paper.

Apropos of exogenetic evolution, Professor Niko Tinbergen, who was unfortunately unable to be present at the meeting, expressed some doubt about the total reversibility I claimed for it, and in particular about my contention that human beings could revert to the Stone Age in one generation if the cultural nexus between one generation and the next were to be wholly severed. The matter can only be resolved by a purely notional experiment in which we are to imagine the development and likely fate of a community consisting of a number of human babes reared as if by magic on a desert island – kept alive, indeed, but without the benefit of any of that inheritance which is passed on from one generation to the next by precept, books or word of mouth. Without that inheritance – without formal schooling or any indoctrination in any of the useful arts – would they not regenerate among themselves into something like a palaeolithic culture?

This experiment becomes more and more vague and unsatisfactory the more one thinks about it, so there is no point in going on about it. I admit, though, that the 'Stone Age' was a rhetorical over-simplification; the point I really want to make, which I think most people concede, was the *reversibility* of exogenetic evolution.

Another point made by Tinbergen was that we should seek analogies no less often than homologies – meaning by analogies 'similarities convergently evolved in less closely related species in adaptation to similar niches'. I accept this criticism completely, and smiled at the justice of the rebuke given to someone trained in a school of zoology for which the notion of 'homology' was the central – even the energizing – concept of the whole of biology.

I am not very well up in the literature of ethology and was not therefore surprised to learn several examples unfamiliar to me of exogenetic evolution or cultural transmission among animals. Tinbergen, in particular, referred me to 'acculturation' in Japanese macaques and referred me to pp. 243–244 of Hinde, 1974; in addition, Mr Gady Katzir called my attention to a description of a cognate phenomenon in Marais, 1969.

Amid the general discussion that followed my paper I reminded the members of the symposium that molecular genetics began with the complete solution, by intent and single-minded research, of the phenomenon of pneumococcal transformation, and I went on to ask if there were not some comparable phenomenon in ethology, the complete interpretation of which would have an effect analogous to that of an aircraft's breaking cloud; I referred to the transformation of male into female sexual behaviour and vice versa by the injection of hormones characteristic of the opposite sex – phenomena of which a number of specially apt examples were given by Lehrman (1961). A complete interpretation of any one of these phenomena, I argued, might have an effect on ethology comparable with the effect of O. T. Avery's work on the subsequent growth of molecular genetics. In the discussion which followed it became clear that members of the conference were not at all sure that any such parallel between ethology and the growth of molecular biology could be drawn.

It may, in any case, turn out that the revolution I have in mind will be conceptual rather than one that turns upon the analysis of behavioural phenomena. Commenting on my paper both Hinde and Bateson professed the rather strong feeling that the Lorenzian distinction between instinctual and learned behaviour is no longer useful: 'all behaviour is both genetically and environmentally influenced; all learned behaviour is genetically influenced'.

Methodologically speaking, I think this is a rather weak declaration and part of it, at least, is tautologous; obviously no behavioural performance could take place without the physical and physiological means to execute it – there must be nerves and muscles and other apparatus of the kind we normally think of as being 'laid on' by developmental processes. It could be said of *every* character trait whatsoever that its determination is partly natural and partly nurtural, yet we do know of character differences that are wholly genetic in determination in the usual sense of the term – e.g. a human being carrying the blood group gene associated with the Group A will be of Group A in almost any environment that is capable of sustaining life. It is an old, vexed question, and it will not be solved in the context of behaviour until Aubrey Manning's ambition for the foundation of the genetics of behaviour is realized.

REFERENCES

Barnett, S. A. (1973). *Homo docens. Journal of biosocial Science*, **5**, 393–403.

Darwin, C. (1871). *The Descent of Man.* John Murray: London.

Hinde, R. A. (1970). *Animal Behaviour: A Synthesis of Ethology and Comparative Psychology.* McGraw-Hill: New York.

Hinde, R. A. (1974). *Biological Bases of Human Social Behaviour.* McGraw-Hill: New York.

Hinde, R. A. (1976). The use of differences and similarities in comparative psychopathology. In *Relevance of the Psychopathological Animal Model to the Human*, Proceedings of 2nd International Symposium, Kittay Scientific Foundation. Plenum Press: New York, in press.

Hinde, R. A. & Stevenson Hinde, J. (ed.) (1973). *Constraints on Learning: Limitations and Predispositions.* Academic Press: London & New York.

Kuo, Z. Y. (1930). Genesis of the cat's behaviour towards the rat. *Journal of comparative Psychology*, **11**, 1–35.

Lehrman, D. S. (1961). Gonadal hormones and parental behaviour in birds and infra-human mammals. In *Sex and Internal Secretion*, ed. W. C. Young. Williams & Wilkins: Baltimore.

Marais, E. (1969). *The Soul of the Ape.* Blond: London. 2nd edn (1973). Penguin: Harmonsworth.

Medawar, P. B. (1957). *The Uniqueness of the Individual.* Methuen: London.

Medawar, P. B. (1960). *The Future of Man.* Methuen: London.

Medawar, P. B. (1972). *Technology and Evolution.* In *Technology and the Frontiers of Knowledge.* The Frank Nelson Doubleday Lectures, 1972–73. Doubleday: New York.

Montague, M. F. A. (ed.) (1968). *Man and Aggression.* Oxford University Press: New York & London.

Seligman, M. E. P. & Hager, J. L. (ed.) (1972). *Biological Boundaries of Learning.* Appleton-Century-Crofts: New York.

Whitehead, A. N. (1911). *An Introduction to Mathematics.* Home University Library: London.

18

Ethology in a changing world

N. TINBERGEN

By juggling with half-vacancies, as we soon shall be juggling again...
Medawar, 1972

(Ethology has) had a far-reaching influence on such medical disciplines as social medicine, psychiatry, and psychosomatic medicine.
Cronholm, 1974[1]*

OPTIMISM AND REALISM

The Editors have invited us to be 'forward looking' and, using a familiar metaphor, to pay special attention to 'growing points'. In the garden of the behavioural sciences there is certainly no dearth of such growing points. But just as the forward looking gardener is concerned about the optimal conditions for growth – sunlight, warmth, water and nutrients – so the growing points of our science need their essential supplies: manpower of behaviour students and the resources needed for their employment and for their training.

How likely is it that these conditions will indeed be met, and what can we ethologists ourselves do to ensure that our science shall continue to flourish? Because of the gradual amalgamation between ethology in its old restricted sense and neurophysiology, many parts of psychology, ecology and the study of evolution and genetics, I shall use the term 'ethologists' in the wide sense of biologically oriented students of behaviour.[2]

In speculating about and planning for the future, not only of our science but of science in general, we have to keep the following points in mind.

When, as we have to do, we take the past as a guideline for the future, we have to realise that the past of our science covers only a very short span of time. In this short history we have witnessed an unprecedented growth, almost an explosion, of the behavioural sciences. There have been setbacks – the depression of the early thirties, the Second World War, and the recent reduction of funds. Yet until quite recently, optimism with regard to the future

* Superscript numbers refer to endnotes on p. 526.

seemed quite justified, and in particular the large postwar generation of ethologists tend to consider the vigorous growth of the last few decades as representative. In fact it may have been, and in my opinion has been, quite exceptional, and on grounds of cool, sober considerations of probability alone, further growth, stagnation at the present level, and even a reduction of resources must be considered equally probable. In fact it would seem highly likely that the steep upward curve, not only in manpower but also in the (real) cost per worker (which has also risen spectacularly) will flatten out, or even turn into a decline. The last few decades are indeed an exceptionally unreliable basis for the prediction of the future.

To see the problem in perspective we have to realise that all scientific endeavour is a function of attitudes in society as a whole. This is certainly true of 'pure' or exploratory science, but to a lesser degree it also holds for the executive, applied branches, the various technologies. In particular exploratory science, the intellectual game of 'science for the sake of science' flourishes only in rich countries and in affluent periods, i.e. when and where, apart from and beyond the resources needed for sheer survival (food, water, housing, clothing, communication, health, education, defence), there is a surplus of manpower and cash that can be allocated to the 'luxury' occupation of being curious. This means that, when trying to forecast the future of our (as of any) science we shall have to make inspired guesses about the future of society as a whole.

At the same time, scientific endeavour is not merely a reflection of the spirit and the affluence of a society; it can also in its turn influence the attitude of society by enhancing the status, prestige, or esteem which it enjoys. This esteem, which determines for each science what part of the overall budget, of government and private foundations combined, shall be allocated to it, is in its turn, conversely, determined by how interesting and above all how useful it is seen to be.

Thus there is a two-way traffic between society and science (and for that matter all types of intellectual activity). I shall try to sketch what trends I think can be discerned in both aspects of this reciprocal relationship between society and science, and what moral we can draw from this analysis of the future of our own science. I shall have to do this in the briefest possible form.

CULTURAL EVOLUTION – A MIXED BLESSING

It is now becoming generally recognised, at least by those who compare man with other animals, that the uniqueness of our species can be characterised best by the general statement that our unprecedented mental abilities have enabled us to embark, over and above genetic evolution, on a fundamentally new type of evolution, commonly called cultural or psychosocial.[3] New evidence on the past of the hominids, and of man as their only surviving species, keeps pushing back in time the beginning of this 'experiment of nature'.[4] Approximately two million years ago, real men began to transfer typically acquired behaviour programming ('experience' or 'knowledge') from each generation to the next. This enabled them to change, rather than merely accept, their environment. The process started very slowly, but has steadily gathered momentum, until in the present times each new generation grows up in, and has to adjust to, a world that is very different indeed from that in which its parents have lived.[5] The preparation of tools of increasing sophistication, the controlled use of fire, intelligent planning of communal hunting expeditions, the building of houses, the making of clothes, the switch to agriculture and animal husbandry etc. evolved through a series of inventions which, as history proceeded, assumed more and more the scope of real saltations; of major, sudden changes. Saltations of equivalent impact are impossible in genetic evolution because they produce unviable embryos. But 'great leaps forward' such as those mentioned, and later the invention of writing and printing, that of the steam engine, the harnessing of electricity, the construction of the internal combustion engine, the use of nuclear energy, the proliferation of the mass media, and the application of new educational methods (often labelled with the emotionally, ideologically loaded term of 'brainwashing') have so far not been lethal thanks to our unique behavioural flexibility. How thoroughly our world has changed in the last few decades can naturally be realised best by those who have known the world of the twenties and, a little further back, that before the First World War, a world with hardly any motor cars, without airliners, without radio, let alone television; a world in which at least some governments worried about the decline of populations, a Western world that lived to a large extent on the produce of colonies, a world in which food surpluses were burned to prevent a drop in prices.

Until recently the vast majority of people living in the Western world have (as we now begin to see, naively) seen the cultural evolution as unmitigated progress, as a movement not only towards greater affluence but also towards greater happiness. However much we can learn from the study of the many great civilisations in recorded history that have flowered and declined, it is the Western civilisation of today that concerns us most, not only because it has produced our science, but also because it is now, admittedly with regional variations, dominating the whole world. Conversely, it is the very growth of science that now makes us look more detachedly at ourselves and that begins, hopefully just in time, to alert us to the fact of life that every form of success carries its penalties, that every blessing is mixed, and that the cultural evolution too is beginning, or rather *has already begun*, to produce an (extremely complex) set of negative feedbacks. That we have for so long failed to notice this (in spite of the writings of Malthus (1817), Osborne (1963), Vogt (1948), Rachel Carson (1963) and others) is because of our one-sided appreciation and our complacent acceptance of the blessings of our civilisation: of reduced infant mortality, of increased affluence, of our 'spiritual life', and last but not least of science itself.

The practical result of science and technology has been a growing mastery of our habitat. By progressively changing our environment we have been trying to have, and for quite a time actually have had, the best of two worlds. Medical technologies have been highly successful in reducing to a minimum the massive weeding-out of comparative failures that occurs in every generation of every species of animal and plant, and which alone has ensured continued adaptedness, i.e. viability. Agricultural and physico-chemical technologies have increasingly modified our environment to meet our growing needs. It deserves emphasising that these 'needs' have grown not only in proportion to the population growth, but in addition in proportion to what our industries offer us. Beyond wanting the essentials for a happy and healthy life, almost all people develop an insatiable, almost exponentially growing greed for ever more. This combination of over-reproduction (through postponed mortality) and growing consumption, even per head, is in principle impossible on finite resources. Even if some non-renewable resources are at least in theory still over-abundant, others are not, or they are out of reach; and the growing production of renewable resources such as food and water is at or very near the point of being overtaken by the growth of what the world's population wants. In the present context I can do no more than refer to the writings of e.g. Borgström (1969), Brown (1974*a*, *b*), Commoner (1966, 1972), Ehrlich & Ehrlich (1972) Ehrlich, Ehrlich & Holden (1973), Meadows, Meadows, Randers & Behrens (1972), and Ward & Dubos (1972), who have in my opinion made a far more convincing case than optimists such as Beckerman (1974) and Maddox (1972). It cannot be repeated too often that the great question is not *whether* but *when* population 'crashes' will come, what shape they will take, and how massive they will be. At present, famines, epidemics of newly evolved strains of parasites and massive unemployment in exploding populations are the most immediate threats to health and to survival itself. Needless to say that concern for this makes sense only as long as no government is foolish enough to unleash nuclear war.

Without entering into the details of the innumerable aspects of the 'human predicament', I must emphasise one fundamental aspect of the situation: one single lethal negative feedback can undo all the gains of the cultural evolution, and can even threaten our very survival.

LOSS OF VIABILITY

However incomplete and in many respects tentative the evidence about new, man-made environmental pressures is in the material sphere, and about somatic response to them, there seems to be no doubt that on balance and contrary to superficial signs our species is in a phase of disadaptation, of loss of viability. The new pressures or stressors* fall roughly into two categories – over-exploitation and depletion of a number of non-renewable and even of renewable vital resources, and accumulation of harmful waste products, which leads to pollution of land, water and air. Most writers on the subject have concentrated their attention on this 'material/somatic' form of disadaptation, probably because they are more tangible and invite measurement. The well-known Meadows models are attempts to provide a framework for quantitative description of these relationships. Without suggesting that we should not measure what we can, I should like to point out that a number of possibly vital aspects of disadaptation, either demonstrated or as yet merely suspected, cannot as yet be expressed in figures. The list of questions still requiring a qualitative answer is almost endless. Often material and psychosocial aspects are jointly involved. I am thinking of such questions as: what is a sufficient and balanced diet in particular for pregnant women and children; how do modern farming methods affect the soil structure; what exactly caused the dwindling of grain reserves after they were shipped to countries in need; what makes farmers abandon their land; what exactly makes many malcontents turn to the use of force, etc. In the overall picture of consistent population growth, over-exploitation of vital resources and pollution it is still too early to claim that we have identified, categorised and measured the relevant pressures, let alone that we have explained their influences on health and life.

The difficulty facing quantification is even greater in the field of psychosocial pressures. Not surprisingly in view of the complexity of human behaviour, these are even less well understood, yet they may well ultimately be more decisive for our future well-being.

PSYCHOSOCIAL PRESSURES

Disadaptation in the psychosocial sphere expresses itself in a bewildering variety of effects. For attempts to design a framework of classification of psychosocial stressors and their deleterious effects I must refer to recent

* Through an unfortunate semantic slip in the past the word 'stress' has, since the pioneering work of Selye, been misapplied to what is in reality reaction-to-stress, while the word 'stressor' is applied to outside influences eliciting stress. The terms stressor and pressure, used in the medical and biological literature, respectively, have the disadvantage of suggesting the idea of a bundle of separate outside stimuli, whereas we are in reality faced by a web with many interconnections.

publications in the field of stress research, e.g. Levi & Andersson (1975). This field is still very much in the exploratory phase, that of paving the way for much needed, multi-disciplinary research. To review the vast and scattered evidence and half-evidence on this subject is far beyond my competence, but for my present purpose, that of alerting my fellow ethologists to the need for entering this field, it is sufficient to point to at least some problems to which they could make useful contributions.

As I have said already, the demarcation between material/somatic and psychosocial effects is by no means sharp. To begin with, some material changes in our habitat affect mental health. The toxic metals are one example; for several of them lead to serious behaviour disorders (Tucker, 1972). Starvation in childhood is another; and recently many aspects of nutrition are receiving close and worried attention (Pauling, 1968).

Conversely, a great deal is now known about 'psychological' pressures that cause 'psychosomatic' diseases: duodenal ulcers, forms of asthma, coronary thrombosis and other heart diseases, hypertension, and many other often lethal illnesses have a considerable 'psychogenic' component.

Least understood and least open to measurement, but probably still greatly under-rated are behaviour disorders resulting from new social circumstances. The complex of sciences covered by the term psychopathology attempts to understand and cope with this vast category, but – again, not surprising in view of our lack of understanding of human behaviour – they must, with few incidental exceptions, still be considered as being in the stage of an art or a pre-science. It is in this field that ethology can contribute more than it has so far been doing. It can do so in a variety of ways. One potentially fruitful procedure is the detailed comparison of social conditions in those populations, small, scattered and dwindling fast, which can be considered to have deviated least from the way of life of ancestral man, with conditions under which populations in highly industrialised and urbanised countries live at present.[6] A second line of attack is to study specific behaviour disorders and trace their relation to contemporaneous social conditions. In both these procedures the influence and usefulness of an ethological approach seems to me to become already evident. Thirdly, the comparative ethologist can derive much information and inspiration from detailed studies of healthy as compared with unhealthy consequences of different living conditions for animals. In this field it is equally important to study species such as some primates, that are at the same time related to us and in their social organisation convergent to primitive man (Hinde, 1974), as species such as the lion, the wolf and the spotted hyaena, that are less closely related to us but are in their social organisation equally or even more strikingly convergent (e.g. Kruuk, 1972; Mech, 1970; Schaller, 1972). That in both cases extrapolation of results will always have to be done with the greatest caution goes without saying.

Although we will still have a long way to go before the many disciplines concerned shall be able to give a coherent account of the new psychosocial pressure system, let alone of its damaging effects on our behaviour, a few aspects can be mentioned that are serious candidates for a place in the 'web of stressors'. Some of these affect mainly the adults of modern populations, some the children and subadults, and some both, or the relationship between young and adult.

In assessing this kind of influence two points are worth remembering. First, our criteria for 'normal' are basically derived from the present distribution of characteristics; a person considered normal in our society might well appear highly abnormal to a Rip van Winkle of 100, even 50 years ago. Secondly, the stressors I discuss do not act in isolation, but usually together with a number of others.

For adult members of our society the growth of populations united in enormous states, the resulting anonymity of society, particularly in the large cities and megalopolises, as well as the sheer complexity of human relations, have aspects that can stretch individual adjustability beyond the tolerable. The view is gaining support, especially among anthropologists and ethologists, that our social behaviour has initially been adapted to conditions that prevail in small, closed in-groups in which every member knew each other member individually and which had relatively rare contacts with strangers (Lee & DeVore, 1968). Modern urban society – and with the explosion of means of communication and transport many 'rural' societies have acquired urban traits – bombards most of its members with an avalanche of contacts with strangers, and that this represents a strain on our social behaviour seems to me beyond doubt. Paradoxically, the anonymity of urban life also produces the opposite: in modern cities there are remarkably many lonely people.

The increased complexity of social relations, made possible by the vastly increased amount of information that reaches the individual, appears also to become excessive and harmful in many ways. The most alert and resistant individuals react to this 'information input overload' (Lipowski, 1973) by cutting too short the time needed for thinking before deciding on action, and in the process many of them develop psychosomatic illnesses or signs of mental strain. The less resilient person may react by cutting himself off from a great deal of information and is for this reason bound to behave suboptimally on many occasions.

Another set of pressures is related to the nature of work. For a large proportion of the population the relation between effort and reward is very different from what it was originally. Much work in modern industrialised society is extremely monotonous and boring. In addition the reward is often hardly related to the work itself; there is little connection between the work done on the assembly line and the final product. This has at least two aspects:

the modern worker often makes components rather than finished products, and the pay packet comes at the end of at least a week. The emphasis has shifted from the immediate satisfaction and pride of the hunter and even of the farmer and of the craftsman to a reward which has little to do with the nature of the work done. Under such circumstances any break from the monotony is welcome. The under-stimulation of the assembly line worker can exist side by side with over-stimulation, for instance from the mass media, from the immediate environment, from attendance at and betting on games, etc. In this context the pressure caused by change in conditions (as for instance required by 'mobility of the labour force') can be an additional factor. And even more difficult to bear is of course lasting unemployment, which seems to be an inevitable consequence of capital-intensive industrialisation. Every year the number of jobless school-leavers is increasing, in particular in the developing countries. To see this merely as consequences of temporary economic recessions is to ignore the deeper, ecological and social causes.

Yet other sources of dissatisfaction are related to the expectations with regard to the 'standard of living' to which I referred above. Industrialised, commercialised societies are in a sense self-perpetuating in that they produce goods, which, once available, tend to become 'necessities of life'. This, together with the urge to 'keep up with the Joneses' inevitably creates a greed for steadily more.

Closely connected with this, and with the anonymity of our society, is the hypertrophy of competitiveness. While life in primitive societies, and undoubtedly in our small ancestral in-groups, required a balanced mixture of competition and cooperation, many circumstances seem now to combine to reward over-competitiveness; in a complex, impersonal aggregation the victors in the 'rat race' do not directly see the harm they inflict on others. A contributory source of over-competitiveness is the unfair distribution of income among the members of many societies, and between nations.

The very high pace of change in modern society also disrupts the relations between the generations. In the past the 'generation gap' was, as in animals, a passing phase in the life of each generation, which ended with the young, after having found their place in the community, on the whole conforming with the adult standards. Now each new generation grows up in a society that is drastically different from that into which their parents' generation was integrated in their youth. Under those circumstances one can hardly expect the young to conform – rather the opposite: since the new society has so many harmful aspects it is not only to be expected but also to be hoped that the young want to change it. In view of the immense complexity of society it is not to be expected either that this rejection by the young expresses itself in mature, constructive behaviour; rebellion almost always begins with destruction. Yet underneath this negative, rejecting behaviour of the younger generation there is almost invariably a wish, often a craving, for an overall worthwhile aim in life. In the secularised atmosphere of today such an overall ideal is sought more and more in practical, material spheres, in ideologies and in nationalism.

The basis for dissatisfaction and for socially disruptive behaviour is laid early in life. Bronffenbrenner (1974) has called attention to a phenomenon which he calls the virtual break-up of family life in American society. Since what happens in the USA today tends to happen here tomorrow, we should be well advised to pay close attention to these developments in the States. There, the weakening of the family bond is not compensated for by a strengthening of what could be called the neighbourhood group, in which even young children form part of a closely knit in-group of known individuals of all ages. Rather the opposite happens: neighbourhood groups too are disappearing.

Nor do schools in general provide what the family and 'inner circle' used to offer. Many educationalists (e.g. Holt, 1970) have pointed out that institutionalised teaching-by-instruction elicits in many pupils either boredom or apprehension – reactions which cannot but predispose children for later rebellion. In fact the rejection of schools by the pupil population and the number of 'educationally subnormal' children are already worrying the authorities in more than one British city and has been called 'epidemic'.

The discrepancy between on the one hand conditions in rapidly growing and increasingly urbanised populations (of which I have mentioned only a few), and on the other the conditions that we desire, causes increased dissatisfaction and growing competitiveness, and so creates a breeding ground for the use of force, not only between but also within nations. Already there is in many Western societies a climate of continuous tension, which leads to ill-health and unhappiness, even to increased mortality. J. Eyer has pointed out that official mortality figures for the USA show an increase in mortality over the last decade, most pronounced for the age group of 15–24 years; this mortality increase is largely a result of homicide, suicide, drug addiction and heart failure.[7]

These few probes into new, deleterious psychosocial aspects of our new society must suffice to alert us to the seriousness and to the complexity of the situation. The various fields of psychosocial medicine and of sociology that are trying to understand these new trends have not yet produced a coherent analysis, and I repeat that here is a field in which ethologists can be of assistance. A combination of knowledge of man's original, deep-rooted behavioural equipment for life and of his living conditions, and of a better understanding of the causes of behavioural and other 'psychogenic' disorders will be a prerequisite if only for a sound diagnosis of the ills of our society. This itself is only a first step, a precondition for the design of a coherent system of remedial measures. In spite of a considerable amount of incidental evidence we are still far from having a comprehensive picture of psychosocial pressures, let alone of the changes to be made.

When I contrast modern Western civilisation with man's 'original', 'precultural' conditions and adaptedness I am of course not forgetting that the two extremes are linked by a very long, gradual and tortuous history. But just because this history has been so long, and above all so gradual, and because its pace has only recently stepped up so dramatically, the juxtaposition of the two extremes in time and space can alert us to many aspects of our present condition which have crept in on us and have therefore escaped attention.

MAN'S AGONISM TO MAN

At the risk of stating the obvious, a few words must be said on one of the most strikingly maladapted social traits of our species: the tendency, clearly of very old origin, to resort not only to force when dissatisfied, frustrated or threatened (which we share with many animals) but to the distortion of this system which makes men kill other men, and culminates in organised warfare. Whereas archaeologists have unearthed evidence of incidental killing of individuals, the oldest archaeological evidence of real warfare is from approximately 8000 years B.C.: the (pre-agricultural) city of Jericho, the oldest known city in the world, has been found to be surrounded by strong defensive walls (Kenyon, 1969). While the most spectacular and most typically human form of intra-specific agonism is war between large population groups, homicide can best be considered in a wider aspect, to include all forms of actual killing, even within well-organised groups – at an individual level – and between, for example, economically, racially, religiously or ideologically separate subgroups. And we should of course not confine our attention to actual killing – there are a variety of less extreme forms of the use of force that aim at subduing fellow human beings, or at making them harmless as competitors. As a rule the better organised societies have, at least for long periods, kept such intra-group hostile behaviour within bounds by cultural counterpressures, by penal codes enforced by some system of police and justice who together maintained law and order.

Our present agonistic behaviour is in my opinion unlikely to be a purely cultural phenomenon; its roots must be sought in our animal ancestry, for it is obviously highly environment resistant. I am convinced that it will soon be more generally recognised that primitive man had already long ago evolved the machinery for various forms of agonistic interaction. Life in our original hunter–gatherer societies must for instance have incorporated forms of sexual competition, forms of a peck order of some kind, including a leader–follower relationship, and also agonistic tendencies of each group towards other in-groups – a point that has still not received the attention it deserves. The indications for these various forms of agonism come not only from studies of contemporaneous hunter–gatherer societies[6] (who would never have evolved inhibitory rules if agonism had not made them necessary) but also from the comparative study of species of mammals which, occupying a niche similar to ours, characterised by hunting, gathering and dwelling in small groups, have convergently developed similar social systems.[8] The issue, so hotly disputed in some treatises on 'aggression' in man, of whether or not, or to what extent our present agonistic behaviour is purely reactive in character or is at least to a certain extent fed by an internal urge which can build up and overspill (Lorenz, 1966; Hinde, 1967, 1974; Tinbergen, 1968), seems to me less relevant than the fact that we have inherited the potential, in the form of complex behaviour mechanisms whose details need not concern us here, for using force in a number of diverse situations. It seems to be more than likely in view of our knowledge of still existing hunter–gatherer societies that, in the long past, most agonistic classes, between individuals and between groups, were of the relatively harmless type of 'fight–flight' balance that characterises similar intra-specific frictions in the vast majority of higher animals in which killing is relatively rare. For a long time the step towards actual killing must have been prevented by the evolution of protective, cultural codes. But modern man, i.e. man from at least 10000 years ago, has taken the disastrous step to war by using his unique capacity for foresight and experience, and recognising that under certain circumstances killing does pay, because a dead man will not return to fight again.

How this knowledge has come to over-ride on so many occasions our old, equally deep-rooted inhibitory systems based on fear and on response to appeasement behaviour, is another aspect of the problem on which biologically oriented behaviour students might well help to throw more light. Recorded history seems to indicate two main sources of mass-killing: fear of being dominated or even killed by members of a hostile force; and lust for conquest, robbery and subjugation of societies seen or suspected to be either richer, or weaker, or both. For an understanding of the question of 'why killing?' I suggest that the following idea, which to my knowledge has not been mentioned, may be helpful. It is the application to our various forms of agonistic behaviour of the concept, coined long ago by Lorenz, of the *Werkzeughandlungen*, of motor patterns that, in a hierarchically organised behaviour machinery, can be employed by a number of different major behaviour subsystems (Lorenz, 1939). The clearest example of this is locomotion, which is used in any functional system that requires movement from one place to another. It seems to me clear that aggressive behaviour – whether motivated by an 'urge' or of a merely reactive nature – is also used by many species, and certainly by man, in the service of a number of different functional and motivational systems. Aggressive and defensive behaviour, the use of force, forms part not only of the systems involved in intra-specific encounters, but also in our defence against predators and parasites and, in predatory species, in feeding behaviour. But while *intra*-specific agonistic behaviour has always evolved together with inhibitory mechanisms such as flight and response to appeasement, the *inter*-specific forms of agonistic behaviour such as for instance killing a poisonous snake, have had either very weak inhibitory mechanisms or none at all. Now, while in animals the various major functional systems usually exclude each other mutually, so that for instance a raptor will either hunt, or fight a rival, or defend itself and its brood against larger predators or against parasites, a man engaged in war or other forms of intra-specific killing seems to be able to fuse these different motivations into one super-motivation: the enemy is to the warrior not merely another human being; he is at the same time a dangerous predator, a parasite, and/or an obstacle to be removed. It seems to me that such 'supermotivations' are also at work in other forms of typically human behaviour – in religion, in art, and certainly in science. I believe that something of this kind is just as important an aspect of warfare in the widest sense as for instance improved mass communication, sophisticated indoctrination, the use of long-range weapons, increased population pressure and other factors that have been discussed by various writers.

Until relatively recently, it was not really biologically correct to call war a form of disadaptation; it never did more than slow down or set back temporarily the overall population growth, and throughout our recent history conquerors have on the whole profited from the outcome of war. But the two great World Wars, and certainly the realisation of what an all-out nuclear war would carry in its wake, have made the majority of people aware of the new situation, i.e. that wars, apart perhaps from wars of liberation, are, even in practical terms, decidedly harmful to both sides; in fact war is now one of the major aspects of our present disadaptation. Apart from bringing unimaginable misery to millions, war no longer pays. In particular nuclear war, even if waged one-sidedly by one of the super-powers, would cripple the entire population of the '*victor*'. This incidentally seems to me to be a fundamentally new aspect of the armaments race; a new kind of inhibition which, we must and can hope, will prevent a nuclear catastrophe. It is not unthinkable that future historians will consider organised mass-killing as a temporary deviation which evolved some 10000 years or more ago and stopped at about the present time.

LAISSEZ-FAIRE OR CULTURAL ENGINEERING?

From what I have said so far, in however sketchy a form, about some aspects of the human condition it will be clear that I am convinced by what I have seen, heard and read, that our species, in spite of all its cultural achievements, has now reached a stage at which a variety of dangerous pressures threaten to overwhelm all the advantages that our cultural evolution has for so long carried with it. We are entering one of the great, if not the greatest, discontinuities of our entire history. As has been pointed out time and again, the crisis which we begin to experience is so different from earlier, more local crises, in that it affects the Earth as a whole and in that it is now evolving so rapidly, that we must consider the present situation as fundamentally different from those earlier crises. It would be a deliberate cut-off from the mounting evidence to this effect if we would ignore the many warnings that, if present trends in our ecology and in our social organisation were to continue, we have to expect mortality on a greatly increased scale. What seems to me an even worse prospect is that, in the wake of famines, epidemics, social pressures and wars, large numbers of people who will survive will be somatically and mentally damaged for life.

For a frightening example of what can happen to a society under extreme stress I recommend a close study of Turnbull's (1973) account of his experiences among the Ik. Forceful eviction of this small tribe from their ancestral hunting grounds into a mountainous region which, even if these unfortunate people had had the agricultural knowhow, could not possibly yield a minimum diet, led to starvation, and – particularly instructive (as well as horrifying) to behaviour students – a complete breakdown of even the most deep-rooted form of altruism: the care of their own children. In this dying society literally everyone tries to care for him- or herself only, and children are left to fend for themselves at the age of three years or thereabout, long before they could possibly be materially, let alone psychologically independent. It would in my opinion be a dangerous delusion to believe that this kind of lethal deterioration could not happen to people at present living in the most affluent and civilised societies – the lessons from the Nazi concentration camps have taught us differently.

The extreme complexity of the situation should not paralyse us. However far we still are from a detailed diagnosis, not to speak of an overall and detailed plan for attacking our troubles at the roots, we can already identify quite a number of harmful pressures, conditions that are already now damaging us. The population growth *will* have to be slowed down, and ultimately stopped if not reversed, though with allowances for regional differences; the production of food and other essentials of life *will* have to be both stepped and stopped; the already known sources of material and psychosocial pollution *will* have to be curbed. To many of us biologists these statements are so obvious as to be almost commonplace, but to those of us who have taken part in recent conferences and symposia, and who have scanned the literature, all the way from the highly technical to more popular discussions, it must have become clear that the vast majority even of otherwise educated people are not informed about the fundamental fact of our overall disadaptation. Not until many more people become aware of the dangers threatening us can we expect the leaders of our society to turn their attention to possible remedial measures. In the meantime, scientists of many kinds will not only have to continue their attempts at analysis, but also try to work out possible remedies. The main contribution that the biological sciences can and must make is to show the core of our difficulties to be a change for the worse in our environment on the one hand, and in its demands on our phenotypic adjustability on the other a change in this *relationship* between our environment and ourselves (Wilson, 1975).

Once this is realised, it will also become clear that remedial measures will have to involve a two-pronged attack; on the one hand one that aims at making our environment more suitable to us, and on the other one that has to try to make the citizens of the future more able to cope with their habitat. This will be an awesome task, of redirecting our technologies, of finding ways for recycling of the masses of waste products that are not merely unused but act as pollutants; of curbing our population growth; above all of making people want to build a different society. Ethology is well placed to play a part in re-shaping both environment and our own society because, after a phase in which it stressed mainly the genetic aspect of behaviour programming, it is now moving to a position where it begins to map the interplay between our genetic blueprint and phenotypic flexibility, and to spot the pressures which overstretch even our exceptional adjustability.

The safest way of achieving re-adaptation would of course be a return to an earlier phase of our history, but such putting-back-the-clock is clearly neither possible nor desirable. There simply is no way back. Our re-adaptation will have to lead to a new level of adaptedness, and it will have to be achieved by rational, cultural means, for the time available is infinitely shorter than genetic re-adaptation would require.

Both with respect to our habitat and to our behaviour this task of biological engineering will have to be done in two phases. First, the most damaging among the already known new pressures, both ecological and social, will have to be reduced. In many spheres such 'symptom treatments' could start already now; in fact on a small, haphazard scale they have already been started. Some of these symptom treatments are: halting the population growth; ensuring a fairer distribution of the necessities of life; reduction of many forms of over-exploitation and tapping instead renewable energy, e.g. from the sun and from wave action; reduction of physico-chemical and psychogenic pollution; the pruning of many excessive luxury industries and in general a thorough re-orientation of our technologies, etc. Whereas with respect to the size of human populations we must aim at a 'no growth' state, the question 'growth or no growth of our technologies?' is usually wrongly put. Growth, i.e. further evolution, there must be, but this growth must be of a qualitative kind, aiming at a global self-support system. This ultimate aim of designing a new type of society faces us with a challenge of a magnitude that, even if we were to start now, would require more time than we have available; therefore the symptom treatments will have to be started in order to enable us to buy time for the design of a new environment. Both stages will involve drastic changes in our system of values, and whatever the changes we shall have to make will undoubtedly require a considerable lowering of our living standard instead of, as so unrealistically desired by so many, a further increase of our (material) standard of living. I am often told that this view of the future is utopian. I submit that this is reversing the positions; it is those who believe that we can with impunity continue on our present course who live in a dream world. Nor is it impossible to start symptom treatments now; however fragmented the many sciences involved are at the moment, we are not totally ignorant – as I have said, we cannot only identify some of the worst stressors, we can also suggest countermeasures. Finally, we can begin an effective advocacy for measures known to be necessary.

The second arm of the attack, the 'production' of a new type of citizen, of people who have not only the expertise but above all the motivation, the imaginativeness and the flexibility of mind to take part in all this work will require the collaboration of many as yet disparate sciences, first of all those concerned with child development and education. As I have argued elsewhere, it is both possible and desirable to experiment, partly on old principles, partly on a new level, with a 'biologically more balanced form of education', i.e. one that places less emphasis on instruction by adults, and that gives more scope for playful, exploratory and imitative self-teaching (Tinbergen, 1975). Experiments such as are now being conducted in countries such as Russia, China, Cuba, and Israel deserve close attention.

People who call themselves realists often reject these ideas because they are 'revolutionary'. But we shall have to realise that re-evolution is inevitable, and that the choice is merely between rational, peaceful re-evolution and a forceful, bloody revolution. We must attempt to work for as bloodless a revolution as possible because history teaches us that explosive, bloody revolutions inevitably begin with premature destruction.

PROBLEMS OF IMPLEMENTATION

Even if and when we will be in a position to present a full analysis of our predicament and recommendations for the reconstruction of society that will be required, there are likely to be formidable obstacles in the way of political implementation. It will be good to remember that, while intellectuals can rarely influence political decision-making directly, they can have considerable influence in helping to create a climate of opinion which leads to new decisions. We shall have to point out that time presses; and also that a number of remedial measures will at best act with a long delay. For example, many of the exploding populations in the world contain a large cohort of young citizens who are certain to reproduce. The main obstacle will be the unwillingness and/or inability of politicians and other leaders of society to act with long-term aims rather than 'the fast buck' in mind. In most Western societies, any long-term proposals which involve, as they will have to, a drastic tightening of our belts would at the moment mean political suicide. It will therefore be an essential part of our task to inform, in as simple and clear terms as possible, the general public, and so to influence public opinion until it becomes a political factor to be reckoned with. Just as J. B. Watson applied behaviourism to advertising, so modern behaviour students and ecologists will have to consider it part of their task as citizens to explain to the population at large what is actually happening. In this exercise in persuasion the method most likely to succeed is in principle simple: we shall have to appeal, by cool reasoning, to the deep-rooted motivations that are part of our genetic heritage. Knowledge of these mainsprings of human action can come from ethology, psychology and psychopathology, from the study of educational successes and failures, and from our growing knowledge of the behaviour systems of 'pre-cultural man', that we still carry with us.

When it comes to designing a re-adaptation programme that appeals to our fundamental motivations, it looks as if we have a choice between two alternatives:

(1) Whatever our present theories and speculations are about the mechanisms of 'aggression' in man, it is a simple fact of life, of which we are already seeing new manifestations, that, when frustrated or provoked, we all tend to use force. Even seemingly mild and as yet restrained forms of incipient anarchy are visible in all Western societies: unrest and delinquency, wildcat strikes (as well as 'wildcat working'), racial and other tensions between different subgroups within nations are already leading to, on occasion hardly containable, violence. It will be one of our tasks to make clear that, although resort to force is our most 'ready' response, it almost always leads to short-sighted solutions, to situations in which some sections of populations

subdue others, to slavery and, in the present international climate, ultimately to war.

(2) The other alternative is a policy which appeals to long-term self interest, including the interest of everyone's children and grandchildren – to at least this deeply rooted and most 'stubborn' form of altruism. A combined appeal to this motivation, and also to fear, will have to supplement appeals to our moral sense, for with the majority of people, moral considerations alone will not be sufficiently effectual in the face of the severe pressures to come. For political use the analogy with the paying of insurance premiums might well be effective. Most motorists are perfectly willing – admittedly under social pressure – to pay the premium for 'third party risk'. This type of short-term sacrifice for the mere *possibility* of a bad accident is directly comparable, though on a small scale, to the high taxes, the massive efforts and the considerable lowering of the living standard that will be essential. As in many forms of insurance and long-term investment the basic issue, in purely rational terms alone, is one between short-term gain with long-term deleterious effects on the one hand, and short-term sacrifices for the sake of long-term advantages on the other. However tangled and ill-understood the web of our overall disadaptation is, the core of the disadaptation issue, and certainly many of the symptoms visible now, can be understood by most.

Paradoxically, at least at first glance, the issue has one aspect which gives reason for hope that the problem will be recognised in the prosperous and powerful industrialised countries who now live in abundance. This is that a number of the most dangerous new pressures, such as many forms of pollution, but also a shortage of energy and even of (health-giving) food, will hit just those most advanced countries hardest. If this is realised in these countries, if the real causes are understood, and if their members are seen to take seriously their own population growth, their own energy problems, their own pollution and their own social injustices, and are seen to do something about them, the obviously strong suspicions of the Third World about our ulterior motives might be overcome. And that the governments of industrialised and overpopulated countries will gradually begin to think and act in the desirable directions may well be more likely than it seems now, since so many of the younger generation in these 'advanced' countries are beginning to rebel against their own societies.

TASKS FOR ETHOLOGY

What, in the light of this thumbnail sketch of the human predicament, can we ethologists hope to contribute? It will be clear that I believe that for the purpose of diagnosis of the present status of our species and of the dangers inherent in an unchanged continuation of our present course, the insights of 'whole animal biology', particularly of ecology and ethology, have to be taken into account. It is the behavioural ecologists who, in collaboration with a variety of human sciences, are in the best position to show that we are faced by unprecedented problems of adaptedness, of dis- and of re-adaptation. On many aspects of these problems experimentation will be impossible, unethical, or both. But the human scene is so varied, and so much in flux anyway, that clinical research in the widest sense, and the interpretation of 'natural experiments' can, together with animal studies, carry us a long way. It is in the tasks of *diagnosis* and of providing guidelines with respect to the overall *aims* for the future that the expertise of whole-animal biologists, including those studying man, will be required. For designing the *means whereby* the hoped-for revolution can best be attained, the know-how of those studying individual and social *mechanisms* of human conduct will have to be called upon. In particular a better knowledge of the interplay of reason and our deep-rooted, typically human motivations will be essential – knowledge, in other words, of our 'true nature', which includes that of both our largely genetic programming and the genetically imposed potential and the limitations of our flexibility. There are signs that economic and social planners are beginning to see that they will need this better understanding of human behaviour; we shall have to try to help them as well as we can.

In attempts to pave the way for re-adaptation we shall have to rely to a great extent on the principle of 'redirection' of our motivations, of employing reason to influence people's choice of action. It is another deep-rooted attribute of man that he tends to unite in warding off a common enemy. If we could steer our fears, our hostility, and our considerable potential for social cooperativeness towards attacking, not our fellow-men, but the negative feedbacks of our cultural evolution, we might with luck succeed in a united 'attack on our disadaptation'. In such an undertaking, the loss of, yet the longing for, an ideal in the younger generation could and must be turned to advantage. They are both the rulers and the ruled of the future; as teachers we academics can help them in developing a constructive attitude.

I realise full well that I am advocating a swing of animal ethology towards a more applied course. But it must by now be clear to many scientists that 'applied' science can be intellectually just as stimulating as 'pure' exploratory research. Even if this were not the case, the seriousness of our own situation is such that, unless we address ourselves to these vital problems, our very survival, and certainly our welfare, *and with this our own science*, will be in serious danger. In a sick, greatly impoverished and damaged society there will be no place for, no resources to spare and no interest in attending, to those 'growing points' of our science that have been discussed in such persuasive and stimulating ways in many of the other contributions to this volume. There is no lack of plans for future research, nor is there among the younger generation a shortage of keen and gifted recruits. If our science is to flourish, it must be seen not only to plan future ethological exploration, but also to work towards convincing our fellow-men of the necessity for its further growth. As part of this it will be the task of all the sciences concerned to work towards further integration and towards a more general awareness that the behavioural sciences are not a dispensable luxury, but an essential part of our overall effort to ensure a healthier future for society, in which alone man's highest mental potential can be fully realised.

REFERENCES

Beckerman, W. (1974). *Defence of Growth.* Cape: London.

Borgström, G. (1969). *Too Many.* Collier–Mcmillan: New York.

Bronffenbrenner, U. (1974). *Two Worlds of Childhood.* Penguin: Harmondsworth.

Brown, L. R. (1974*a*). *In the Human Interest.* Norton & Co: New York.

Brown, L. R. (1974*b*). *By Bread Alone.* Praeger: New York & Washington.

Carson, R. (1963). *Silent Spring.* Hamish Hamilton: London.

Commoner, B. (1966). *Science and Survival.* Viking Press: New York.

Commoner, B. (1972). *The Closing Circle.* Cape: London.

Ehrlich, P. R. & Ehrlich, A. H. (1972). *Population, Resources, Environment.* Freeman: San Francisco.

Ehrlich, P. A., Ehrlich, A. H. & Holden, J. P. (1973). *Human Ecology.* Freeman: San Francisco.

Hinde, R. A. (1967). Review of Lorenz, on aggression, *New Society,* **9**, 302–304.

Hinde, R. A. (1974). *Biological Bases of Human Social Behaviour.* McGraw-Hill: New York.

Holt, J. (1970). *How Children Fail.* Penguin: Harmondsworth.

Kenyon, K. M. (1969). *Archaeology in the Holy Land.* Benn: London.

Kruuk, H. (1972). *The Spotted Hyena.* University of Chicago Press: Chicago.

Lee, R. B. & DeVore, I. (ed.) (1968). *Man the Hunter.* Aldine: Chicago.

Levi, L. & Andersson, L. (1975). *Psychosocial Stress: Population, Environment and Quality of Life.* Spectrum Publs. Inc.: New York.

Lipowski, Z. J. (1973). Affluence, information inputs and health. *Social Science and Medicine,* **7**, 517–529.

Lorenz, K. (1939). Vergleichende Verhaltensforschung. *Verhandlungen der Deutschen zoologischen Gesellschaft, Supplement,* **12**, 69–102.

Lorenz, K. (1966). *On Aggression.* Methuen: London.

Maddox, J. (1972). *The Doomsday Syndrome.* Macmillan: London.

Malthus, T. R. (1817). *An Essay on the Principle of Population.* Murray: London.

Meadows, D. H., Meadows, D. L., Randers, J. & Behrens, W. W. III. (1972). *The Limits to Growth.* Earth Island Ltd: London.

Mech, L. D. (1970). *The Wolf*. Natural History Press: New York.

Medawar, P. B. (1972). In Tinbergen, N., *The Animal in its World*, p. 10. Allen & Unwin: London.

Osborne, F. (1963). *Our Crowded Planet*. Allen & Unwin: London.

Pauling, L. (1968). Orthomolecular psychiatry. *Science, Washington*, **160**, 265–271.

Schaller, G. B. (1972). *The Serengeti Lion*. University of Chicago Press: Chicago.

Tinbergen, N. (1968). On war and peace in animals and man. *Science, Washington*, **160**, 1411–1418.

Tinbergen, N. (1975). The importance of being playful. *Times Educational Supplement*, 10 January 1975, pp. 19–21.

Tucker, A. (1972). *The Toxic Metals*. Pan/Ballantine: London.

Turnbull, C. M. (1973). *The Mountain People*. Cape: London.

Vogt, W. (1948). *Road to Survival*. Sloane & Ass.: New York.

Ward, B. & Dubos, R. (1972). *Only One Earth*. Deutsch: London.

Wilson, E. O. (1975). *Sociobiology, the New Synthesis*. Belknap Press: Cambridge, Mass.

NOTES

1 Cronholm, B. (1974). *Les Prix Nobel en 1973*, p. 29. Norstedt & Soner: Stockholm. Flattering though this citation is, we shall have to acknowledge, with G. P. Baerends (*Nederlands Tijdschrift voor Geneeskunde*, **117**, p. 1726, 1973) that it expresses an expectation rather than an established fact, and that future developments will have to justify the Nobel Committee's high hopes.

2 By this I mean that behaviour be treated as an observable, tangible phenomenon; that a clear distinction be made between what is observed and what is inferred; and that equal attention be given to the 'four why's' (Tinbergen, N. (1963). On aims and methods of ethology. *Zeitschrift für Tierpsychologie*, **20**, 410–433).

3 See, for instance: Dubos, R. (1968). *So Human an Animal*. Scribner: New York. Dobzhansky, T. (1971). *Mankind Evolving*. Yale University Press: New Haven. Huxley, J. S. (1944). *Man in the Modern World*. Mentor: New York. Medawar, P. B. (1961). *The Future of Man*. Mentor: New York. Morris, D. (1967). *The Naked Ape*. Cape: London. Thorpe, W. H. (1975). *Animal Nature and Human Nature*. Methuen: London. Tinbergen, N. (1972). Functional ethology and the human sciences. *Proceedings of the Royal Society of London* B, **182**, 385–410.

4 Campbell, B. G. (1974). *Human Evolution*. Aldine: Chicago. For a recent, non-technical, clearly written account of the pre-history of man see the series *The Emergence of Man*, Time–Life Inc.: New York.

5 On the survival value of ontogenetic flexibility see: Bruner, J. S. (1972). *Nature and uses of immaturity. American Journal of Psychology*, **27**, 687–708.

6 Coon, C. S. (1974). *The Hunting People*. Cape: London. Lee, R. B. & DeVore, I. (ed.) (1968). *Man the Hunter*. Aldine: Chicago. Turnbull, C. M. (1974). *The Forest People*. Cape: London. Gould, R. A. (1969). *Yiwara: Foragers of the Australian Desert*. Collins: London.

7 Eyer, J. (personal communication). Based on figures supplied by the US Department of Health, Education and Welfare.

8 See Kruuk, 1972; Mech, 1970; & Schaller, 1972. The main pressures that have forced some kind of inter-group territoriality on these species are (1) the need for an intimate, slowly acquired knowledge of innumerable details of the hunting range; and (2) the need to reserve a minimum food supply for the members of each group – both extremely strong pressures.

EDITORIAL: 8

A few years ago the possibility that ethologists could make a significant contribution to the study of human behaviour was often doubted. We believe that such doubts have now largely been dispelled. Ethological work concerned with the study of man requires, however, some reorientation, for methods and concepts useful for studying fish, say, are not equally useful for studying man. At the same time, specialisation exclusively on *human* ethology could be unduly restrictive.

One reason for this is as follows. When the biological sciences were blossoming in the nineteenth and early twentieth centuries, it was convenient to divide up the subject matter on phyletic grounds. The obvious route for specialisation was along the lines of the natural classification. Gradually, however, as the growth of knowledge demanded further specialisation, a new type of division arose. Investigators became interested in particular aspects of the groups they studied – there were specialists in the classification of butterflies, the behaviour of birds, the control of population in lemmings. But the solution of such problems required investigators to enlarge their scope beyond the phyletic group on which they were focussed. Butterfly systematics required principles similar to those used in bird systematics; the same techniques and generalisations would serve for some aspects of insect and bird behaviour; the ecology of lemmings had something in common with that of crossbills. In short, investigators became primarily systematists, comparative psychologists, or ecologists, and only secondarily specialists in a particular phyletic group.

Carving up science along phyletic lines does, therefore, seem retrogressive. Without denying that man is special, and deserves a special focus, no ethologist can afford to be deprived for long of the benefits of a comparative approach. To some extent this is already exemplified in studies of non-human primates. Through limiting their vision by phyletic boundaries, primatologists have too often tackled issues with which ornithologists were already highly experienced. The same could easily happen to human ethologists. Even though the current spirit is one of exploration, the camp of human ethologists with its shining new banner could easily become a defensive fortress. And, before long, as the banner becomes a little tattered, the fortress could become a prison. We believe it is healthier to keep research focussed on problems rather than on species – even one as important to mankind as man.

The value of drawing on a wide body of literature, not all of it concerned with man, is apparent in the first three chapters in this section. Many of the themes occurring in earlier sections recur in these chapters, but there are also a number of new ones. For instance, the value of a functional approach is emphasised by Blurton Jones. However, as he stresses, in so far as human behaviour is adapted the origins of that adaptedness may lie neither far back in the evolution of the species nor in the development of the individual but in that of the culture. It is usually easy to distinguish factors contributing to evolutionary change from the immediate determinants of individual behaviour. But cultures can change relatively rapidly, and the causes of cultural change are not so easy to distinguish from the causes of events within the culture. Whilst it is the case that causal and functional approaches can be complementary and inter-fertile, in the study of cultural differences it is especially necessary to be clear-headed about their relations to each other. Some of the authors cited by Blurton Jones as giving minor or non-existent roles to ecological or biological factors are quite simply concerned with other questions. Blurton Jones, indeed, points out (p. 436) that 'Their structuralist generalisations seem to concern the immediate contemporary motivational or cognitive state of people... [they] may be making no statement about long-term influences.' We would add that just where a functional approach can and cannot aid understanding of immediate causes is yet to be determined. In the case of cultural change, it will require understanding not only of how changes are initiated, but also of how changes once initiated become accepted and transmitted to succeeding generations.

The dynamic stability of relationships, discussed by Hinde & Stevenson-Hinde, also raises issues that scarcely arose in early chapters. So far as behaviour is concerned, at any rate, a relationship can be studied only by observing interactions of limited duration. However, a relationship is itself extended in time, and its description requires reference not only to the nature of the interactions that occur, but to how they are patterned in time. The study of inter-individual relationships thus demands synthesis as well as analysis.

In many respects, the approach of Hinde & Stevenson-Hinde is complementary to that of Dunn. Her analysis is primarily concerned with whether differences in the social environment or in social behaviour at one stage affect later development. Her evidence suggests that considerable self-regulation may occur. If so, the processes involved could well be similar to those required for dynamically stable social relations.

In his overview, Medawar raised issues which demand careful thought even if many ethologists will not agree with all the points he makes. He suggests that ethologists need to discover a key experimental preparation which would unlock the door to dramatic understanding of behavioural mechanisms. With the infinite diversity of nature always beckoning ethologists to explore fresh fields, Medawar's point is a useful reminder that a general advance in understanding can come from focussing research effort on a particular and judiciously chosen topic. Much progress in ethology has come from a relatively few preparations that have been analysed in depth – stickleback courtship, herring gull reproductive behaviour, the following response of precocial birds, to mention a few. But at the same time broader comparative studies have been essential both to establish the generality of the findings of detailed analyses and to reach for conclusions about the evolution of behaviour not attainable in any other way. A balance between these two approaches must be achieved.

Tinbergen's plea that ethologists should put a major part of their effort into solving some of the urgent practical problems now facing mankind will strike a chord in the minds of many readers. Certainly we are wholly in sympathy with his view. But he is of course not implying that ethologists should give up fundamental work. The present strength of ethology was derived from studies of other animals; and an *exclusive* focus on problems of immediate practical importance could deprive ethology of the possibility of theoretical growth. Furthermore, the conceptual structure necessary for the practical problems of the 1990s will not necessarily arise from examination of the problems of today. Thus the urgent need for a concerted attack on applied problems should not weaken the vigour of research on basic theoretical and conceptual issues. And here the contribution of comparative studies is not limited by the extent to which animals resemble man. Comparative studies can be of value just because animals are *different* from man. In some instances this value arises merely because they are simpler, and provide material on which we can try out new concepts or approaches without the complexity of the human case. In others it lies in the fact they they lack man's most special attributes – verbal language and the cultural differentiation that depends upon it. Separation of cultural and biological factors in human social structure is almost impossible, but comparison of man with other primates, in which the cultural factors are less important, may be of considerable help.

Conclusion – on asking the right questions

The contributors to this volume represent diverse interests within ethology and, as we hoped, they all had their own ideas about its growing points. Furthermore, we should reiterate that not all aspects of ethology are represented – for instance, no chapter is concerned with the physiological analysis of mechanisms, or with comparative studies sketching the evolutionary history of behaviour. For these reasons, it would be inappropriate for us to specify the areas or approaches that seem to us most likely to be fertile in the immediate future.

But wherever valuable progress seems most possible, it will not be achieved unless the questions asked are clearly specified. In the history of ethology, perhaps one of the most important milestones was the formulation by J. S. Huxley and more definitively by Tinbergen (1951, 1963) of the four questions that can be separated out from a generalised interest in a behavioural problem – the questions of immediate causation, ontogeny, function and evolutionary history. Only too often statements about, say, the evolution of behaviour had been muddled with an explanation of its ontogeny: Tinbergen's analysis did much to dispel the confusion.

Whilst recognising that those questions are logically distinct, ethologists have also recognised that they are inter-fertile. For instance, a classification of behaviour in terms of function often corresponds with a classification in terms of common causal factors. From an evolutionary point of view this is not hard to understand. The efficiency of the underlying mechanisms may be greatly enhanced if functionally related activities share causal factors, the mechanisms involving perhaps hierarchical organisation. However, the correspondence is not always close, as Beer (1963) demonstrated in his analysis of black-headed gull (*Larus ridibundus*) reproductive behaviour. Conclusions about causation must ultimately rest on analysis of antecedent events and conclusions about function on analysis of consequences. In the chapters in this book, the nourishing influence of one perspective on another is frequently recognised. On this issue McFarland took rather an extreme position, arguing that analysis of a system controlling behaviour would be virtually impossible without knowledge of its survival value. Not all members of the conference agreed with him, but everybody accepted that a functional account of behaviour can usefully restrict the variety of mechanisms which may plausibly be proposed, though in the human case the issues are especially complex.

Tinbergen's questions can also be mutually fertile in another, rather different, way. When questions are formulated, the investigator must take a decision as to the degree of quantitative precision at which he is aiming. On the whole, increased precision can be obtained only at the cost of generality. While each type of behaviour of every species requires investigation as a distinct problem the scientist must ultimately seek for generalisations, not for a limitless array of particulars. Clearly a balance must be struck. In practice, however, studies of antecedent conditions generally demand detailed analyses of particular instances with a precision limited only by considerations of economy and the tools available. The generality of the conclusions which can be drawn are often assessed from comparative studies, usually pursued with the quite different goal of understanding how behaviour has evolved. It is in part because ethologists have been simultaneously engaged in detailed studies of limited scope and wide-ranging surveys of broader areas that they have not become lost in the morass of detail into which attempts to understand the behaviour of organisms could so easily lead them.

Another decision concerns the level of analysis at which the questions should be aimed. At one time ethologists saw analysis of causal questions at the behavioural level as paving the way for physiological analysis. For example, Tinbergen (1951, p. 5) wrote 'it is our job to carry our analysis from these high levels down to the level of the neurophysiologist, the sense physiologist, and the muscle physiologist'. Not only has this task proved much more difficult than it then seemed, but increasingly it is being argued that a specifically behavioural analysis is a proper goal in its own right, whether or not it leads to further understanding in terms of neural events. This view is implicit in practically every chapter of this book. But here again, the recognition that questions can properly be posed at each level should not be taken to imply that once holes are found for pigeons they can no longer coo to each other. It is not only that behavioural analysis can lead to analysis of neural events, but also that neurobiology should become increasingly fruitful in suggesting directions for behavioural studies. While time has increasingly justified study at particular levels of integration, it has not lessened the need for providing links between levels.

Linking levels nearly always implies asking increasingly refined questions – in other words, analysis from the compound to the more particulate.

Understanding of the complexities of behaviour always requires fragmentation of the problems, and progressive understanding at successively finer levels of analysis. But an understanding of component processes does not necessarily imply understanding of the whole. At each level are to be found emergent properties which are not apparent in the products of analysis: collections of neurones, or the pattern of interactions within inter-personal relationships, have properties not present in the components. Furthermore, progressive analysis of one aspect of a problem can lead to neglect of others. Analysis is thus never sufficient, but must be constantly accompanied by re-synthesis to assess the relations between the parts and to check the adequacy of the initial analysis. Ethologists have often said this in the past, but it can perhaps never be said too often. Within ethology the issue is illustrated by the study of 'imprinting'. Naturalistic observations of the parent–offspring relationship, and of how that relationship could go wrong, led to detailed studies of one particular type of response within that relationship – following in controlled situations with inanimate models substituting for parents. The word 'imprinting' came to mean the learning that occurred in the context of the following response, and only in the last few years has the nature of imprinting come to be studied in the context of the formation of the developing relationship – with benefit for the understanding both of the learning process and the developing relationship.

A related issue concerns the extent to which, in framing their questions, ethologists should aim towards some degree of formalisation. Fed by natural history, and wary of the spectre of learning theories sterilised by over-formalisation, ethology has tended to eschew the seemingly rigid constraints of formal theory building. The search has been instead for broad conceptual frameworks to order the wealth of evidence. But the time is clearly coming when something more than this will be required. Some of the chapters in this book, for instance McFarland's and Bertram's, indicate that formalisation within limited areas is already worthwhile.

Another way in which the questions asked by ethologists are slowly changing concerns the extent to which they are concerned with types of *behaviour*. A decade or two ago the majority of studies were concerned with 'the stimuli releasing this type of response', or 'the evolution of that type of behaviour'. Such a focus on types of behaviour is perhaps to be understood in terms of the early fertility of the concept of Fixed Action Patterns and the classification of activities into 'instincts'. More recently it has been possible to discern an increasing emphasis on faculties or properties of behaviour that bridge the conventional categories. Two obvious examples are the way in which studies of imprinting in the context of the following response have become linked to studies of perception (e.g. Sluckin, 1972), and studies concerned with the problem of motivation have led to concepts like time-sharing (e.g. McFarland, 1974). It seems likely that this trend will continue, and will be facilitated by the growing interest of ethologists in the nature and bases of individual differences.

This is important also because it will facilitate another tendency which has been becoming more and more conspicuous recently – the tendency of ethologists to be concerned with questions within the boundaries of other disciplines. To some extent, this has always been true – links with psychology, ecology, evolutionary studies, genetics and physiology have long been fostered, and have indeed been a continuous vitalising influence. Even so, the boundaries of the province of ethology were more or less apparent. But this book provides examples of the eclectic manner in which ethologists are now straying even farther from what could be called classical ethology. The cost of such willingness to engage in dialogues all round can be a loss of identity – and tortuous attempts to define who is and who is not an ethologist. But does that matter? The nature of the specifically ethological contribution ceases to be an issue as disciplines join forces in coming to grips with common problems.

REFERENCES

Beer, C. G. (1963). Incubation and nestbuilding behaviour of black-headed gulls. III. The pre-laying period. *Behaviour*, **21**, 155–176.

McFarland, D. J. (1974). Time-sharing as a behavioral phenomenon. *Advances in the Study of Animal Behavior*, **5**, 201–225.

Sluckin, W. (1972). *Imprinting and Early Learning*, 2nd edn. Methuen: London.

Tinbergen, N. (1951). *Study of Instinct.* Clarendon Press: Oxford.

Tinbergen, N. (1963). On aims and methods of ethology. *Zeitschrift für Tierpsychologie*, **20**, 410–433.

P.P.G.B.
R.A.H.

Index